Developments in Volcanology 3

EXPLOSIVE VOLCANISM

DEVELOPMENTS IN VOLCANOLOGY

Further titles in this series

1 *H. TAZIEFF and J.C. SABROUX*
   FORECASTING VOLCANIC EVENTS
2 *S. ARAMAKI and I. KUSHIRO (Editors)*
   ARC VOLCANISM

DEVELOPMENTS IN VOLCANOLOGY 3

EXPLOSIVE VOLCANISM

edited by

MICHAEL F. SHERIDAN

Department of Geology, Arizona State University, Tempe, Arizona, U.S.A.

and

FRANCO BARBERI

Istituto di Mineralogia e Petrologia, Universita di Pisa, Pisa, Italy

reprinted from Journal of Volcanology and Geothermal Research, vol. 17 (1–4)

ELSEVIER Amsterdam — Oxford — New York — Tokyo   1983

ELSEVIER SCIENCE PUBLISHERS B.V.
Molenwerf 1
P.O. Box 211, 1000 AE Amsterdam, The Netherlands

Distributors for the United States and Canada:

ELSEVIER SCIENCE PUBLISHING COMPANY INC.
52, Vanderbilt Avenue
New York, NY 10017

Library of Congress Cataloging in Publication Data
Main entry under title:

Explosive volcanism.

(Developments in volcanology ; 3)
Papers from a workshop held in Italy, May 1982 and sponsored by the National Science Foundation (U.S.) and Consiglio nazionale delle ricerche (Italy).
"Reprinted from Journal of volcanology and geothermal research, v. 17 (1-4)."
Bibliography: p.
Includes index.
1. Volcanism--Congresses. I. Sheridan, Michael F. II. Barberi, Franco. III. National Science Foundation (U.S.) IV. Consiglio nazionale delle ricerche (Italy) V. Journal of volcanology and geothermal research. VI. Series.
QE521.5.E96 1983   551.2'1   83-16579
ISBN 0-444-42251-X (v. 3)

ISBN 0-444-42251-X (Vol. 3)
ISBN 0-444-42235-8 (Series)

© Elsevier Science Publishers B.V., 1983
All rights reserved. No part of this publication may be reproduced, stored in a retrieval system or transmitted in any form or by any means, electronic, mechanical, photocopying, recording or otherwise, without the prior written permission of the publisher, Elsevier Science Publishers B.V., P.O. Box 330, 1000 AH Amsterdam, The Netherlands

Printed in The Netherlands

CONTENTS

Preface VII

Hydrovolcanism: basic considerations and review
  M.F. Sheridan (Tempe, AZ, U.S.A.) and K.H. Wohletz (Los Alamos, NM, U.S.A.) 1

Mechanisms of hydrovolcanic pyroclast formation: grain-size, scanning electron microscopy, and experimental studies
  K.H. Wohletz (Los Alamos, NM, U.S.A.) 31

Ignimbrite types and ignimbrite problems
  G.P.L. Walker (Honolulu, HI, U.S.A.) 65

Computer simulation of transport and deposition of the Campanian Y-5 ash
  W. Cornell, S. Carey, and H. Sigurdsson (Kingston, RI, U.S.A.) 89

Explosive activity associated with the growth of volcanic domes
  C.G. Newhall (Vancouver, WA, U.S.A.) and W.G. Melson (Washington, D.C., U.S.A.) 111

A volcanologist's review of atmospheric hazards of volcanic activity: Fuego and Mount St. Helens
  W.I. Rose, R.L. Wunderman, M.F. Hoffman, and L. Gale (Houghton, MI, U.S.A.) 133

The past 5,000 years of volcanic activity at Mt. Pelée, Martinique, (F.W.I.): implications for assessment of volcanic hazards
  D. Westercamp (Fort de France, Martinique, F.W.I.) and H. Traineau (Oreleans, Cedex, France) 159

Application of computer-assisted mapping to volcanic hazard evaluation of surge eruptions: Vulcano, Lipari, and Vesuvius
  M.F. Sheridan and M.C. Malin (Tempe, AZ, U.S.A.) 187

Geologic, volcanologic, and tectonic setting of the Vico-Cimino area, Italy
  F. Sollevanti (Milano, Italy) 203

Structure and evolution of the Sacrofano-Baccano caldera, Sabatini volcanic complex, Rome
  D. DeRita, R. Funiciello (Roma, Italy), U. Rossi (Pisa, Italy), and A. Sposato (Italy) 219

A general model for the behavior of the Somma-Vesuvius volcanic complex
  R. Santacroce (Pisa, Italy) 237

The A.D. 472 "Pollena" eruption: volcanological and petrological data for this poorly-known, Plinian-type event at Vesuvius
  M. Rosi and R. Santacroce (Pisa, Italy) 249

The Phlegrean Fields: structural evolution, volcanic history, and
  eruptive mechanisms
    M. Rosi (Pisa, Italy), A. Sbrana, and C. Principe (Milano, Italy)..   273
The Phlegrean Fields: magma evolution within a shallow chamber
    P. Armienti, F. Barberi (Pisa, Italy), H. Bizouard, R. Clocchiatti
    (Orsay, France), F. Innocenti (Pisa, Italy), N. Metrich (Orsay,
    France), M. Rosi (Pisa, Italy), and A. Sbrana (Milano, Italy)......   289
Evidence for magma mixing in the surge deposits of the Monte
  Guardia sequence, Lipari
    R. DeRosa (Cosenza, Italy) and M.F. Sheridan (Tempe, AZ, U.S.A.)...   313
Evolution of the Fossa cone, Vulcano
    G. Frazzetta (Catania, Italy), L. LaVolpe (Bari, Italy), and
    M.F. Sheridan (Tempe, AZ, U.S.A.)..................................   329
Recent volcanic history of Pantelleria: a new interpretation
    Y. Cornette (Gif sur Yvette, France), G. Crisci (Cosenza, Italy),
    P.Y. Gillot (Gif sur Yvette, France), and G. Orsi (Napoli, Italy)..   361
Origin and emplacement of a pyroclastic flow and surge unit at
  Laacher See, Germany
    R.V. Fisher (Santa Barbara, CA, U.S.A.), H.-U. Schmincke, and
    P. Van Bogaard (Ruhr, West Germany)................................   375
Volcanism in the eastern Aleutian arc: Late Quaternary and Holocene
  centers, tectonic setting and petrology
    J. Kienle and S.E. Swanson (Fairbanks, AK, U.S.A.).................   393
Large-scale phreatomagmatic silicic volcanism: a case study
  from New Zealand
    S. Self (Arlington, TX, U.S.A.)....................................   433
Index..   471

PREFACE

The problems of explosive volcanism have been the focus of intense research during the last decade .Prior to the eruptions of Mount St.Helens in 1980 and El Chichon in 1982 a number of models were developed to explain the phenomena of Plinian eruption columns,Strombolian explosions,column collapse and runout of pyroclastic flows,and emplacement of pyroclastic surges.Also during this period was a growing realization of the strong effect produced by the interaction of external water with magma to produce hydrovolcanic explosions.

Studies of explosive volcanism are important for both basic research into geologic processes and practical application of volcanic phenomena.Models of magma evolution in shallow chamber generally require an interpretation of samples taken from explosive products.Likewise,the evaluation of the potential volcanic risk at active volcanoes and the exploration for geothermal energy in volcanic regions require an understanding of the processes of explosive volcanism and the interaction of subsurface water with magma.

A workshop held in Italy was jointly sponsored by National Science Foundation (U.S.A.) and Consiglio Nazionale delle Ricerche (Italy) in May 1982,in order to bring together researchers concerned with these problems.As a result of this workshop,21 papers were collected for inclusion in this special volume dedicated to the problems of explosive volcanism.The papers are arranged in three groups: general topics,volcanic risk,and case studies.Because half of the workshop time was devoted to examination of field problems in the Roman,Campanian,and Aeolian regions,many of the topical papers concern these areas.

Besides the sponsoring agencies,the compilation of this book depended on the dedicated work of many individuals.More than 50 people reviewed the manuscripts and greatly improved the quality of this work through their comments.Mary Rose prepared all of the manuscript on a word processor and provided critical editorial revisions of the manuscripts.Joan Bahamonde helped with the photographic layout, and early preparation of the texts.Tom Moyer provided final editing of the manuscripts and checked for correct pagination.

Michael F.Sheridan and Franco Barberi

HYDROVOLCANISM: BASIC CONSIDERATIONS AND REVIEW

MICHAEL F. SHERIDAN[1] and KENNETH H. WOHLETZ[2]

[1] Department of Geology, Arizona State University, Tempe, AZ 85287 (U.S.A.)
[2] ESS1, Los Alamos National Laboratory, Los Alamos, NM 87545 (U.S.A.)
(Received December 7, 1982; revised and accepted April 9, 1983)

ABSTRACT

Sheridan, M.F. and Wohletz, K.H., 1983. Hydrovolcanism: basic considerations and review. In: M.F. Sheridan and F. Barberi (Editors), Explosive Volcanism. J. Volcanol. Geotherm. Res., 17: 1-29.

Hydrovolcanism refers to natural phenomena produced by the interaction of magma or magmatic heat with an external source of water, such as a surface body or an aquifer. Hydroexplosions range from relatively small single events to devastating explosive eruptive sequences. Fuel-coolant interaction (FCI) serves as a model for understanding similar natural explosive processes. This phenomena occurs with magmas of all compositions.

Experiments have determined that the optimal mass mixing ratio of water to basaltic melt for efficient conversion of thermal energy into mechanical energy is in the range of 0.1 to 0.3. For experiments near this optimum mixture, the grain-size of explosion products is always fine (less than 50 µm). The particles generated are much larger (greater than 1-10 mm) for explosions at relatively low or high ratios. Both natural and experimental pyroclasts produced by hydroexplosions have characteristic morphologies and surface textures. SEM micrographs show that blocky, equant grain shapes dominate. Glassy clasts formed from fluid magma have low vesicularity, thick bubble walls, and drop-like form. Microcrystalline essential clasts result from chilling of magma during or shortly following explosive mixing. Crystals commonly exhibit perfect faces with patches of adhering glass or large cleavage surfaces. Edge modification and rounding of pyroclasts is slight to moderate. Grain surface alteration (pitting and secondary mineral overgrowths) are a function of the initial water to melt ratio as well as age. Deposits are typically fine-grained and moderately sorted, having distinctive size distributions compared with those of fall and flow origin.

Hydrovolcanic processes occur at volcanoes of all sizes ranging from small phreatic craters to huge calderas. The most common hydrovolcanic edifice is either a tuff ring or a tuff cone, depending on whether the surges were dry (superheated steam media) or wet (condensing steam media). Hydrovolcanic

0377-0273/83/$03.00 © 1983 Elsevier Science Publishers B.V.

products are also a characteristic component of eruption cycles at polygenetic volcanoes. A repeated pattern of dry to wet products (Vesuvius) or wet to dry products (Vulcano) may typify eruption cycles at many other volcanoes. Reconstruction of eruption cycles in terms of water-melt mixing is extremely useful in modeling processes and evaluating risk at active volcanoes.

INTRODUCTION

Hydrovolcanism refers to volcanic phenomena produced by the interaction of magma or magmatic heat with an external source of water, such as a surface body or an aquifer (MacDonald, 1972; Sheridan and Wohletz, 1981). Hydroexplosion (Ollier, 1974; Schmincke, 1977) is an analogous term for explosive activity caused by this process. Hydromagmatic processes could even occur within deep (a few km) hydrothermal zones related to plutonic bodies. Stable isotope studies are important for determining the origin of water concentrated in explosive products erupted from the tops of large magma chambers (Forrester and Taylor, 1972; Kalamarides, 1982). This is especially true for silicic, caldera-forming eruptions (Christiansen and Blank, 1972; Lipman and Friedman, 1975; Hildreth, 1981).

Surficial hydroexplosions range from relatively small phreatic events, through common base-surge phenomena, up to devastating eruptions like the 1982 eruption of El Chichon. They may consist of a single explosion that opens a vent, hydrovolcanic pulses interspersed with purely magmatic activity, or a long series of steam-and-ash jets typical of sustained eruptions. Because water is plentiful near the surface of the earth, its relationship to erupting magma must play an important role in volcanism. For this reason it is useful to consider volcanism within a continuum that ranges from purely magmatic processes as one end member to steam eruptions at the other.

The controls of hydroexplosions are poorly understood at the present. Because the observed contact of magma with water at the earth's surface or beneath the sea does not always lead to explosive activity (e.g., Moore, 1975; Shepherd and Sigurdsson, 1982), some may be skeptical of the potential explosive energy of such a system. A general model and specific definitions for hydrovolcanic phenomena are lacking. Continuing experiments on water-melt interaction and careful observations of hydromagmatic eruptions and products should eventually lead to a better working model.

Hydrovolcanism affects all shallow (< 200 m) subaquatic volcanoes and most subaerial vents. Tuff cones and tuff rings (Heiken, 1971; Wohletz and Sheridan, 1983), which both result from this process, are second in global abundance only to scoria cones among pyroclastic vents (Green and Short, 1971). Hydrovolcanic explosions are also common activity on stratovolcanoes and calderas.

ENVIRONMENTS OF HYDROVOLCANISM

Hydrovolcanism encompasses all environments where the intermixing of water and magma produce explosive volcanic phenomena or extensive brecciation of rock and magma. Some of the important specific environments include: deep submarine, shallow submarine, littoral, lacustrine, phreatic, and subglacial. Hydromagmatism refers to a general process rather than to a specific event type or class of geologic situation. In many cases the specific hydrologic environment leading to an hydroexplosion is difficult to determine from surficial geologic data. However, the deposits may provide good evidence for the interaction of external water with melt. We concur with Schmincke (1977) that terms like hydromagmatism, hydrovolcanism, or hydroexplosion are preferred for situations where a strong interaction of water and magma can be proven, regardless of whether or not the source of the external water is known.

The subaqueous environment includes all activity beneath a standing body of water. Products from this environment have been termed subaquatic (Sigvaldason, 1968) or aquagene (Carlisle, 1963). Included in this category are submarine, littoral, sublacustrine, or other specific cases.

Submarine events (Bonatti, 1967) occur within deep (greater than 200 m) saline water (Honnorez and Kirst, 1975). The most common occurrence for this type of volcanism is at oceanic spreading centers and on large submarine volcanoes that form flat-topped guyots (Cotton, 1969) consisting of pillow basalts and hyaloclastites.

Littoral refers to near shore and shallow (less than 200 m) subaqueous activity (Wentworth, 1938). Common constructs include littoral cones that form where lava enters the sea (Moore and Ault, 1965; Fisher, 1968) or tuff cones near the shoreline such as Diamond Head and Koko craters in Hawaii (Wentworth and Winchell, 1947). Pseudocraters, such as those at Myvatn in Iceland, represent a type of littoral activity where a lava flowed into a fresh-water lake (Rittman, 1938).

Phreatic (Greek word for well, see Macdonald, 1972) refers to the eruption from the phreatic zone (ground water) of vaporized water and solid materials without juvenile clasts (Ollier, 1974). Included in this category are hydrothermal explosion craters such as those at Yellowstone National Park (Muffler et al., 1971) or New Zealand (Nairn and Wirdadiradja, 1980) which were produced by steam explosions at the top of hydrothermal systems. The deposits generally consist of massive explosion breccias that contain hydrothermally altered blocks in a clay matrix.

The term phreatomagmatic was used by Stearns and MacDonald (1946) in reference to explosions resulting from the conversion of groundwater to steam by ascending magma. It has been used for shallow lakes and submarine explosions as well. Because this is a common type of terrestrial hydrovolcanic environment, this term frequently occurs in the literature. The products are water, steam, brecciated country-rock, and must include juvenile clasts. Tuffs with a wide range of bedding structures are the common products.

Subglacial phenomena occur where magma is erupted beneath a glacier (Noe-Nygaard, 1940). In addition to deposits from massive floods (jökullaups), thick accumulations of pillow basalts, pillow breccias, massive palagonite tuffs, and stratified palagonite tuffs construct table mountains (stapi) or ridges (mobergs) above their vents (Sigvaldason, 1968; Ollier, 1974).

A MODEL FOR WATER-MELT INTERACTION

The relationship of explosive energy to crater size and particle velocity (and hence distribution of ejecta fragments) is generally expressed in terms of scaling laws. Considerable effort has been directed toward this problem with respect to large planetary impacts (Gault et al., 1963; Stöffler et al., 1975; Oberbeck, 1975). However, explosive phenomena occur over a wide range of time scales. Because volcanic explosions take place at a slower rate than thermochemical or thermonuclear explosions, scaling laws developed for hypervelocity impacts cannot be directly used to calculate the explosive energy from volcanic crater size or distribution of products. Perhaps dimensional analysis of theoretical scaling laws (Housen et al., 1983) will eventually prove appropriate for volcanic explosions. An alternative method is to extrapolate data from water/melt experiments (Wohletz and McQueen, 1981; Wohletz and Sheridan, 1982) to the scale of volcanic hydroexplosions.

NATURE OF THE PHYSICAL PHENOMENA

Hydrovolcanic eruptions can be considered to be the natural equivalent of a class of physical processes termed fuel-coolant interactions (FCI) by investigators of large industrial explosions. See Colgate and Sigurgeirsson, (1973) and Peckover et al., (1973) for applications of this theory to volcanic phenomena. FCI involves the contact of two fluids, the fuel having a temperature above the boiling point of the coolant (Board et al., 1974; Buchannan, 1974; Board and Hall, 1975; Frohlich et al., 1976; Drumheller, 1979; Corradini, 1981). The interaction generally results in vaporization of the coolant and chilling or quenching of the fuel. This process has attracted considerable interest because the vaporization often occurs at explosive rates. Examples from industry include destructive explosions at foundries where molten metal accidentally contacts water. Recent investigations have attempted to predict conditions that would lead to an FCI in the event of a nuclear core meltdown. These studies (Sandia Laboratories, 1975) were conducted to determine the controlling factors of FCI so that nuclear plants can be designed to prevent explosions.

An explosive FCI rapidly converts thermal energy to mechanical energy with a heat transfer rate greatly in excess of normal boiling by several orders of magnitude (Witte et al., 1970). The rapid vaporization of large volumes of water by magma in volcanic regimes and consequent expansion results in explosive yields that can reach one-quarter to one-third that of an equivalent mass of TNT.

The process of rapid heat transfer is periodic: pulses are separated by millisecond or shorter intervals. Initially a small volume of water is vaporized due to contact with the melt. At this stage the dominant effect of the vaporization energy is to fragment the melt which results in an increased surface area of contact between water and the melt (Corradini, 1981). The larger area of melt/water contact in turn promotes further vaporization of water that, through this feedback process, rapidly increases the total mechanical energy (P∆V) of the system.

When the total vaporization energy exceeds the limit of containment, the system explodes (Fig. 1) with the rapidly expanding vapors propelling the entrained melt fragments, sometimes with pieces of the containment chamber. At this stage the dominant effect of the vaporization energy is to accelerate the particles into the surrounding lower pressure space. If unmixed magma and water remain in the system after the initial explosion, a regular influx of melt and

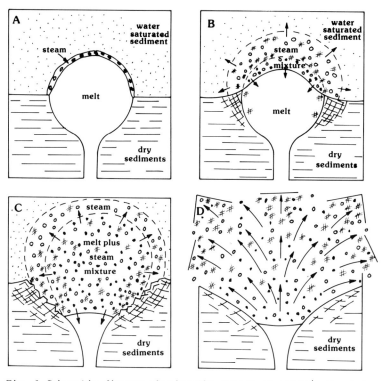

Fig. 1 Schematic diagram showing the stages of water/melt mixing within a multi-layered medium. A. Emplacement of melt into contact with water-saturated sediments. A thin vapor film develops along the contact. B. Pulsating increases in the high-pressure steam volume within the aquifer. Possible local brecciation of the country rock at this stage. C. Large-scale water/melt interaction. Mixing of country rock, steam, and melt. D. Explosive rupture of the confinement chamber.

water into the zone of mixing could lead to a sustained period of discrete explosions. Non-equilibrium thermodynamics and shock-wave physics must be considered in the analysis of this complex vaporization process. Wohletz (this volume) applies this cyclic model to develop an hypothesis on the formation of hydrovolcanic ash.

Many types of FCI phenomena have been produced under controlled conditions at Los Alamos National Laboratory (Wohletz and Sheridan, 1981; 1982; Wohletz and McQueen, 1981). The magnitude of observed explosivity varies from sporadic, pulsating ejection of large, centimeter-sized melt fragments to supersonic bursts of millimeter-sized fragments in billowing envelopes of wet condensing steam. Non-exlosive chilling and fragmentation of melt also occurred. This spectrum of experimental FCI phenomena correlates with the large-scale volcanic phenomena described as Strombolian, Surtseyan, and submarine (Walker and Croasdale, 1971).

Explosive phenomena are difficult to quantify because of the dependence of their effects on a time scale. Typically, explosive energy is scaled to a specified mass of TNT, the destructive energy of which is produced by hot, rapidly expanding gases. These expanding gases drive a shock front (the detonation wave) that causes thermal combustion of the explosive material. The rate of gas evolution depends on the velocity of the detonation wave through the material. Detonation wave velocities in high explosives are on the order of 10^3-10^4 m-s^{-1}. In contrast, the vaporization of water in FCI systems is triggered by a much slower shock wave. The propagation of such acoustic waves in systems composed of a mixture of liquid, vapor, and solids is in the range of 10^2 m-s^{-1} or less (Kieffer, 1977). Because hydrovolcanic eruptions release energy at slower rates they have less destructive potential than high explosives of equal energy. For this reason, hydrovolcanic craters can not be easily scaled to the same energy function as craters produced by high explosives.

An important aspect of FCI theory is that calculation of the explosive energy is complicated by the non-equilibrium effects of transition boiling (Buchanan and Dullforce, 1973) and superheating (Reid, 1976). These latter factors are strongly influenced by the geometry of the contact, containment pressure in the mixing zone, mass ratio of water to melt, and the temperature difference between the water and melt.

The main physical steps in an FCI cycle are summarized below:
(1) Initial contact of melt with water.
(2) Creation of a semi-insulating, superheated vapor film along the contact of the two fluids.
(3) Repeated collapse and expansion of the vapor film due to a complex balance of kinetic energy between the surrounding liquid and the film.
(4) Progressive fragmentation of the melt surface due to the kinetic energy generated by collapse and expansion of the vapor film.
(5) Increased surface contact area due to melt fragmentation and mixing with water.

(6) Increased conductive heat transfer rates with concurrent increases in film mass and energy.
(7) Vapor explosion if vaporization energy increases beyond the confinement strength (which includes surface tension effects).

At stage 7 the system can cycle back to stage 1, provided that more water and melt are available for mixing. The collapse of the steam envelopes in stage 3, which is required for explosive mixing, occurs on a time-scale of milliseconds. This collapse is inhibited by the presence of a non-condensible gas or solidification of the melt (Corradini, 1981).

DISCUSSION OF NATURAL HYDROVOLCANIC PHENOMENA

Hydroexplosions are characterized by the production of great quantities of steam and fragmented magma that are ejected from the vent in a series of eruptive pulses. Jaggar (1949) was an early advocate of the idea that many volcanic eruptions were the result of rapid vaporization of meteoric water by magma. Since the description of the 1924 eruption of Kilauea Volcano in Hawaii (Jaggar and Finch, 1924), several eyewitness accounts of other volcanic steam explosions have provided valuable information on the nature of this phenomena. These studies include the birth of maar volcano Nilahue in Chile (Muller and Vehl, 1957), the eruption of Capelinhos in the Azores (Tazieff, 1958; Servico Geologicos de Portugal, 1959), the birth of Surtsey in Iceland (Thorarinsson, 1964), and the eruption of Taal Volcano in the Philippines (Moore et al., 1966). Some important observations of the above studies include:
(1) The explosions were all periodic or pulsating.
(2) Hydroexplosions occur directly after water pours into the vent.
(3) The amount of water entering the vent and the apparent depth of explosions greatly affect the manner of pyroclast ejection.
(4) Base surges are produced.

Essentially all of the classical eruption types (Mercalli, 1907), as well as some more recently recognized types (Walker, 1973), can contain at least a small hydromagmatic component. Surtseyan activity (Thorarinsson, 1964; Walker and Croasdale, 1971) is dominantly hydromagmatic, producing mainly pyroclastic surges with minor ash or lapilli falls. Vulcanian activity (Mercali and Silvestri, 1891) has recently been shown to have a strong hydrovolcanic component (Schmincke, 1977; Nairn and Self, 1978; Frazzetta et al., this volume). Large phreatoplinian explosions produce a wide dispersal of hydrovolcanic products (Self and Sparks, 1978). The Plinian activity of Vesuvius characteristically finishes with surges and lahars (Sheridan et al., 1981; Santacroce, this volume). Strombolian activity may alternate between ash-fall and cinder production and surge clouds (Walker and Croasdale, 1971), although hydrovolcanic products are not common. Even Hawaiian volcanoes occasionally emit steam blast eruptions such as the 1790 and 1924 eruptions of Kilauea (Jaggar and Finch, 1924).

The phenomenology of volcanic steam explosions is similar in many respects to

that of underwater or underground chemical or nuclear explosions (Glasstone, 1962). However, repetition of explosions from the same vent complicate the analysis of volcanic eruptions. Both types of explosions produce the following physical phenomena:
(1) Ground seismic events that can fracture the country rock near the explosion.
(2) Atmospheric acoustic events that include shock waves (Nairn, 1976; Livshits and Bolkhovitinov, 1977; and Nairn and Self, 1978). Adiabatic cooling by rarefaction behind the shock, or refraction of light at the shock front (Perret, 1912), may form visible condensation fronts that move through the atmosphere away from the vent.
(3) An ejection plume dominantly composed of steam and clasts. The emission of accidental (country rock) and juvenile fragments usually excavates a crater, or enlargens a conduit/vent system, that is surrounded by an ejecta ring.

An ejecta plume can be considered to be comprised of two vertical components (Sparks and Wilson, 1976; Wilson, 1976); a gas-thrust region and a convective thrust region. In some cases an additional horizontal component forms a base surge or a pyroclastic surge. According to Wilson (1976), the gas thrust region is characterized by rapid deceleration of erupted materials whereas the convective thrust region receives a buoyant uplift due to heating of entrained air by the hot pyroclasts.

Coarse-grained ejecta in the vertical eruption plume move in dominantly ballistic trajectories, modified to various degrees by aerodynamic drag. Fine-grained ejecta, in contrast, are entrained in an expanding buoyant gas cloud. Their rise, fall, or lateral movements depend on the density of the cloud relative to the surrounding air. Clots of ejecta form ballistic jets, characteristic of buried explosions, with cypressoid or cock's tail shape. Variations in the amount of ejecta and steam in different eruption pulses are common in hydromagmatic explosions. Calculations of ejecta dynamics, based on ballistic theory, yield velocities ranging from tens to several hundreds of meters per second (Lorenz, 1970; Fudali and Melson, 1972; Nairn, 1976; Steinberg and Babenko, 1978; Self et al., 1980).

The most devastating component of a hydrovolcanic eruption plume is a base surge (Moore et al., 1966; Waters and Fisher, 1971; Moore and Sisson, 1981). Two mechanisms of surge formation from a nuclear detonation have been proposed (Young, 1965): (1) directed blast related to overturning of the crater rim during excavation, and (2) bulk subsidence of material falling out of the vertical explosion plume. A related mehcanism is the lateral movement of the entire eruption cloud due to gravitational instability (Waitt, 1981; Malin and Sheridan, 1982). As the surge moves outward the grain concentration in the basal layer increases whereas upward streaming vapors elutriate fine ash into a buoyant overriding cloud (Wohletz and Sheridan, 1979).

A poorly understood aspect of hydroexplosions is the mode of contact between magma and water. Most obvious is the direct pouring of water into an open vent or the movement of magma into a standing body of water. In other cases maar volcanoes have erupted where only ground water was present. However, water-saturated country rock generally contains insufficient volatiles in the pore spaces for the maximum explosive mixing ratios (30 to 50 percent).

Many maar volcanoes occur along faults, suggesting that the rise of magma into a fracture-controlled zone in the aquifer may upset the hydrologic conditions sufficiently to cause a hydroexplosion. Consider a tabular-shaped body of magma moving upward through a system of fissures. Hydrostatic pressure could drive the water into the dilated fault zone during the intrusion. Water within the system may be locally heated enough to initiate small vapor explosions. Such explosions near the surface could fracture the country rock and excavate a crater increasing the magma-water contact area. Eruptions that begin this way first eject an explosion breccia composed of dominantly fragmented country rock with subordinant juvenile material (Kienle et al., 1980; Wohletz and Sheridan, 1983). As the size of the mixing zone grows, the eruptive energy ($P \Delta V$) progressively increases due to the greater volumes of water contacting the magma. By this process, eruptions may evolve from a Strombolian type with a low rate of transfer of thermal to mechanical energy to a Surtseyan type with a high efficiency of energy transfer.

A relatively deep (a few kilometers) magma chamber may also experience a sudden influx of water, as is the case for Plinian eruptions of Vesuvius (Sheridan et al., 1981; Santacroce, this volume). This could occur when pressure within the chamber becomes less than hydrostatic pressure in the aquifer toward the end of the Plinian stage, causing implosion of the chamber roof and walls. The pore water in the surrounding rocks would be at high pressures, as expected in low permeability zones adjacent to a heat source (Delaney, 1982). The sudden pressure differential could cause brecciation of the chamber walls and flooding of the chamber interior with water and accidental blocks leading to hydroexplosions.

Water could also be gradually excluded from the erupting magma interface during the course of the eruptive cycle, as was the case during the emergence of Surtsey from the sea. At the initial contact of magma with water on the sea floor, relatively quiet melt fragmentation produced pillow lavas and hyaloclastites. When the magma erupted through this pile near the surface of the sea, violent Surtseyan activity ensued. As the vent moved above sea level the activity changed to Strombolian accompanied by passive lava emission (Thorarinsson et al., 1964).

EXPERIMENTAL MODELING OF HYDROVOLCANISM

The first experiments that attempt to quantify the phenomenology of hydroexplosions were conducted at Los Alamos National Laboratory (Wohletz and McQueen, 1981; Wohletz and Sheridan, 1981; 1982). Early experiments demonstrated the feasibility of using large quantities of thermite (100 kg) to simulate magma in various configurations with water. The thermite reaction is highly exothermic:

$$3Fe_3O_4 + 8Al = 4Al_2O_3 + 9Fe + \Delta H \qquad (1)$$

Thermite melt is similar to basaltic magma with respect to viscosity, density, and crystallization behavior. However, its enthalpy is about three times greater. In our experiments the excess enthalpy was used to produce a silicate melt by mixing quartz sand with the thermite (magnetite plus aluminum) in the explosion device. The controlled explosions ejected melt particles in a cloud of steam. Detailed descriptions of these experiments include characterization of the experimentally formed ash (Wohletz, this volume), experimental description and energy calculations (Wohletz and McQueen, 1981), and applications to planetary problems (Wohletz and Sheridan, 1981).

Previous and current experimental studies on FCIs have a strong bearing on the interpretation of our work. Our experimental configuration allowed quantitative measurements of the explosive phenomena using high speed cinematography of the ejecta plume as well as pressure and temperature records of the chamber. Approximately 90 kg of melt was produced in the upper chamber of the device. This melt penetrated an aluminum partition and contacted the water in the lower chamber. Vaporization caused a rapid (< 1 second) rise in the pressure sufficient to excede the burst limit (70 bars) of the vent seal. Pressure histories within the confinement vessel for various experiments showed spikes exceeding 350 bars that lasted less than one second, oscillating responses from 40 to 150 bars over periods of seconds, and a sustained response of several seconds exceeding 350 bars.

Melt fragments enclosed in a steam envelope were explosively ejected from the device. The sizes of melt fragments varied from micrometers to centimeters in diameter, ejection velocities from 10 to over 100 m-s^{-1}, and steam temperatures from that of condensing steam to highly superheated, expanding steam (up to at least 500°C). Particle paths followed both ballistic trajectories and turbulent, horizontally-directed flow lines.

The results of this work (Fig. 2) show that the energy of hydroexplosions depends strongly upon the mass ratio of interacting water and magma within the vent, as well as the confining pressure and geometry of the contact. Explosive efficiency is manifested by the fine-grained ejecta, superheated steam, and surging flow of materials from the orifice. Maximum explosivity for thermite experiments, measured as the conversion of efficiency of thermal to mechanical energy, occurs for ratios between 0.3 and 0.7. Because of the difference in enthalpy, these values correspond to ratios between 0.1 and 0.3 for basaltic magmas.

CHARACTERISTICS OF THE GRAINS

Clast Grain size

Abundant grain size data on surge deposits has been collected in recent

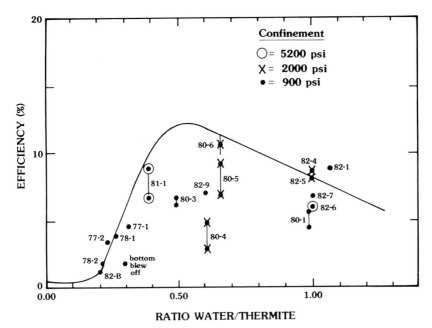

Fig. 2 Efficiency vs. water/melt ratio.

years, including information from the Azores (Walker and Croasdale, 1971), California (Crowe and Fisher, 1973), Arizona (Sheridan and Updike, 1975), southwestern U.S.A. (Wohletz and Sheridan, 1979), Idaho (Womer and Greeley, 1980), Sabatini, Italy (De Rita, et al., 1982), Mount St. Helens, Washington (Hoblitt et al., 1981), Japan (Yokoyama, 1982), Laacher See, Germany (Fisher et al., this volume), and Vulcano, Italy (Frazzetta et al., this volume). Some general observations are possible from the above data.

Walker (1971) has shown that pyroclastic-flow and pyroclastic-fall deposits occupy relatively discrete fields on a plot of median size vs sorting, but sample from surge deposits fall between the fall and flow fields. However, when considered alone, pyroclastic-surge deposits occupy a relatively well-defined field on a plot of median size and sorting (Fig. 3). Where samples are selected from different depositional units for comparison (see: Hoblitt et al., 1981; Frazzetta et al., this volume; Fisher et al., this volume), there is an apparent correspondence of grain-size distribution to bedform, as pointed out by Sheridan and Updike (1975) and Wohletz and Sheridan, (1979). Grain size gradually increases from beds with accretionary lapilli, through sandwave, massive, and planar surge beds, to lapilli-fall and breccia horizons.

The correlation of grain size with eruptive and transport mechanisms is complicated by the common incorporation of two or more discrete size populations within a single bed. Thus, moment analysis of total sample populations cannot strictly represent grain-size data of surge deposits. The separation of

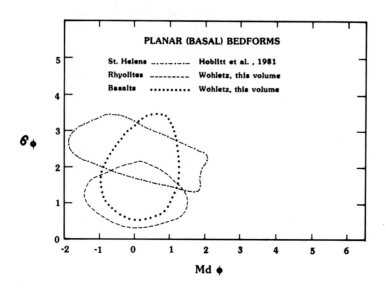

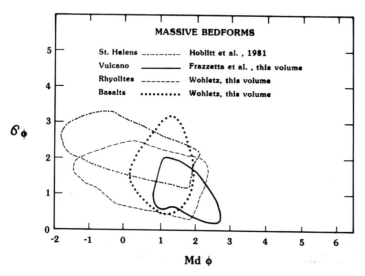

Fig. 3 Median diameter (M_ϕ) vs. sorting (σ_ϕ) for pyroclastic surge deposits.

composite particle-size data of surge deposits into their component subpopulations is a problem yet to be satisfactorily addressed. One method that has been used to distinguish subpopulations is factor analysis (Sheridan and Updike, 1975). Other techniques, such as isolating subpopulations by fitting polymodal distributions to spline functions, may also prove useful.

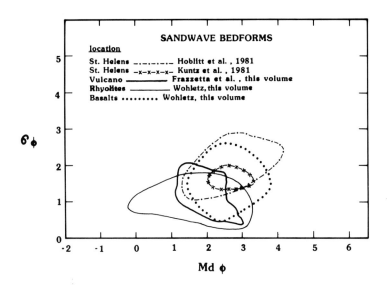

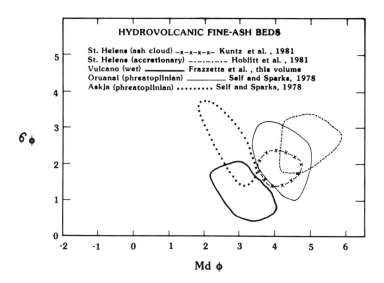

Particle morphology (SEM)

Hydrovolcanic deposits contain grains that have been significantly affected by a variety of processes related to their generation, transport, and alteration. Hence grain surface features may record a wealth of information about the history of the particles. Typically, clasts of different types and histories are brought together in hydrovolcanic deposits. Because of the complexity of these deposits, a methodology for the study of pyroclasts should be adopted that will optimize retrieval of this information (Sheridan and Marshall, 1982).

The distinctive shapes of pyroclasts resulting from volcanic steam explosions has been observed in many studies, most notably those of Heiken (1972; 1974). More recent studies of glassy hydrovolcanic clasts (Honnorez and Kirst, 1975; Wohletz and Krinsley, 1982) have found a wide range of particle shapes and textures than can be genetically interpreted. The common shapes of glass pyroclasts can be ascribed to varying energies and modes of contact of water with magma (Wohletz, this volume): blocky-equant (Fig. 4A), moss-like (Fig. 4B), plate-like (Fig. 4C), and drop or spherical (Fig. 4D).

Crystalline pyroclasts likewise have distinctive features (De Rita et al., 1982), although the origin of crystal surface textures is more obscure. The most common crystalline morphology are blocky grains bounded by large cleavage surfaces with smaller step fractures which are also related to cleavages (Fig. 4E). Where abundant water is present in the vent perfect crystals or perfect crystals coated with vesiculated glass are produced (Fig. 4F).

Besides the general shape of the particles, fine-scale surface features, readily observable by a scanning electron microscope, provide information on the depositional and diagenetic history of the clasts. Features related to grain collisions during transport include: adhering particles, edge modifications, grooves, scratches, step-like fractures, and dish-shaped fractures. Alteration features include vesicle fillings, skin cracks, and microcrystalline encrustations.

Thus, particle morphology reveals not only the mechanism for initial fragmentation of the melt, but also the amount and type of abrasion that occured during transport and deposition. Even post-depositional processes, such as palagonitization and lithification, are recorded. The percentage of broken, angular ash and fused or drop-like surfaces increases (but the vesicularity decreases) with increasing amounts of external water interaction during the eruption (Wohletz and Sheridan, 1982; Wohletz and Krinsley, 1983).

Alteration of grains

Glassy pyroclasts sampled from hydrovolcanic deposits display a variety of post-depositional alteration features. Outcrops of strongly altered deposits are usually orange to tan or brown in color and well-lithified, even in young deposits. In contrast, the color of outcrops composed of unaltered materials usually indicates their original composition (i.e. black to gray for basalts; white to tan for rhyolites), and the beds are generally unconsolidated.

Chemical alteration of the glass may be due to: 1) reaction with gases or fluids in the vent or hydration and diagenesis during initial cooling of the deposit, 2) post-emplacement hydrothermal activity, or 3) subsequent groundwater reactions. These three origins can be distinguished by the respective field geometries of their alteration zones: 1) bedform dependence (wet versus dry emplacement), 2) near-vent proximity (alteration independent of bedform or

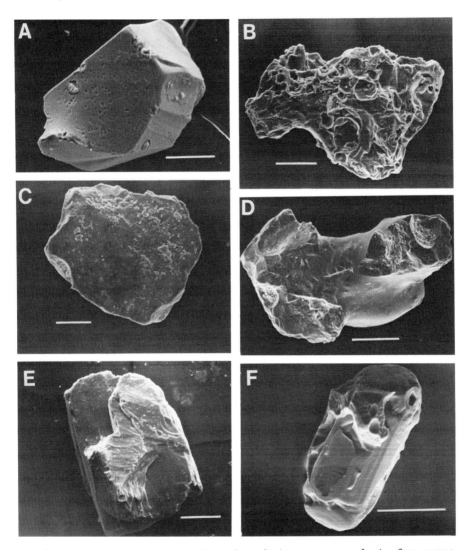

Fig. 4 SEM micrographs of typical surface features on pyroclasts from surge deposits. Glassy fragments: A. Blocky equant (Vulcano). B. Moss-like (Vulcano). C. Plate-like (Vulcano). D. Drop or fused (Taal). Crystalline pyroclasts: E. Blocky grains with large cleavage faces (Vulcano). F. Perfect crystals with adhering layer of vesiculated glass (Lipari).

horizontality), or 3) water table dependency (horizontally controlled alteration distribution). The common alteration of hydrovolcanic ash is palagonitization (Bonatti, 1965; Hay and Iijima, 1968; Honnorez, 1972; and Jakobsson, 1978) during which various elements are mobilized and redistributed on grain surfaces as a complex intergrowth of clays and zeolites.

Observation of textures on glassy particles using a scanning electron microscope reveals several stages of alteration which correspond in part to the degree of alteration and consolidation observable in outcrop. Stage 1 is marked by development of microlites of clays and zeolites within vesicle hollows or other indentations on particle surfaces. Stage 2 is noted by the formation of hydration cracks on vesicle surfaces. A thin (1 to 5 μm) hydrated skin may detach from the underlying glass between these cracks yielding an appearance similar to surface dessication textures in mud. Stage 3 alteration is characterized by the presence of hydration cracks over much of the particle surface. Overgrowths of palagonite as crusts or aggregates of fine-grained, crystalline or non-crystalline material are also abundant in this stage of alteration.

The surface chemistry of grains determined during electron microbeam analysis is also useful for distinguishing fresh and altered surfaces. Energy dispersive spectral analysis (EDS) provides adequate quantitative data for assessment of the degree of alteration. Chemical signatures change with the inferred degree of melt-water interaction from samples of tuff cone and tuff ring deposits. Strong systematic variations in iron, silica, alumina, sodium and potassium are noted between fresh and altered glass. As silica increases for these natural samples, the other elements decrease. A similar trend was found to be related to the water content of experimentally produced palagonite (Furness, 1975).

DEPOSITS

Deposits formed by hydromagmatic eruptions display a great variety of textures and structures because of the wide range of environments and explosivity of the water/melt interaction. These deposits contain primary ejecta that range from primitive to highly-evolved magma compositions. Their vents are associated with all types of feeding systems. The unifying factor for all is the contact of external water with magma that leads to fragmentation and dispersal of the clasts.

The main hydromagmatic eruption phenomena include Surtseyan, phreatoplinian, pyroclastic surge and fall, and subaqueous (including subglacial) processes. Surge phenomena (Young, 1965; Moore, 1967; Fisher and Waters, 1970) result from a range of conditions (Sheridan and Wohletz, 1981) including dry (superheated steam media) to wet (condensing steam media). The basal part of a surge may be dilute near the vent but grain concentration increases with runout distance (Wohletz and Sheridan, 1979; Fisher, 1979). The overriding cloud is generally dilute and very hot in the dry surges. Subaqueous clastic eruptions produce a mixture of particles within a water matrix that is transported at near ambient conditions. The particle concentration within such flows is extremely variable ranging from dense to dilute. Pyroclasts from phreatoplinian explosions (Self and Sparks, 1978; Self, this volume) rise in dilute, buoyant plumes to great heights in the atmosphere. The broad dispersal of clasts by lateral wind currents produces deposits at nearly ambient temperatures.

Their deposits form widespread, thin sheets of fine ash. Proximal deposits range from surge beds to stratified ash and layers with accretionary lapilli are common. Climbing megaripples and inversely graded planar stratification have been reported. The deposits show little downwind decrease in median grain size. The known vents are silicic calderas which contain lakes, the suspected source of external water.

Dry-surge deposits form thin sheets composed of unconsolidated, well-stratified deposits (Fisher and Waters, 1970). The three common types of bedforms (sandwave, massive, and planar) have a stochastic relationship that allows the definition of specific facies that are related to the distance from the source (Wohletz and Sheridan, 1979). On steep slopes proximal to the vent typical sandwave beds may be cut by U-shaped channels (Fisher, 1977) that are filled by massive density-flow deposits. Near-vent explosion breccias may cover large impact sags formed by deformation of underlying plastic layers. Many large blocks, however, are carried by the surge currents and are matrix-supported with no underlying depressions. Beds with abundant accretionary lapilli extend to medial distances where lensoid massive beds are common. Distal planar beds lack cross-stratification, but display reverse grading due to their emplacement by grain flow.

Wet surge beds typically form thick near-vent, accumulations that are strongly indurated by secondary minerals formed in the warm damp ash shortly after deposition. The beds are generally thick, massive to planar types with indistinct stratification. Mudflow and sheetwash deposits are common. Large-scale slumps and megaripples due to post-deposition deformation are common on steep slopes. Beds of vesiculated tuffs (Lorenz, 1974) and accretionary lapilli occur in most deposits. Layers plastered onto cliff faces (Heiken, 1971), trees or buildings (Waters and Fisher, 1971) attest to the cohesion provided by condensed water on grain surfaces.

Subglacial and subaqueous aquagene deposits (Carlisle, 1963) have some textural features in common with surge beds. Shallow water deposits are generally well-stratified and consist of massive and cross-stratified deposits of sand-sized clasts (Sigvaldason, 1968). The size of the fragments increases in deeper water with the progressive appearance of pillow lavas. Channels formed by turbidity flows of breccia are common on the flanks of larger volcanoes. Foreset breccia layers deposited at the angle of repose may surround some vents. Fluidization pipes and hydrobreccias are common in the vicinity of feeder dikes within volcanic constructs. Thick sections of relatively uniform hyaloclastite may form in some cases, and in others pillows may be supported in a palagonite matrix. Deposits of acid composition occur in both the submarine

(Pichler, 1965) and subglacial (Furnes et al., 1980) environments, although mafic deposits are much more common.

TYPES OF VOLCANOES AND SCALE OF THE PHENOMENA

Tuff rings and tuff cones (Heiken, 1971; Macdonald, 1972) are the most common landforms created by hydroexplosions. An understanding of these monogenetic volcanoes forms the basis for the interpretation of more complex eruption cycles in polygenetic volcanoes. Tuff rings have low topographic profiles and gentle external slopes whereas tuff cones have high profiles and steep outer slopes (Wohletz and Sheridan, 1983). Both are small volcanoes (less than 5 km diameter) and contain relatively large craters (Darwin, 1844). If the floors extend below the original ground surface they may be called maars (Ollier, 1967; Lorenz et al., 1970; 1973). Tuff rings are more commonly associated with maars than tuff cones. Their craters generally broaden and deepen as eruption progresses, leading to collapse, slump structures, and near vertical bedding inside their topographic rims (Heiken, 1971). Strongly asymmetrical deposits may be due to a change of vent location, multiple vents with different production rates, or strong prevailing winds. Crater rims may occur along crests of beds with quaquaversal dips, parallel collapse scarps, or at the intersection of adjacent craters.

The pyroclastic deposits surrounding hydrovolcanic vents range in morphology from steep-sided (30 to 35 degrees) cinder cones with small apical craters through tuff cones with moderate slopes (25 to 30 degrees) and much larger craters to tuff rings with very gentle slopes of 2 to 15 degrees. The avalanche slopes of cinder cones are due to steep angles of repose for centimeter-sized, rough-surfaced cinders. The difference in slope of tuff rings and tuff cones is due to the cohesion of the wet ash that constructs the latter structure. In addition, tuff rings rarely have rim deposit thicknesses that exceed 50 m, whereas those for tuff cones generally exceed 100 m in thickness.

The morphology of pyroclastic deposits surrounding hydrovolcanic vents is useful in determining the nature of the eruptions that produced those deposits. Tuff-ring deposits indicate high-energy surge eruptions in which mobile clouds of pyroclasts are transported relatively far from the vent. Clasts in these deposits are fine-grained and the abundant sandwave structures indicate high-energy transport. In contrast, tuff-cone deposits generally extend less than one crater diameter from the crater rim. Their tephra are relatively coarser than those of tuff-ring deposits and lapilli- or ash-fall beds are more abundant. These indicators, as well as the strong lithification due to wetness of emplacement, suggest that tuff cones result from low-energy surge and fall eruptions.

Small craters or pits are typical features of explosive activity in fumarolic geothermal areas (Muffler et al., 1971). Phreatic explosion pits can also occur where hot pyroclastic flows cover standing bodies of water (Rowley et al., 1981; Moyer, 1982). Such craters range in diameter from a few tens of meters to greater than one kilometer. They may be surrounded by a tuff cone, tuff ring, or a thin blanket of non-juvenile explosion breccia. Pseudocraters may also form by the explosive vaporization of water trapped beneath lava flows that enter water as those near Myvatn, Iceland (Rittman, 1938). These features are generally less than several hundred meters in size.

Subglacial volcanoes, common in Iceland (Walker and Blake, 1965) and Antarctica (LeMasurier, 1972), are steep-sided mountains composed of pillow lavas, pillow breccias, hyaloclastite breccias, and bedded tuffs similar in appearance to surge deposits (Jones, 1966; Sigvaldason, 1968). Some have flat-topped surfaces, with or without a vent cone, due to subaerial extrusion of lava above the level of the ice (Walker, 1965). Subaqueous volcanoes like Surtsey have an analogous structure and morphology (Kjartansson, 1966), as supported by the geologic descriptions of guyots (Christiansen and Gilbert, 1964; Moore and Fiske, 1969; Bonatti and Tazieff, 1970).

Hydrovolcanic phenomena are common at polygenetic volcanoes as well, although they seldom produce such distinctive landforms as with the monogenetic types. Their expression varies from occasional steam-blast explosions, such as the 1790 and 1924 events at Hawaii (Jaggar and Finch, 1924), to regular incorporation into the pattern of activity, as with the Plinian eruptions at Vesuvius (Sheridan et al., 1981; Santacroce, this volume) and Vulcano (Frazzetta et al., this volume). Because their deposits are relatively thin and similar in appearance to some water-laid tuffs, hydrovolcanic deposits have not been widely recognized on large volcanic structures. However, they merit much more attention because of their significance for volcanic risk evaluation and for interpretation of the role of external water in the general behavior of the volcano.

These hydrovolcanic landforms demonstrate the variety of eruption phenomena that results from the interaction of water and magma. Because water is abundant near the surface of the earth, hydrovolcanism is a likely occurrence at most volcanoes. The interaction of water and magma is also common in the subsurface, as evidenced by the brecciation of dikes and intrusions (Delaney and Pollard, 1981). Peperite (Macdonald, 1939) is produced by the shallow injection of basaltic magma into muds (Williams and McBirney, 1979).

HYDROVCLCANIC CYCLES

Hydrovolcanic phenomena occur in such a regular pattern at some volcanoes that they can be integrated into typical eruptive cycles. The interface of magmatic and hydrologic systems at central volcanoes remains fairly constant over long periods of time. A volcanic cycle can be considered as a sequence of events that follows a recognizable pattern with definable starting and ending points. Cycles may represent a period of days, as with the Plinian activity of Vesuvius, or a few centuries, as with the Fossa activity at Vulcano (Frazzetta et al., this volume). For many volcanoes a sufficient repose period exists between cycles for a soil horizon to develop at the boundary. At some central volcanoes a similar cycle may repeat several times throughout their history. Other volcanoes may exhibit an alternation of cycle types throughout its history. In general, a cycle follows a predictable pattern to its close unless the volcano-tectonic situation changes. In the case of volcanoes with a strong tendency for hydrovolcanic involvement, a cycle may record either an increase in interaction of magma with external water or a decrease in this process.

The A.D. 79 Plinian eruption of Vesuvius (Sheridan et al., 1981; Sigurdsson et al., 1982; Santacroce, this volume) is an excellent example of an eruption cycle that exhibits an increase in the hydromagmatic component with time. The eruption starts with a Plinian-fall layer overlain by stratified-fall deposits that are interspersed with surge beds. The next sequence of beds consists of pumice and ash flows, separated by surge horizons. The uppermost unit is a "wet" surge deposit with abundant accretionary lapilli.

The eruptions from Fossa of Vulcano (Frazzetta et al., this volume) represent typical examples of the reverse type of cycle that shows a decrease in hydromagmatic component with time. These cycles begin with lahars or wet surges and proceed to dry surge beds. These deposits are overlain by pumice-fall beds and the sequence is capped by a cycle-ending lava flow.

The above examples suggest that several textural indicators can be placed in an order of increasing interaction of external water with magma in order to define the progress of the cycle: lava flows, lapilli fall, stratified lapilli fall, dry surge with cross-laminations, accretionary lapilli, vesiculated tuffs, wet surge, lahar, pillows, and hyaloclastite. This data can be combined with the experimental results of Wohletz and McQueen (1981) to map the water to melt ratios during the cycle. An example of this technique for the Vesuvius and Vulcano types is shown in Fig. 5.

SUMMARY

Hydrovolcanism is a common natural phenomena that occurs in every volcanic setting. Its role is generally underestimated in volcanic eruptions. Fuel coolant interactions (FCI) are an industrial analog of natural hydroexplosions that serve as a model for hydrovolcanism. The explosive mixing of water and magma produces very fine melt fragmentation (10 to 50 µm). Experiments at Los Alamos National Laboratory have duplicated most of the phenomenology associated with hydrovolcanic explosions. The optimum mixing ratio of water to basaltic magma for efficient transfer of thermal energy to mechanical work is 0.1 to 0.3 mass fraction. Pyroclast morphology, bedding structures, and deposit morphology are distinctive and can be used to estimate the water to melt mixing ratio of the eruption (Fig. 6). The change in the character of the deposit throughout an eruption cycle can be used to interpret the behavior of the volcano and to evaluate potential hazards related to future activity.

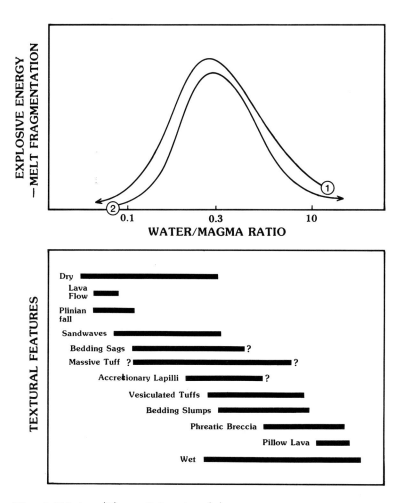

Fig. 5 Plinian (2) vs. Vulcanian (1) eruption cycles plotted on an energetics diagram. Various textures associated with water/melt ratio ranges shown below.

This paper has benefited through extended field discussions of hydrovolcanic products at Italian volcanoes with a number of people, especially: Franco Barberi, Mauro Rosi, Roberto Santacroce, Luigi La Volpe, Giovanni Frazzetta, Donatella De Rita, Gian Zuffa, Rosana De Rosa, Renato Funiciello, and Tom Moyer. The experimental work at Los Alamos National Laboratory was done with the cooperation of Robert McQueen. The SEM work was done at Los Alamos National Laboratory and Arizona State University. We would like to thank Stephen Self and Grant Heiken for critically reading the manuscript. The research was partially supported by NASA grant NAGW-245 and NSF grant INT-8200856.

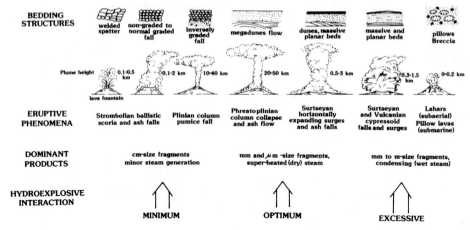

Fig. 6 The systematics of hydrovolcanic activity.

REFERENCES

Board, S.J. and Hall, R.W., 1975. Thermal explosions at molten tin/water interfaces. In: J.R. Okendon and W.R. Hodgkins (Editors), Moving boundary problems in heat flow diffusion. Clarendon Press, Oxford, pp. 259-269.

Board, S.J., Hall, R.W., and Hall, R.S., 1974. Fragmentation in thermal explosions. Inst. J. Heat Mass Transfer, 17: 331-339.

Bonatti, E., 1965. Palagonite, hyaloclastites and alteration of volcanic glass in the ocean. Bull. Volcanol., 28: 1-15.

Bonatti, E., 1967. Mechanisms of deep sea volcanism in the South Pacific. In: P.H. Ableson (Editor), Researches in Geochemistry, 2: 453-491. John Wiley & Sons, New York, N.Y.

Bonatti, E. and Tazieff, H., 1970. Exposed guyot from the Afar Rift, Ethiopia. Science, 168: 1087-1089.

Buchanan, D.S., 1974. A model for fuel-coolant interactions. J. Phys. D.: Appl. Phys., 7: 1441-1457.

Buchanan, D.S. and Dullforce, T.A., 1973. Mechanism for vapor explosions. Nature, 245: 32-34.

Carlisle, D., 1963. Pillow breccias and their aquagene tuffs: Quandra Island, British Columbia. Amer. J. Sci., 71: 48-71.

Christiansen, M.N. and Gilbert, C.M., 1964. Basaltic cone suggests constructional origin of some guyots. Science, 143: 240-242.

Christiansen, R.L. and Blank, H.R., Jr., 1972, Volcanic stratigraphy of the Quaternary rhyolite plateau in Yellowstone National Park. U.S. Geol. Survey Prof. Paper, 729-B, 18 pp.

Colgate, S.E., and Sigurgeirsson, H., 1973. Dynamic mixing of water and lava. Nature, 244: 552-555.

Corradini, M.L., 1981. Analysis and modelling of steam explosion experiments. Sandia Natl. Labs., SAND80-2131, NUREG/CR-2072, pp. 114.

Cotton, C.A., 1969. The pedestals of oceanic volcanic islands. Geol. Soc. Amer. Bull., 80: 749-760.

Crowe, B.M. and Fisher, R.V., 1973. Sedimentary structures in base-surge deposits with special reference to cross bedding, Ubehebe Craters, Death Valley, California. Geol. Soc. Amer. Bull., 84: 663-682.

Darwin, C.R., 1844. Geologic observations on the volcanic islands visited during the voyage of HMS Beagle, with brief notes on the geology of Australia and the Cape of Good Hope, being the second part of the geology of the voyage of the Beagle. London.

Delaney, P.T., 1982. Rapid intrusion of magma into wet rock: groundwater flow due to pore-pressure increases. Jour. Geophys. Res., 87: 7739-7756.

Delaney, P.T. and Pollard, D.D., 1981. Deformation of host rock and flow of magma during growth of minette dikes and breccia-bearing intrusions near Ship Rock, New Mexico. U.S. Geol. Survey Prof. Paper 1202, 61 pp.

De Rita, D., Sheridan, M.F., and Marshall, J.R., 1983. SEM surface textural analysis of phenocrysts from pyroclastic deposits at Baccano and Sacrofano volcanoes, Latium, Italy. In: W.B. Whalley and D.H. Krinsley (Editors), SEM in Geology. Geoabstracts, Norwich, England, in press.

Drumheller, D.S., 1979. The initiation of melt fragmentation in fuel-coolant interactions. Nucl. Sci. Eng., 72: 347-356.

Fisher, R.V., 1968. Puu Hou littoral cones, Hawaii. Geol. Rundschau, 57: 837-864.

Fisher, R.V., 1977. Erosion by volcanic base-surge density currents: U-shaped channels. Geol. Soc. Am. Bull., 88: 1287-1297.

Fisher, R.V., 1979. Models for pyroclastic surges and pyroclastic flows. J. Volcanol. Geotherm. Res., 6: 305-318.

Fisher, R.V. and Waters, A.C., 1970. Base-surge bed forms in maar volcanoes. Am. J. Sci., 268: 157-180.

Fisher, R.V., Schmincke, H.U., and Van Bogaard, P., 1983. Origin and emplacement of a pyroclastic flow and surge unit at Laacher See, Germany. In: M.F. Sheridan and F. Barberi (Editors), Explosive Volcanism. J. Volcanol. Geotherm. Res., 17: (this volume).

Forrester, R.W. and Taylor, H.P., Jr., 1972. Oxygen and hydrogen isotope data on the interaction of meteoric ground water with a gabbro-diorite stock, San Juan Mountains, Colorado. Int. Geol. Cong., 24th, Montreal, Geochem. Proc., 10(24): 254-263.

Frazzetta, G., La Volpe, L., and Sheridan, M.F., 1983. Evolution of the Fossa cone, Vulcano. In: M.F. Sheridan and F. Barberi (Editors), Explosive Volcanism. J. Volcanol. Geotherm. Res., 17: (this volume).

Frölich, G., Müller, G., and Unger, G., 1976. Experiments with water and hot melts of lead. J. Non-Equilib. Thermodyn., 1: 91-103.

Fudali, A.F., and Melson, W.G., 1972. Ejecta velocities, magma chamber pressures, and kinetic energy associated with the 1968 eruption of Arenal Volcano. Bull. Volcanol., 35: 381-401.

Furnes, H., 1974. Volume relations between palagonite and authigenic minerals in hyaloclastites, and its bearing on the rate of palagonitization. Bull. Volcanol., 38: 173-186.

Furnes, H., Fridleifsson, I.B., Atkins, F.B., 1980. Subglacial volcanoes: on the formation of acid hyaloclastites. J. Volcanol. Geotherm. Res., 8: 137-160.

Gault, D.E., Shoemaker, E.M., and Moore, H.J., 1963. Spray ejecta from the lunar surface by meteoroid impact. NASA Tech. Note, D-1767.

Glasstone, S., 1950. (Editor), The effects of atomic weapons, Los Alamos, New Mexico. U.S. Atomic Energy Comm., Los Alamos Sci. Lab., N.M., 156 pp.

Green, J. and Short, N.M., 1971. Volcanic landforms and surface features, a photographic atlas and glossary. Springer-Verlag, New York, pp. 522.

Hay, R.L. and Iijima, A., 1968. Nature and origin of palagonite tuffs of the Honolulu group on Oahu, Hawaii. Geol. Soc. Amer. Mem., 116: 331-376.

Heiken, G., 1971. Tuff rings: examples from the Fort Rock-Christmas Lake Valley basin, south-central Oregon. J. Geophys. Res., 76: 5615-5626.

Heiken, G., 1972. Morphology and petrography of volcanic ashes. Geol. Soc. America Bull., 83: 1961-1988.

Heiken, G., 1974. An atlas of volcanic ash. Smithsonian Contrib. Earth Sci., no. 12, pp. 101.

Hildreth, W., 1981. Gradients in silicic magma chambers: implications for lithospheric magmatism. J. Geophys. Res., 86: 10153-10192.

Hoblitt, R.P., Miller, C.D., and Vallance, J.W., 1981. Origin and stratigraphy of the deposit produced by the May 18 directed blast. In: P.W. Lipman and D.R. Mullineaux (Editors), The 1980 eruptions of Mount St. Helens, Washington. U.S. Geol. Survey Prof. Paper 1250: 401-420.

Honnorez, J., 1972. La Palagonitisation. Vulkaninstitut Immanuel Friedllaender, No. 9. Birkhauser Verlag, Basel., 131 pp.

Honnorez, J. and Kirst, P., 1975. Submarine basaltic volcanism: morphometric parameters for discriminating hyaloclastites from hyalotuffs. Bull. Volcanol., 39: 441-465.

Housen, K.R., Schmidt, R.M., and Holsapple, K.A., 1983. Crater ejecta scaling laws: fundamental forms based on dimensional analysis. J. Geophys. Res., 88: 2485-2499.

Jaggar, T.A., 1949. Steam blast volcanic eruptions. Hawaiian Volcano Obs., 4th Spec. Rept., pp. 137.

Jaggar, T.A. and Finch, R.H., 1924. The explosive eruption of Kilauea in Hawaii, 1924. Am. J. Sci., 8: 353-374.

Jakobsson, S.P., 1978. Environmental factors controlling the palagonitization of Surtsey tephra, Iceland. Geol. Soc. Denmark Bull. 27: 91-105.

Jones, J.G., 1966. Intraglacial volcanoes of southwest Iceland and their significance in the interpretation of the form of the marine basaltic volcanoes. Nature, 212: 586-588.

Kalamarides, R.I., 1982. $\delta^{18}O$ fractionation among phases and evaluation of contamination of the Kiglapait layered intrusion. Am. Geophys. Union Trans., EOS, 63(45): 1151.

Kieffer, S.W., 1977. Sound speed in liquid-gas mixtures: water-air and water-steam. J. Geophys. Res., 82: 2895-2904.

Kienle, J., Kyle, P.R., Self, S., Motyka, R.J., and Lorenz, V., 1980. Ukinrek Maars, Alaska, I. April 1977 eruption sequence, petrology and tectonic setting. J. Volcanol. Geotherm. Res., 7: 11-37.

Kjartansson, G., 1966. A comparison of table mountains in Iceland and the volcanic island of Surtsey of the south coast of Iceland. Natturufraedingurinn, 36: 1-34.

Lemasurier, W.E., 1972. Volcanic record of Cenozoic glacial history of Marie Byrd Land. In: R.J. Adie (Editor), Antarctic Geology and Geophysics. Universitetsforglaget, Oslo. pp. 251-260.

Lipman, P.W. and Friedman, I., 1975. Interaction of meteoric water with magma: an oxygen-isotope study of ash-flow sheets from southern Nevada. Geol. Soc. Amer. Bull., 86: 695-702.

Livshits and Bolkhovilinov, 1977. Weak shock wave in the eruption column. Nature, 267: 420-421.

Lorenz, V.W., 1970. Origin of Hole-in-the-ground, a maar in central Oregon. Unpub. NASA Rept., 113 pp.

Lorenz, V.W., 1974. Vesiculated tuffs and associated features. Sedimentology, 21: 273-291.

Lorenz, V.W., 1973. On the formation of maars. Bull. Volcanol., 37: 183-204.

Lorenz, V.W., McBirney, A.R., and Williams, H., 1970. An investigation of volcanic depressions. Part III, maars, tuff rings, tuff cones, and diatremes. Houston, NASA Prog. Rept. NGR 38003012, pp. 198.

Macdonald, G.A., 1939. An intrusive peperite at San Pedro Hill, Calif. Univ. Calif. Pub., Bull. Geol. Sci., 24: 329-338.

Macdonald, G.A., 1972. Volcanoes. Prentice-Hall, Inc., N.J., 510 pp.

Malin, M.C. and Sheridan, M.F., 1982. Computer-assisted mapping of pyroclastic surges. Science, 217: 637-639.

Mercali,, G., 1907. I vulcani attivi della terra. Ulrico Hoepli, Milan, 421 pp.

Mercali, G. and Silvestri, O., 1891. Le eruzioni dell'isola di Vulcano, incominciate il 3 Agosto 1888 e terminate il 22 Marzo 1890. Relazione Scientifica, 1891 Ann. Uff. Cent. Metr. Geodin., 10(4): 1-213.

Moore, J.G., 1967. Base surge in recent volcanic eruptions. Bull. Volcanol., 30: 337-363.

Moore, J.G., 1975. Mechanism of formation of pillow lava. Amer. Sci., 63: 269-277.

Moore, J.G. and Ault, M.V., 1965. Historic littoral cones in Hawaii. Pacific Sci., 19: 3-11.

Moore, J.G. and Fiske, R.S., 1969. Volcanic substructure inferred from dredge samples and ocean bottom photographs, Hawaii. Geol. Soc. Amer. Bull., 80: 1191-1202.

Moore, J.G. and Sisson, T.W., 1981. Deposits and effects of the May 18 pyroclastic surge, In: P. W. Lipman and R. R. Mullineaux (Editors), The 1980 eruptions of Mount St. Helens. U.S. Geol. Survey Prof. Paper, 1250: 421-438.

Moore, J.G., Nakamura, K., and Alcaraz, A., 1966. The 1965 eruption of Taal Volcano. Science, 151: 995-960.

Moyer, T.C., 1982. Deposits of the phreatic-explosion pits of the 18 May, 1980 Mount St. Helens pyroclastic flows. Am. Geophys. Union. EOS, 63(45): 1141.

Muffler, L.J.P., White, D.E., and Truesdell, A.H., 1971. Hydrothermal explosion craters in Yellowstone National Park. Geol. Soc. America Bull., 82: 723-740.

Muller, G. and Vehl, G., 1957. The birth of Nilahue, a new maar type volcano at Rinihue, Chile. Int. Geol. Cong. 25th, Mexico 1956, Rept. sec. 1, 375-396 pp.

Nairn, I.A., 1976. Atmospheric shock waves and condensation clouds from Ngauruhoe explosive eruptions. Nature, 259: 190-192.

Nairn, I.A. and Self, S., 1978. Explosive eruptions and pyroclastic avalanches from Ngauruhoe in February, 1975. J. Volcanol. Geotherm. Res., 3: 39-60.

Nairn, I.A. and Wirdadiradja, S., 1980. Late Quaternary hydrothermal explosion breccias at Kawerau geothermal field, New Zealand. Bull. Volcanol., 43: 1-14.

Noe-Nygaard, A., 1940. Subglacial volcanic activity in ancient and recent times. Folia Geogr. Danica, 1: no. 2.

Oberbeck, V.R., 1975. The role of ballistic erosion and sedimentation in the lunar stratigraphy. Rev. Geophys. Space Phys., 13: 337-362.

Ollier, C.D., 1967. Maars: their characteristics, varieties and definitions. Bull. Volcanol., 31: 232-247.

Ollier, C.D., 1974. Phreatic eruptions and maars. In: L. Civetta et al., (Editors), Physical Volcanology. Elsevier, New York, N.Y., pp. 289-310.

Peckover, R.S., Buchanan, D.J., and Ashby, D.E.T.F., 1973. Fuel-coolant interactions in submarine volcansim. Nature, 245: 307-308.

Perret, F.A., 1912. Eruptions of Vesuvius, 1906. Am. J. Sci., 4: 329-333.

Pichler, H., 1965. Acid hyaloclastites. Bull. Volcanol., 28: 293-310.

Reid, R.G., 1976. Superheated liquids. Amer. Sci., 64: 146-156.

Rittman, A., 1938. Die Vulkane am Myvatn in Nordost-Island. Bull. Volcanol., 4: 3-38.

Rowley, P.D., Kuntz, M.A., and Macleod, N.S., 1981. Pyroclastic flow deposits. In: P.W. Lipman and D.R. Mullineaux (Editors), The 1980 Eruptions of Mount St. Helens, Washington, U.S. Geol. Survey Prof. Paper, 1250: 489-512.

Sandia Laboratories, 1975. Core meltdown experimental review. SAND74-0382, pp. 472.

Santacroce, R., 1983. A general model for the behavior of the Somma-Vesuvius volcanic complex. In: M.F. Sheridan and F. Barberi (Editors), Explosive Volcanism. J. Volcanol. Geotherm. Res., 17: (this volume).

Schmincke, H.U., 1977. Phreatomagmatische Phasen in quartaren Vulkanen der Osteif. Geol. Jahrbuch., A39: 3-45.

Self, S. and Sparks, R.S.J., 1978. Characteristics of widespread pyroclastic deposits formed by the interaction of silicic magma and water. Bull. Volcanol., 41(3): 196-212.

Self, S., Kienle, J., and Huot, J.P., 1980. Ukinrek maars, Alaska, II. Deposits and formation of the 1977 craters. J. Volcanol. Geotherm. Res., 7: 39-65.

Servico Geologicos de Portugal, 1959. Le volcanisme de l'isle de Faial et l'eruption de volcan de Capelinhos. Mem. 4 (nova ser.), 100 pp.

Shepherd, J.B. and Sigurdsson, H., 1982. Mechanism of the 1979 explosive eruption of Soufriere Volcano, St. Vincent J. Volcanol. Geotherm. Res., 13: 119-130.

Sheridan, M.F. and Updike, R.G., 1975. Sugarloaf Mountain Tephra: A Pleistocene rhyolitic deposit of base-surge origin in northern Arizona. Geol. Soc. Amer. Bull., 86: 571-581.

Sheridan, M.F. and Wohletz, K.H., 1981. Hydrovolcanic explosions: the systematics of water-pyroclast equilibration. Science, 212: 1387-1389.

Sheridan, M.F. and Marshall, J.R., 1983. SEM examination of pyroclastic materials: basic considerations. Scanning Electron Microscopy, in press.

Sheridan, M.F., Barberi, F., Rosi, M., and Santacroce, R., 1981. A model for the eruptions of Vesuvius. Nature, 289: 282-285.

Sigurdsson, H., Cashdollar, S., and Sparks, R.S.J., 1982. The eruption of Vesuvius in A.D. 79: reconstruction from historical and volcanological evidence. Am. Jour. Archeol., 86: 39-51.

Sigvaldason, G., 1968. Structure and products of subaquatic volcanoes in Iceland. Contr. Mineral. and Petrol., 18: 1-16.

Sparks, R.S.J. and Wilson, L., 1976. A model for the formation of ignimbrite by gravitational column collapse. Geol. Soc. London J., 132: 441-451.

Stearns, H.T. and MacDonald, G.A., 1946. Geology and groundwater resources of the island of Hawaii. Hawaii Div. Hydrol. Bull., 67: 13-49.

Steinberg, G.S. and Babenko, J.I., 1978. Experimental velocity and density determination of volcanic gasses during eruption. J. Volcanol. Geotherm. Res., 3: 89-98.

Stöffler, D., Gault, D.E., Wedekind, J., and Polkowski, G., 1975. Experimental hypervelocity impact into quartz sand: distribution and shock metamorphism of ejecta. J. Geophys. Res., 80: 4062-4077.

Tazieff, H.K., 1958. L'eruption 1957-1958 et la tectonique de Fail (Azores). Soc. Belg. Geol. Bull., 67: 13-47.

Thorarinsson, S., 1964. Surtsey, the new island in the North Atlantic. Reykjavik, Almenna Bokofelagid, pp. 47.

Thorarinsson, S., Einarsson, T., and Sigvaldason, G., 1964. The submarine eruption of the Vestmann Islands, 1963-63. Bull. Volcanol., 27: 437-446.

Waitt, R.B., 1981. Devastating pyroclastic density flow and attendant air fall of May 18 -- stratigraphy and sedimentology of deposits. In: P.W. Lipman and D.R. Mullineaux (Editors), The 1980 Eruptions of Mount St. Helens, Washington. U.S. Geol. Survey Prof. Paper, 1250: 439-460.

Walker, G.P.L., 1971. Grain size characteristics of pyroclastic deposits. J. Geol., 79: 696-714.

Walker, G.P.L., 1973. Explosive volcanic eruptions - A new classification scheme. Geol. Rundsch., 62: 431-446.

Walker, G.P.L. and Blake, D.H., 1966. The formation of a palagonite breccia mass beneath a valley glacier in Iceland. Quart. J. Geol. Soc. London, 122: 45-61.

Walker, G.P.L. and Croasdale, R., 1971. Characteristics of some basaltic pyroclasts. Bull. Volcanol., 35: 303-317.

Waters, A.C. and Fisher, R.V., 1971. Base surges and their deposits: Capelinhos and Taal Volcanoes. J. Geophys. Res., 76: 5596-5614.

Wentworth, C.K., 1938. Ash formations of the island of Hawaii. Hawaii Volcano Observ. 3rd Spec. Rept., Honolulu, Hawaiian Volcano Res. Assoc., pp. 173.

Wentworth, C.K. and Winchell, H., 1947. Koolau basalt series, Oahu, Hawaii. Geol. Soc. Amer. Bull., 58: 49-78.

Williams, H. and McBirney, A., 1979. Volcanology. Freeman, Cooper & Co., San Francisco. 397 pp.

Wilson, L., 1976. Explosive volcanic eruptions - III. Plinian eruption columns. Geophys. J. Roy. Soc., 45: 543-556.

Witte, L.C., Cox, J.E., and Bouvier, J.E., 1970. The vapor explosion. J. Metals, 22: 39-44.

Wohletz, K.H., 1983. Mechanism of hydrovolcanic pyroclast formation: Size, scanning electron microscopy, and experimental studies. In: M.F. Sheridan and F. Barberi (Editors), Explosive Volcanism. J. Volcanol. Geotherm. Res., 17: (this volume).

Wohletz, K.H. and Sheridan, M.F., 1979. A model of pyroclastic surge. In: C.E. Chapin and W.E. Elston (Editors), Ash-flow Tuffs. Geol. Soc. America Spec. Paper, 180: 177-194.

Wohletz, K.H. and Sheridan, M.F., 1981. Rampart crater ejecta: experiments and analysis of melt/water interactions. NASA Tech. Memor. 82385: 134-136.

Wohletz, K.H. and McQueen, R., 1981. Experimental hydromagmatic volcanism. Amer. Geophys. Union Trans., EOS, 62(45): 1085.

Wohletz, K.H. and Sheridan, M.F., 1982. Melt/water interactions: series II experimental design. NASA Tech. Memor., 84211: 169-171.

Wohletz, K.H. and Krinsley, D.H., 1983. Scanning electron microscopic analysis of basaltic hydromagmatic ash. In: W.B. Whaley and D.H. Krinsley (Editors), SEM in Geology. Geoabstracts Norwich, England, (in press).

Wohletz, K.H. and Sheridan, M.F., 1983. Hydrovolcanic explosions II. Tuff rings and tuff cones. Am. J. Sci., in press.

Womer, M.B., Greeley, R., and King, K.S., 1980. The geology of Split Butte -- a maar of the south-central Snake River Plain, Idaho. Bull. Volcanol., 43: 453-471.

Yokoyama, S., 1980. Base surge deposits in Japan. In: S. Self and R.S.J. Sparks (Editors), Tephra Studies. D. Reidel Pub. Co., Dordrecht, Holland. pp. 427-432.

Young, G.A., 1965. The physics of the base surge. White Oak, Maryland, U.S. Naval Ordinance Lab., NOLTR64-130, pp. 284.

MECHANISMS OF HYDROVOLCANIC PYROCLAST FORMATION: GRAIN-SIZE, SCANNING ELECTRON MICROSCOPY, AND EXPERIMENTAL STUDIES

KENNETH H. WOHLETZ

Earth and Space Division, Los Alamos National Laboratory, Los Alamos, NM 87545 (U.S.A.)

(Received September 15, 1982; revised and accepted December 20, 1982)

ABSTRACT

Wohletz, K.H., 1983. Mechanisms of hydrovolcanic pyroclast formation: grain-size, scanning electron microscopy, and experimental studies. In: M. F. Sheridan and F. Barberi (Editors), Explosive Volcanism. J. Volcanol. Geotherm. Res., 17: 31-63.

Pyroclasts produced by explosive magma/water interactions are of various sizes and shapes. Data from analysis of over 200 samples of hydrovolcanic ash are interpreted by comparison with experimentally produced ash. Grain size and scanning electron microscopy (SEM) reveal information on the formation of hydrovolcanic pyroclasts. Strombolian explosions result from limited water interaction with magma and the pyroclasts produced are dominantly centimeter-sized. With increasing water interaction, hydrovolcanism increases in explosivity to Surtseyan and Vulcanian activity. These eruptions produce millimeter- to micron-sized pyroclasts. The abundance of fine ash (<63 µm diameter) increases from 5 to over 30 percent as water interaction reaches an explosive maximum. This maximum occurs with interactions of virtually equal volumes of melt and water.

Five dominant pyroclast shape-types, determined by SEM, result from hydrovolcanic fragmentation: (1) blocky and equant; (2) vesicular and irregular with smooth surfaces; (3) moss-like and convoluted; (4) spherical or drop-like; and (5) plate-like. Types 1 and 2 dominate pyroclasts greater than 100 µm in diameter. Types 3 and 4 are typical of fine ash. Type 5 pyroclasts characterize ash less than 100 µm in diameter resulting from hydrovolcanic fragmentation after strong vesiculation.

Fragmentation mechanisms observed in experimental melt/water interactions result from vapor-film generation, expansion, and collapse. Fragments of congealed melt are products of several alternative mechanisms including stress-wave cavitation, detonation waves, and fluid instability mixing. All result in rapid heat transfer. These mechanisms can explain the five observed

0377-0273/83/$03.00   © 1983 Elsevier Science Publishers B.V.

pyroclast shapes. Stress-wave fracturing (cavitation) of the melt results from high pressure and temperature gradients at the melt/water interface. Simultaneous brittle fracture and quenching produces Type 1 pyroclasts. Type 2 develops smooth fused surfaces due to turbulent mixing with water after fracture and before quenching. Fluid instabilities promote turbulent mixing of melt and water and produce fine ash. This kind of fragmentation occurs during high-energy explosions. The increased melt surface area due to fine fragmentation promotes high-efficiency heat exchange between the melt and water. Shapes of resulting pyroclasts are determined by maximum surface area (Type 3) and surface tension effects (Type 4). Type 5 pyroclasts result from nearly simultaneous vesicle burst and melt/water fragmentation.

INTRODUCTION

The modeling of pyroclastic rock formation is of prime importance, not only in consideration of the hazards of explosive volcanism, but also in understanding the role of volatiles in the chemical and physical evolution of magmas. The volatile species in this study is water and its equation of state under extreme conditions of pressure and temperature determines not only explosive energy but also affects post-eruptive alteration of pyroclastic deposits, an important aspect in the formation of many soils.

Two dominant mechanisms of pyroclast formation are generally considered important: (1) the magmatic mechanism which involves vesicle nucleation, growth, and disruption of the magma by exsolution of volatiles from the melt during its rise and decompression, and (2) the hydrovolcanic mechanism (see Sheridan and Wohletz, this volume) which operates during contact of melt with external water at or near the surface of the earth (marine, lacustrine, fluvatile, ground, or connate water).

Magmatic fragmentation has recently been reviewed and analyzed by Sparks (1978). Earlier informative considerations of bubble growth in magmas include Verhoogen (1951) who discussed the disruption of magma by bubble coalescence, McBirney and Murase (1970) who demonstrated the effect of gas pressure within bubbles exceeding the surface tension of the magma, and Bennett (1974) who discussed the role of expansion waves in a vesiculating magma.

In contrast to magmatic fragmentation which has been studied from both theoretical approaches and experimental evidence, hydrovolcanic fragmentation has been investigated largely by field observation. Hyaloclastite is the term suggested by Honnorez and Kirst (1975) for glass found with pillow basalts produced by non-explosive quenching and fracturing of basaltic glass whereas hyalotuff is used for explosive fragmentation of glass due to phreatomagmatic eruptions. Hyaloclastite may also be formed at depths greater than 500 m on seamounts where hydrostatic pressure is great enough to prevent vesiculation (McBirney, 1963). Consideration of the explosive mechanism and observation of blocky, equant glass shapes has resulted in the general conclusion that hydrovolcanic (phreatomagmatic) ash is formed by thermal contraction and shattering of glass (von Waltershausen, 1853; Peacock, 1926; Fisher and Waters, 1969; and Heiken, 1971). This conclusion is supported by Carlisle (1963) who observed non-vesicular teardrop- and spindle-shaped hyaloclasts and experimentally produced curved splinters and pointed chips of sideromelane. Also, Honnorez and Kirst (1975) observed blocky grain shapes in blast furnace slags quenched by water.

Walker and Croasdale (1971), Heiken (1972, 1974), and Honnorez and Kirst (1975) used optical microscopy and scanning electron microscopy (SEM) to characterize hydrovolcanic ash produced in Surtseyan eruptions and submarine extrusion of pillow basalts. Heiken (1972, 1974) presented the most extensive SEM study of pyroclast shapes and found a marked difference in grain morphology between magmatic and phreatomagmatic ashes. Equant, blocky shapes and curviplanar surfaces with a paucity of vesicles resulted when fragmentation was caused by thermal contraction due to rapid chilling. Conversely, the flat, elongate, and pyramidal shapes and the drop-like shapes that Heiken (1972) described were variations due to magma chemistry and vesicle abundance.

Wohletz and Krinsley (1982) studied glassy basaltic pyroclasts and distinguished a number of textural features related to fragmentation mechanism from those caused by transport abrasion and secondary alteration. The frequency of broken planar surfaces decreases and the abundance of vesicle surfaces increases with decreasing energy of emplacement and with increasing median grain size of the deposit. This relationship is complicated, however, by transport processes which cause pyroclasts to increase in both roundness and number of small broken surfaces with increasing transport duration in pyroclastic surges. Many hydrovolcanic pyroclasts also have surfaces covered with fine adhering dust. This characterizing feature, the result of the cohesiveness of larger hydrovolcanic particles with the fine fraction contribution, however, may be confused or obscured by secondary alteration products such as the fine clay, opal, and zeolite materials of palagonite. In fact, strong alteration may completely obscure primary grain morphology. Since hydrovolcanic ash may be quickly altered because of its emplacement by steam-rich eruptions, care must be taken to distinguish between primary and alteration morphologies.

The size distributions of hydrovolcanic pyroclasts are strongly controlled by eruption energy which, in turn, determines dispersal mechanisms and resulting deposit types. Numerous published size analyses of pyroclasts have shown the relationship of particle size to mechanisms of transport and deposition. Walker (1971) demonstrated the size characteristics of flow and fall tephra by plotting the sorting coefficient versus median diameter. Later Walker (1973) showed that phreatomagmatic pyroclasts have a much higher degree of fragmentation than do magmatic ones.

An important consideration is that both the magmatic and hydrovolcanic fragmentation mechanisms may operate simultaneously during eruption. This situation is illustrated by Self and Sparks (1978) for phreatoplinian silicic eruptions in which the magma is initially disrupted by exsolution and expansion of magmatic volatiles producing a relatively coarse-grained population of pyroclasts followed by further fragmentation (fine-grained pyroclasts) due to explosive interaction with water (Tazieff, 1968). Theoretical consideration of experimental fragmentation mechanisms suggests that stress waves produced by high-pressure vaporization of water at the magma-water interface may induce vesiculation in the melt.

The samples discussed in this report are basaltic and silicic pyroclasts and their collection localities are summarized in Table 1. These samples were taken from tuff cones and tuff rings (Heiken, 1971; Wohletz and Sheridan, 1982), the characteristic vent types of hydrovolcanism. These volcanoes consist of pyroclastic fall, surge, and flow deposits. There is increasing evidence that portions of some large ignimbrites surrounding calderas (Self and Sparks, 1978) and some ash flows erupted from stratovolcanoes (Sheridan et al., 1981) are also of hydrovolcanic origin. Samples of these are not included in this study but they may yield additional information.

FUEL-COOLANT INTERACTION THEORY

Fuel-coolant interaction (FCI) explosions result from the interaction of a hot fluid (fuel) with a cold fluid (coolant) whose vaporization temperature is below that of the former. FCIs have attracted considerable investigation in the realm of small-scale laboratory experiments and in theoretical developments (Sandia Laboratories, 1975). Accidental contact of molten materials with water at foundries has produced violent explosions approaching maximum thermodynamic yield as discussed by Lipsett (1966) and Witte et al. (1970). Analysis of the debris revealed that the source of explosive energy is not due to chemical reactions, but rather is due to rapid (millisecond) heat transfer from the melt to the water which produced explosive vaporization and production of fine-grained debris. Because theoretical models of this process have been given in detail elsewhere (Corradini, 1981a; Drumheller, 1979; Buchanan, 1974), only a qualitative discussion is presented here. Although laboratory experiments develop only low thermodynamic efficiencies (0.1 to 10 percent), valuable information about the explosive heat-transfer mechanism has been obtained (Board and Hall, 1975; Dullforce et al., 1976; Fröhlich et al., 1976; Nelson and Duda, 1981). Larger-scale experiments conducted in the field (Buxton and Benedict, 1979) and those discussed here (Wohletz and McQueen, 1981), however, have produced higher efficiencies. The field experiments do pose problems in quantification of mechanical energy which, when divided by the total thermal energy, gives the efficiency. Essentially there are two explanations of FCI explosions, one is superheating and homogeneous nucleation of water, and the other is pressure-induced detonation. Superheat vaporization (Reid, 1976) occurs after relatively slow heating of water into a metastable state. During this process, the water temperature increases past the vaporization point and is limited by the homogeneous nucleation temperature at which all the water

TABLE 1

Hydrovolcanic pyroclast sample sources

Location	Composition	Vent type	Deposit type
Pinacate Mexico	Basalt	maar with tuff ring and tuff cone	Surtseyan fall, surge
Hopi Buttes Arizona	Basalt	diatreme	Vulcanian, vesiculated tuff, surge
Kilbourne Hole New Mexico	Basalt	maar with tuff ring	Surtseyan surge, fall, lahar
Zuni Salt Lake New Mexico	Basalt	maar with tuff ring	Surtseyan surge, Strombolian fall
Koko Crater Hawaii	Basalt	tuff cone	Surtseyan fall and surge
Taal Volcano Philippines	Basalt	tuff ring	Surtseyan fall and surge
Surtsey Iceland	Basalt	tuff ring	Surtseyan fall and surge
Ubehebe California	Basalt	maar with tuff ring	Surtseyan fall, surge and explosion breccia
Inyo-Mono Craters California	Rhyolite	tuff rings	Phreatoplinian fall and surge, phreatic breccia
Panum Crater, California	Rhyolite	tuff ring	Phreatoplinian fall, surge, and flow
Clear Lake California	Basalt through dacite mixed magma	tuff ring and fall blankets	Surtseyan and Phreatoplinian fall and surge
Vulcano Italy	Trachyte and Rhyolite	tuff cone	Vulcanian and Surtseyan surge, fall, flow, and lahar

vaporizes instantaneously. The maximum measured superheat for water is 280°C at one atmosphere (Apfel, 1972) and increases with increasing pressure. Pressure-induced detonation (Drumheller, 1979) requires some physical disturbance initially fragmenting the melt and thereby increasing the contact surface area with water which, in turn, greatly increases heat-transfer rates. It is likely that both of these mechanisms operate during high-efficiency explosions. Heat transfer during superheating is still poorly understood and requires assessment of non-equilibrium thermodynamics, which may be a very important consideration in explosive heat-transfer process which occurs at rates three orders of magnitude greater than those in normal boiling (Witte et al., 1970).

The fragmentation-vaporization process has been shown experimentally to be a cyclic process of vapor film generation and collapse. The energy of this collapse is partially cycled back into the system, generating new contact surfaces so that the system is self-sustaining. The steps in this feedback process are included by models of Buchanan (1974), Board et al. (1974), and Corradini (1981b). They present the following cycle:

Stage 1. The initial contact of melt with water creates a vapor film at the interface. This stage includes the rise of magma into a zone of near surface water or water-saturated, unconsolidated materials.

Stage 2. The vapor film expands to the limit of condensation and then collapses. This expansion and collapse may occur several times until the energy of the collapse is sufficient to fragment the melt.

Stage 3. The penetration into or mixing of the collapsed film with the melt increases surface area.

Stage 4. Rapidly increasing heat transfer takes place as the water encloses melt fragments.

Stage 5. Formation of a new vapor film as water is suddenly vaporized by superheating. This new film expands and the process cycles back to stage two.

This feedback process allows a small vaporization zone to grow in size by many cycles until an explosion occurs. However, subsequent vapor collapses may be of limited strength so that only a non-chilled, heat conductive surface area is maintained. By this means a coarse, melt breccia is formed.

The expansion and collapse of a film jacket has been documented by high-speed cinematography by Nelson and Duda (1981). They and other investigators conclude that melt fragmentation by film collapse can occur by several mechanisms: (1) Axisymmetric collapse produces a water jet which penetrates and fragments the melt. (2) Symmetric or asymmetric film collapse allows the water to impact the melt surface. This impact generates a stress wave that is of sufficient energy to cavitate the melt. Kazimi (1976) showed that stress waves can also be formed by the violent film expand/collapse oscillation. (3) Trigger-induced film collapse causes liquid-liquid contact between water and melt which, in turn, causes rapid fuel fragmentation by Taylor instabilities. The rapid high-presure steam generation then causes further melt framentation. (4) Instability fragmentation due to the relative velocities of the melt and water occurs due to the passage of a shock wave. Both Rayleigh-Taylor and Kelvin-Helmholtz fluid instabilities develop where the lighter fluid, water, is accelerated into the melt (Board et al., 1975).

In all except the last of the above fragmentation mechanisms, vapor-film collapse is required as is also the case in explaining the small-scale laboratory tests (Nelson and Buxton, 1978). Corradini (1981b) shows that the presence of either non-condensible gasses or high ambient pressures can suppress a vapor explosion. Both of these conditions are tigger-related, that is, a quick fluctuation in either may initiate film collapse and explosion. Buchanan and Dullforce (1973) describe transition boiling effects which may cause a sudden drop in heat-transfer rates and thereby trigger collapse.

The hydrovolcanic mechanism of explosive fragmentation has been noted by many workers since Fuller (1931) discussed the aqueous chilling of basalt and Jaggar (1949) wrote his detailed observations of the 1924 eruption of Kilauea Volcano. Many descriptions of the activity and resulting deposits deal with the base surge (Moore, 1967; Fisher and Walters, 1970). Sheridan and Wohletz (1981) outlined the controlling factors of hydrovolcanic eruptions using the results of field studies, laboratory analysis, and experimental investigation. Colgate and Sigurgeirsson (1973), Buchanan and Dullforce (1973), and Peckover et al. (1973a and 1973b) demonstrated the strong similarity of submarine volcanic explosions to FCIs. Although the mechanism of fuel (melt) fragmentation and controls of explosiveness are not completely understood, applications to volcanic phenomena have yielded the following conclusions (Wohletz, 1980). Explosive energy, generally expressed as a scaled quantity, is measured as the efficiency of conversion of the melt's thermal energy to mechanical energy. The mechanical energy results from flash vaporization of water due to rapid energy transfer from the melt by superheating. The mechanical energy produced is then partitioned into several dominant modes including: fragmentation energy, particle kinetic energy, seismic energy, and acoustic energies. The efficiency of this process is dominantly a function of the mass-flux ratio of melt and water into the zone of interaction and the confining pressure on that zone. Fig. 1 summarizes experimental and field studies (Wohletz and McQueen, 1981; Wohletz and Sheridan, 1982). The explosive energy curve can be subdivided into regions of Strombolian activity (mass ratios of 0 to 0.1), Surtseyan activity (ratios 0.1 to 1.00), and submarine extrusion of pillow basalt (greater than 3.00).

PYROCLASTIC SIZE ANALYSIS

A hydrovolcanic eruption consists of numerous bursts, each resulting from a water/melt interaction. The size analysis presented here assumes that individual fall and surge beds (or bedding sets) represent the deposit resulting from one eruptive burst. Samples were collected from deposits within one crater radius of the crater rim. With exception of sandwave beds, samples were taken from individual bedding layers. Single sandwave beds were found usually to be less than 1 cm thick, so samples were taken from bedding sets several centimeters in thickness. The samples were chosen to investigate the variation of size characteristics among bedforms and, therefore, the relative degree of

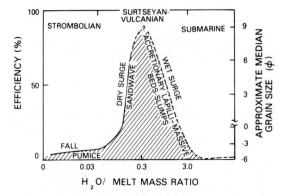

Fig. 1 Plot of conversion efficiency of thermal energy to explosive mechanical energy versus water to melt mass ratio ($\rho_{H_2O} \simeq 1.0$ g-cm^{-3}). The curve is adapted from Wohletz (1980) and also shows approximate, median grain size of melt fragments. Optimal efficiencies are expressed as percent of thermodynamic maximum. Onset of vapor explosion occurs at ratios near 0.1 and reaches a maximum near 0.3. Maximum explosive interactions produce pyroclastic surges of dry, superheated steam that deposit dune or sandwave beds. Ratios above the maximum result in wet (condensing steam) surge eruptions that deposit massive surge and flow beds. Wet surge deposits commonly are associated with accretionary lapilli, lahars, and soft-sediment bedding deformations.

fragmentation produced by each eruptive burst. This method is not strictly meaningful for fall deposits because of the strong dependence of their grain size on distance from the vent (Walker, 1971). Grain-size distributions of surge bedforms, however, do not appear to be so sensitive to distance of transport.

Figs. 2 and 3 are plots of sorting versus median diameter for 127 basaltic pyroclasts and 80 silicic samples, respectively. The plots of sandwave, massive, planar, and fall samples are delineated showing distinct fields of median size as a function of bedform. The size fields are less clearly defined for silicic pyroclasts than for basaltic pyroclasts. This result reflects upon the lower ability of high-viscosity melts to explosively mix with water. For both basaltic and silicic compositions, median sizes are: fall, 1100-2000 μm; planar surge, 750-1600 μm; massive surge, 370-650 μm; and sandwave surge, 150-300 μm.

Most size distributions of hydrovolcanic pyroclasts are polymodal. Depending upon deposit type, modes generally occur in both the coarse ash and fine ash (<63 μm) divisions which result from: (1) the degree of explosive fragmentation, and (2) subsequent sorting due to transport by inertial and viscous flow processes. Fig. 4 demonstrates the strong increase in fine ash abundance going from fall and planar surge deposits to massive and sandwave surge deposits.

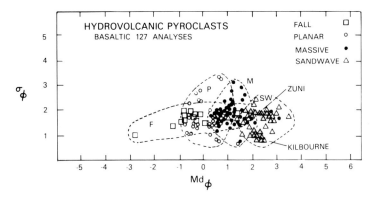

Fig. 2 Plot of standard deviation (σ_ϕ) versus median diameter (Md_ϕ, $\phi = -\log_2$ mm) for basaltic pyroclasts. Analyses of fall (F) and planar (P), massive (M), and sandwave (SW) surge deposits plot in distinct fields. Variation in sorting between sandwave deposits of Zuni Salt Lake and Kilbourne Hole result from a strong contribution of lithic (quartz sand) material at Kilbourne Hole.

SCANNING ELECTRON MICROSCOPY ANALYSIS

Following the method of Sheridan and Marshall (1982), samples were inspected with a binocular microscope to distinguish glass, lithic, and crystal constituents. Samples were cleaned using dilute HCl with ultrasound and grains were mounted individually on metal stubs. Sizes investigated include the 250 to 500 µm range and a fine fraction less than 43 µm in diameter. The SEM was operated in both the secondary electron mode at 15 keV and the backscatter mode when sample charging prevented adequate imaging. Surface charging was found to be prevalent on those grains that were most altered. Highly irregular surfaces,

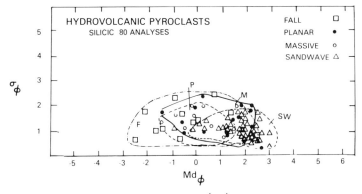

Fig. 3 Plot of standard deviation (σ_ϕ) versus median diameter (Md_ϕ) for silicic hydrovolcanic pyroclasts. Distinction of analyses of various bedforms is less apparent than for those of basaltic pyroclasts due mainly to the eruptive contribution of magmatic gasses.

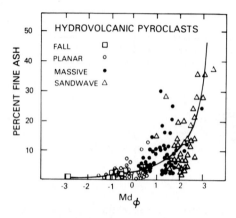

Fig. 4 Plot of percent fine ash (<63 μm diameter) versus median diameter showing an exponential increase for massive and sandwave deposits. The abundance of fine ash is a measure of explosive violence and the degree of water interaction for hydrovolcanic deposits.

prevented deposition of a uniform gold conductive coat. Altered surfaces were easily distinguished using energy dispersive spectral analysis.

Five dominant pyroclast types were distinguished from SEM images: (1) blocky, equant shapes; (2) vesicular, irregular shapes with rounded, fluid-formed surfaces; (3) moss-like, convoluted shapes; (4) spherical to drop-like shapes; and (5) platy shapes. More than one pyroclast type is commonly present in samples. However, one type generally characterizes the coarse or the fine fraction.

Type 1 pyroclasts (Fig. 5) are the most frequently observed shapes of coarse (>63-μm-diameter) hydrovolcanic ash. These "chunky" shapes are found in compositions ranging from basaltic to rhyolitic. Typically, vesicle surfaces are rare and are cut by curviplanar fracture surfaces. The equant, blocky surfaces are smooth and surface irregularities where present are due to abrasion and alteration features, or vesicle embayments. Surfaces of silicic ash are commonly slab-like. They are elongated in two dimensions and shortened in the other due to a foliation formed during injection prior to fragmentation. Pyramidal shapes are also typical expressions of Type 1 pyroclasts, which when elongated in one or two dimensions, resemble pointed chips or splinters.

Type 2 pyroclasts (Fig. 6) also are evident in coarse fractions and have surfaces controlled by vesicle walls. Flat breakage surfaces with distinct corners are absent. Vesicle edges are rounded and smoothed and overall grain shape is irregular. The smooth, curved surfaces between vesicles are lumpy and appear fused and fluid-formed. These pyroclasts have only been found in basaltic compositions and are especially abundant in Surtsey tephra where copious amounts of water had access to the vent as evidenced by periodic eruptions of water-pyroclast slurries (Thorarinson, 1966).

Type 3 pyroclasts (Fig. 7) are found only in the fine fraction (<63 μm diameter) of basalts. These moss-like, convoluted shapes have highly irregular

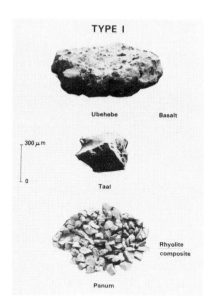

Fig. 5 SEM photomicrographs of Type I pyroclasts showing blocky shapes with curviplanar surfaces.

surfaces formed by several or more globular masses attached together. The appearance of vesicle-like embayments is due to the tortuous convolution of the composite surface, which shows smaller attached globules. The overall appearance of grains is that of delicate, interconnected structures. Some grains of this type show fluid-form connections between larger masses. The high surface area and delicate shape resembles moss.

Type 4 pyroclasts (Fig. 8) also are found in the fine fraction of basalts. This type shows roughly spherical or drop-like boundaries with smooth, curved surfaces. These pyroclasts rarely exist as separate particles. They are attached to larger blocky grains or are agglutinated to form botryoidal surface encrustations. Drop-like forms are elongated and may be broken showing vesicular interiors.

Type 5 pyroclasts (Fig. 9) include plate-like or crescent shapes. Surfaces are smoothly curved or irregular, the latter formed in magma with abundant microlites. These shapes show at least one curved surface that formed a wall of a vesicle bubble whose diameter was greater than that of the grain. This type is typical of the fine fraction of vesicular magmas. Chips and splinters of bubble walls have characteristic sharp edges.

Since hydrovolcanic pyroclasts show a multitude of grain shapes, the five types discussed above were chosen as useful divisions to characterize shapes that grade from one into another. It is likely that further investigation of

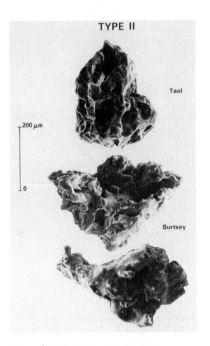

Fig. 6 SEM photomicrographs of Type II pyroclasts showing vesicular, irregular shapes with smooth, fluid-form surfaces.

magmas with unusual compositions and crystallinities will show that more types can be delineated. Many of the shapes characteristic of the fine fraction require a high-resolution stage which permits clear images of magnification in excess of 50,000 times. Samples containing pyroclasts less than a few microns in diameter have revealed quite a different spectrum of shapes (included in Types 3 and 4) than those observed in coarser ash.

EXPERIMENTAL RESULTS

The experimental basis for hydrovolcanic (water/melt) theory is from work performed over the last seven years at Los Alamos National Laboratory. The experimental apparatus and methodology are discussed in detail by Wohletz (1980), and Wohletz and McQueen (1981, in press). Concurrently and independently, research at Sandia National Laboratories by Corradini (1981a), Buxton and Benedict (1979), and Nelson and Duda (1981) investigated nuclear reactor core melt down. In both studies, the molten material used was thermite (iron oxide and aluminum).

Quartzo-feldspathic sand added to thermite in my volcano experiments produced a silicate melt approximating basaltic compositions, density, viscosity, and

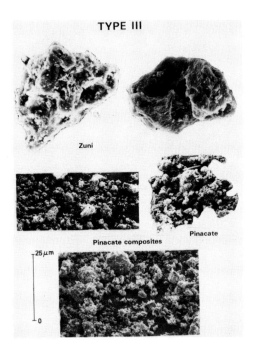

Fig. 7 SEM photomicrographs of Type III pyroclasts showing high surface-area, moss-like shapes.

phase relations. This melt was brought into contact with water inside a confinement chamber with monitoring of pressure and temperature. Venting of the chamber upon explosion was documented by high-speed cine cameras.

Variation of experimental design, melt/water contact geometry, water-to-melt mass ratio, and confinement pressure has made possible the investigation of a wide variation of explosive energy. Highly explosive experiments produced fine-grained ejected debris less than 50 μm in median diameter and less explosive interactions resulted in centimeter-sized debris. Although recovery of explosion debris is difficult because of its relatively wide dispersal away from the experimental device, size as well as SEM studies of the debris has aided in understanding the melt-fragmentation mechanism (Buxton and Benedict, 1979; Corradini, 1981a). High-speed cinematography of molten metal dropped into water (Nelson et al., 1980) also aided in the understanding of the fragmentation mechanism. The experimental debris investigated by SEM in this study shows many similarities to hydrovolcanic ash and is described below.

The experimental fallout debris was collected on polyethylene sheets. Additional debris from directed blasts was trapped using metal blocks set on the

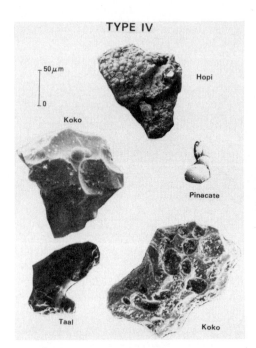

Fig. 8 SEM photomicrographs of Type IV pyroclasts showing spherical particles and drop-like shape both unattached and attached to larger particles.

sheets which served as barriers to the ejecta as it moved horizontally from the confinement chamber. Debris recovered by these means ranged in particle diameter of less than one micron to a maximum of nearly one centimeter and most was less than 50 µm in diameter.

Fig. 10 consists of micrographs of the dominant form of artificially formed explosion debris. These shapes are characterized by "moss-like" convolute shapes showing high surface area. Most surfaces are smooth and rounded forming globules and "chunks" fused together into particles usually less than 20 µm in maximum dimension. Deep embayments separate globular lobes and tiny spheres are attached on some particles. Jagged edges are rare except where ribbon-like or spongy material forms the particle lobes.

Less abundant particles have spheroidal and drop-like shapes (Fig. 11) ranging from nearly 20 to 150 µm in long dimension. Spheres commonly have smaller attached chunky and spheroidal debris. Sphere surfaces are smooth, but show intricate internal patterns of interlocking plates and elongate crystals reminiscent of Widmänstätten texture. Spheres may be attached to other debris but usually occur as separate particles. Some spheres appear to be partially

Fig. 9 SEM photomicrograph of Type V pyroclasts showing plate-like shapes and curved bubble wall shards.

hollow and have openings on their surface which reveal chunky debris inside. Drop-like shapes are usually elongated spindles with smooth surfaces and attached chunky debris.

The largest type of experimental debris shows blocky, equant shapes (Fig. 12) ranging in size from 40 to 200 μm in long dimension. Most surfaces are smooth and curviplanar; however, some show pitting and small attached particles. These forms, which show the most similarity to coarse hydrovolcanic pyroclasts, are relatively rare.

Plate-like shapes (Fig. 13) are broken pieces of curved bubble walls and angular flat chips. Although vesicles are rare in the fine experimental debris, they are common in the centimeter-sized ejecta (Fig. 14) which bears strong resemblence to basalt scoria. Gasses trapped in the thermite melt form vesicles of 0.1 to 1.0 cm in diameter.

Although thermite melt varies considerably in chemical composition from basaltic melts, its similar viscosity, density, and surface tension make it a reasonable model. Table 2 gives representative energy dispersive spectral analysis (EDS) of several melt particles. Due to incomplete mixing of the reactants during melting, the products show variation in chemistry on a fine

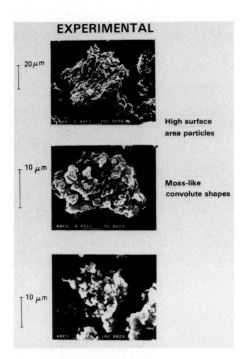

Fig. 10 SEM photomicrographs of experimental debris. These particles are typical of high surface-area convolute shapes that dominate most samples.

scale. In this section the thermite explosive product consists of thin, lath-like crystals enclosed in a quenched groundmass. The crystals are birefringent and the quenched groundmass is opaque in plane polarized light. The texture is microcrystalline variolitic.

DISCUSSION OF FRAGMENTATION MECHANISMS

Experimentally produced ash shows a strong similarity in size and shape when compared to hydrovolcanic ash. Considering the limitations imposed due to scaling and differences in bulk chemistry of the thermite from that of basalt, the following discussion of pyroclast formation illustrates the significance of size and shape.

Pyroclast Size

The dominant form of heat transfer from the melt to the water in melt/water interaction is assumed to be conductive. The difference in temperature and the contact surface area between the melt and water are important parameters in determining the rate of heat transfer and the explosive efficiency. From

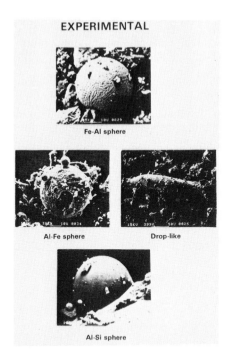

Fig. 11 SEM photomicrographs of experimental spherical and drop-like particles. Notice the intricate pattern of crystal growth on the upper sphere and the hollow center on the sphere pictured left center.

small-scale experiments, Buxton and Benedict (1979) have quantitatively shown the decreasing fragment size with increasing explosive efficiency of thermite and water. Large-scale experiments (Wohletz and McQueen, in press) have qualitatively shown that highly explosive "Surtseyan" interactions result in micron-size fragmentation and surge dispersal of the melt, whereas less explosive "Strombolian" interactions produce millimeter- and centimeter-size fragments dispersed by fallout.

The size break between surge and fall ashes occurs near 1 mm which corresponds to the distance of penetration of a thermal wave into magma in less than 10^{-1} second. Fig. 15 is a plot of cooling time, t_c, and surface area of spherical pyroclasts versus grain diameter; t_c is calculated by two methods:

$$t_c = \frac{d^2}{D} \tag{1}$$

$$t_c = \frac{d^2}{8k} \tag{2}$$

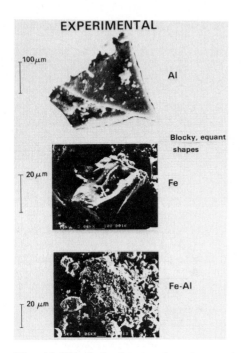

Fig. 12 SEM photomicrographs of experimental blocky or "chunky" debris. Smooth surfaces characterize most samples although some show abundant adhering fine dust.

(from Colgate and Sigurgeirsson (1973) and Sparks (1978) respectively), where d is the depth of penetration of a thermal wave (grain radius), D is the thermal conductivity, $k(5 \times 10^{-3}$ cal-cm^{-1}-s^{-1}-deg^{-1} for basalt) divided by the heat capacity, C_V(0.25 cal-cm^{-3}-deg^{-1} for basalt). Explosive heat-transfer times decrease over seven orders of magnitude from seconds to microseconds as grain diameter decreases from millimeters to microns. Concurrently, the specific surface area of particles increases nearly 6,000 times as melt is fragmented to micron size. The efficiency of conductive heat transfer, a function of both surface area and heat transfer time increases dramatically with increasing melt fragmentation.

Heat transfer from pyroclasts to a surrounding vapor film (Fig. 16) can be evaluated by considering heat flow from a spherical body. Integration of the conductive heat-flow equation for spherical coordinates yields Q, the rate of heat transfer expressed for unit area:

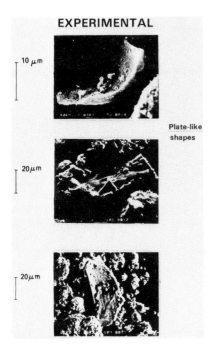

Fig. 13 SEM photomicrograph of experimental plate-like shapes. The upper photo shows a broken bubble wall that formed a vesicle from trapped gasses in the melt. Other shapes appear to have been "peeled" or cavitated as quenched skin from the surface of the melt.

$$Q = \frac{4\pi \ k(T_1 - T_2)R_1R_2}{A(R_2 - R_1)} \tag{3}$$

where k is the conductivity of steam, R_1 and R_2 the radii of the melt sphere and the surrounding vapor film (measured from the melt sphere center) and T_1 and T_2 their respective temperatures, and A the contact area. The thermal conductivity of steam can be approximated as the linear function of temperature $6.22\times10^{-5}+2.43\times10^{-7}[T(°C)-127]$ cal-cm^{-1}-s^{-1}-deg^{-1} from values given by Weast (1977) and that of basalt is 5×10^{-3} cal-cm^{-1}-s^{-1}-deg^{-1}. Assuming a constant vapor film thickness of one tenth that of the melt sphere diameter and the temperature gradient over the film to be 20 degrees, Q increases from 6.1×10^{-2} cal-cm^{-2}-s^{-1} to 7.6×10^2 cal-cm^{-2}-s^{-1} as the particle diameter decreases from about 1 cm to 10 μm.

Fig. 17 is a plot of data from Buxton and Benedict (1979) showing melt-fragment median diameter versus the experimental explosion efficiency. Median fragment diameters of 2 mm or greater were recovered from non-explosive

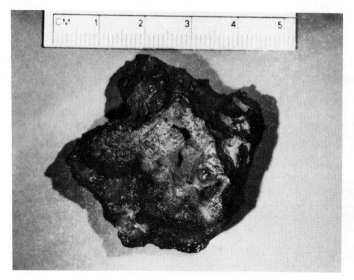

Fig. 14 Photograph of a piece of experimental scoria. The vesicles and jagged edges as well as similar density and color make this sample difficult to distinguish from basalt scoria.

TABLE 2

Representative chemical analyses* of thermite melt debris

Oxide	1[+]	2	3	4	5
SiO_2	10.1	10.7	14.3	36.6	18.6
TiO_2	-	-	2.3	1.4	1.9
Al_2O_3	31.2	25.1	11.4	34.4	42.5
FeO	57.9	64.2	57.0	17.3	23.3
MgO	-	-	6.4	3.7	5.8
MnO	-	-	1.6	1.0	1.3
CaO	0.5	-	2.1	1.7	1.9
Na_2O	-	-	3.1	2.0	3.2
K_2O	0.3	-	2.0	2.0	1.9

*Standardless energy dispersive spectral analyses (EDS)
[+]1 - Iron-aluminum sphere, large
 2 - Iron-aluminum sphere, small
 3 - Blocky iron particle
 4 - Coating on iron particle
 5 - Iron-aluminum spindle

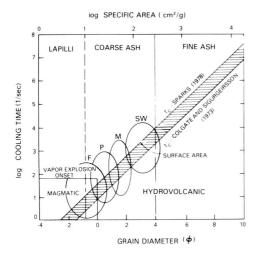

Fig. 15 Plot of penetration time of a thermal wave into a spherical pyroclast versus grain diameter. This time decreases exponentially with grain size and is calculated by the two similar formulae from Sparks (1978) and Colgate and Sigurgeirsson (1973). The specific area of grain surfaces increases exponentially with decreasing grain size. High surface area and short thermal equilibration times 10^{-1} to 10^{-2} s are required for vapor explosions. This boundary is near that for the division of lapilli and coarse ash (2 mm), and for basalts the division is between dominantly magmatic Strombolian and hydrovolcanic Surtseyan eruptions. The fields of grain size are shown for fall and surge deposits and demonstrate a two order of magnitude increase in surface area for hydrovolcanic over non-hydrovolcanic pyroclasts.

experiments. Fig. 17 also shows the calculated rate of conductive heat transfer, Q, from Eq. (3) versus fragment size, R_1, showing the range of values obtained for steam at temperatures between 300 and 1180°C heated by basaltic melt droplets at 1200°C. The surface area dependency of conductive heat flow upon fragment size demonstrates an exponential relationship of explosive efficiency to degree of fragmentation. Initial vapor expansions produce tensile stresses and fluid instabilities that increasingly fragment the melt resulting in an exponential increase in explosive, conductive heat transfer. The partitioning of vaporization energy into melt fragmentation and ejection modes is dependent upon the density, viscosity, surface tension, and yield strength of the steam and the melt, their seismic and acoustic velocities, and two phase flow complexities.

The partial effect of the steam film thickness (R_2) upon heat transfer can be derived from Eq. (3) to give a conductive factor:

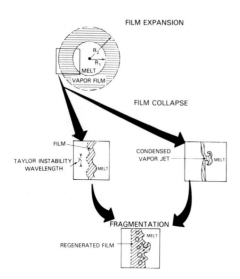

Fig. 16 Diagrammatic illustration of the film collapse model showing fragmentation during collapse by Taylor instabilities and water jetting.

$$dQ]_{R_1} = - dR_2/(R_2-R_1)^2 \ .$$

The above equation is a hyperbolic curve with an inflection near a point where the steam film radius, R_2, is twice that of the melt sphere radius, R_1. The change in heat conductance due to the conductive factor is shown for different values of R_2 in Fig. 18. Q increases logarithmically as the steam jacket R_2 decreases in thickness below a value equal to nearly 2 to 5 times that of the thickness R_1 (the depth of the thermal wave penetration in the melt). Above this value, Q remains constant as R_2 varies. During collapse of the film jacket, Q increases to the point where the condensed phase is instantaneously vaporized forming an expanding steam jacket. During expansion, Q decreases to a minimum point where the bubble becomes unstable again and collapses.

Drumheller (1979) approaches the problem of vapor-film collapse by calculating the energy and work of vapor and liquid water surrounding molten iron spheres at 1600°C. These quantities are dominantly functions of the film and melt sphere radii, rate of condensation, and vapor densities. The functions are integrated over time to obtain the equations of motion which, when evaluated with heat conductivity requirements, predict the following: For 5- and 10-mm-diameter melt spheres, the collapsing film reaches impact velocity peaks of 3 to 7 ms^{-1} and impact pressures of 5 to 10 MPa for film thicknesses of about 0.05 to 0.1 mm. The functions expressing these values are strongly damped with increasing ambient pressure. Pressure waves generated in the melt sphere reach peak values approaching 25 MPa on microsecond time scale. These values when

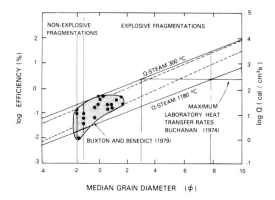

Fig. 17 Plot of conversion efficiency of thermal to mechanical energy versus median grain size of debris formed by experiments. Points shown are from medium scale (10-20 kg thermite) experiments (Buxton and Benedict, 1979) and fall on an extrapolated trend (dashed lines) that intersects maximum thermodynamic efficiency near 10 ϕ. Below -1.0 to -1.5 ϕ fragmentations of thermite due to contact with water were non-explosive. Also plotted are values of heat-transfer rates, Q, versus grain size showing an exponential increase with decreasing grain size. The maximum measured laboratory rates corresponds to fragmentation sizes between 3 ϕ (125 µm) and 8 ϕ (4 µm), which are observed in small-scale laboratory (~1 g) experiments. Values of Q are calculated for thermal gradients between basalt at 1200°C and steam at 1180°C and 300°C.

considered with surface tension effects are great enough to cause fragmentation of the melt (Galloway, 1954) thereby increasing the surface area by over two orders of magnitude.

Corradini (1981a) models FCI experiments using a thermal fragmentation mechanism. In this model near fuel-coolant contact during film collapse generates Taylor instabilities and high-pressure vapor at the contact. The instabilities fragment the melt which is subsequently quenched during convective mixing and heat transfer. This model predicts the generation of a high-pressure wave (150 MPa) propagating at 90 ms^{-1}. Critical to these calculations are those of equilibrium pressure via a Redlich-Kwong equation of state. These pressures cause acceleration of the fuel giving rise to instabilities. The Taylor wavelength, a function of film thickness, melt and vapor densities, and melt-acceleration determines the surface area increase.

Board et al. (1975) show that the structure of FCIs may produce a detonation wave. High pressures observed in laboratory experiments and fast propagation of these pressure waves suggest the development of shock waves. The front of the wave shatters the melt. Vaporization due to fuel-coolant thermal equilibrium occurs directly behind the shock front and the expanding vapors drive and

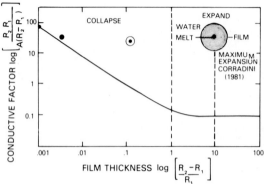

Fig. 18 Plot of a conductive factor versus vapor film thickness, R_2, normalized to a constant melt sphere radius, R_1. As the film collapses to small thicknesses, the conductive heat transfer from the melt to the film increases rapidly until an additional volume of water is vaporized. The film then expands (Corradini, 1981a) while conducting progressively less energy until heat transfer to the vapor eventually reaches a minimum. At that point heat energy is lost from the vapor to the surrounding fluid, vapor expansion slows, and condensation occurs causing the film to collapse. This process may be repeated many times on a millisecond scale. Each collapse fragments more melt resulting in larger heat transfers and greater volumes of vaporized water. The process will continue until vapor expansion is greater than confining pressure and the system explodes.

maintain the front. This mechanism is analogous to detonation of chemical explosives in which the generated gas drives the detonation wave. Board et al. (1975) calculate that pressures as high as 1.5×10^3 MPa (15 kbar) could be generated by this mechanism in large systems. The passage of the shock front also produces large velocities of the fuel relative to the coolant. These velocities are sufficient to develop both Rayleigh-Taylor and Kelvin-Helmholtz fluid instabilities which cause melt fragmentation (Theofanus, 1979; Berenson, 1961).

Shock propagation may also initiate axisymmetric vapor film collapse causing water jets to penetrate the melt. In this model, Buchanan (1974) calculates the energy of film collapse, and the penetration depth, heat transfer, and fragmentation size of the water jet. After each collapse and fragmentation, more water is vaporized and the cycle repeated. In his model, Buchanan (1974) shows that the first cycle of vapor collapse produces a film with an energy of 1.2×10^{-4} J and thickness of 6.6×10^{-4} m in 2.6×10^{-3} s. This cycle fragments a very small amount of melt (2.2×10^{-11} kg) and produces a peak pressure of 0.33 MPa. After six cycles and an elapsed time of 0.21 s, the vapor film is 1.3 m thick with collapse energy of 9.8×10^5 J. At this point 0.17 kg of melt has been fragmented and a peak pressure 662 MPa is produced by the feedback mechanism. This model also predicts a strong damping affect due to the system ambient pressure and illustrates the remarkable efficiency of FCIs.

With these discussions in mind let us return to the problem of hydrovolcanic fragmentation. The pyroclast size-distributions of the various bedforms shown in Figs. 2 and 3 pose some difficulty for interpretation. This difficulty lies in the fact that pyroclastic deposits generally become finer grained with distance from the vent. For sandwave beds, Sheridan and Updike (1975) found median diameter to vary nearly two phi-size units with distance from the vent. This variation, however, did not show a systematic "fining" of grain size. Data given by Wohletz and Sheridan (1979) suggest that overall, the median diameter of surge tephra varies less than one phi-size unit within a given bedform at varying distances from the vent. Although this problem has not been studied systematically, the following interpretation of size data is put forth based upon models of surge eruption and emplacement. An FCI model for surge eruption predicts that the most explosive interactions produce the greatest steam-to-pyroclast volume ratio and the finest grain sizes. Using the grain-size data in Fig. 4, this model indicates a likelihood of sandwave deposition if highly explosive eruptions produce surges, which is in agreement with experimental observations (Wohletz and McQueen, 1981). Independent of the FCI model, Wohletz and Sheridan (1979) suggested that sandwave, massive, and planar bed forms are deposited from surges of decreasing void space (decreasing steam-to-pyroclast volume ratio). Assuming the validity of these models, variation of grain size among surge bedforms at near-vent localities reflects a fluctuation in the explosivity of the eruptions that produced the tephra. In this reasoning, highly explosive eruptions result in emplacement of dominantly fine-grained sandwave surge deposits while massive and planar surge deposits of coarser grain sizes are emplaced after less explosive bursts.

Pyroclast Shape

The shape of the fragments produced is complexly dependent upon the physical properties of the melt and rate of heat energy release as mentioned earlier. Fragmentation may be of a brittle, ductile, or viscous nature depending upon the viscosity, surface tension, and yield strength of the melt. Each of these deformation modes will produce a distinct fragment shape. The dominant mode of deformation, as evidenced by shape, may be related to one of the fragmentation mechanisms outlined above.

Brittle and ductile fractures depend upon the strength of the melt. Cavitation due to brittle fracture may result from stress waves propagating into the melt forming regions of tension and compression. If the strain rate exceeds the melt bulk modulus, then brittle failure will occur with increasing tendency as confining pressure decreases. Fractures will propagate at an angle less than 45° to the direction of compression or extension. This mechanism could explain the formation of blocky pyroclasts Type 1 and Type 2. Fig. 19 illustrates this model. Quenching and solidification during and after brittle fracture preserves blocky shapes with curvi-planar surfaces. More ductile behavior would result in irregular or elongated fragments. If solidification and formation of a quenched

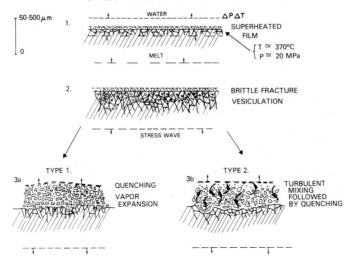

Fig. 19 The collapse of a superheated vapor film or the explosive expansion of the film will produce stress waves in the melt. If these exceed the bulk modulus of the melt and it fractures brittly, blocky Type 1 or Type 2 pyroclasts may form.

crust is not complete after fracture, subsequent movement of fragments out of the zone of interaction forms smooth, fluid-like surfaces on fragments (Type 2).

Fluid instabilities resulting from either water jet penetration of the melt or Taylor and Kelvin-Helmholtz mechanisms result in turbulent mixing of the melt and water. Viscous deformation of the melt dominates and fluid-form shapes result. Repeated vapor-film collapses generate high surface-area particles on a millisecond time scale. When heat-transfer rates due to high surface area result in vapor generation at pressures greater than the confining system pressure, vapor explosion occurs. Moss-like, Type 3 (Fig. 20) pyroclasts result from viscous effects of the melt whereas spherical or drop-like shapes, Type 4 (Fig. 21) are due to the dominance of surface tension effects. These instabilities are highly probable features at the contact of the melt and water because of the density difference between the two fluids, accelerations due to vaporization, hydrostatic head, and explosion shock waves as well as the seismic disturbances that exist in a volcano.

Type 5 pyroclasts are most typical in silicic deposits and characterize phreatoplinian eruptions (Self and Sparks, 1978). The burst of vesicle bubbles fragments the rising magma and propels shards into a zone of mixing with external water. The following rapid vaporization fractures the shards and produces stress waves in the melt that may enhance vesiculation (Bennett, 1974)

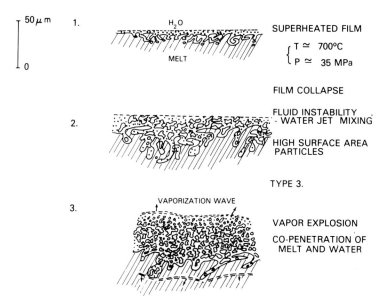

Fig. 20 Fluid instabilities can form at the contact of water with melt. These instabilities are of a Taylor type, Rayleigh-Taylor and Kelvin-Helmholtz type, or axisymmetric film collapse that causes water jets to penetrate the melt. The result is high surface-area fragmentation, rapid heat exchange, followed by vapor explosion. The explosion then generates new contact areas and/or a shock wave that fragments more melt and acts as a detonation wave. This process requires abundant water to be present.

as shown in Fig. 22. A compressive pressure wave propagating into a melt may reflect off physical boundaries due to vent geometry or density differences. Once the pressure wave reflects, it becomes negative (tensile stress) and cavitation proceeds behind the wave by formation of vesicles of exsolving melt volatiles (Corradini, 1981a). In this manner, vesiculation waves form and, if the volatile content is great enough, may cause initial fragmentation of the melt.

The fact that Types 3 and 4 pyroclasts are typically much finer grained than are Types 1 and 2 suggests that they result from higher explosion efficiencies. It follows that fragmentation due to fluid instabilities results in more complete mixing of melt and water with higher heat-transfer rates. Thus, the fine ash fraction (<63 μm diameter) indicates strong water interaction. This feature is illustrated by grain-shape correlation with deposit type. Phreatic explosion breccias, ash falls, and planar surge deposits show mostly Type 1 pyroclasts whereas sandwave and massive surge beds, accretionary lapilli beds, and vesiculated tuffs have a strong contribution of Types 3 and 4.

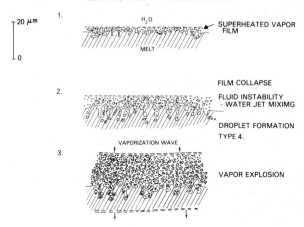

Fig. 21 Fluid instabilities will form spherical or drop-like shapes of melt if viscosity is low and surface tension affects are strong.

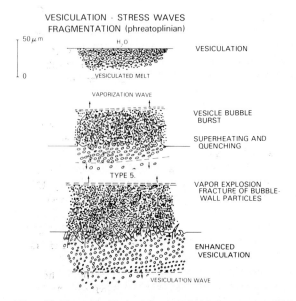

Fig. 22 Phreatoplinian fragmentation occurs when stress waves fracture bubble-wall shards formed by vesicle bubble burst. This requires contact of a vesiculating, ash-producing melt with external water during eruption and results in production of fine ash consisting of tiny plate-like forms.

CONCLUSIONS

The ash formation mechanisms discussed in this paper have both a theoretical and an experimental basis in the wealth of literature on FCIs. Comparison of ash debris from experiments that model hydrovolcanic explosions with natural samples has allowed development of several theoretical models of hydrovolcanic explosions. Although there are physical limitations to comparison of man-made metallic melts with magma, experiments demonstrate that the fragmentation mechanisms are basically the same for a wide range of melt compositions.

Shapes of experimentally produced ash particles are characterized as blocky and equant, spherical and drop-like, mossy aggregates, and the plate-like broken bubble walls. These shapes, as well as their sizes, show a strong resemblance to those of ash produced by hydrovolcanic eruptions. Comparison of volcanic ash shapes and sizes to those of artificial ash gives insight into the hydrovolcanic fragmentation mechanism.

Since rapid vaporization of water is the driving mechanism of hydrovolcanic explosions, study of the pyroclastic material generated provides information on the efficiency of the heat-transfer process. Heat transfer is dominantly conductive and requires large surface areas to reach explosive rates. Size studies of experimental debris show decreasing grain size with increasing explosive efficiency. The maximum size limit of experimental explosion debris is 2 to 3 mm with non-explosive debris being coarser. The lower boundary of size is in the submicron range. Similarly, explosive hydrovolcanic debris shows median grain sizes less than 1 or 2 mm. Size distributions do not appear to depend upon ejected volumes, but depend upon the fragmentation mechanism that occurs on a millimeter and smaller scale. For this reason, scaling is not critical when comparing experimental debris to volcanic ash.

The explosive contact of water with melt begins with the formation and collapse of steam films on the melt surface. This process is cyclic on a micro- or millisecond time scale and results in the generation of fluid instabilities at the contact, water jet penetration of the melt, and stress waves propagating into the melt. These mechanisms fragment the melt thereby increasing surface area and heat-transfer rates. Critical to the vaporization of water are the effects of superheating and detonation waves. Superheating is a process involving non-equilibrium heat transfer and homogeneous nucleation causing instantaneous vaporization. It is still poorly understood but it results in explosive efficiencies several orders of magnitude higher than those of normal boiling processes. Detonation waves are shock waves that propagate through the melt causing fragmentation of the melt and mixing with water by fluid instabilities and vapor film collapse. These waves are sensitive to the system size and permit interaction of large volumes of melt and water in a short time span.

Various debris shapes produced in melt-water interactions reflect fragmentation by brittle failure due to stress waves and viscous melt-water mixing by fluid instabilities. The latter of these two fragmentation mechanisms appears to be the result of the most efficient explosive interactions. Hence, the production of fine-grained debris with high surface area and fluid shapes is predicted for highly explosive eruptions of fluid basalt. However, melt viscosity and strength strongly affects the tendency for development of instabilities. Therefore, brittle fragmentation dominates for intermediate and silicic melts.

Future development of a quantitative model of hydrovolcanic pyroclast formation could approach the problem of the actual amounts of external water and magma involved in explosive heat exchange during eruption. This empirical value, calculated for experiments by Corradini (1981a), may be quantitatively correlated to rates of ash production and emplacement mode. Also of importance is the partitioning of vaporization energy into fragmentation and ejection modes. These and other quantitative treatments will be the subject of future studies.

ACKNOWLEDGEMENTS

Robert McQueen provided constant support for several years in design and implementation of experiments. His scientific support and interest have made this work possible. The initial SEM work on experimental debris was completed with the help of Mike Sheridan and Rosanna DeRosa. Their ideas and encouragement are greatly appreciated. Grant Heiken provided many stimulating discussions on ash formation. Lloyd Nelson and others at Sandia National Laboratories provided encouragement as well as expert advice on FCI theory. Robert Raymond and Ronald Gooley provided technical assistance for the high resolution microscopy. I thank George Walker, Aaron Waters, and Grant Heiken for their helpful reviews of this manuscript. NASA grants NSG-7642 and NAWG-245 and funds from the Los Alamos Institutional Supporting Research and Development supplied partial support for this work.

REFERENCES

Apfel, R. E., 1972. Water superheated to 279.5°C at atmospheric pressure. Nature, 238: 63-64.
Bennett, F. D., 1974. On volcanic ash formation. Amer. J. Sci., 274: 648-661.
Berenson, P. J., 1961. Film boiling heat transfer from a horizontal surface. J. Heat Trans., August: 351-356.
Board, S. J. and Hall, R. W., 1975. Thermal explosions at molten tin/water interfaces. In: J. R. Okendon and W. R. Hodgkins (Editors), Moving Boundary Problems in Heat Flow Diffusion. Clarendon Press, Oxford, pp. 259-269.
Board, S. J., Hall, R. W. and Hall, R. S., 1975. Detonation of fuel coolant explosions. Nature, 254: 319-321.
Board, S. J., Farmer, C. L. and Poole, D. H., 1974. Fragmentation in thermal explosions. Inst. J. Heat Mass Trans., 17: 331-339.

Buchanan, D. J., 1974. A model for fuel-coolant interactions. J. Phys. D: Appl. Phys., 7: 1441-1457.

Buchanan, D. J. and Dullforce, T. A., 1973. Mechanism for vapor explosions. Nature, 245: 32-34.

Buxton, L. D. and Benedict, W. B., 1979. Steam explosion efficiency studies. Sandia National Laboratories, SAND79-1399, NUREG/CR-0947, pp. 1-62.

Carlisle, D., 1963. Pillow breccias and their aquagene tuffs, Quadra Island, British Columbia. J. Geol., 71: 48-71.

Colgate, S. A. and Sigurgeirsson, T., 1973. Dynamic mixing of water and lava. Nature, 244: 552-555.

Corradini, M. L., 1981a. Analysis and modelling of steam explosion experiments. Sandia National Laboratories, SAND80-2131, NUREG/CR-2072, pp. 1-114.

Corradini, M. L., 1981b. Phenomenological modelling of the triggering phase of small-scale steam explosion experiments. Nucl. Sci. Eng., 78: 154-170.

Drumheller, D. S., 1979. The initiation of melt fragmentation in fuel-coolant interactions. Nucl. Sci. Eng., 72: 347-356.

Dullforce, T. A., Buchanan, D. J. and Peckover, R. S., 1976. Self triggering of small-scale fuel-coolant interactions: I experiments. J. Phys. D.: Appl. Phys., 9: 1295-1303.

Fisher, R. V. and Waters, A. C., 1969. Bedforms in base surge deposits: Lunar implications. Sci., 165: 1349-1352.

Fisher, R. V. and Waters, A. C., 1970. Base surge bedforms in maar volcanoes. Amer. J. Sci., 268: 157-180.

Fröhlich, G., Müller, G. and Unger, G., 1976. Experiments with water and hot melts of lead. J. Non-Equilib. Thermodyn., 1: 91-103.

Fuller, R. E., 1931. The aqueous chilling of basaltic lava on the Columbia River Plateau. Amer. J. Sci., 21: 281-300.

Galloway, W. J., 1954. An experimental study of acoustically induced cavitation in liquids. J. Acoust. Soc. Amer., 26: 849-857.

Heiken, G. H., 1971. Tuff rings: examples from the Fort Rock-Christmas Lake Valley, south-central Oregon. J. Geophys. Res., 76: 5615-1626.

Heiken, G. H., 1972. Morphology and petrography of volcanic ashes. Geol. Soc. Amer. Bull., 83: 1961-1988.

Heiken, G. H., 1974. An atlas of volcanic ash. Smithsonian Contrib. Earth Science, 12: 1-101.

Honnorez, J. and Kirst, P., 1975. Submarine basaltic volcanism: morphometric parameters for discriminating hyaloclastites from hyalotuffs. Bull. Volcanol., 32: 441-465.

Jaggar, T. A., 1949. Steam blast volcanic eruptions. Hawaiian Volcano Observatory, 4th Spec. Report, pp. 1-137.

Kazimi, M. S., 1976. Acoustic cavitation as a mechanism of fragmentation of molten droplets in coolant liquids. MIT Report, CO-2781-6TR.

Lipsett, S. G., 1966. Explosions from molten materials and water. Fire Tech., May: 118-126.

McBirney, A. R., 1963. Factors governing the nature of submarine volcanism. Bull. Volcanol., 26: 455-469.

McBirney, A. R. and Murase, T., 1970. Factors governing the formation of pyroclastic rocks. Bull. Volcanol, 34: 372-384.

Moore, J.G., 1967. Base surge in recent volcanic eruptions. Bull. Volcanol., 30: 337-363.

Nelson, L. S. and Duda, P. M., 1981. Steam explosion experiments with single drops of CO_2 laser-melted iron oxide. Trans. Amer. Nucl. Soc., 38: 453-454.

Nelson, L.S., Buxton, L.D. and Planner, H.N., 1980. Steam explosion triggering phenomena, Part 2: Corium-A and Corium-E and oxides of iron and cobalt studied with a floodable arc-melting apparatus. Sandia Laboratories, SAND79-0260, NUREG/CR-0633.

Nelson, L. S. and Buxton, L. D., 1978. Steam explosion triggering phenomena: stainless steel and corium-E similants studied with a floodable arc melting apparatus. Sandia Laboratories, SAND77-0998, NUREG/CR-0122.

Peacock, M. A., 1926. The basic tuffs. In: G. W. Tyrell and M. A. Peacock (Editors), The petrology of Iceland. Royal Soc. Edinburgh Trans., 45: 51-76.

Peckover, R. S., Buchanan, D. J. and Ashby, D. E. T. F., 1973a. Fuel-coolant interactions in submarine volcanism. Nature, 245: 307-308.

Peckover, R. S., Buchanan, D. J. and Ashby, D. E. T. F., 1973b. Fuel-coolant interactions in submarine volcanism. Culham Laboratory Publ., Abingdon, pp. 1-348.

Reid, R. C., 1976. Superheated liquids. Amer. Sci., 64: 146-156.

Sandia Laboratories, 1975. Core-meltdown experimental review. SAND74-0382, pp. 1-472.

Self, S. and Sparks, R. S. J., 1978. Characteristics of widespread pyroclastic deposits formed by the interaction of silicic magma and water. Bull. Volcanol., 41-3: 196-212.

Sheridan, M. F., Barberi, F., Rosi, M., and Santacroce, R., 1981. A model for Plinian eruptions of Vesuvius. Nature, 289: 282-285.

Sheridan, M. F. and Marshall, J. R., 1982. SEM examination of pyroclastic materials: basic considerations. SEM Inc., (in press).

Sheridan, M. F. and Updike, R., 1975. Sugarloaf Mountain tephra - a Pleistocene rhyolite deposit of base-surge origin. Geol. Soc. Amer. Bull., 86: 571-581.

Sheridan, M. F. and Wohletz, K. H., 1981. Hydrovolcanic explosions, the systematics of water-pyroclast equilibration. Science, 212: 1387-1389.

Sparks, R. S. J., 1978. The dynamics of bubble formation and growth in magmas: a review and analysis. J. Volcanol. Geotherm. Res., 3: 1-37.

Tazieff, H., 1968. Sur le mécanisme des eruptions basaltiques sous-marines a faibles profundeurs et la genese d'hyaloclastites associeês. Geol. Rund., 57: 955-966.

Theofonus, T. G., 1979. Fuel-coolant interactions and hydrodynamic fragmentation. Proceedings of the Fast Reactor Safety Meeting, Seattle, Washington.

Thorarinsson, S., 1966. Surtsey the new island in the North Atlantic. Almenna Bokafelagid, Reykjavik, pp. 1-47.

Verhoogen, J., 1951. Mechanics of ash formation. Amer. J. Sci., 249: 729-739.
von Waltershausen, W. S., 1853. Uber die vulkanischen gesteine in Sizilien und Island und ihre submarine umbildung. Gottingen.
Walker, G. P. L., 1971. Grain-size characteristics of pyroclastic deposits. J. Geol., 79: 696-714.
Walker, G. P. L., 1973. Explosive volcanic eruptions - a new classification scheme. Sond. Geol. Rund., 62: 431-446.
Walker, G. P. L. and Croasdale, R., 1971. Characteristics of some basaltic pyroclastics. Bull. Volcanol., 35: 305-317.
Weast, L. C., 1977. CRC handbook of chemistry and physics 58th. CRC Press Inc. Cleveland, Ohio.
Witte, L. C., Cox, J. E. and Bouvier, J. E., 1970. The vapor explosion. J. Metals, 22: 39-44.
Wohletz, K. H., 1980. Explosive hydromagmatic volcanism. Ph.D. Thesis, Arizona State University, Tempe, Arizona, pp. 1-303.
Wohletz, K. H. and Krinsley, D. H., 1982. Scanning electron microscopic analysis of basaltic hydromagmatic ash. In: B. Whaley and D. Krinsley (Editors), Scanning Electron Microscopy in Geology. Geo. Abstracts, Inc., Norwich, England.
Wohletz, K. H. and McQueen, R. G., 1981. Experimental hydromagmatic volcanism. Amer. Geophys. Union Trans., EOS, 62(45): 1085.
Wohletz, K. H. and McQueen, R. G., in press. Experimental studies of hydromagmatic volcanism. In: F. R. Boyd (Editor), Explosive volcanism: inception, evolution, and hazards. Studies in Geophysics, National Academy Sciences.
Wohletz, K. H. and Sheridan, M. F., 1979. A model of pyroclastic surge. In: C.E. Chapin and W.E. Elston (Editors), Ash-flow Tuffs. Geol. Soc. Amer. Spec. Paper, 180: 177-193.
Wohletz, K. H. and Sheridan, M. F., 1982. Hydrovolcanic explosions II: evolution of tuff cones and tuff rings. Amer. J. Sci., (in press).

IGNIMBRITE TYPES AND IGNIMBRITE PROBLEMS

GEORGE P. L. WALKER

Hawaii Institute of Geophysics, 2525 Correa Road, Honolulu, Hawaii 96822

(Received August 19, 1982; Revised and accepted February 5, 1983)

ABSTRACT

Walker, G.P.L., 1983. Ignimbrite types and Ignimbrite Problems. In: M. F. Sheridan and F. Barberi (Editors), Explosive Volcanism. J. Volcanol. Geotherm. Res., 17: 65-88.

A spectrum of ignimbrite emplacement types exists, ranging from the "conventional" high-aspect ratio (H.A.R.I.) type, emplaced relatively quietly and passively in valleys, to the low-aspect ratio (L.A.R.I.) type, emplaced cataclysmically. Features of the L.A.R.I., such as a remarkable ability to scale mountains and cross open water and a strong fines-depletion of part of their deposits, stem from a high flow velocity which may result from an extremely high magma discharge rate. Being less rare than large-volume H.A.R.I. eruptions covering the same area, L.A.R.I. eruptions are a much more immediate volcanic hazard. Being thin and inconspicuous, a L.A.R.I. may easily be overlooked when determining the past record of a volcano.

Another ignimbrite spectrum depends on variations in particle viscosity during emplacement and extends from the low-grade (water-cooled?) ignimbrite which is totally non-welded even if >50 m thick, to the high-grade (superheated?) one which is densely welded even if <50 m thick. Problems of air-cooling and water-cooling of ash flows need to be tackled, and it may be necessary to recognize strongly cooled ash flows which were emplaced in part at <100°C.

One problem of ignimbrite eruptions is the origin of the extensive associated ash fall, comparable in volume to the ignimbrite. This ash may be: (a) pre-ignimbrite Plinian pumice; (b) co-ignimbrite ash, containing material lost from both the eruptive column and ash flow; (c) phreatoplinian, due to the entry of significant amounts of water into the vent; (d) phreatoplinian co-ignimbrite, due to explosions at rootless vents where ash flows enter water from land.

Another problem is the origin of associated well-sorted and sometimes wavy-bedded deposits. These deposits may be from: (a) base surges, related to either the collapsing column or entry of water to the vent; (b) base surges, due to explosions at rootless vents where ash flows enter water from land; (c) fines depletion in, and deposition from, the strongly fluidized head of the ash flow; (d) standing waves in a high-velocity ash flow; (e) pyroclastic surges springing from the ash flow; (f) superficial turbulence in the topmost fractions of the ash flow as it comes to rest.

Major problems concern the relationship between pyroclastic surges and flows, the ability of one to change into the other, and the distinction between their deposits. Thus, the May 18th 1980 "directed blast" of Mount St. Helens is widely regarded as a surge, yet produced deposits having many characteristics of a L.A.R.I.. Understanding the behaviour of the fine ash and dust fraction is thought to be critical to the solution of these problems.

INTRODUCTION

The unravelling of the origin of ignimbrites, or ash-flow tuffs as the welded ones are often called, is one of the outstanding success stories of modern volcanology. This work focused attention on features such as the welding and crystallization zonations shown by ignimbrite sheets, and the association of major sheets with major calderas (e.g., Marshall, 1935; Smith, 1960a, 1960b; Ross and Smith, 1960; Boyd, 1961). It was recognized how voluminous some ignimbrite units are and, incidentally, partly resolved the dispute on whether granites can have a magmatic origin. More recent studies are using ignimbrites to explore physico-chemical conditions and magma zonations within magma chambers (e.g., Hildreth, 1979, 1981; Smith, 1979).

An important direction of recent and current research is focusing attention on the emplacement mechanisms of ignimbrites, and it is this direction which is reviewed in the present paper. This approach includes studies on the field relations of young ignimbrites and their response to the pre-existing topography; it also places much reliance on granulometric and component analyses to document features such as fines depletion and crystal concentration in pyroclastic deposits as a means of understanding the physical processes which operate to produce them. These studies have been co-ordinated with fluidization experiments and numerical modelling towards the same end (e.g., Wilson, 1980).

This direction of research is actively being pursued because it is so abundantly clear that there is still much that is not understood. The present article is therefore a progress report, discussing concepts still only partially propounded or not yet fully tested, and serves as a guide to some of the "grey" areas of incomplete knowledge. The research is, *inter alia*, radically changing the diagnostic criteria for the different pyroclastic types, and is very relevant to volcanic hazards evaluation.

Few words in volcanology have had so chequered a history and been used in so many different ways as "ignimbrite", which is why "ash-flow tuff" was introduced in its place. "Ignimbrite" is, however, a convenient word and here it is defined as a pyroclastic deposit or rock body, made predominantly from pumiceous material, which shows evidence of having been emplaced as a concentrated hot and dry particulate flow. "Ash flow" is here retained to denote the pyroclastic flow when it is moving. As with many definitions in geology, it is difficult to allow for all contingencies by devising an all-embracing definition; thus, as discussed later, there are grounds for including as "ignimbrites" some deposits which may not have been hot when they were emplaced.

Ignimbrites are associated with nearly half of the world's volcanoes. They embrace all but the most mafic compositions. They occur in geological formations of all ages. They range in size over at least five orders of magnitude, and the largest are the greatest eruptive units known, the formation of which would be cataclysmic events on a scale too big to be envisaged clearly by modern man.

HIGH- AND LOW-ASPECT RATIO IGNIMBRITES

A convenient non-genetic means of describing the overall geometry of rock units is by means of the "aspect ratio", applicable, for example, to extrusive lava bodies, pillows as seen in cross-sections of pillow lava, and deformed juvenile clasts (fiamme) in welded tuffs. The aspect ratio is V/H, where V is a vertical dimension (e.g., the average thickness), and H is a horizontal dimension (e.g., the diameter of a circle covering the same areal extent as the rock unit).

It is known that V/H for ignimbrites, as for lava extrusions, covers a wide range in values (Fig. 1); it varies from about 1/400 for the Valley of Ten Thousand Smokes (V.T.T.S.) ignimbrite to about 1/100,000 for the Taupo and Koya ignimbrites (Ui, 1973; Walker et al., 1980a).

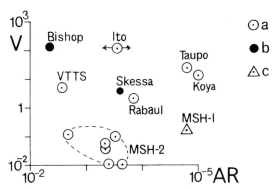

Fig. 1 Plot of volume (V, in km^3) against aspect ratio (AR) for a number of ignimbrites and related deposits. a - mainly non-welded; b - mainly welded; c - relatively poor in pumice; VTTS - Valley of Ten Thousand Smokes; MSH - Mount St. Helens (1 - "directed blast" of May 18, 1980, from Moore and Sisson 1981; 2 - post climactic 1980 flows, from Rowley et al., 1981). The arrows for Ito express uncertainty, whether to take the area known to be occupied by ignimbrite, or the area enclosed by an envelope drawn around the most outlying outcrops.

Variations in the aspect ratio of ignimbrites appear to be correlated with other significant differences, notably in the response of the ash flow to the pre-existing topography (Fig. 2). High-aspect ratio flows such as the V.T.T.S. respond passively to the topography. They follow and infill existing valleys, and the distance they travel is greatest along the main vallleys. Low-aspect ratio flows such as Taupo, on the other hand, are emplaced in such an active way that only a minimal control is exercised by the existing topography. They tend to travel about the same distance radially outwards in all directions from the vent, irrespective of the topography. The Taupo flow, for example, covered a near-circular area and reached about the same distance after crossing several substantial mountain ridges as it did after travelling in other sectors over nearly level ground (Fig. 2).

There are striking contrasts in the extent to which the flows surmount the topography. Thus, the high-aspect ratio V.T.T.S. flow is virtually confined to two valleys, and the "high-tide mark", which is locally found on valley sides (Fenner, 1925) rises to only 50 m above the general level of the flow and may adequately be accounted for by the volume reduction accompanying compaction (W. Hildreth, pers. comm.). The Taupo flow evidently surmounted and flowed over virtually the entire topography and recorded the fact by draping the landscape with a thin, but more or less continuous "ignimbrite veneer deposit", as well as filling small valley-bottom ponds of the more familiar type (Walker et al., 1981a). The smooth top surface of the V.T.T.S. deposit falls gradually and uniformly in level away from source at 1° to 2°. In contrast, the innumerable valley ponds in the Taupo ignimbrite occur at many different levels and decrease in level outwards from source only where the pre-existing topography also decreases in level. The Ito ignimbrite, which in many respects resembles Taupo, shows up-vent-sloping depositional ramps in some valleys (Suzuki and Ui, 1982).

The evident ability of low-aspect ratio flows to cross high mountain ridges -- the Ito flow crossed passes up to 720 m high to reach valleys on the far side (Yokoyama, 1974) -- led to the belief that active ash flows travel in a highly expanded condition, having a low ratio of gas to particulate material. This expanded "ash cloud" was then able to overtop and pour over mountain ridges. The author considers that this mechanism cannot be sustained, on the grounds that the resulting ignimbrite lacks any significant normal grading or internal stratification, and has retained a high content of fine particles (Walker, 1981a).

The evidence of high mobility and a remarkable ability to surmount topographic obstacles as outlined above indicates that low-aspect ratio ignimbrites are emplaced at a high flow velocity. For the Taupo flow the velocity is calculated, from the height climbed, to be not less than 200 m s^{-1}. The high flow velocity in turn is correlated with a high discharge rate estimated at about 10^8 m^{3}s^{-1} of pumice (Wilson and Walker, 1981), in contrast with 2×10^5 m^{3}s^{-1} for the V.T.T.S. flow (Curtis, 1968). Another general line of supporting evidence for a high discharge rate is the remarkably wide dispersal of the Plinian (ultraplinian) pumice fall, which was erupted

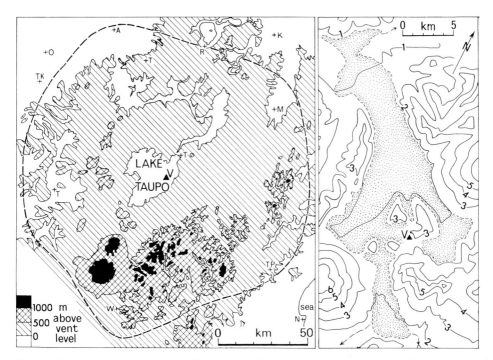

Fig. 2 Contrasted response to the topography of a low-aspect ratio ignimbrite (left) and a high-aspect ratio ignimbrite (right); note the 6-fold difference in scale of the two maps. V - vent.
Left: the Taupo ignimbrite (Walker et al., 1981a); the dashed line is an envelope enclosing all the known outcrops. Shading indicates land higher than the level of Lake Taupo; there is evidence that the vent was near lake level at the time of the eruption.
Right: the Valley of Ten Thousand Smokes ignimbrite (Fenner 1925; Curtis 1968). Contour interval to 300 ft. (= approx 100 m).

immediately before the Taupo ignimbrite. This dispersal also indicates a high energy discharge rate (Wilson et al., 1978; Settle, 1978).

A number of other features of low-aspect ratio ignimbrites may be correlated with the high flow velocity (Fig. 3). One is the conspicuous development of fines-depleted "layer 1" deposits (F.D.I.) underlying normal ignimbrite; attributed to the highly fluidized condition of the flow head resulting from a high gas throughput. Most of this gas was presumably ingested air. The fact that these deposits are best developed where the flow traversed forests suggests that surface roughness promoted air ingestion, and that gases generated by the heating and combustion of ingested macerated vegetation may also have made a

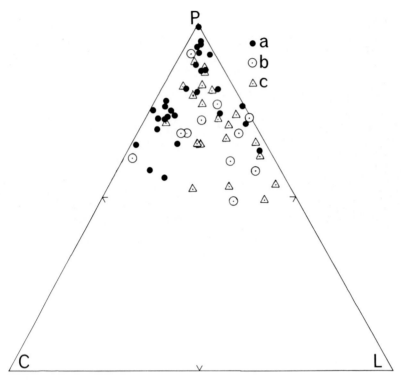

Fig. 3 Composition plot of 62 ignimbrites showing weight of pumice plus shards (P), free crystals (C), and lithics (L), for different petrological types: a - rhyolite; b - dacite, andesite; c - trachyte, phonolite. One sample from each ignimbrite was taken at random, the percentage of components in classes 1/4 mm and coarser was determined, and material finer than 1/4 mm was assumed to consist of pumice. In all samples pumice is the main constituent, but either crystals or lithics can reach about 40%. The content of free crystals is generally significantly higher than the magmatic content.

significant contribution (Walker et al., 1980b). The fines-depleted layer 1 deposits include the pumice-rich F.D.I. variant attributed to forward jetting of parts of the flow head (Walker et al., 1981b; Wilson and Walker, 1982), and the more common "heavies-enriched" ground layer variant attributed to sedimentation of heavy particles in the flow head (Walker et al., 1981c). Various segregation features are also well developed including a pronounced upward concentration of light particles which were translated by laminar flowage into a pronounced lateral variation (Walker and Wilson, 1982).

One corollary of the foregoing is that, if ash flows can deposit well-sorted fines-depleted material and produce landscape-mantling veneer deposits, the diagnostic criteria for the various pyroclastic types need to be revised: good sorting and mantle bedding are not attributes of fall deposits alone. Again, it is now known that accretionary lapilli similar to those in rain-flushed ash falls occur in many ignimbrites. Care is also needed when distinguishing between welded tuffs of air fall and flow origin. In consequence, diagnostic criteria must continually be updated as pyroclastic types become better characterized.

Another corollary of the foregoing is that the distance travelled by a low-aspect ratio ash flow is set by the quantity of material in the flow. This is because all the time on its journey outwards from the vent the flow loses material as it deposits layers of fines-depleted and veneer-type material on the ground over which it passes, and as dust escapes to the air. The flow travels outwards until all the material has been used up. A crude analogy may be drawn to a rolled carpet, which, when given a lateral push, unrolls to form a layer on the ground, and continues moving until it is fully unrolled.

A third corollary is that, because parts of a L.A.R.I. represent deposits from an ash flow and do not consist of the ash flow itself in the position where it came to rest, their thickness and grain size at any given point may bear little relationship to the thickness or grain size of the ash-flow when it passed over that point.

The commonly held view of ignimbrite formation envisages the ash flow as coming to rest en masse, so that the thickness and structure of the resulting ignimbrite are similar to those of the ash flow. In these respects, an ignimbrite is thus similar to a lava flow. This view is probably valid for some ignimbrites, but it is manifestly not valid for the L.A.R.I.. Thus, a veneer deposit a few centimeters thick may record the passage of an ash flow many tens of meters deep, and a well-sorted and strongly fines-depleted layer 1 may be deposited from an ash flow which was itself unsorted and carried a high content of fine material.

The volcanological literature has tended to focus attention on the thick ignimbrites at the expense of the thin ones. The idea that a pyroclastic layer only a few centimeters thick (as found in some L.A.R.I. examples) may have an ash flow origin is still a somewhat novel one. Very few examples of ignimbrites under 1 m thick have yet been documented, but such thin flows seem nevertheless to be very common. Thus, upwards of ten separate flows ranging from 0.1 to 1.5 m thick are found toward the top of the Mangaone Plinian pumice deposit in New Zealand. The Vulsini D ignimbrite in Italy (Sparks, 1976) and the Rio Caliente ignimbrite in Mexico (Wright, 1982) are examples of the common multiple type of ignimbrite which is built of an accumulation of many individual, thin flow units.

HAZARDS FROM LOW-ASPECT RATIO IGNIMBRITES

Hazard from eruptions of small-volume L.A.R. flows is more immediate than that from eruptions of large-volume H.A.R. flows of the more "conventional" type which have the same destructive potential. One reason is that eruptions of

L.A.R. flow are much less rare: there have been not less than five in New Zealand, Kyusyu and Rabaul in the past 6000 years. Furthermore, the disastrous eruptions of Mt. Pelée 1902 and Mt. St. Helens 1980 produced some deposits which, though too poor in pumice to be called ignimbrite, are closely akin to L.A.R.I.s. The L.A.R.I., being thin, inconspicuous, and readily eroded, may easily be overlooked when making a stratigraphic study of the past record of a volcano. The study of L.A.R.I.s is still in an early stage and the full capabilities of volcanoes to produce them are not yet established. Eruption of a moderately large-volume L.A.R.I. comparable to deposits at Taupo or Koya would undoubtedly be a major disaster to any country, and because of the large quantity of released ash would have repercussions throughout the hemisphere.

To reiterate, the thickness of a deposit at any point is no guide to the thickness of the ash flow that passed that point and hence to the destructive potential of the flow. This is abundantly clear at Mt. St. Helens and St. Pierre (Martinique) where massive demolition of forest or buildings took place yet where the accumulated thickness of the deposit is only tens of centimeters.

HIGH- AND LOW-GRADE IGNIMBRITES

It is a familiar fact that one ignimbrite may be totally non-welded whereas another having the same thickness may be densely welded more or less from bottom to top. It is convenient to designate as "low-grade" those ignimbrites which are totally non-welded even where they are 50 m or more thick, and as "high-grade" those which are densely welded even where they are less than 5 m thick. The dependence of welding on the viscosity of the non-crystalline juvenile particles at the time of deposition is well established from the numerical modelling of Kono and Osima (1971) and Riehle (1973). What is uncertain is the reason why in some ignimbrites the particle viscosity at that time was more than about 10^{15} poises, whereas in others having a similar composition the viscosity was less than about 10^{10} poises.

A first possible explanation for this variation is that the grade reflects the initial temperature of the magma immediately prior to eruption. It is a fact that many high-grade ignimbrites are aphyric or nearly so, indicating that the magma was near the liquidus temperature. Examples are the Skessa tuff in Iceland (Walker, 1962) and the Walcott tuff in Idaho, both of which have a phenocryst mineralogy indicative of relatively hot and dry magmas. However, exceptions occur: the Cimino ignimbrite in Central Italy has a fairly high grade but carries nearly 50% of crystals. Conversely, some ignimbrites are practically aphyric yet are of low-grade type; an example is the fairly low-grade Rio Caliente ignimbrite in Mexico.

A second possible explanation is that low-grade ignimbrites suffered appreciable cooling during eruption. An example is the Rabaul ignimbrite which had a high magmatic temperature when quenched (Heming, 1974; Heming and Carmichael, 1973), yet is of fairly low-grade type. The eruptive column is one place where cooling might occur, mainly by thermal exchange with indrawn air.

Sparks and Wilson (1976) found that ignimbrites preceded by a Plinian phase tend to be less welded than those not preceded, and suggested that the former resulted by collapse from a high column and were more cooled. Cooling by as much as 300°C could occur in the collapsing column (Sparks et al., 1978).

Another place where heat loss could occur is from the head of the ash flow. Boyd (1961) calculated that heat loss from the active flow is small, but loss could be considerable in strongly fluidized examples where there is a high throughput of ingested air in the head. Any surface water would also contribute to the cooling, but ingested vegetation might act in the opposite direction and contribute some heat if sufficient oxygen for combustion could gain access to it. The informally named "Morrinsville ignimbrite" in New Zealand (Walker and Wilson, 1982) shows some evidence for having been cooler at its far distal end after travelling 200 km than it was nearer source, but the author does not know of any well documented example of an ignimbrite that markedly changes grade laterally.

The view could be taken that these various kinds of heat loss are inadequate to account for the observed variations in grade, and recourse then be taken to more drastic solutions: the high-grade ignimbrites stem from superheated magma, and low-grade ignimbrites develop where the magma has been water-cooled. Consider the second. Phreatomagmatic eruptions are those in which sufficient water (either surface water or groundwater) enters the vent to modify significantly the character of the eruption and the nature of the resulting ash deposit (see Sheridan and Wohltez, this volume). It is not the absolute flux of water into the vent that is important: it is the ratio of the water to magma flux. Thus, Plinian eruptions produce coarse pumice deposits and are generally regarded as characterizing dry vents. There may, however, be a considerable influx of water into the vent, but the Plinian charcter is preserved if the water flux is small in comparison with the magma flux. A phreatomagmatic eruption is one in which the water flux is sufficiently large relative to the magma flux to have a significant effect on the eruption.

Some low-grade ignimbrites may have come from an eruption column in which significant water-cooling of the juvenile material occurred at a water flux which was insufficient to generate a phreatoplinian column. One gram of water heated to steam at 100°C can cool 10 g of rhyolitic magma by 250°C, sufficient to increase the magma viscosity by more than three orders of magnitude and produce a significant decrease in grade of the resulting ignimbrite. The Oruanui ignimbrite in New Zealand is a possible example (Self, this volume). It lacks the usual field evidence for a high temperature (e.g., welding or the presence of carbonized plant remains), and its association with a particularly widespread phreatoplinian ash (Self and Sparks, 1978) testifies to the involvement of water in the eruption.

Generally the larger pumice clasts in ignimbrite lack jointing, and in this respect they contrast with pumice-fall deposits in which each of the larger pumice clasts readily breaks along its own system of cooling joints. The explanation is that the pumice clasts in ignimbrite were enclosed in a hot matrix and cooled slowly together with the matrix, whereas in a fall deposit each pumice clast cooled more or less independently of the others in the air.

In some ignimbrites there is clear evidence that the larger clasts were much hotter than the matrix. Thus in the 1883 ignimbrite of Krakatau, the larger pumice show a pink thermal coloration, contrasting with the uniformly white or grey color of the matrix, and each pumice lump has its own system of cooling joints showing that it cooled by loss of heat into the matrix. Breadcrust blocks, such as occur in part of the Rio Caliente Ignimbrite (Walker et al., 1981c) and in several New Zealand ignimbrites, carry a similar connotation: the breadcrusted structure was caused by cooling due to a loss of heat into a cooler matrix. Whether the matrix in the Krakatau example was water-cooled is speculative, but the possibility that it was so cooled clearly exists.

Many high-grade ignimbrites show rheomorphic structures, normally confined to the basal part of rather thick units. Rheomorphism results when the ash particles coalesce to form a reconstituted lava body which then flows downslope. The finest rheomorphic structures in thin eruptive units are shown not by ignimbrites, but by welded tuffs of fall origin, exemplified by those of Pantelleria (Wright, 1980; Wolff and Wright, 1981); and the reason for this is clear. For an ash flow to be generated at the site of column collapse, the magma viscosity must be sufficiently high that the particles do not stick together, otherwise a welded ash-fall tuff forms around the vent. If indeed the particle viscosity is sufficiently low, the whole mass flows away from the vent as a lava flow, as often happens at the base of lava fountains in Hawaii. The point is that if the ash particles had been sufficiently fluid to flow as a thin rheoignimbrite, it is unlikely that an ash flow would have formed in the first place. Basaltic ignimbrites are scarce for the same reason: basaltic magma in general, is too fluid to permit ash flows to form.

THE LAHAR-IGNIMBRITE BOUNDARY

There is a semantic problem attached to certain particulate flows which lack evidence for having been hot when emplaced. Conventionally, these would be termed lahars or mudflows although it is possible that their mobility was due to fluidization by gas. An example is the Morrinsville ignimbrite which in some distal outcrops shows the pink thermal coloration common to many non-welded ignimbrites but in the most distal outcrops lacks this feature. This ignimbrite shows a gradual and progressive change in characteristics out to these distal parts and there is no reason to suspect that the continuous phase was anything but gas, albeit more or less cold (Walker and Wilson, 1982).

It is here proposed that usage of the terms "ash flow" and "ignimbrite" should logically be extended to embrace all examples in which there is reason to believe that the continuous phase was gas, the kind of evidence in favor of gas being the common presence of gas elutriation pipes and fines-depleted layer 1 deposits. "Lahar" and "mudflow" should be restricted to examples where there is reason to believe that the continuous phase was liquid water, the kind of evidence being the capacity of lahars (because of their strength) to carry coarse lithic debris, the inverse coarse-tail grading sometimes shown by this debris, and the common presence of vesicles in the muddy matrix.

CO-IGNIMBRITE AND RELATED ASH-FALLS

Recent studies of young pyroclastic deposits have shown that major ignimbrites are characteristically associated with ash-fall deposits of comparable volume to the ignimbrite. Examples are the Campanian Tuff (Italy), the Toba Tuff (Sumatra), the Bishop Tuff (California), the Ito and Koya ignimbrites (Japan), the Los Chocoyos ignimbrite (Guatemala), and the Rotoiti and Oruanui ignimbrites (New Zealand), all of which have ignimbrite and associated ash-fall volumes of tens of hundreds of cubic kilometers. The ash-fall deposits are characteristically widely dispersed, examples being the Toba Ash found to 2500 km west of Sumatra (Ninkovich et al., 1978), and deep sea ashes found 2000 km east of New Zealand (Ninkovich, 1968) which are probably related to major Quaternary ignimbrites in the Taupo Zone.

Impressed with the evidence for crystal enrichment in ignimbrites (Hay, 1959; Lipman, 1963; Walker, 1971) which could be accounted for by the selective loss of vitric ash, Sparks and Walker (1977) proposed that the "lost" vitric material resided in the associated widely dispersed ash-fall beds, and that these ashes of "co-ignimbrite" type were thus complementary to ignimbrite. From more recent work it has appeared that, while this interpretation may be valid in some instances, there is another way to account for crystal enrichment in ignimbrite, and there are several possible alternative origins for the ash-fall.

First, an alternative way of accounting for the "lost" vitric ash, demonstrated at Taupo (Walker and Wilson, 1982), is that it was removed from the proximal ignimbrite (to produce the strong crystal enrichment observed there), and is contained in the predominantly very fine and crystal-poor distal ignimbrite. The mechanism is a simple one: light vitric material is preferentially concentrated towards the top of the moving ash flow, and by laminar flowage this top part came to travel farther and was translated to the distal part of the ignimbrite. It is not known how common this mechanism is, but the distal part being in general thin and less likely to be welded, tends to have a lower survival potential than the relatively crystal-rich main part of an ignimbrite and may thus be lost or escape notice. The distal part of the Morrinsville ignimbrite is so fine and lacking in evidence for having been hot that it could easily fail even to be identified as ignimbrite.

Second, one alternative origin for the extensive ash-fall is that it is the fine distal part of a pre-ignimbrite Plinian pumice fall. Most studies of Plinian pumice deposits (with the notable exception of some Icelandic studies, Thorarinsson, 1954, 1981; Thorarinsson et al., 1959; Larsen and Thorarinsson, 1977; Persson, 1966a, 1966b, 1967) have concentrated on the well preserved and impressively coarse near-vent parts. Mass budget studies from crystal contents have revealed that Plinian deposits may be much more voluminous and may include a much more extensive fine distal part than was generally thought (Walker, 1980, 1981b). The ash fall associated with the Bishop Tuff (W. Hildreth, pers. comm.) and one lobe of the ash associated with the Los Chocoyos ignimbrite (W. I. Rose, pers. comm.) agree more closely in chemical composition with the pre-ignimbrite Plinian pumice than with the ignimbrite itself.

Another alternative origin for the ash-fall is that it is of phreatoplinian type. There is evidence that the Oruanui and Rotoehu ashes are of this type, and their remarkably wide dispersal shows that the phreatomagmatic explosions were of quite exceptionally great power. The former are regarded as originating at the primary vent (Self and Sparks, 1978), and the latter at rootless vents where pyroclastic flows entered water from land (Walker, 1979). The rate of decay in thickness and grain size of both ashes is so low as to make location of the vent position extremely imprecise and hence identification of the origin uncertain. The presumed position of the Oruanui primary vent is submerged beneath Lake Taupo, and of the Rotoehu is deeply buried beneath younger volcanic rocks.

The origin of the great paroxysmal explosions of Krakatau in 1883 is still in doubt. It is known that each major explosion coincided with the formation of an ignimbrite flow unit (Williams, 1941; Self and Rampino, 1981). One possibility is that the explosions resulted from the entry of ash flows from land into the sea, as was postulated for the Rotoehu Ash.

It may be significant that most of the major ignimbrites which are associated with a widespread ash-fall deposit occur near the sea or a large lake and have calderas now flooded by the sea or lake, examples being Krakatau, Aira (source of the Ito ignimbrite), Kikai (source of the Koya Tuff), and Taupo (source of the Oruanui Ash). All of these are situations where the extensive ash fall could result from the interaction of magma with water.

HAZARDS FROM EXTENSIVE ASH-FALL

The extensive ash fall generated in a major ignimbrite eruption might cause more disruption than the ignimbrite itself because of the great area encompassed. In known examples the ash-fall exceeds 10 cm thick over 10^5 to 10^6 km^2. For example, the Oruanui ash (Self and Sparks, 1978) covered half of New Zealand to this thickness, and the Akaroa ash (Machida and Arai, 1978) covered half of Japan. Such an ash fall would ruin crops and deprive grazing animals of their food and drink over an enormous area.

The occurrence of accretionary lapili in such ashes indicates that they commonly fall in a damp or wet condition. Being cohesive and having a density when wet typically of 1500 $kg-m^{-3}$, a modest ash thickness might cause the collapse of most buildings and endanger the lives of those sheltering inside. Herein lies the main hazard to life unless people take positive steps to remove the ash while it accumulates. Note that civil defense authorities tend to recommend people to seek shelter in times of ash fall which may be unsound advice. Ash falling in a damp and hence cohesive condition would also bring down telephone and electric power lines and make roads temporarily impassable, thereby causing a complete breakdown of communications.

ABILITY TO CROSS OPEN WATER?

The 1883 ignimbrite of Krakatau is several meters thick on Sertung, an island 20 km from the presumed vent position. How the ignimbrite reached Sertung is uncertain. The sea is shallow and possibly the ignimbrite formed a temporary "causeway" to the island. Alternatively, a fast-moving flow might have been able to displace the water or flow over the sea floor.

Evidence is not lacking that some ash flows were able to cross significantly wider and deeper stretches of open sea. One, the 6000-year old Koya flow (Ui, 1973), which is widely spread in southern Kyusyu, originated in the Kikai islands 40 km offshore (Tadahide Ui, pers. comm.). Two of the Aso 4 flows (Watanabe, 1978), which originated at Aso caldera, crop out on offshore islands or extend to Honsyu, implying an ability to cross more than 40 km of sea.

The entry of ash flows from land into water thus appears, in different situations, to have followed three quite different courses: the ash-flow plunged into water to generate a pyroturbidite (e.g., the Roseau Ash: Carey and Sigurdsson, 1980), exploded to generate extensive phreatoplinian ash-falls (Walker, 1979), or traversed the water to deposit ignimbrite on land beyond. Conditions which favour these courses have not yet been explored.

THE PYROCLASTIC FLOW-PYROCLASTIC SURGE PROBLEM

One of the unresolved problems of volcanology concerns the exact relationship between pyroclastic flows and pyroclastic surges, and how to distinguish reliably between their deposits. The following explores these problems, starting with the premise that pyroclastic flows are concentrated particulate systems in which the continuous phase (gas) is volumetrically equal or subordinate to particulate material, and pyroclastic surges are dilute and turbulent particulate systems in which gas greatly predominates. The character of the resulting deposits reflects these differences. Thus, the high particle concentration in pyroclastic flows inhibits sorting or the loss of fine particles, whereas the low particle concentration in pyroclastic surges may allow sorting and the ready escape of the fine particles so that the resulting deposits are strongly fines-depleted (Fig. 4).

Base surges commonly develop in explosive eruptions in which the convective plume is not sufficiently powerful to carry aloft a major proportion of the pyroclasts. Column collapse therefore occurs. They often develop in phreatomagmatic eruptions where much of the thermal energy that might contribute to a convective plume is dissipated by converting water to steam, and also in phreatic or Vulcanian eruptions where a major proportion of the ejecta are cold and have no thermal energy to contribute. Not all base surges are caused by eruptive column collapse. Others may be caused by very shallow and strongly

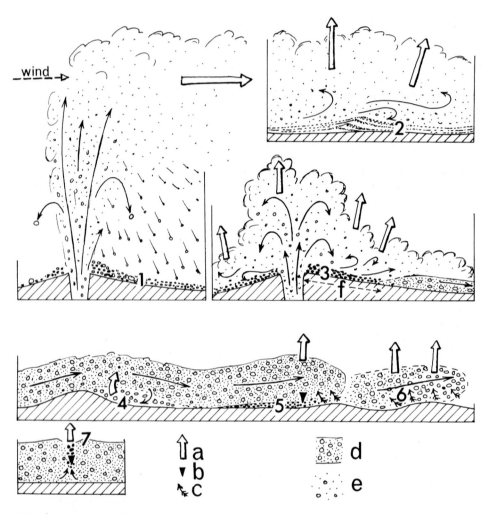

Fig. 4 Schematic views showing seven situations which produce fines depletion and good sorting of pyroclastic deposits. 1 - pyroclastic fall; 2 - pyroclastic surge; 3 - lithic lag breccia (co-ignimbrite lag-fall deposit); 4 - coarse pumice lee-side lens; 5 - ground layer forming at flow head; 6 - fines-depleted ignimbrite formed by forward jetting from flow head; 7 - elutriation pipe; a - loss of fines; b - sinking of dense particles; c - strong ingestion of air; d - high-particle concentration flow (i.e., pyroclastic flow); e - low particle concentration flow; f - deflation interval (the interval required to deflate a dilute turbulent flow to a high-particle concentration pyroclastic flow).

divergent, in part outwardly-directed, explosions. Perhaps the eruptive column in some is diverted laterally by the "stoppering" effect of a heavy load of debris already aloft above the vent.

It became realized following the 1965 eruption of Taal volcano that the deposits of base surges commonly show dunes and wavy bedforms (Moore et al., 1966; Moore, 1967) and dune-type bedding found in pyroclastic deposits (e.g., Fisher and Waters, 1970; Waters and Fisher, 1971; Crowe and Fisher, 1973) has since come to be regarded as diagnostic of a base-surge origin. This prompts three cautionary comments. One, lenticular or wavy bedforms and a form of cross-bedding can be generated by other processes operating in ash flows. Two, similar bedforms can be generated by wind or running water in non-volcanic sedimentary processes, and could indeed be primary bedforms if a strong wind is blowing during ash deposition. Three, not all base-surge deposits are wavy bedded: some are planar bedded (Sheridan and Updike, 1975; Wohletz and Sheridan, 1979).

Lenticular and wavy bedforms in ash flows develop in fast-moving flows, either downflow from topographic elevations where the flow, due to its momentum, leaps over the ground (Fig. 5), or where standing waves develop in a flow travelling at a very high velocity over a planar surface (Walker et al., 1980c). These are situations where a local low particle concentration exists in vortices in or below the flow, and a strongly fines-depleted pumice deposit accumulates there. The resulting coarse pumice lee-side lenses are quite distinct in their grain-size characteristics from base-surge deposits.

An outstanding problem concerns the nature of the nuee ardente. It is well established that the deposits of the nuee ardente include true pyroclastic flows. Commonly the resulting flow deposits are very coarse and consist of dense material, which implies that these particular flows owed their mobility to the height and steepness of the volcano slope, and that grain flow may have been a more important flow mechanism than fluidization. These pyroclastic flows tend to be channelled by valleys, often stand on moderate slopes of 5° or more and at their fartherest extension have steep flow fronts of lobate form. A seared zone often extends a short distance beyond and on either side of such flows and may result from an accompanying pyroclastic surge. The deposits in these seared zones, being thin and rapidly eroded, have seldom been described. One view is that the searing is due to an ash-cloud surge -- a hot ash-cloud fed from, riding over, becoming detached from, and extending beyond the pyroclastic flow (Fisher, 1979). Another view is that searing is due to surges emitted from the flow margin (Rose et al., 1976).

In major "directed blast" eruptions, as at Mt. Pelee on May 8th and 20th, 1902, and Mt. St. Helens on May 18th, 1980, thin deposits which may include a fines-depleted lower part are spread widely as a blanketing layer over the landscape. These deposits have generally been attributed to powerful pyroclastic surges (Fisher et al., 1980; Hoblitt et al., 1981; Moore and Sisson, 1981; Smith and Roobal, 1982). For the following reasons this interpretation is here questioned. One, internal bedding is absent from or is very ill-defined in

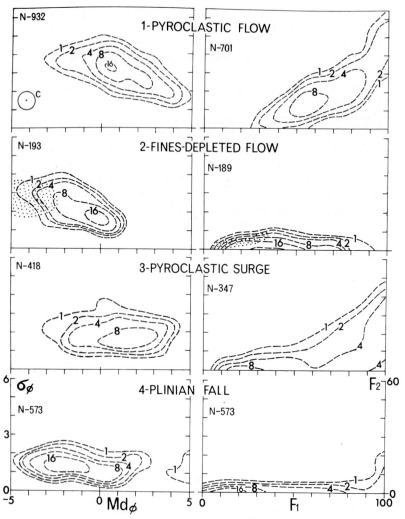

Fig. 5 Contoured fields (left) of σ_ϕ - graphic standard deviation against Md_ϕ - median size, and (right) of F_2 - wt.% finer than 1.6 mm, against F_1 - wt.% finer than 1 mm, for pyroclastic flows and three kinds of fines-depleted pyroclastic deposits. The contours are based on the percentage of samples at any point contained by a circle of size C centered on that point. N - number of samples. Plot number 2 includes various kinds of fines depleted flow deposits associated with ignimbrite: ground layers, fines-depleted ignimbrite, and elutriation gas pipes. The dotted area contains lithic lag breccias (co-ignimbrite lag falls) not included in the contoured field. Note that 2, 3 and 4 overlap, showing that grain size parameters alone are not good criteria of origin.

the relatively coarse basal fines-depleted layer (layer A1 of Waitt, 1981). Two, there are many places where the main layer of the deposit (layer A2 of Waitt, 1981) is found both in valley pond situations and on gently sloping valley sides. Three, although layer 2 at Mount St. Helens elsewhere often shows a dune type bedding, this bedding is faint and the variance between co-existing beds is much less than in the deposits of a powerful surge. The author's opinion is that the distance of travel (nearly 30 km from vent at Mount St. Helens) is unduly great for a low-concentration cloud moving against air resistance and depending on internal turbulence to maintain particles in suspension. Sparks et al. (1978) have demonstrated how rapidly particles settle out from such a cloud.

In its landscape-mantling form and fines-depletion shown by some layers, the directed-blast deposit closely resembles a L.A.R.I., suggesting that the "blast" was a high-velocity ash flow. Having a high velocity, ingestion of air in the flow head (aided at both Mt. Pelee and Mt. St. Helens by the dense vegetation) caused strong fines depletion of head (layer 1) deposits, and dune bedding was caused by local turbulence. The narrow marginal seared zone may however result from a pyroclastic surge. The outer margin of this zone shows evidence at Mount St. Helens for the rising of the "blast cloud" as, with loss of particulate material and heating of entrapped air, its density fell below ambient, and it is easy to envisage this happening to a fast-moving ash flow at the stage when the material is largely used up and is all contained in the head.

It seems possible to recognize a spectrum of pyroclastic flows in nuees ardente and "directed blasts" analogous with the spectrum of ignimbrite types (Fig. 6), ranging from the weakly-emplaced, stubby and lobate flow of ill-sorted material exemplified by Ngauruhoe, 1975 (Nairn and Self, 1978) to the violently emplaced low-aspect ratio flow exemplified by Mt. Pelee 1902 and Mount St. Helens 1980.

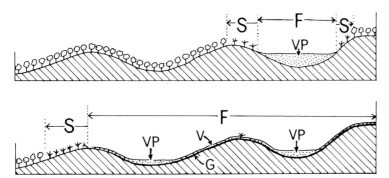

Fig. 6 Schematic view contrasting a high-aspect ratio pyroclastic flow or nuée ardente deposit (above) with a low-aspect ratio flow or nuée ardente deposit (below). F - extent of pyroclastic flow deposits; S - seared zone: pyroclastic surge; VP - valley pond deposit; V - veneer deposit; G - fines-depleted deposits underlying normal flow material.

Returning now to ignimbrites, Sparks et al. (1973), impressed by the frequent occurrence of a particular sequence of erupted layers among the products of many ignimbrite eruptions, proposed a standard sequence including a layer 1 or "ground surge" deposit underlying the ignimbrite.

Layer 1 was loosely defined, and it is now apparent that it embraced a variety of different kinds of deposits. At Taupo the layer 1 deposits are strongly fines-depleted and include "fines-depleted ignimbrite" and also a ground layer. Both are interpreted as deposits from the pyroclastic flow head, their abrupt and near-planar upper boundary against normal ignimbrite being due to shearing as they were over-ridden by the main mass of the flow. There are other possible origins for layer 1 deposits and each deposit must be assessed separately on each ignimbrite by applying the appropriate tests. On present knowledge, a ground layer as found at Taupo is particularly common among layer 1 deposits, and there is merit in restricting the designation "layer 1" to deposits such as these which are an integral part of the ignimbrite.

Some ignimbrites are overlain by wavy-bedded deposits, and there are several ways in which such deposits may have formed. In one proposed mechanism the bedded material resulted from an ash-cloud surge (Fisher, 1979) which emanated from the top of the pyroclastic flow. That it did not originate independently at the main ignimbrite vent may be shown by its outwardly increasing thickness. The thickness alone may, however, not be a reliable indicator since it may be slope-dependent. In another mechanism, the bedded material is of base-surge origin formed as a result of secondary explosions at rootless vents where the ignimbrite entered water. A feature of May 18, 1980 Mount St. Helens flow deposits is the considerable number of rootless vents that developed on them, and the large size of the secondary explosions some of which generated base surges.

Dune-like bedforms interpreted to result from the pyroclastic surge mechanism are seen also in the main part of the "directed blast" deposit of Mount St. Helens and are apparently unrelated to rootless vents. A grain size study (Walker, in prep.) shows that they are less coarse and have less variance than deposits of powerful base surges: the turbulence that generated them was relatively weak. This leads to the alternative interpretation that they were developed in the topmost few tens of centimeters of the loose ash-flow material during minor turbulence at the time of its deposition. This turbulence was caused by the surface roughness where a thin and fast-moving ash flow traversed a mountainous area diversified by a chaotic mass of tree stumps and felled trees.

SIGNIFICANCE OF FINES DEPLETION

A critical element in current discussions about the relationship between pyroclastic surges and pyroclastic flows is whether deflation of a surge is capable of generating a flow. The problem is seen to be one of retaining, or generating, enough of the fine dust on which fluidization of the flow depends.

A highly turbulent environment like that of a violent pyroclastic surge exists at the site of column collapse (Sparks et al., 1978), and as the expanded particle/gas mixture moves outwards from there, deflation to generate a high-particle-concentration mix occurs as gasses escape. Coarse lithic-rich lag breccias which accumulate within the deflation interval, consisting of debris too heavy to be carried far, are strongly fines depleted.

The gasses which escape while deflation proceeds inevitably carry off much fine ash and dust, and the crucial factor determining whether or not a pyroclastic flow then forms is whether there is a sufficient quantity of fines in the resulting high-cencentration mix. Factors which favour the retention or generation of fines favour the formation of pyroclastic flows: a high discharge rate and particle concentration, a low gas content in the column, and hot magmatic eruptive conditions in which fines are rapidly replenished by the continual bursting of vesicles as well as being generated by attrition during flow. Sparks et al. (1978) express the view that continual fines generation is essential, and the author concurs except where column collapse takes place on a steep cone and this requirement is relaxed as grain-flow then becomes a significant mobilising mechanism.

Factors which favour the loss of fines favour the formation of pyroclastic surges instead of flows: a low discharge rate, a high gas content, and a high content of more or less cold pyroclasts with a limited capacity to replenish lost fines. Note that pyroclastic surges are particularly common in eruptions that are dominated by copious amounts of non-magmatic steam, and where a major proportion of the pyroclasts are quenched or consist of reworked cold debris.

The highly expanded environment of pyroclastic surges is regarded as one in which deflation to produce a high particle-concentration basal ash flow is very unlikely because of the strong fines depletion that accompanies deflation. On the other hand, there are no such constraints inhibiting the reverse process: the addition of sufficient gas to a pyroclastic flow may very readily convert it to a pyroclastic surge. Such surges stemming from pyroclastic flows are judged to be responsible for the seared zone that often margins a flow.

CONCLUSIONS

Recent progress in the study of the emplacement and related features of ignimbrites has enabled a facies model to be developed, and is beginning to sort out the inter-related factors of discharge rate, eruptive column height, gas content, flow volume, vesiculation and fluidization state, and effects of ground roughness, which determine whether or not ash flows are generated in eruptions, and (if they are generated) then control the overall structure of ignimbrite sheets. This research has also defined a number of problems for which solutions are needed, such as the nature of interactions between ash flows and water, and the reliable distinction between the products of ash flows and the products of pyroclastic surges.

This article takes stock of what is known, and also of what needs to be known before ignimbrites are well understood. Hawaii Institute of Geophysics contribution No. 1321.

REFERENCES

Bond, A., and Sparks, R.J.S., 1976. The Minoan eruption of Santorini, Greece. J. Geol. Soc. London, 132: 1-16.

Boyd, F.R., 1961. Welded tuffs and flows in the rhyolite plateau of Yellowstone Park, Wyoming. Geol. Soc. Am. Bull., 72: 387-426.

Carey, S.N., and Sigurdsson, H., 1980. The Roseau Ash: deep-sea tephra deposits from a major eruption on Dominica, Lesser Antilles arc. J. Volcanol. Geotherm. Res. 7: 67-86.

Crowe, B.M., and Fisher, R.V., 1973. Sedimentary structures in base-surge deposits with special reference to cross-bedding, Ubehebe Craters, Death Valley, California. Geol. So. Am. Bull., 84: 663-682.

Curtis, G.H., 1968. The stratigraphy of the ejecta from the 1912 eruption of Mt. Katmai and Novarupta, Alaska. Geol. Soc. Am. Mem., 116: 153-210.

Fenner, C.N., 1925. Earth movements accompanying the Katmai eruption. J. Geol., 33: 193-223.

Fisher, R.V., and Waters, A.C., 1970. Base surge bed forms in maar volcanoes. Am. J. Sci., 268: 157-180.

Fisher, R.V., 1979. Models for pyroclastic surges and pyroclastic flows. J. Volcanol. Geotherm. Res., 6: 305-318.

Fisher, R.V., Smith, A.L., and Roobal, M.J., 1980. Destruction of St. Pierre, Martinique, by ash-cloud surges, May 8 and 20, 1902. Geology, 8: 472-476.

Fisher, R.V., Smith, A.L., Wright, J.V., and Roobal, M.J., 1980. Ignimbrite veneer deposits or pyroclastic surge deposits? Nature, 286: 912.

Hay, R.L., 1959. Formation of the crystal-rich glowing avalanche deposits of St. Vincent. B.W.I. J. Geol., 67: 540-562.

Heming, R.F., and Carmichael, I.S.E., 1973. High-temperature pumice flows from the Rabaul caldera, Papua, New Guinea. Contr. Miner. Petrol., 38: 1-20.

Heming, R.F., 1974. Geology and petrology of Rabaul Caldera, Papua, New Guinea. Geol. Soc. Am. Bull., 85: 1253-1264.

Hildreth, W., 1979. The Bishop Tuff: evidence for the origin of compositional zonation in silicic magma chambers. Geol. Soc. Am. Spec. Pap., 180: 43-75.

Hildreth, W., 1981. Gradients in silicic magma chambers: implications for lithospheric magmatism. J. Geophys. Res., 86: 10153-10192.

Hoblitt, R.P., Miller, C.D., and Vallance, J.W., 1981. The 1980 eruptions of Mount St. Helens, Washington. Origin and stratigraphy of the deposit produced by the May 18 directed blast. U.S. Geol. Surv. Prof. Pap., 1250: 401-419.

Kono, Y., and Osima, Y., 1971. Numerical experiments on the welding process in the pyroclastic flow deposits. Bull. Volcanol. Soc. Japan., 16: 1-14.

Kuntz, M.A., Rowley, P.D., Kaplan, A.M., and Lidke, D.J., 1981. The 1980 eruptions of Mount St. Helens, Washington. Petrography and particle-size distribution of pyroclastic-flow, ash-cloud, and surge deposits. U.S. Geol. Surv. Prof. Pap., 1250: 525-539.

Larsen, G., and Thorarinsson, S., 1977. H_4 and other acid Hekla tephra layers. Jokull, 27: 28-46.

Lipman, P.W., 1963. Mineral and chemical variations within an ash-flow sheet from Aso caldera, southwestern Japan. Contr. Miner. Petrol., 16: 300-327.

Machida, H., and Arai, F., 1978. Akahoya ash -- a Holocene widespread tephra erupted from the Kikai caldera, South Kyusyu, Japan. Quat. Res., 17: 143-163.

Marshall, P., 1935. Acid rocks of Taupo-Rotorua volcanic district. Trans. R. Soc. N.Z., 64: 323-375.

Moore, J.G., Nakamura, K., and Alcaraz, A., 1966. The 1965 eruption of Taal Volcano. Science, 151: 955-960.

Moore, J.G., 1967. Base surge in recent volcanic eruptions. Bull. Volcanol. 30: 337-363.

Moore, J.G., and Sisson, T.W., 1981. The 1980 eruptions of Mount St. Helens, Washington. Deposits and effects of the May 18 pyroclastic surge. U.S. Geol. Surv. Prof. Pap., 1250: 421-438.

Nairn, I.A., and Self, S., 1980. Explosive eruptions and pyroclastic avalanches from Noauruhoe in February 1975. J. Volcanol. Geotherm. Res., 3: 39-60.

Ninkovich, D., 1968. Pleistocene volcanic eruptions in New Zealand recorded in deep sea sediments. Earth Planet. Sci. Lett., 4: 89-102.

Ninkovich, D., Sparks, R.S.J., and Ledbetter, M.T., 1978. The exceptional magnitude and intensity of the Toba eruption, Sumatra: an example of the use of deep-sea tephra layers as a geological tool. Bull. Volcanol., 41: 286-298.

Persson, C., 1966a. Försök till tefrokronologisk datering av nagra svenska torvmossar. Geol. Fören. Förh. Stockholm, 88: 361-394.

Persson, C., 1966b. Undersökning av tre sun asklager pa islànd. Geol. Fören. Förh. Stockholm, 88: 500-519.

Persson, C., 1967. Försök till tefrokronologisk datering; tre norska myrar. Geol. Fören. Förh. Stockholm, 89: 181-197.

Riehle, J.R., 1975. Calculated compaction profiles of rhyolite ash-flow tuffs. Geol. Soc. Am. Bull., 84: 2193-2216.

Rose, W.I., Pearson, T., and Bonis, S., 1976. Nuee ardente eruption from the foot of a dacite lava flow, Santiaguito volcano, Guatemala. Bull. Volcanol., 40: 23-38.

Ross, C.S., and Smith, R.L., 1960. Ash-flow tuffs: their origin, geologic relations, and identification. U.S. Geol. Surv. Prof. Pap., 336.

Rowley, P.D., Kuntz, M.A., and Macleod, N.S., 1981. The 1980 eruptions of Mount St. Helens. Pyroclastic flow deposits. U.S. Geol. Surv. Prof. Pap., 1250: 489-512.

Self, S., 1983. Large-scale phreatomagmatic silicic volcanism: a case study from New Zealand. In: M.F. Sheridan and F. Barberi (Editors), Explosive Volcanism. J. Volcanol. Geotherm. Res., 17, (this volume).

Self, S., and Sparks, R.S.J., 1978. Characteristics of widespread pyroclastic deposits formed by the interaction of silicic magma and water. Bull. Volcanol., 41: 196-212.

Self, S., and Rampino, M.R., 1981. The 1883 eruption of Krakatau. Nature. 294: 699-704.

Settle, M., 1978. Volcanic eruption clouds and the thermal output of explosive eruptions. J. Volcanol. Geotherm. Res., 3: 309-324.

Sheridan, M.F., 1979. Emplacement of pyroclastic flows: a review. Geol. Soc. Spec. Pap., 180: 125-136.

Sheridan, M.F., and Updike, R.G., 1975. Sugarloaf Mountain tephra - a Pleistocene rhyolitic deposit of base surge origin. Geol. Soc. America Bull. 86: 571-581.

Sheridan, M.F., and Wohletz, K.H., 1983. Explosive hydrovolcanism: basic considerations and review. In: M.F. Sheridan and F. Barberi (Editors), Explosive Volcanism. J. Volcanol. Geotherm. Res., 17, (this volume).

Smith, A.L., and Roobal, M.J., 1982. Andesitic pyroclastic flows, in R.S. Thrope (ed.), Orogenic Andesites, John Wiley, pp. 415-433.

Smith, R.L., 1960a. Ash flows. Geol. Soc. Am. Bull., 71: 795-842.

Smith, R.L., 1960b. Zones and zonal variations in welded ash flows. U.S. Geol. Surv. Prof. Pap., 354-F: 149-159.

Smith, R.L., 1979. Ash-flow magmatism. Geol. Soc. Am. Spec. Pap., 180: 5-27.

Sparks, R.J.S., Self, S., and Walker, G.P.L., 1973. Products of ignimbrite eruptions. Geology, 1: 115-118.

Sparks, R.S.J., 1976. Stratigraphy and geology of the ignimbrites of Vulsini volcano, Central Italy. Geol. Rundsch., 64: 497-523.

Sparks, R.S.J., and Wilson, L., 1976. A model for the formation of ignimbrite by gravitational column collapse. J. Geol. Soc. London, 132: 441-451.

Sparks, R.S.J., and Walker, G.P.L., 1977. The significance of vitric-enriched air-fall ashes associated with crystal-enriched ignimbrites. J. Volcanol. Geotherm. Res., 2: 329-341.

Sparks, R.S.J., Wilson, L., and Hulme, G., 1978. Theoretical modeling of the generation, movement, and emplacement of pyroclastic flows by column collapse. J. Geophys. Res., 83: 1727-1739.

Suzuki, K., and Ui, T., 1982. Grain orientation and depositional ramps as flow direction indicators of a large-scale pyroclastic flow deposit in Japan. Geology, 10: 429-432.

Thorarinsson, S., 1954. The eruption of Hekla 1947-1948. II, The tephra-fall from Hekla on March 29th., 1947. Visindafelag Islendinga, H.F. Leiftur, Reykjavik. 68 pp.

Thorarinsson, S., Einarsson, T., and Kjartansson, G., 1959. On the geology and geomorphology of Iceland. Geogr. Annal. Stockholm, 41: 135-169.

Thorarinsson, S., 1981. Greetings from Iceland. Ash-falls and volcanic aerosols in Scandinavia. Geogr. Annal. Stockholm, 63-A: 109-118.

Ui, T., 1973. Exceptionally far-reaching, thin pyroclastic flows in southern Kyusyu, Japan. Bull. Volcanol. Soc. Jpn., 18: 153-168.

Waitt, R.B., 1981. The 1980 eruptions of Mount St. Helens, Washington. Devastating pyroclastic density flow and attendant air fall of May 18 -- Stratigraphy and sedimentology of deposits. U.S. Geol. Surv. Prof. Pap., 1250: 439-458.

Walker, G.P.L., 1962. Tertiary welded tuffs in eastern Iceland. Q.J. Geol. Soc. London., 118: 275-293.

Walker, G.P.L., 1971. Crystal concentration in ignimbrites. Contr. Miner. Petrol. 36: 135-149.

Walker, G.P.L., 1979. A volcanic ash generated by explosions were ignimbrite entered the sea. Nature, 281: 642-646.

Walker, G.P.L., 1980. The Taupo Pumice: products of the most powerful known (ultraplinian) eruption? J. Volcanol. Geotherm. Res., 8: 69-94.

Walker, G.P.L., Heming, R.F., and Wilson, C.J.N., 1980a. Low-aspect ratio ignimbrites. Nature, 283: 286-287.

Walker, G.P.L., Wilson, C.J.N., and Froggatt, P.C., 1980b. Fines-depleted ignimbrite in New Zealand -- the product of turbulent pyroclastic flow. Geology, 8: 245-249.

Walker, G.P.L., Heming, R.F., and Wilson, C.J.N., 1980c. Ignimbrite veneer deposits or pyroclastic surge deposits? Reply. Nature, 286: 912.

Walker, G.P.L., 1981a. Generation and dispersal of fine ash and dust by volcanic eruptions. J. Volcanol. Geotherm. Res., 11: 81-92.

Walker, G.P.L., 1981b. The Waimihia and Hatepe Plinian deposits from the rhyolitic Taupo Volcanic centre. N.Z. J. Geol. Geophys., 24: 305-324.

Walker, G.P.L., Wilson, C.J.N., and Froggatt, P.C., 1981a. An ignimbrite veneer deposit: the trail-marker of a pyroclastic flow. J. Volcanol. Geotherm. Res., 9: 409-421.

Walker, G.P.L., Self, S., and Froggatt, P.C., 1981b. The ground layer of the Taupo ignimbrite: a striking example of sedimentation from a pyroclastic flow. J. Volcanol. Geotherm. Res., 10: 1-11.

Walker, G.P.L., Wright, J.V., Clough, B.J., and Booth, B., 1981c. Pyroclastic geology of the rhyolitic volcano of La Primavera, Mexico. Geol. Rundsch., 70: 1100-1118.

Walker, G.P.L., and Wilson, C.J.N., 1982. Lateral variations in the Taupo ignimbrite. J. Volcanol. Geotherm. Res. (in press).

Watanabe, K., 1978. Studies on the Aso pyroclastic flow deposits in the region to the west of Aso Caldera, Southwest Japan, I: Geology, Mem. Fac. Ed. Kumamoto Univ., Nat. Sci. No. 27: 97-120.

Waters, A.C., and Fisher, R.V., 1971. Base surges and their deposits: Capelinhos and Taal Volcanoes. J. Geophys. Res., 76: 5596-5614.

Williams, H., 1941. Calderas and their origin. Univ. Calif. Pubs. Geol. Sci., 25: 239-346.

Wilson, C.J.N., 1980. The role of fluidization in the emplacement of pyroclastic flows: an experimental approach. J. Volcanol. Geotherm. Res., 8: 231-249.

Wilson, C.J.N., and Walker, G.P.L., 1981. Violence in pyroclastic flow eruptions, in: Tephra Studies (S. Self and R.S.J. Sparks, Editors). Reidal, Dordrecht, pp. 441-448.

Wilson, C.J.N., and Walker, G.P.L., 1982. Ignimbrite depositional facies: the anatomy of a pyroclastic flow. J. Geol. Soc. London. (in press).

Wilson, L., Sparks, R.S.J., Huang, T.C., and Watkins, N.D., 1978. The control of volcanic column heights by eruption energetics and dynamics. J. Geophys. Res., 83: 1829-1836.

Wohletz, K.H., and Sheridan, M.F., 1979. A model of pyroclastic surge. Geol. Soc. Am. Spec. Pap., 180: 177-194.

Wolff, J.A., and Wright, J.V., 1981. Rheomorphism of welded tuffs. J. Volcanol. Geotherm. Res., 10: 13-34.

Wright, J.V., 1980. Stratigraphy and geology of the welded air-fall tuffs of Pantelleria, Italy. Geol. Rundsch., 69: 263-291.

Wright, J.V., 1982. The Rio Caliente ignimbrite: analysis of a compound intraplinian ignimbrite from a major Late Quaternary Mexican eruption. Bull. Volcanol., 44: 189-212.

Yokoyama, S., 1974. Mode of movement and emplacement of Ito pyroclastic flow from Aira caldera, Japan. Sci. Kyoiku Daigaku, C, 12: 17-62.

COMPUTER SIMULATION OF TRANSPORT AND DEPOSITION OF THE CAMPANIAN Y-5 ASH

WINTON CORNELL, STEVEN CAREY and HARALDUR SIGURDSSON

Graduate School of Oceanography, University of Rhode Island, Kingston, R.I. 02881 (U.S.A.)

(Received May 26, 1982; revised and accepted January 5, 1983)

ABSTRACT

Cornell, W., Carey, S., and Sigurdsson, H., 1983. Computer simulation of transport and deposition of the Campanian Y-5 Ash. In: M.F. Sheridan and F. Barberi (Editors), Explosive Volcanism. J. Volcanol. Geotherm. Res., 17: 89-109.

Analyses of grain-size and modal composition of the Campanian tuff ash layer (Y-5) from 11 deep-sea cores have been carried out. This layer represents ash fall that has been correlated with the 38,000 y.b.p. Campanian ignimbrite (Thunell et al., 1979), a deposit formed by the largest eruption documented in the Mediterranean region during the late Pleistocene (Barberi et al., 1978). The bulk deposit is bimodal in grain-size and dominated by glass shards. The calculated mean grain-size of the coarse mode of the individual size distributions decreases with distance from the source and progressively approaches a near-constant fine mode of approximately 13 microns. Distal samples are unimodal in grain-size.

These data combined with a set of vertical profiles of wind (10 year average) have been used as input to a computer model that simulates fallout of tephra. Modelling indicates that the downwind variation of grain-size of the coarse mode can be accurately reproduced with transport of ash between 5 and 35 km. The observed fine mode of the deposit cannot, however, be generated by transport of ash as individual particles at these elevations. Such transport would result in deposition of virtually all of the fine ash beyond the studied area. Deposition of fine ash within the studied distance of 1600 km from source can only occur by fallout as particle aggregates from a high eruption plume or as individual particles from co-ignimbrite ash clouds with a maximum elevation of 3 km. The large volume of ash in the fine mode (>70 wt.%) and the irregularity in azimuth of low-level winds argue against major low-level transport of co-ignimbrite ash. Rather, the ash may have been derived from both a plinian eruption column and high-altitude clouds of co-ignimbrite ash, with settling of fine ash as particle aggregates.

0377-0273/83/$03.00   © 1983 Elsevier Science Publishers B.V.

INTRODUCTION

One aim in the study of tephra deposits is the reconstruction of ancient explosive eruptions. Such reconstructions strive to determine the properties and dynamics of the eruption column and transport processes on the basis of the observed properties of the tephra deposit. The May 18, 1980 eruption of Mount St. Helens provided an opportunity to develop and test a computer model of tephra fallout constrained by observations of eruption column height, elevation of major ash transport, lateral spreading of the eruption plume, atmospheric wind profile, fallout area, volume, grain-size distribution, and modal composition of the deposit (Carey and Sigurdsson, 1982). The computer model accurately simulates the downwind variation in grain-size, thickness and composition of the Mount St. Helens deposit as far as 440 km from source. The model also reproduces the thickness maximum of the deposit 325 km from source and shows that fallout of particle aggregates has an important influence on grain-size characteristics, variation in model composition, and thickness of the deposit. The deposition of particle aggregates produced a major fine mode in the deposit.

In this paper we develop a quantitative model of the formation of the Campanian ash layer (Y-5), which has a volume two orders of magnitude greater than the Mount St. Helens 1980 ash-fall deposit. The ~38,000 y.b.p. Campanian ash layer originated from a major eruption in the Phlegraean Fields caldera (Rosi et al., this volume) and consists of both an ash-fall deposit mantling the central and eastern basins of the Mediterranean Sea and a major ignimbrite on land (Barberi et al., 1978). The Campanian tephra, identified in piston cores from the Mediterranean, is correlated by composition of glass with the Campanian ignimbrite on land (Thunnell et al., 1979). Sparks and Huang (1980) have shown that the ash-fall layer is bimodal in grain size and that the median diameter of the coarse mode decreases steadily with increasing distance from source. In contrast, the median size of the fine mode changes little with distance from the source. They proposed that a Plinian eruption column supplied tephra composing the coarse mode but ash clouds rising from pyroclastic flows produced the fine mode. We evaluate various scenarios for transport and deposition of the Campanian tephra, including fallout from a single-stage eruption column with and without particle aggregation and transport as low-altitude co-ignimbrite ash.

DISTRIBUTION

The locations of cores containing the Campanian Y-5 ash are shown in Fig. 1 and range from 430 to 1535 km from source. The isopach map of the deposit is well constrained in the eastern Mediterranean, because of the good coverage of core sites. Distribution and thickness of the tephra in the proximal area is virtually unknown due to limited coring in the Ionian, Tyrrhenian and Adriatic Seas. The lack of core sites in the northern and central Aegean Sea precludes determination of the north boundary of the layer in the distal area. There is,

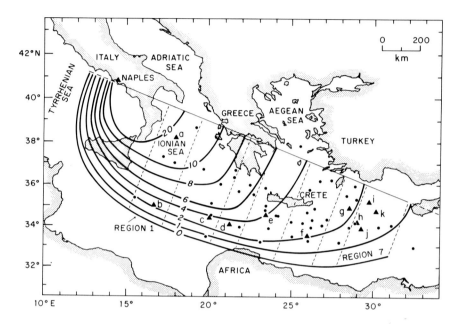

Fig. 1 Isopach map of the Y-5 ash layer derived from distribution of eastern Mediterranean piston cores. Core locations indicated by solid circles. Triangles indicate samples used for grain-size analysis in this study. Subrectangular areas refer to regions 1 to 7 in Table 1. Isopachs are in centimeters.

evidence of thinning of the deposit to the north. Isopach contouring, based on available piston core data thus cannot uniquely define a maximum thickness axis. The 115° azimuth of the fallout axis shown in Fig. 1 is at best approximate. We thus consider only the southern half of the deposit.

We divide the half-ellipse of the mapped deposit into 7 regions from the proximal to distal areas in order to reconstruct the bulk grain-size distribution in the study area and calculate tephra volume. The seven regions (Table 1) represent an area of 689,000 km^2 and a tephra volume of 37 km^3 which is roughly half the total ash-fall deposit. By comparison, the volume of the ignimbrite on land is estimated to be 80 km^3 (Rosi et al., this volume). The composite volume thus exceeds 150 km^3.

Samples and Techniques

Samples were obtained from the core collections of the University of Rhode

TABLE 1

Volume and areal extent of tephra within regions of the half-ellipse isopach of the Campanian layer (Fig. 1)

Region	area km^2	volume km^3
1	108.767	12.57
2	154,402	11.43
3	74,848	3.94
4	109,783	4.03
5	100,901	2.49
6	54,143	1.00
7	85,921	1.10
Total half-ellipse	688,765	36.56
Total deposit	1.38x10^6 km^2	73.1 km^3

Island and Lamont-Doherty Geological Observatory. Channel samples were taken of the Y-5 layer in 11 cores (Table 2). Only the ash layer in core RC9-191 exhibited stratigraphic layering (Sparks and Huang, 1980). Ash coarser than 25 microns in diameter was wet-sieved at one-half phi intervals (phi units are -log(base 2) of size in mm; Inman, 1952). Finer ash was analysed with a Particle Data electro-resistance size analyzer (Muerdter et al., 1981). The proportions of modal components (glass, lithics, felsic and mafic minerals) were determined by counting particle types in each of the one-half phi size intervals down to 25 microns. The size fraction less than 25 microns was assumed to have the same modal proportions as the 25 micron size class.

GRAIN-SIZE DISTRIBUTION

In general, the Y-5 tephra samples are poorly sorted, bimodal, and exhibit a coarse mode that progressively decreases in grain-size with distance from source (Fig. 2) in accordance with the findings of Sparks and Huang (1980). For each sample, the dividing point between a fine and coarse mode was chosen as the size class with the minimum weight percent between the two respective modes. The mean for the coarse mode (calculated by the method of moments) decreases steadily from about 200 microns at 430 km to 30 microns at 1535 km from source (Fig. 3). The mean of the fine mode shows much less variation, ranging from about 15 microns at 430 km to 10 microns at 1535 km (Fig. 3). The tephra layer is unimodal in only two cores in the distal region and this may be related to reworking, bioturbation, or merging of the two grain-size modes. The broad geographic spread and limited number of samples do not allow us to study across-lobe sorting by low-level winds. This effect has been shown to have produced variations in grain-size characteristics across ash-fall lobes from Mount St. Helens (Sarna-Wojcicki et al., 1981). Thunnell et al., (1979)

TABLE 2

Locations of cores sampled for Y-5 ash layer

ID on Fig. 1	Core no.	Latitude	Longitude	Distance from vent (km)	Thickness of layer (cm)	Depth of layer in core (cm)
a	RC9-191	38°10'	18°00'	430	4.0	306-310
b	LYNCHII-3	35°02'	16°42'	660	6.0?	417-423
c	TR171-21	34°27'	20°08'	860	2.5	22-24.5
d	TR171-22	34°06'	21°22'	945	3.0	35-38
e	RC9-183	34°35'	23°25'	1025	3.0	118.5-121.5
f	TR171-27	33°50'	26°00'	1275	1.5	111.5-113
g	TR172-24	34°53'	28°28'	1405	1.5	152-153.5
h	TR172-22	35°19'	29°01'	1420	2.0	143-145
i	TR172-11	34°08'	28°59'	1450	0.8	99.4-100.2
j	TR172-12	33°54'	29°16'	1510	0.8	551.2-552
k	TR172-19	34°43'	30°09'	1535	1.0	126.5-127.5

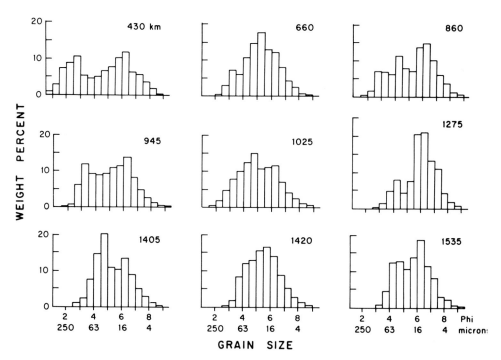

Fig. 2 Grain-size distributions of the bulk Y-5 ash layer at various distances downwind from the Campanian source area.

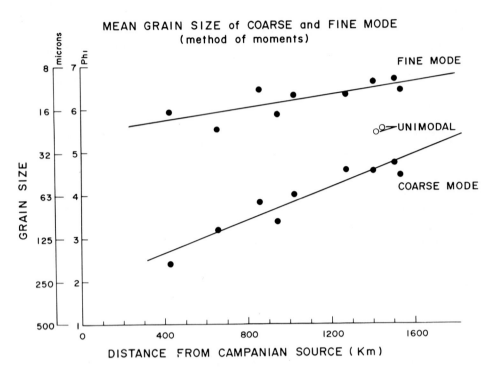

Fig. 3 Mean grain-size of the coarse and fine mode of the Y-5 ash layer as a function of distance from the Campanian source area, in kilometers. Means calculated by the method of moments (Folk, 1968). Open symbols indicate unimodal samples.

recognize, however, a decrease in grain-size from south to north for the Y-5 ash.

The grain-size distribution of the Campanian tephra contrasts with that of other studied layers of Italian origin. The Y-5 layer and three other tephra layers in core RC9-191, 430 km from source, are compared in Fig. 4. The other layers are unimodal, have modes near 19 microns, and lack a coarse mode, which suggests that they fell from less voluminous plumes.

The grain-size data for each of the seven regions (Fig. 1) have been averaged and weighted according to the mass in each region in order to obtain a bulk grain-size distribution of the deposit within the study area (Fig. 5). This distribution does not include the proximal deposit (0-400 km), which contains the majority of the coarse material, and thus underestimates the coarse component. The bulk grain-size distribution is bimodal, with modes at 3 to 3.5 phi (125-88 microns) and 6 to 6.5 phi (16-11 microns). More than half (57%) of the ash is finer than 5 phi (32 microns). Similarly, 52% of the bulk grain-size distribution of the May 18, 1980 Mount St. Helens deposit is finer than 5 phi

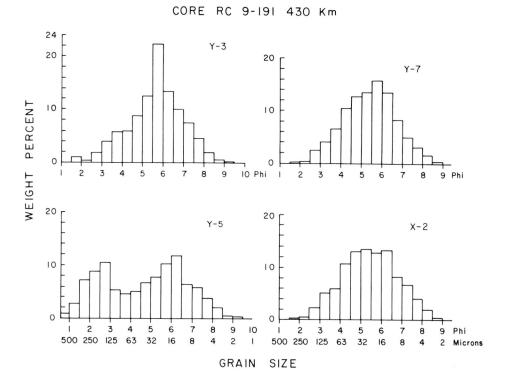

Fig. 4 Grain-size distributions of the Y-3, Y-7, Y-5, and X-2 ash layers in core RC9-191 (430 km from source).

(Carey and Sigurdsson, 1982).

MODAL CONSTITUENTS

Glass shards and pumice fragments make up 70 to 90% of the tephra (Fig. 6). The glass shards are flat, gently curved, or Y-shaped (Fig. 7), and about 5 to 10 microns in thickness. The habit indicates an origin as bubble walls, resulting from extreme vesiculation of a relatively low-viscosity magma. The tephra is of trachytic composition and calculated viscosities (Shaw, 1972) for this magma with 1% H_2O are 6.2×10^6 and 1.8×10^5 poise at 850° and 1020°C respectively. In comparison, the dacitic Mount St. Helens magma was erupted as microvesicular pumices. Calculated viscosities for the Mount St. Helens magma with 1% H_2O are 4.4×10^8 and 2.0×10^7 at 830° and 950°C respectively.

In addition to platy glass shards, the Campanian tephra contains highly

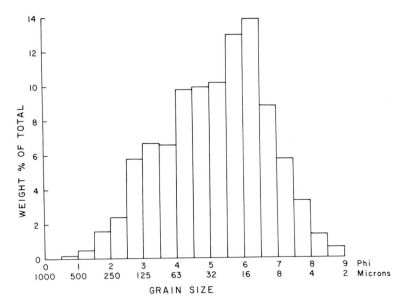

Fig. 5 Total grain-size distribution of the Y-5 ash layer based on samples between 430-1535 km from source.

elongate pipe-vesicular pumice fragments (Fig. 7), a variety of lithic fragments, and crystals of sanidine, plagioclase, pyroxene, and phlogopite. The relative proportions and downwind variation in these component abundances are shown in Fig. 6.

DISPERSAL MODEL OF CAMPANIAN TEPHRA

The atmospheric dispersal of tephra from the Campanian eruption has been quantitatively modelled using a revised version of the computer model described in Carey and Sigurdsson (1982). In the revised program, terminal settling velocities of tephra components are calculated as a function of shape and density at each one vertical kilometer increment of the fall trajectory, beginning at the initial elevation of transport. The model incorporates equations that take into account the vertical variations in atmospheric density, viscosity, and temperature (R. S. J. Sparks and L. Wilson, written communication).

It has been shown that morphology significantly effects the terminal settling velocity of a particle (Walker et al., 1971; Wilson and Huang, 1979). To compensate for this effect we introduced a factor that adjusts the terminal settling of a particle, calculated by equations of Sparks and Wilson (written

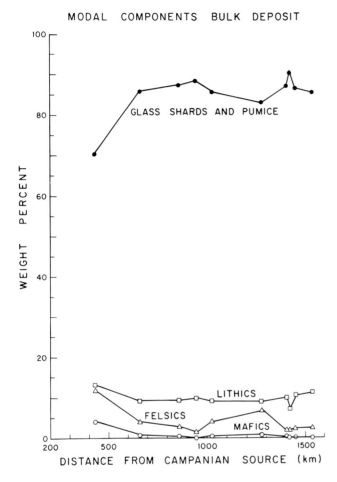

Fig. 6 Variation in modal composition of the bulk Y-5 ash layer as a function of distance from a Campanian source area.

communication). This factor, based on the experimental work of Wilson and Huang (1979), depends on particle diameter. It is only applied to particles with a diameter greater than 44 microns, because of the diminishing effect of shape on fall velocities for smaller particles.

Particle density is an additional variable that complicates the calculation of terminal settling velocities. Densities can be assumed to be essentially constant for felsic and mafic crystals but for pumice and glass shards the density varies with grain-size. Because many of the glass shards with diameters less than 125 microns in size are poorly vesicular and platy, we have assigned

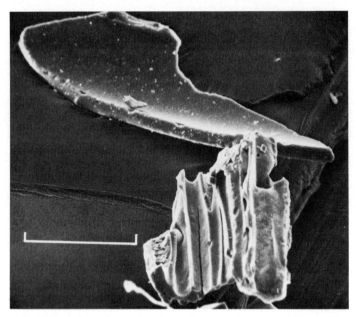

Fig. 7 Scanning electron photomicrograph of a bubble-wall and pipe-vesicular glass shard from the Y-5 ash layer in core RC9-191 (430 km from source). Scale bar is 100 microns in length.

them a density of 2.3 g-cm^{-3} (the density of a trachytic liquid of Y-5 composition). Glass particles with a diameter between 125 and 180 microns were assigned a density of 1.72 g-cm^{-3} (75% of magma density). Glass particles larger than 180 microns were assigned a density of 1.15 g-cm^{-3} (50% of magma density).

Modelling of ash transport for the Campanian eruption requires some assumptions about the vertical profile of wind. We assume that the atmospheric conditions prevailing at the time of the Campanian eruption are analogous to those occuring in the Italian region today. The 38,000 y.b.p. age of the Y-5 falls close to the isotope stage 3 (Thunnell et al., 1979; Cita and Ryan, 1978), an interstadial. The coincidence of the age, however, with the cooling side of stage 3 may suggest a better comparison to glacial conditions. The assumption of equivalence of present and past atmospheric conditions, while clearly imperfect, nonetheless provides a starting point in that present-day meterological observations can put quantitative limits on various models of tephra dispersal.

The 10-year data set of meteorlogical measurements collected over Brindisi, Italy (N40°30', E17°57') can be used to assess the present day atmospheric structure and variability. The most frequent wind direction between 2 to 34 km

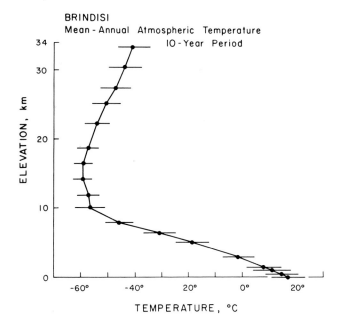

Fig. 8 Mean annual atmospheric temperature profile (10-year period) recorded at Brindisi, Italy.

elevation is rather constant for much of the year (Sept.-May) with an azimuth of 265° (W-SW) (Figs. 8 and 9). In summer (June-Aug.) a significant change of wind azimuth occurs above 14 km, when the dominant direction is from the east (90°). Below 2 km there is considerable seasonal variability in wind direction. Low-level winds are from the WSW and SW for most of the year (Sept.-May), but from NW in summer (June-Aug.). The intensity of the most frequently occurring wind direction varies both as a function of altitude and season (Fig. 10). Each seasonal wind profile is typically S-shaped with maxima near 8 km (20-30 m/s) and 34 km (25-35 m/s).

Linear regression equations were calculated for the major segments of the seasonal-wind intensity profiles and the mean-temperature profiles (Table 3). These equations were incorporated in computer models to simulate dispersal and deposition of Campanian tephra during the four seasons of the year. Input to the model consists of the bulk grain-size distribution of the eruption plume, and the total modal proportions.

RESULTS

The computer model has been used to evaluate the most reasonable range of

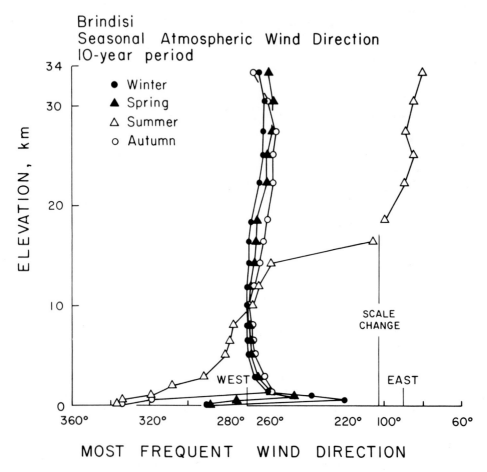

Fig 9. Seasonal atmospheric wind direction based on the most frequently occurring azimuth (10-year period) recorded at Brindisi, Italy.

eruption column heights, wind conditions and processes of tephra transport that can account for the observed bimodal grain-size distribution and variation in modal composition in the Y-5 ash layer. The observed decrease of the mean grain-size of the coarse mode with distance away from source (Figs. 2 and 3) suggests that the coarse tephra fell as individual particles during atmospheric transport. As such, the coarse mode may provide constraints on the altitude of tephra transport. To this end, we used the seasonal wind data (Fig. 10) to constrain the upper level of major ash transport.

The observed grain-size relations of the coarse mode are reproduced fairly

TABLE 3

Seasonal wind velocity equations used in the FALLOUT model

Season	Elevation interval (km)	Equation*	Correlation coefficient
Winter	>18	V=1.3509 H-10.938	0.96
	18-8	V=41.031-1.4088 H	0.98
	< 8	V=2.9245 H+5.166	0.99
Spring	>18	V=1.1909 H-12.512	0.94
	18-8	V=40.520-1.6712 H	0.97
	< 8	V=2.5930 H+4.294	0.99
Summer	>16	V=0.2458 H+2.546	0.79
	16-10	V=66.1041-3.6455 H	0.99
	<10	V=2.5228 H+2.7144	0.99
Fall	>18	V=1.2216 H-15.114	0.98
	18-10	V=33.9024-1.4068 H	0.95
	<10	V=1.6168 H+5.856	0.99

*-Derived from 10-year seasonal upper-wind averages collected for Brindisi, Italy. Data Processing Division - USAFETAC - Air Weather Service (MAC). Asheville, N.C. V is expressed as wind velocity in m/s and H in kilometers.

well by models which place upper limits of ash transport at 25 to 40 km for autumn, 20 to 35 km for winter, and 25 to 35 km for spring (Figs. 11 and 12). A lower limit of ash transport of 5 km was used for all three of these solutions. Transport of the coarse ash during summer can not have occurred because of the reversal of wind direction above 14 km during that season (Fig. 9). A summer eruption would therefore result in deposition of the tephra in the western Mediterranean. The model for ash fall during spring gave the best fit to the observed data for the coarse mode, with 67% of the actual data points (observed means) falling between the curves predicted for an upper level of ash transport at 25 to 35 km (Fig. 12). This modelling of settling of individual particles indicates that ash finer than 22 microns should not fall within a distance of 1600 km from source; thus the ash comprising the fine mode (13 microns) should not be deposited in this area.

The occurrence of a nearly constant fine mode in the deposit (Figs. 2 and 3) suggests that an additional mechanism is needed to explain its deposition. This relationship could be due to either: (1) particle aggregation of fine ash in the eruption plume (e.g. Brazier et al., 1982; Carey and Sigurdsson, 1982) or (2) co-ignimbrite ash fall, as proposed by Sparks and Walker (1977). These two hypotheses have been tested by the computer model.

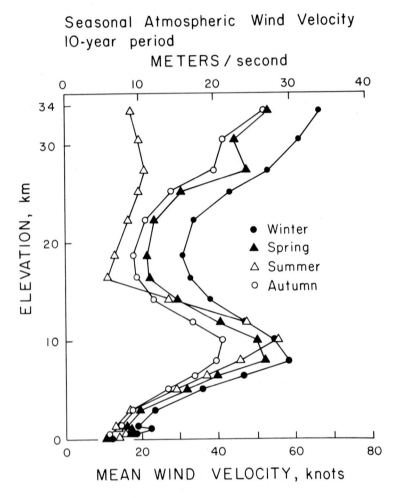

Fig. 10 Seasonal vertical variation in mean wind velocity along the most frequently occurring wind azimuth recorded at Brindisi, Italy.

Particle aggregation was modelled using the spring wind profile, which gave the best fit in modelling of the coarse mode. Two models produced a mean for the aggregated fine mode similar to the observed mean. In the first model of aggregation 50% of the 63-44 micron ash, 75% of the 44-31 micron ash and 100% of the less than 31 micron ash are treated as aggregate particles with a diameter of 200 microns and a density of 0.2 g-cm^{-3} (model 1; Fig. 12). This model produced a fine mode with slight decrease of mean grain-size away from source (Fig. 12). In the second model of aggregation, 25% of the 31-22 micron ash, 75%

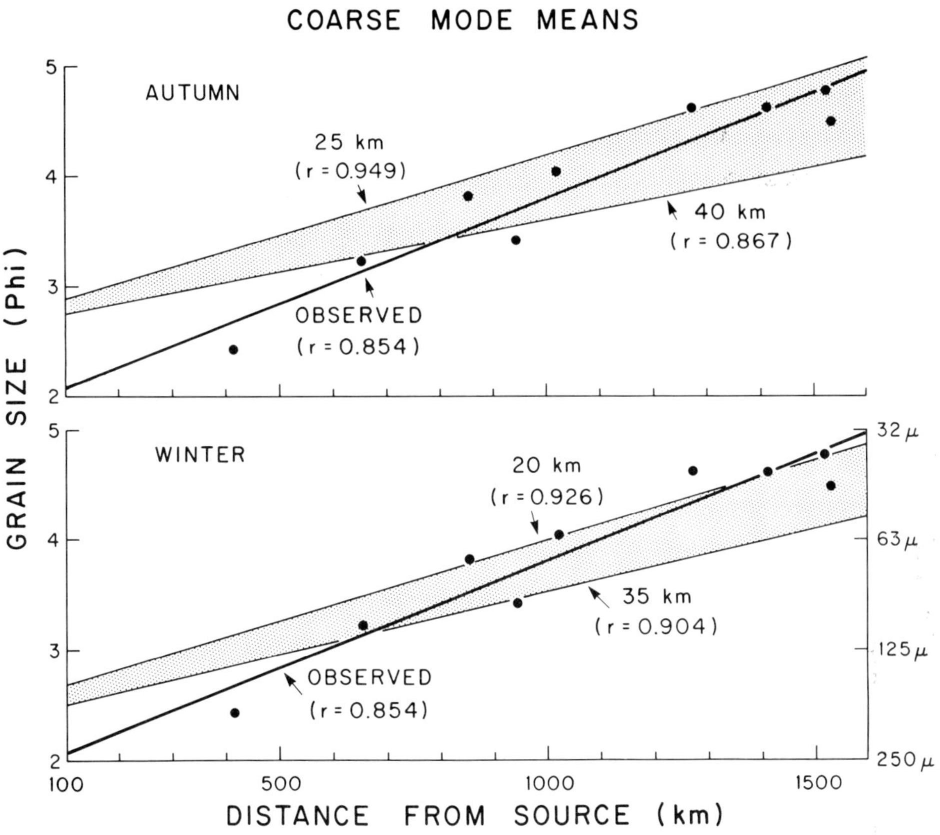

Fig. 11 Comparison of the FALLOUT-predicted and observed coarse mode means of the Y-5 ash layer as a function of distance from source for the autumn (upper part of figure) and winter (lower part of figure) seasonal wind data. Means calculated by the method of moments. Shaded areas are linear regressions through FALLOUT-predicted coarse mode means with maximum ash injection heights at 25 and 40 km for the autumn and 20 and 25 km for the winter. The lower limit of injection is 5 km. Observed coarse mode means for the Y-5 layer are shown as solid circles. The r values refer to correlation coefficients for linear regressions.

of the 22-16 micron ash and 100% of the less than 16 micron ash (model 2; Fig. 12) are treated as aggregates of 200 microns diameter with a density of 0.2 g-cm^{-3}. This model produced a fine mode with constant mean grain-size over the 1600 km of ash dispersal. In the second model, the fine-ash mode does not show a decrease of mean grain-size with distance from source, but matches the observed, nearly constant, fine mode. Agreement is also reasonable between the

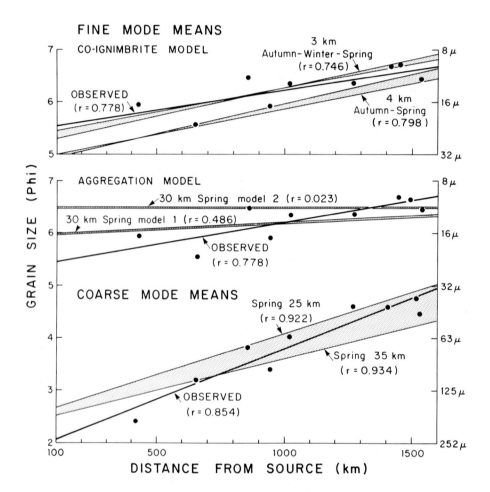

Fig. 12 Comparison of the FALLOUT-predicted and observed coarse and fine mode means of the Y-5 ash layer as a function of distance from source. Shaded areas are defined by linear regressions through the FALLOUT-predicted means for the coarse and fine mode with maximum ash injection heights at 25 and 35 km for the spring (lower part of the figure; coarse mode), 30 km for spring (fine mode aggregation models 1 and 2; middle part of the figure) and 3 or 4 km for the autumn, winter, and spring (fine mode co-ignimbrite model; upper part of the figure). Observed coarse and fine mode means are shown as solid circles. The r values refer to correlation coefficients for linear regressions.

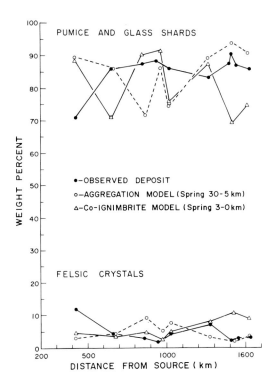

Fig. 13 Comparison of the FALLOUT-predicted and observed component proportions of the bulk Y-5 ash layer as a function of distance from source. Two types of models are presented: the co-ignimbrite and particle aggregation models for spring wind conditions at the 3 and 30 km elevation maxima, respectively.

predicted and observed proportions of modal components (Fig. 13) for both models.

To test if the fine mode could be produced by co-ignimbrite ash fall from low elevations (Sparks and Walker, 1977; Sparks and Huang, 1980) we modelled the dispersal of the bulk grain-size distribution at elevations of 5-0 km, 4-0 km and 3-0 km, with the spring, fall, and winter winds data, with the tephra settling as individual particles. The summer wind profile was not considered, as it has a polarity opposite to that of the deposit. The modelling shows that mean grain-size of the fine ash of the modelled deposit accords well with that of the observed deposit in the case of ash transport at 0 to 3 km height (Fig. 12). Agreement between the predicted and observed proportions of modal components is also reasonable in this model (Fig. 13). The model is, however, quite sensitive to height of dispersal and a change to 4 km or higher results in a predicted fine mode which is much coarser than observed (Fig. 12).

DISCUSSION

Sparks and Huang (1980) present a volcanological model to account for the occurrence and behavior of the coarse and fine grain-size modes of the Y-5 deposit. They suggest that two sources of ash fall existed during the eruption. The first was a high-altitude Plinian eruption column which provided the majority of the coarse mode. Dispersal of the tephra by upper-level winds resulted both in strong sorting and the transport of relatively large particles great distances from source. A second source was attributed by them to ash clouds generated from the top of pyroclastic flows, characterized by fine grain-size and depletion of dense components such as crystals and lithics (Sparks and Walker, 1977). These co-ignimbrite ash clouds are envisaged to rise to great heights in order to account for the lack of lateral grain-size variation in the fine mode of the deposit. However, because the base of these clouds was near ground level, deposition of fine ash could also occur close to source.

The results of our modelling of the Y-5 layer enables a rigorous evaluation of the Sparks and Huang (1980) model. We show that the coarse mode of the deposit can be accurately modelled with the present-day spring wind profile with 25-35 km as the upper limit of significant ash transport. These conditions reproduce the downwind variation of the mean grain-size of the coarse mode. The results agree with the Plinian eruption column phase envisaged by Sparks and Huang (1980). However, the modelling also demonstrates that if all ash fell as individual particles, no significant deposition of ash less than 22 microns in size would occur within the 1600 km fallout area, and thus a fine mode would not be generated under these conditions.

Our modelling of ash deposition indicates that a relatively constant fine-ash mode (Figs. 2 and 3) can, in theory, be produced over a large area by either deposition from low-level co-ignimbrite processes or the aggregation of particles. If the fine-ash mode is generated solely by deposition of material from co-ignimbrite ash clouds, then the modelling demonstrates that these clouds could not have exceeded three kilometers in height. The co-ignimbrite modelling indicates, furthermore, that no ash coarser than 32 microns (3 to 4 km upper limit of transport) can be contributed by the co-ignimbrite ash fall to the deposit at distances greater than 430 km (first sample locality). Sparks and Huang (1980) had suggested that part of the coarse mode was derived from the co-ignimbrite ash fall.

The process of particle aggregation is an alternative to low-level co-ignimbrite ash transport for the generation of a pervasive fine-ash mode. Carey and Sigurdsson (1982) suggest that particle aggregation caused the "premature" fallout of fine ash and the production of a fine-ash mode in the May 18, 1980 Mt. St. Helens tephra. Modelling of aggregate transport for the Y-5 layer shows that a fine-ash mode resembling the observed mode could be produced with 200 micron diameter particle aggregates transported at the same altitude as the coarse mode.

The model of particle aggregation is, in our opinion, more likely to account for the grain-size distribution in the Y-5 layer, rather than low-level transport of co-ignimbrite ash. The three kilometer maximum elevation imposed by the model for dispersal of co-ignimbrite ash-fall seems low when the scale of the Campanian eruption is considered. Our data for the bulk grain-size distribution of the sampled deposit indicates that over 70% of the ash fallout consists of the fine mode. If transport and deposition of much of the fine mode was solely from co-ignimbrite ash clouds, then this implies that most of the deposit was transported below three kilometers. Furthermore, the prevailing direction of low-level winds over Italy and their highly variable direction over the Mediterranean as a whole, makes it extremely unlikely that low-level transport can produce such a large volume and polarized deposit. We do not argue that co-ignimbrite ash clouds cannot be an important source of fine ash. In the light of modelling presented in this paper, we believe, however, that the two-source model presented by Sparks and Huang (1980) is not sufficient to explain quantitatively the grain-size relations and transport of the Y-5 ash.

The eruption column of the Campanian event was more likely a complex combination of a high-altitude Plinian column and high-altitude co-ignimbrite ash clouds. Deposition of fine ash, both at proximal and distal sites, occurred as a result of the low level base of the co-ignimbrite clouds and particle aggregation of ash throughout the eruption column and volcanic plume as a whole.

CONCLUSIONS

The 38,000 y.b.p. Campanian eruption produced the most extensive ash-fall layer in the Mediterranean Sea. Deposition of tephra covered an area in excess of 1.4×10^6 km^2 with a minimum volume of 73 km^3 (tephra). At distances up to 1535 km from source, the ash displays bimodal grain-size distribution, and a coarse mode that migrates towards a relatively constant fine mode (13 microns) with increasing distance from the source.

Computer modelling of the grain-size features of the deposit indicates that downwind variation of the grain-size of the coarse mode can be accurately reproduced with transport of ash between 5 and 35 km elevation. However, the observed fine mode of the deposit can not be generated by transport of ash as individual particles at these elevations but would result in transport of virtually all of the fine ash beyond the studied area.

Deposition of fine ash within 1600 km can only occur by either "premature" fallout of fine ash as particle aggregates from a high eruption plume or as individual particles from co-ignimbrite ash clouds with a maximum elevation of 3 km. The low level (<3 km) constraint imposed by the modelling of co-ignimbrite ash transport, the large volume of ash in the fine mode (>70% of the deposit) and the irregularity of azimuth of low-level winds all argue against low-level co-ignimbrite ash transport and indicate that particle aggregation strongly influenced the origin of the observed fine-ash mode of the Y-5 layer. The earlier depositional model of two volcanological phases for the Y-5 ash layer (Sparks and Huang, 1980) appears to be inadequate as a quantitative explanation of the grain-size features of the Y-5 deposit. Instead the eruption column was more likely a complex combination of material from both Plinian and ignimbrite activity. Fine ash was deposited as a result of aggregation of ash particles which were carried to great height. The ash may have been derived from both the Plinian eruption column and high-altitude co-ignimbrite clouds.

ACKNOWLEDGEMENTS

We are grateful to R.S.J. Sparks and L. Wilson for supplying us with unpublished information on settling velocity equations. The paper benefited greatly by a thorough review by R. B. Waitt and K. H. Wohletz. This work was supported by National Science Foundation grant EAR-82-05955.

REFERENCES

Barberi, F., F. Innocenti, L. Lirer, R. Munno, T. Pescatore and R. Santacroce, 1978. The Campanian ignimbrite: a major prehistoric eruption in the Naples area (Italy). Bull. Volcanol. 41: 1-22.

Brazier, S., Carey, S.N., Davis, A., Sigurdsson, H., and Sparks, R.S.J.. Fallout and deposition of volcanic ash during the 1979 explosive eruption of the Soufriere of St. Vincent. Submitted to Nature.

Carey, S.N. and Sigurdsson, H., 1982. The influence of particle aggregation on the deposition of distal tephra from the 18 May 1980 eruption of Mt. St. Helens volcano. J. Geophys. Res., 87: 7061-7072.

Cita, M.B., and Ryan, W.B.F., 1978. The deep-sea record of the eastern Mediteranean in the last 150,000 years. Thera and the Aegean World I. In: C. Doumas (Editor), Thera and the Aegean World. London.

Farrand, W.R., 1977. Occurrence and age of Ischia tephra in Franchithi cave, Peleponnesos, Greece, Geol. Soc. Amer. Abstr. with Progr. 9: 971.

Folk, R.L., 1968. Petrology of sedimentary rocks. Hemphill's. Austin, Texas. 170 pp.

Inman, D.L., 1952. Methods of describing the size distribution of sediments. Jour. Sed. Petrol., 22: 125-145.

Keller, J., 1982. Quaternary tephrochronolgy in the Mediterranean region. In: S. Self and R.S.J. Sparks (Editors), Tephra Studies. D. Reidel, Dordrecht. pp. 227-244.

Muerdter, D., Dauphin, J.P., and Steele, G., 1981. An interactive computerized system for grain-size analysis of silt using electro-resistance. Jour. Sed. Petrol., 51: 647-650.

Rosi, M., Sbrana, A., and Principe, C., 1982. The Phlegraean Fields: structural evolution, volcanic history and eruptive mechanism. In: M.F. Sheridan and F. Barberi (Editors), Explosive Volcanism. J. Volcanol. Geotherm. Res., 17: (this volume).

Sarna-Wojcicki, A.M., Shipley, S., Waitt, R.B. Jr., Dzurisin, D., Wood, S.H., 1981. Areal distribution, thickness, mass, volume, and grain-size of air-fall ash from the six major eruptions of 1980. In: P.W. Lipman and D.R. Mullineaux (Editors), The 1980 eruptions of Mount St. Helens, Washington. U.S. Geol. Survey Prof. Paper 1250: 577-600.

Shaw, H.R., 1972. Viscosities of magmatic silicate liquids: an empriical method of prediction. Am. J. Sci., 272: 870-893.

Sparks, R.S.J. and Walker, G.P.L., 1977. The significance of vitric-enriched air-fall ashes associated with crystal-enriched ignimbrites. J. Volcanol. Geotherm. Res., 2: 329-341.

Sparks, R.S.J. and Huang, T.C., 1980. The volcanological significance of deep-sea ash layers associated with ignimbrites. Geol. Mag., 117: 425-436.

Thunnell, R., Federman, A., Sparks, R.S.J., and Williams, D.F., 1979. The age, origin, and volcanological significance of the Y-5 ash layer in the Mediterranean. Quat. Res., 12: 241-253.

Walker, G.P.L., Wilson, L., Bowell, E., 1971. Explosive volcanic eruptions I: The rate of fall of pyroclasts. Geophy. J.R. Astr. Soc. 22: 377-383.

Wilson, L. and Huang, T.C., 1970. The influence of shape on the atmospheric settling velocity of volcanic ash particles. Earth Planet. Sci. Lett., 44: 311-324.

EXPLOSIVE ACTIVITY ASSOCIATED WITH THE GROWTH OF VOLCANIC DOMES

C. G. NEWHALL[1] and W. G. MELSON[2]

[1] U.S. Geological Survey, Vancouver, WA 98661 (U.S.A)
[2] Smithsonian Institution, Washington, DC (U.S.A)

(Manuscript received December 13, 1982)

ABSTRACT

Newhall, C.G. and Melson, W.G., 1983. Explosive activity associated with the growth of volcanic domes. In: M.F. Sheridan and F. Barberi (Editors), Explosive Volcanism. J. Volcanol. Geotherm. Res., 17: 111-131.

Domes offer unique opportunities to measure or infer the characteristics of magmas that, at domes and elsewhere, control explosive activity. A review of explosive activity associated with historical dome growth shows that:
(1) explosive activity has occurred in close association with nearly all historical dome growth;
(2) whole-rock SiO_2 content, a crude but widely reported indicator of magma viscosity, shows no systematic relationship to the timing and character of explosions;
(3) the average rate of dome growth, a crude indicator of the rate of supply of magma and volatiles to the near-surface enviornment, shows no systematic relationship to the timing or character of explosions; and
(4) new studies at Arenal and Mount St. Helens suggest that water content is the dominant control on explosions from water-rich magmas, whereas the crystal content and composition of the interstitial melt (and hence magma viscosity) are equally or more important controls on explosions from water-poor magmas.

New efforts should be made to improve current, rather limited techniques for monitoring pre-eruption volatile content and magma viscosity, and thus the explosive potential of magmas.

INTRODUCTION

The papers in this collection describe examples of explosive volcanism in many geologic settings. This contribution reviews a subset of explosive activity -- magmatic explosions that have occurred in association with historic dome growth.

0377-0273/83/$03.00   © 1983 Elsevier Science Publishers B.V.

Because a dome is a part of a magma body exposed at the ground surface, some of the parameters affecting explosions from that magma can be measured or relatively easily inferred (e.g., melt composition, crystallinity, temperature, and viscosity). A dome may develop if the viscosity of a lava is at least 10^8 poises; less viscous lava will form a lava flow (Walker, 1973). Lithostatic or "magmastatic" pressure must be low in and just beneath domes, simply because they are at the surface. Volatile concentrations must also be low, as high volatile concentrations in viscous melts at the surface will normally lead to explosions rather than dome growth. Because characteristics of the magma in domes can be relatively well determined, domes offer opportunities for studying controls of explosive volcanism.

We have identified eruptions in which both dome growth and explosive activity have occurred, referring to Simkin and others (1981), the Catalogue of Active Volcanoes of the World (International Association of Volcanology and Chemistry of the Earth's Interior, 1951-present), the Bulletin of Volcanic Eruptions (Volcanol. Soc. Japan, 1961-present), and detailed accounts of specific eruptions. Historically reported dome growth and the type and timing of explosive activity related to dome growth are listed in Table 1. As in any compilation of information about historical volcanic activity, one finds a large number of poorly described examples and a small number of well-described examples. Explosive activity has occurred within 1 year before or after the start of over 95% of historically reported episodes of dome growth (Table 1). This explosive activity has varied from small, frequent ash emissions to rare, very large eruptions such as that of Katmai and Novarupta in 1912. In some cases explosive activity precedes dome growth, in other cases the two occur concurrently or alternate with each other, and in yet other cases explosive activity follows dome growth, forming a crater in or destroying the new dome.

The approach taken in this paper, to examine the relations of explosions to dome growth rather than vice versa, may seem "backwards" to some, for there is a conventional wisdom that dome growth is usually a late-stage culmination of explosive eruptions, and a sign of "decadence" or decreasing explosive potential (see, for example, Powers, 1916; Williams, 1932). Some domes do indeed fall into this category, but explosive activity is at least as common after dome growth begins as before (Figs. 1 and 2). The perspective taken in this paper is of interest when evaluating hazards associated with dome growth, and may also be the most useful perspective for using growing domes as a laboratory for studying explosive volcanism.

DEFINITIONS

Williams (1932) defined volcanic domes as "steep-sided viscous protrusions of lava forming more or less dome-shaped masses around their vents." This definition, when broadened to include cryptodomes (domes that do not quite reach the surface), is used in most compilations of volcanic data. Williams also distinguished between exogenous, endogenous and plug domes. Those domes that

Table 1: HISTORICAL DOME GROWTH AND ASSOCIATED EXPLOSIVE ACTIVITY

Volcano Name	Older domes? (Yes, No)	Date of activity (year)	Duration (eruption/dome growth) (in wks)	Type of dome growth[1] (Ex,En, Pl,Cr)	Activity associated with dome growth P E N L S	CL/Sub[2]	VEI[3] (max)	Times of explosive activity relative to start of dome growth (in wks)[4] SmEx LgEx DCPf[5]	Volume of new dome[6] (m³×10⁶)	Rate of dome growth[7] (m³×10³) per day	SiO₂ of dome	SiO₂ range, this volcano	Remarks
Kameno Vouno	N?	250 BC			E L		3		60		58-60	56-60	
Santorini		197 BC			?	Sub	4?					63-66	Hiera dome
"	Y	19 AD			E	Sub	3					"	Thia dome
"	Y	46 AD			E	Sub	3					"	
"	Y	726 AD			E	Sub	3					"	
"	Y	1570 AD		En,Ex	E L	Sub	3		50		64-66	"	Mikra Kameni dome
"	Y	1707-11	220/220?	"	E L	Sub	2	0 +2,+8	10	65	65	"	Nea Kameni dome
"	Y	1866-70	250/250?	"	E L	Sub	2	+1:+250	100	57	64-65	"	Georgios,Afroessa,Reka
"	Y	1925-26	20/20	"	E L	Sub	2	0:+7	100	360	64-65	"	Dafni dome
"	Y	1928	7/7	"	E L		2	0:+100	50			"	Naftilos dome
"	Y	1939-41	100/100	"	E L	Sub	2	0	30	43	64-66	"	Triton, Ktenas, Fouque Smith A,B, Reck, Niki
"	Y	1950	3/3		E L	Sub	2	0:+3	ca 1	ca 100	63	"	Liatsikas dome
Borawli (Eth.)		1631?			E								Amado dome
Unnamed (Wadi)		1256	12/12?	Ex	E L		3						
Piton de la Fournaise	N?	1766	2/2		E		2	0				39-50	
"	Y	1791	4/4		E L		2	0:+4				"	
"	Y	1955-57	88/23		E L		2	-36:+33	2.5	ca 1000	48-50	"	
Ngauruhoe	N	1949	3/3	Ex	E N L		2	-.07			57	56-57	tholoid filled crater
Ruapehu		1861	ca 12/	En,Ex P	E		2	+12			60	59-62	
"	Y	1945	38/14	" P	E L	CL	3	0:+23	20	200		"	
Tuluman		1953-57	186/186?	Ex,En P	E⁸ L	Sub	3	+32 +84			69?		
Langila	N?	1973-cont.	450+/	Ex	P E L		3	-250:+140			55		Strombolian explosions after dome growth began
Lamington	Y	1951-56	285/284	En,Ex	E N S		4	+1:283 -1, +1:6? +1:6	1000	500	58-60	55-60	less explos. after 3/51

Note: In the month and a half of high activity following the paroxysmal eruption, explosive outbursts near the base of the dome immediately preceded rapid dome growth. After March 1951, explosive activity diminished (Taylor, 1958).

Table 1, continued

Volcano Name	Older domes? (Yes, No)	Date of activity (year)	Duration (eruption/dome growth) (in wks)	Type of dome growth[1] (Ex,En, Pl,Cr)	Activity associated with dome growth P E N L S CL/Sub[2]	VEI[3] (max)	Times of explosive activity relative to start of dome growth (in wks)[4] SmEx LgEx DCPf[5]	Volume of new dome[6] ($m^3 \times 10^6$)	Rate of dome growth[7] ($m^3 \times 10^3$ per day)	SiO_2 of dome	SiO_2 range, this volcano	Remarks
Victory	Y	late 1800's			E N	2						
Bagana		1948-52	200/100	Ex	E[8]N L	4	-100	small		54	53-58	2 cycles of expl+dome
"	Y	1966-67	90/10	"	E	3	-80			55	"	
"	Y	1972-cont.	500+/	"	E N L	1	-70			56	"	
Puet Sague		1918-21	ca 180		E N	2				hy-ol-hb andesite		nuees=block+ash flows?
Galunggung		1822	.7/	En,Ex	E N	5	-				48-56	1822 dome destr. 1894
	Y	1918	2/2		E	2	-.1:2	CL	10	60	56		1918 dome destr. 1982
Merapi		1786				1		vol. of dome varies			51-58	
"	Y	1797				1		between 1 and 5 mill.			"	
"	Y	1810				1		cu. m., with growth			"	
"	Y	1812-13	ca 52/52			1		balanced by avalanches			"	
"	Y	1820-23	118/			3	-	and nuees ardentes;			"	Hartmann Type D
"	Y	1832-36	180/			3	-	rates of growth vary			"	" Type C
"	Y	1837-38	44/			3	-	between 7000 and			"	" "
"	Y	1846-47	58/			3	+	300000 cu m/day			"	" "
"	Y	1862-69	350/220		E[8]N L	2	+,+++				"	Hartmann Type B
"	Y	1883-84	68/	En,Ex	E[8]N	1	-			55	"	East Dome
"	Y	1887-89		"	E N	3	++ +,+++				"	Hartmann Type B
"	Y	1891-94		"	E N	2	++ +,+++				"	" "
"	Y	1897		"	E[8]N	3	++ -				"	Hartmann Type C
"	Y	1902-08	ca 300	"	E[8]N	3	++ +,+++				"	" Type B
"	Y	1909-18	ca 450	"	E[8]N	2	++ +				"	" Type A
"	Y	1920-23	ca 120	Ex,En	E[8]N	3	++ +,+++				"	" Type B
"	Y	1930-31	46/46	"	E[8]N L	3	+3 +2,+4:45		12	55-56	"	" Type B
"	Y	1933-35	80/80	"	E N	2	-.5 +			55	"	" Type C
"	Y	1939-40	40/40	"	E N	2	-6 0:34,			55	"	
"	Y	1942-43	72/72	"	E[8]N L	3	-8 0:39, 40:64		30		"	Type B
"	Y	1944-45	52/52	"	E N	2	- +				"	Type A
"	Y	1953-54	47/47	"	P E N	1	- +,+++			55	"	Type A
"	Y	1955-58	191/191	"	E N	3	- +,+++			55	"	Type B
"	Y	1967-cont.	800+	"	E[8]N L	3	-13, +104 0:787		100	52-57	"	SiO_2 increasing through this unusually long eruption cycle?
							0:787 +318					Hartmann Type B
							+769?					

Note: Explosive activity at Merapi includes (i) vent-clearing minor explosions, (ii) dome growth, collapse, and small "block and ash" pyroclastic flows as partly degassed melt is remobilized, (iii) moderately large explosions from gas-charged, fresh magma, often with pyroclastic flows from a vertical eruption column, and (iv) minor vertically-directed explosions and "block and ash"-type pyroclastic flows associated with growth and collapse of "fresh" dome. Hartmann (1935) recognized 4 types of activity: Type A with i and ii; Type B with i, ii, iii and iv. Type C with iii and iv, and Type D with iii, with or without iv. Type B is the most common.

Table 1, continued

Volcano Name	Older domes? (Yes, No)	Date of activity (year)	Duration (eruption/ dome growth) (in wks)	Type of dome growth[1] (Ex,En, Pl,Cr)	Activity associated with dome growth P E N L S CL/Sub[2]	VEI[3] (max)	Times of explosive activity relative to start of dome growth (in wks)[4] SmEx LgEx DCPf[5]	Volume of new dome[6] (m³x10⁶)	Rate of dome growth[7] (m³x10³ per day)	SiO_2 of dome	SiO_2 range, this volcano	Remarks
Kelut		1376	1 yr?			3					50-58	
"	Y	1919-20	80/1		P E CL	3	-80, 0:+1	0.1	50		"	
Semeru		1950-64	ca 700	Ex,En	E N L CL	2					47-59	viscous lava on steep slopes produces block and ash flows.
"	Y	1967-cont.	800+	"	E N L	3	0:800 +400 0:800				"	
		Note: During active dome growth, small vertically-directed explosions occur every few hours.										
Lamongan	N?	1898	2?/	Pl?	E L	2				47?	47+	basaltic plug?
Rinjani		1944	30/1		E CL	2	-29,0:1	74	11000	bas and.	51-53	glow; minor explosions
Rokatenda	Y	1928	7/??	En,Ex?	E	2		10	140	58	48-62	
"	Y	1964	a few days?		E?	2?		50	ca 1000	62	"	
Lewotobi Lakilaki		1932-33	80/12	Ex?	E N L S	2	-1:+80	0.5	ca 1	px and.	52-53?	dacite xenoliths
Lewotobi Perempuan		1921	ca 52/		E	2		0.2		bas and.	bas-and.	
Ili Werung	Y	1870		Ex,En?	E	3?	-?,+	50			55+	Ili Werung dome
"	Y	1928		"	E	2	-4,0:200	2			"	tephra on top of dome
"	Y	1948-49	200/30	"	E N	2					"	Ili Gripe dome
Soputan		1915	8/		E L	2	-3, -3,	5		basalt	48-55	Aeseput Weru dome
"	Y	1966-68	120/117	Ex?	E N L	3	0:117 +39	20		48-50	"	two pulses dome growth
Lokon		1971	23/.1		E	2	-18,0:5	0.0006	6		44-67	
"	Y	1976-78+	100+/2?		E N	2	-40, +5:50	0.012	0.8		"	
Tongkoko		1801	less than 1 yr		E L	2		4		px and.	px and.	
"	Y	1880			E	1					"	
Ruang	Y	1856	less than 4 wks		E	2					54-58	as at Merapi, viscous lava forms dome in crater before flowing beyond crater rim
"	Y	1889			E N L	1	0:54 -2	3			"	
"	Y	1904	56/		E	3	-1,0				"	
"	Y	1949	2+/1+		E	2					"	
Api Siau		1970-71	14/2		E N L	2	-12,0:+2 0:+2			52	bas-and	like Merapi Type B, w/early(plug) and later (fresh) dome growth
"	Y	1972-76	245/6		P E[8] L	2	-60,+12 +105					

Note: Pattern seems to include vent-clearing explosions, remobilization of old dome, larger explosions, and late-stage new dome.

Table 1, continued

Volcano Name	Older domes? (Yes, No)	Date of activity (year)	Duration (eruption/ dome growth) (in wks)	Type of dome growth[1] (Ex,En, Pl,Cr)	Activity associated with dome growth P E N L S CL/Sub[2]	VEI[3] (max)	Times of explosive activity relative to start of dome growth (in wks)[4] SmEx LgEx DCPf[5]	Volume of new dome[6] (m³×10⁶)	Rate of dome growth[7] (m³×10³ per day)	SiO₂ of dome	SiO₂ range, this volcano	Remarks
Banua Wuhu	Y	1835	.5?/		E L Sub	2		2?	1000+		px-hb and.	
"		1889			E Sub?			ca 0.1			"	
"	Y	1904-19									"	
Awu		1931	43?/40?	En,Ex	E CL	1	-3	3	ca 10		47-54	
Malupang Warirang		between 1933-36				1						
Hibok-Hibok	Y	1871-75	200/	En	E	2		150			px-ol-hb andes.	Vulcan rose 617m in 3 yr
"	Y	1948-53	253/208	"	E N L	3	-37:199 +132 0:208				"	
Bulusan	Y	1916-22	300/4	En,Pl	E N?L	2	-8,+25:150 0:+4				and.-dac.	
Didicas	Y	1856-60		En	E L Sub	2		ca 100				
"	Y	1952-53	ca 50/	"	E L Sub	2	0:50	ca 100		hb and.	hb and.	" explos. weakened w/time
Chokai		1801	4?/		E	2?	-1.4	75		54	52-55	
Niijima	Y	886 AD?			E N	4		300		76	50-76	
Myojin-sho		1946	few days			2	+	50?			51-68	1st of 2 islands
"	Y	1952-53	55/		E S Sub	3	++ +++	50?	ca 100	63-68	"	2nd of 2 islands; sequence repeated 3x
Komagatake	Y?	1856	one day?		E N	4					58-63	
Usu	Y	1663?		En,Cr	E N	5	0:+3 0:+3	60		71	51-71	Ko-usu lava dome
"	Y	1822	3/3	Cr	E N	4	-2:0 -2:0	150			"	Ogari-yama cryptodome
"	Y	1853	2+/2+	En,Cr	E N	4		25	300	68	"	O-usu lava dome
"	Y	1910	14/12	Cr P	E	2	0:12	70	ca 100	70	"	Meiji-Shinzan cryptod.
"	Y	1944-45	88/88	En,Cr P	E	2	+25:43	ca 70?	50?	68	"	Showa-Shinzan lava+cry
"	Y	1977-cont	210+/210+	Cr P	E	3	+14:64 +1:1				"	Usu-Shinzan cryptodome

Note: earlier pattern for Usu = phreatic to magmatic explosions (some very large), followed by late stage dome growth; recent pattern = cryptodome intrusion, moderate-scale explosions, continued growth of cryptodome or lava dome, sometimes but not always accompanied by small phreatic, phreatomagmatic explosions (Niida and others, 1980).

Tarumai		1867		Ex,En	E	2	-12	20		59	56-61	Lava dome I
"	Y	1909	13/.3		E	4	-3,-1		10000	59	"	Lava dome II
Berg		ca 1940		En,Ex	E		+	ca 20			andes.	Dome has 3 explosion craters; top of dome covered w/ thick layer

Table 1, continued

Volcano Name	Older domes? (Yes, No)	Date of activity (year)	Duration (eruption/ dome growth) (in wks)	Type of dome growth[1] (Ex,En, Pl,Cr)	Activity associated with dome growth P E N L S CL/Sub[2]	VEI[3] (max)	Times of explosive activity relative to start of dome growth (in wks)[4] SmEx LgEx DCPf[5]	Volume of new dome[6] ($m^3 \times 10^6$)	Rate of dome growth[7] ($m^3 \times 10^3$ per day)	SiO_2 of dome	SiO_2 range, this volcano	Remarks
Goraschaia Sopka		1881-83	80+/	En?		1?		50		63	58-63	Dome=last stage of 1881 eruption, but has a crater of unknown age on top of dome
Zavaritzki	Y	betw. 1915-31		En?	E L CL	1				58	52-67	
"		1957		"	E	3					"	
Ushishir		after 1769		En	E			15?		63	56-63	Two domes
Ekarma	Y	1767-69	100/		E	2	0?			60	57-60	
Sinarka	Y	1872-78	300/300		E N L	4	-:0 0:300?	40	50	57-59	56-60	
Harimkotan		1933	14/	En	E N L	3	0:+ -3?	100	1000(min)	59-60	59-67	Pre-dome pumic. pfs= 67% SiO_2; dome cut by 59% SiO_2 lavas
Kresnitzyn Peak	Y	1952	1/	En	E CL	3	- -	30	4000	62	49-65	
Nemo Peak		1906?			E	2				59	54-64	1906 act. Strombolian?
Karymsky		1970-76 ca 300/37			E L	2	-263:-3	0.003	0.01? and.-dac.63-65+			
Bezymianny	Y	1955-cont.1400+/1395+		Ex,En (+Cr?)	E[8]N L	5	0:+17, -5.0, +17, +17: 480 480:1026 1026:1395+	300	40	60-56	56-66	

Note: Stage 1 = premonitory activity; Stage 2, 10/22/55-11/30/55 = strong ash eruptions; Stage 3, 12/1/55-3/29/56 = weak to moderate explosive activity, with dome growth (and cryptodome intrusion?); Stage 4 = paroxysmal eruption of 3/30/56; Stage 5a, 4/56-summer 1965 = episodic dome growth and powerful ash explosions; Stage 5b, summer 1965-1975 = additional but slower dome growth, with moderate-sized explosions; Stage 5c, 1975-present = extrusion of small dome of plastic lava, with pyroclastic flows changing from pumiceous to block-and-ash type. Renewal of activity in 1965 and in 1976, with each new increment of magma slightly more mafic (Bogoyavlenskaya and Kirsanov, 1981).

Sheveluch	Y	1879-83	200/		E[8]N	2	+			52-63		New dome "exploded"
"	Y	1899-98	ca 50/		E	2					"	Central Dome formed
"	Y	1928-30	ca 70/		E N	1					"	
"	Y	1944-50	275/170	En	E N S	2	-55.0 +150:+80; +210 +150	200	200	60	52-63	Suyelich Dome
"	Y	1980	14+/10+	En	E	1		20	280			
Great Sitkin	Y	1945	1/1	En	E L	2		ca 10?	ca 6	57	57+	
"	Y	1974	30/29	En,Ex?	E L	2	-.5:+.5	96	460		"	subglacial

Table 1, continued

Volcano Name	Older domes? (Yes, No)	Date of activity (year)	Duration (eruption/dome growth) (in wks)	Type of dome growth[1] (Ex,En, Pl,Cr)	Activity associated with dome growth P E N L S CL/Sub[2]	VEI[3] (max)	Times of explosive activity relative to start of dome growth (in wks)[4] SmEx LgEx DCPf[5]	Volume of new dome[6] ($m^3 \times 10^6$)	Rate of dome growth[7] ($m^3 \times 10^3$ per day)	SiO_2 of dome	SiO_2 range, this volcano	Remarks
Bogoslof		1796		En,Ex	E L Sub	3?		ca 10		61	46-61	Castle Rock dome
"	Y	1806-23		"	L			ca 10			"	
"	Y	1883-91?	400+/400+	"	E⁸ Sub	3	+3,+350 +400	40	ca 10	52	"	Grewingk Is. Metcalf, McCullogh, and Tahoma Peak
"	Y	1906-10	239/239	"	E⁸ S Sub	3	+239 +40,80	5	3		"	
"	Y	1926-28	60/40	"	E Sub	2	-20	ca 1	4	46	"	decreasing SiO_2 (and volumes?), 1796-1928
Aniakchak		1931	3/		E	3						
Novarupta		1912	ca 1/.5?	En	E N	6	-.5	ca 10	ca 1000	60-75	60-75	zoned, mixed magmas
Trident flow/dome		1953-54	86/50+	Ex,En	E⁸ L	3	0:+86 0,+86	380?	1000			viscous lava
"	Y	1974-75			E	3	+					dome destroyed between summer 1974, summer 1975
Augustine		1883-85+		En	E N L	4	-14,4 -				56-64	
"	Y	1935	20/	"	E L	3	-40:0 -1				"	
"	Y	1963-64	ca 45/5?	"	E N	2	-3,-1 -2.5 0:+2	ca 10	300?		"	
"	Y	1976	13/10	"	E N	4	+8:10 -1 6:10	66	1000	59-64	"	dome vol. incl. block+ ash flows?; mixed magma

Note: general pattern for Augustine is to have minor, vent-clearing explosions, followed by climactic explosions and concluded by dacitic dome growth (Johnston, 1978).

Redoubt		1965-68	165+/short		E	3	-165:0 -8					dome growth= late event
St. Helens	Y	1600+			E N		-,+ -	200		64	50-69	pre-1980 Summit Dome;
"	Y	1842-47+			P E N	3	-,+ 0:+ +	30	40	63	"	Goat Rocks dome
"	Y	1980-cont	133+/133+	Cr,Ex,En	P E N	5	+8:111 8:30			61-63		intermittent activity

Note: general pattern between 6/80 and 10/80 = moderate size explosive eruptions followed within a day or so by dome growth; general pattern since 10/80 = small explosions at or before start of most new episodes of dome growth, followed by non-explosive dome growth. Vigorous gas pulses containing small amounts of ash are common between periods of dome growth.

Barcena	N	1952-53	30/10		E N L	3	-7:0 -7 +7:23	4	57	61	61+	
Colima		1818	2?/	En	E N L	4	-				54-64.	expl. = end of cycle 2 lava = begin cycle 3
"	Y	1869			L	2?					"	
"	Y	1909			P	2?	+				"	dome growth uncertain
"	Y	1913-35	1150/1150+	En	E N L	4?	0:+ -				"	end of cycle 3 (Luhr)
"	Y	1957-67	520/520?	"	E N L	2	+400:520 +300:?				"	
"	Y	1973		"	E	1					"	
"	Y	1975-76	28/28	"	E?N		4?:28	1	5	hb andes		vol. includes lava

Note: Luhr and Carmichael (1980) recognized eruptive cycles beginning with slow, discontinuous piston-like dome growth lasting for 50-140 yrs, largely without explosions, followed by 35-70 years of intermittent explosions and terminated by a large explosive, crater-clearing eruption. Repose periods between historical eruptive periods have been 48, 51 and 138 yrs.

Table 1, continued

Volcano Name	Older domes? (Yes, No)	Date of activity (year)	Duration (eruption/dome growth) (in wks)	Type of dome growth[1] (Ex,En, Pl,Cr)	Activity associated with dome growth P E N L S CL/Sub[2]	VEI[3] (max)	Times of explosive activity relative to start of dome growth (in wks)[4] SmEx LgEx DCPf[5]	Volume of new dome[6] ($m^3 \times 10^6$)	Rate of dome growth[7] ($m^3 \times 10^3$ per day)	SiO_2 of dome	SiO_2 range, this volcano	Remarks
Santa Maria (Santiaguito)		1922-cont	2600/2600+	En,Ex (Pl)	P E L S	3	0:2600 0:2600	700	80	62-65	54-66	dome growth = late phase of large 1902 explos. eruption of Sta. Maria

Note: Rose (1973) recognizes 6 phases of maximum extrusion (1922-25, 1927-35, 1939?-44?, 1948?-52?, 1956-66, and 1970-   ).
General pattern = increased extrusion, increased mild pyroclastic activity, followed by flank lava extrusions when original vent becomes plugged. When lava vents become plugged, plug dome extrusion and pyroclastic activity resume at the original, central vent. Small explosions occur with periods between 20 min and a few hours.

Volcano Name	Older domes?	Date of activity	Duration	Type of dome growth	Activity associated	VEI	SmEx LgEx DCPf	Volume	Rate	SiO₂ dome	SiO₂ range	Remarks
Ilopango (Islas Quemadas)	Y	1879-80	ca 10/10	En	E[8] CL	3	+2:+10 +8	150	2000	67	61-67	
Irazu		1963-65	ca 100/short		P E	3	- -			54-55	54-57	dome = late phase
Dona Juana	Y	1897-1906	450+/450+	En?	E[8]N	4	+ +100			hb-bio andes	and-dac.	dome repeatedly built and destroyed to 1906
Galeras		1924-27	ca 125/ca 80	En	E	3	-45:-17 -45				54-70	2 late pulses of dome growth
Antisana		1801-02	5+/		E L	2					59-74	subglacial
Nevados de Chillan		1906	290/50 or more		E L	2	-36,0:254	1		dacite?	bas-dac	frequent (hourly+) gas eruptions as dome grows
		1973-79			E L	2		1		"	"	
Pelee	Y	1902-05	175/52+	En,Ex	P E N S CL	4	[-2,0:175] +1,16 +25:50			60-64	56-64	SiO₂ incr. 1902-05
"	Y	1929-32	170/162	En	E N S	3	[-14,-10:0] -10:+4 0:+65?			"	"	Nuees less energetic +4 wks onward.
St. Vincent		1971-72	25/25	Ex,En	L CL	1?		80	600	54	50-57	Lava, no explosions
"	Y	1979	21/19	"	P E N L	3	-2:+2 -2			"?	"	
La Palma		1971	3/		E L	2		40?	2000?(min)		43-46	Lava, = dome?
Tristan da Cunha		1961-62	22/		E L	2				55		

119

TABLE 1
Historical dome growth and associated explosive activity. Volcanoes are listed in the order followed in the Catalogue of Active Volcanoes.

Footnotes in Table 1:

1. Ex=exogenous; En=endogenous; Pl=plug; Cr=cryptodome; all as defined in the text. In Ex,En and En,Ex pairs, the dominant mode of growth is indicated first.

2. Categories as listed in the Catalogue of Active Volcanoes and in Simkin and others (1981): P=Phreatic explosion; E=so-called "normal" explosion, vertically-directed and not known to be phreatic; N=Nuée, referring here to all pyroclastic flowage phenomena including pyroclastic flows, pyroclastic surges, and directed blasts; L=Lava flow; S=Spine; CL/Sub=eruption through a crater lake or the sea.

3. VEI=Volcanic Explosivity Index (Newhall and Self, 1982).

4. "-" indicates an explosion before the start of dome growth, "+" indicates an explosion after the start of dome growth. Signs without numerical values indicate the order but not the exact timing of explosions relative to the start of dome growth; explosions in order of their occurrence may appear as --, -, +, ++, +++ and so on. Colons indicate frequent explosions between the two dates shown. Purely phreatic explosions are not listed here.

5. SmEx=Small explosive eruption, VEI 1 or 2; LgEx=larger explosive eruption, VEI 3 or above; DCPf=pyroclastic flow resulting from gravitational collapse of a dome.

6. Volumes are approximate and do not include pyroclastic flows derived from dome unless so stated. Most volumes have been estimated from published dimensions or dimensions read from topographic maps.

7. Rate is an average rate, volume of dome/duration of dome growth; where dome growth is episodic, this rate is significantly less than the maximum rate of dome growth.

8. Newly formed dome was partly or wholly destroyed by a late-stage explosion; this does not include older domes destroyed by renewed explosive activity.

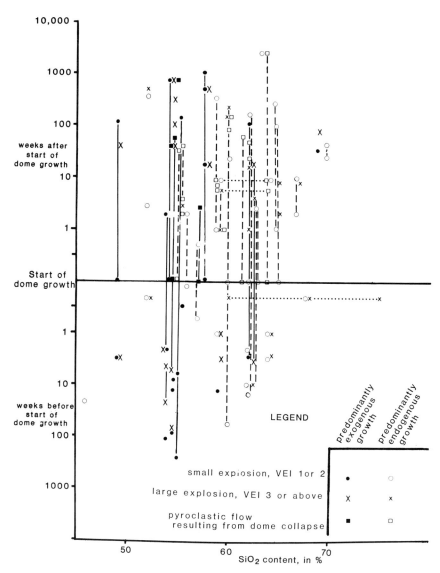

Fig. 1 Types and times of explosive eruptions relative to the start of dome growth vs. whole-rock SiO_2 content of juvenile products. Solid vertical lines indicate frequent explosions associated with exogenous dome growth; dashed vertical lines indicate frequent explosions associated with endogenous dome growth. Dotted horizontal lines indicate variation in the SiO_2 content of juvenile products during a single eruptive sequence. Small explosions = VEI 1 or 2; large explosions = VEI 3 or above. VEI=Volcanic Explosivity Index; (Newhall and Self, 1982).

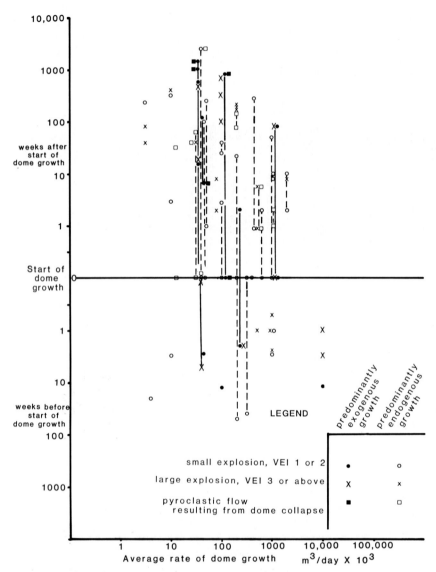

Fig. 2 Types and times of explosions relative to the start of dome growth vs. the average rate of dome growth. Other symbols as in Figure 1. If an average rate of dome growth was not reported in the original account, it was calculated by dividing the volume of the dome by the approximate duration of dome growth; maximum growth rates vary greatly and are commonly much higher than those average rates. If dome volumes or topographic maps were not shown in the original account, volumes have been approximated from available dimensions.

grow by extrusion and piling up of multiple viscous lava flows are "exogenous," domes that grow primarily by expansion from within are "endogenous," and domes that rise as simple pistons from a conduit are "plug domes." There are gradations between all of these types. Some domes exhibit two styles of growth simultaneously (exogenous and minor endogenous growth at Mount St. Helens, 1980-present) (Swanson and others, 1982); others shift from one style to another with time (Merapi, endogenous between 1883 and about 1915; increasingly exogenous from about 1920 to the present). We adopt Williams' distinctions with one additional specification: domes that consist of a single viscous lava flow (Merapi, 1967-present; Soufriere of St. Vincent, 1971, 1979; Trident, 1953) are classed as exogenous because they are more like an individual lava flow of an exogenous dome than like a typical endogenous dome. Distal parts of flows forming exogenous domes may lose thermal and volatile connection with the vent, and this disconnection may affect explosions from such domes.

We distinguish three types of explosive activity that occur in conjunction with dome growth:
(1) relatively small explosive eruptions, with VEI 1 or 2 (VEI=Volcanic Explosivity Index; Newhall and Self, 1982);
(2) larger explosive eruptions (VEI 3 or above); and
(3) pyrocalstic flows resulting from gravitational collapse of domes (nuées ardentes d'avalanche, or so-called "Merapi-type" pyroclastic flows).

For the purposes of discussing magmatic explosions we would like to disregard phreatomagmatic and phreatic explosions (those driven partly or wholly by vaporization of groundwater or surface water). Unfortunately, historical accounts rarely distinguish between phreatic and magmatic activity. Most explosions labeled "phreatic" in the Catalogue of Active Volcanoes are in fact phreatic and are not discussed here, but a significant number of explosions labeled "normal" explosions in the Catalogue have some or even a large phreatic component. Unidentified phreatic explosions in Table 1 contribute substantially to the scatter in Figs. 1 and 2, and to relatively poor correlations between some factors in Table 3. Thus, inferences based on the whole historical data base (Table 1) should be re-examined carefully at volcanoes where phreatic and magmatic activity have been clearly distinguished.

Similar ambiguities arise in historical reports of pyroclastic flows, because reports do not always specify whether a pyroclastic flow was generated by a laterally directed explosion, by column collapse, by lateral expansion of the base of an eruption column, or by an avalanche off of a dome -- distinctions that are important in understanding explosions at domes. These shortcomings in historical records complicate, but in our opinion do not preclude, interpretation of the data in Table 1.

DISCUSSION -- THE WORLDWIDE SAMPLE

Table 1 and the ordinates of Figures 1 and 2 show the timing of various types of explosive activity relative to the start of dome growth. Frequent explosive activity can begin at least 5 years before dome growth begins (Karymsky), and continue for a least 60 years thereafter (Santiaguito). Most explosive activity ceases when dome growth ceases, but in a few cases (Ruapehu, 1945; Lewotobi Lakilaki, 1932-33) explosions continue after domes stop growing. Small explosions are common before, at, and after the onset of dome growth, but are more closely spaced after dome growth begins. Large explosive eruptions also occur before, at, and after the onset of dome growth, but tend to occur at least a week before or several weeks after the onset of dome growth. Dome growth itself is often episodic (as at Santiaguito, Lamington, Bezymianny, Mount St. Helens), and although it is not so indicated in this compilation, explosions during a period of episode dome growth can mark the beginning of each episode of dome growth.

Table 2 contains mean values for the numerical data of Table 1, and the frequency with which various CAVW categories of explosive activity occur in association with endogenous and exogenous dome growth. Table 3 lists correlation coefficients between numerical data of Table 1. Some of the correlations are easily explained, such as that between volume and rate of dome growth. Other correlations, however, raise questions of cause and effect that cannot be resolved from the correlation table alone. As an example of the latter, a good correlation exists between maximum VEI and the duration of dome growth. Is this simply because the longer an eruption lasts, the more likely it is to have a (randomly occurring) large explosion, or does the occurrence of a large explosion cause an eruption to be prolonged, or is there a third factor (such as the size of a shallow magma reservoir) affecting both duration and maximum VEI? To answer questions about cause and effect, and especially about which factors control explosive activity at domes, we plotted the timing and character of explosive activity against two possible controls of that activity (Figs. 1 and 2). Ideally, viscosity (determined in turn by crystallinity, temperature, and chemical composition of the melt), magma supply and extrusion rates, and the magma's volatile budget (supply vs. loss) should be considered as possible controlling variables. In practice, very little of this information is available, and we have selected whole-rock SiO_2 content of domes (Fig. 1) as a crude but frequently reported indicator of viscosity, and average extrusion rate (Fig. 2) as a crude indicator of the rate at which magma moves upward to levels where gases can begin to exsolve.

Neither the characteristics of explosive activity nor its timing relative to dome growth correlate well with the bulk SiO_2 content of domes (Fig. 1). There is a clear tendency for domes with bulk SiO_2 contents less than 55 or 60 percent to grow exogenously, and those with higher silica contents to grow endogenously, but this difference in style of dome growth does not seem to result in (or from) a difference in the style of explosive activity. From Table 3, the correlation between bulk SiO_2 and maximum VEI of eruptions is weaker than we would expect if bulk SiO_2 were a major factor in the magnitude of explosions at domes. Bulk SiO_2 would be a useful indicator of viscosity and explosivity if all other factors were equal; clearly, however, other factors are not equal even during

TABLE 2

Mean-value characteristics of eruptions in which endogenous, exogenous, or undifferentiated dome growth occurs.
N (exogenous) = 67; N (endogenous) = 22; N (undiff.) = 67
Symbols for eruption characteristics as in Table 1.

	Duration of eruption (wks)	Duration of dome growth (wks)	log (volume of dome growth in m^3)	log (rate of dome growth in m^3/day	SiO_2
Endogenous	198	197	7.5	5.2	60
Exogenous	304	271	7.1	5.4	57
Undifferentiated	68	32	6.4	4.7	57

	P	E	N	L	S	Cl/sub	Maximum VEI
Endogenous	8	64	33	24	6	19	2.9
Exogenous	4	22	11	13	0	2	2.6
Undifferentiated	9	60	21	25	3	11	2.3

TABLE 3

Correlation coefficients between possible controls on explosions at domes (duration, volume and rate of dome growth and SiO_2 of juvenile products) and maximum explosivity (maximum VEI).

	Duration of dome growth	Volume of dome growth	Rate of dome growth	SiO_2	Maximum VEI
Duration	1				
Volume	−.04	1			
Rate	−.10	+.59	1		
SiO_2	+.08	+.40	+.19	1	
Max. VEI	+.43	+.32	+.13	+.27	1

dome growth, a relatively homogeneous subset of volcanic activity.

Both large and small explosions are common for a wide range of extrusion rates and in both endogenous and exogenous dome growth (Fig. 2). There is no evidence that major explosive eruptions are influenced by the rate of dome growth. Pyroclastic flows resulting from dome collapse do not notably correlate with high rates of dome growth, but there may be a threshold value below which such pyroclastic flows do not occur. The lowest average rate of dome growth known to lead to collapse and pyroclastic flows is approximately 10^4/m^3/day

(Merapi); a threshold may be higher at domes growing on gentle slopes. The mean volume of domes which partially or wholly collapse to form pyroclastic flows (5×10^7 m^3) is several times greater than the average volume of domes in this compilation (10^7 m^3), but slope angle is critical and some pyroclastic flows can form from domes as small as 10^6 m^3. On large domes, the slope angle of the dome itself may exert an important control on failure.

DISCUSSION -- CASE EXAMPLES

Neither whole-rock SiO_2 nor average rates of dome growth appear to exert a dominant control on explosive activity at domes, so we must look at other, less frequently reported parameters. Crystallinity, matrix glass composition (including pre-eruption volatile content), and temperature exert strong controls on the viscosity and explosive potential of magmas (Bottinga and Weill, 1972; Shaw, 1972; Williams and McBirney, 1979; Marsh, 1981; Melson, 1982a,b). Studies of the products and eruptive style of Arenal and Mount St. Helens offer some useful insights into the influence of crystallinity and matrix glass composition (including volatile content) on explosivity.

In the El Tajo tephra sequence at Arenal Volcano, Costa Rica, two mafic tephras with 52 and 51% bulk silica contents were produced by major explosive activity (Melson, 1982b). The viscosity of these mafic magmas must have been higher than their bulk composition would suggest, and was apparently increased by two factors: (1) the highly evolved character of their matrix glass, and (2) their high crystal content. The SiO_2 contents of their matrix glasses were 58 and 55 percent, respectively. In addition, these mafic magmas had high crystal contents (46 and 54 wt. percent), whereas dacitic magmas of Arenal have less than 5 wt. percent crystals. Together, the silicic matrix and high crystal content raised the viscosities and explosive potential of these mafic magmas dramatically.

Glass inclusions in olivine phenocrysts in the mafic tephras of Arenal are quite similar in composition to the matrix glass of that tephra but have analytical sums of about 96 percent. Following Anderson (1974), the difference between 100 percent and the observed sum is assumed to be volatiles, mostly water. Thus, the mafic magmas of Arenal had about 4 percent dissolved H_2O, equivalent to about 1.2 kilobars of water pressure in a water-saturated melt of andesitic composition at an assumed temperature of 1000 degrees C (Eggler and Burnham, 1973, Fig. 2). This probable high water pressure would contribute significantly to the explosivity of these mafic magmas.

Although not from a dome, this example from Arenal points up the problems inherent in using bulk analyses in getting at some understanding of the physical properties of magmas, and may explain the lack of correlation we have found between explosive and dome-building eruptions and bulk lava compositions.

The suggestion from Arenal that both high crystal content and high pre-eruption volatile content lead to explosive eruptions can be tested at the growing dome of Mount St. Helens. Some specific inferences can be drawn from data about lavas and pumice of Mount St. Helens (Table 4):

(1) Magmas with high pre-eruption volatile contents (samples 3 and 4) contain an average of 40 wt. percent crystals, whereas magmas with lower volatile contents (samples 5, 8, 9, and 10) contain an average of 60 wt. percent crystals. Magma temperatures, as best as can be determined, have not decreased systematically through time. From these two observations we infer that crystallization has been largely isothermal, occurring as the magma has degassed.

(2) Explosively erupted pumice has relatively less fractionated matrix glass (analyses 3a, 4a, 6a, 7a) and higher pre-eruption volatile content (analyses 3b, 4b, 6b, 7b) than does passively extruded dome lava (analyses 5a, 8a, 9a, 10a; 5b, 8b, 9b, 10b). This is particularly well illustrated by samples 7 and 8, representing the explosive and subsequent non-explosive phases of the October 1980 eruption. Parts of the May 18, 1980 cryptodome were highly fractionated (analysis 1a), but had a moderately high pre-eruption volatile content (analysis 1b).

(3) The melt inclusions and matrix glass of the early highly-explosive magmas (pumice of the May 18 and May 25, 1980 eruptions, samples 3 and 4) are essentially identical on an anhydrous basis, crystallization of the melt inclusions occurred before the May 18 and May 25 eruptions.

(4) The overall trend from May 1980 to March 1982 (and probably continuing as of January 1983) has been one of decreasing volatile contents; temporary increases have occurred within this trend, such as the increase in volatile content from the dome of June 12, 1980 (sample 5) to the pumice of July 22, 1980 (sample 6). Over the same period, explosive activity at Mount St. Helens has generally declined, with momentary reversions to explosive activity on July 22 and October 16, 1980, and on March 19, 1982.

From these specific inferences we conclude that at Mount St. Helens, where volatile contents are inferred to be high, magmatic explosions are controlled primarily by volatile content and little or not at all by crystal content. We further conclude that, as at Arenal, bulk compositions of Mount St. Helens samples are of little use in inferring viscosities and other mechanical properties of the magmas unless considered in conjunction with matrix glass composition and crystal content.

MONITORING THE EXPLOSIVE POTENTIAL OF A MAGMA

Of the parameters discussed above, volatile content of the melt appears to be the most reliable indicator of explosive potential at Mount St. Helens and at other volcanoes with water-rich melts. Volatile content can be monitored indirectly by considering:

(1) gas flux (with the assumption that the system is open, so that high concentration in the melt will cause high flux rates) (Casadevall and others, 1983),

(2) ratios between gases with differing solubilities in melts, such as CO_2/SO_2, SO_2/HCl (progressive degassing of a single body of melt will be shown by

Table 4

No.	SiO_2	Al_2O_3	FeO^*	MgO	CaO	K_2O	Na_2O	TiO_2	P_2O_5	MnO	SUM	

1. MAY 18 1980 GRAY DACITE 'BLAST FACIES'. USNM115379-37.
   (62 wt % crystals; temperature = ND)
   a. 79.60 11.76 1.39 .15 .42 3.24 2.55 .32 .12 ND 99.55 MATRIX GLASS, N=2
   b. 73.64 10.97 2.54 .68 .77 2.91 3.44 .47 ND ND 95.42 MELT INCLUSIONS, N=10

2. MAY 18 1980 GRAY DACITE 'BLAST FACIES'. USNM115379-34.
   (58 wt% crystals; temperature = ND)
   a. 77.37 11.73 1.46 .43 .71 2.89 3.55 .34 .19 ND 98.67 MATRIX GLASS, N=9
   b. 68.99 13.49 2.68 .76 1.72 2.22 3.69 .51 .00 .00 94.06 MELT INCLUSIONS, N=32

3. MAY 18 1980 PUMICE LAPILLUS. AF. USNM115230.
   (37 wt% crystals; temperature = 962 C)
   a. 72.66 14.86 2.41 .54 2.64 2.04 4.68 .38 .09 ND 100.30 MATRIX GLASS, N=3
   b. 67.96 14.29 2.05 .54 2.52 1.89 3.27 .31 .09 .00 92.92 MELT INCLUSIONS, N=7

4. MAY 25 1980 PUMICE LAPILLUS. AF. USNM115331A.
   (38 wt% crystals; temperature = 1006 C)
   a. 70.80 14.46 2.40 .49 2.50 1.90 4.59 .39 .08 ND 97.61 MATRIX GLASS, N=4
   b. 67.91 14.14 2.41 .69 2.17 1.91 2.58 .35 .12 .00 92.28 MELT INCLUSIONS, N=14

5. JUNE 12 1980 DOME. USNM115341.
   (63 wt% crystals; temperature = ND)
   a. 76.82 13.07 1.50 .11 1.27 2.52 4.15 .35 .06 ND 99.85 MATRIX GLASS, N=2
   b. 74.46 11.09 2.28 .55 1.28 2.90 3.99 .42 .07 .00 97.04 MELT INCLUSIONS, N=23

6. JULY 22 1980 PUMICE LAPILLUS. PF. USNM115377.
   (60 wt% crystals; temperature = 972 C)
   a. 77.55 12.70 1.69 .24 1.22 2.73 2.56 .40 .09 ND 99.18 MATRIX GLASS, N=10
   b. 70.67 12.71 2.35 .61 1.58 2.38 3.67 .47 .12 .00 94.56 MELT INCLUSIONS, N=24

7. OCTOBER 16 1980 PUMICE LAPILLUS. PF. USNM115418-1-1.
   (59 wt% crystals; temperature = 954 C)
   a. 77.82 11.79 1.53 .19 .81 2.88 3.15 .38 .05 ND 98.60 MATRIX GLASS, N=10
   b. 71.37 11.94 2.63 .54 1.24 2.29 3.74 .41 .12 .00 94.28 MELT INCLUSIONS, N=24

8. OCTOBER 18 1980 DOME. USNM115418-60.
   (58 wt% crystals; temperature = ND)
   a. 77.66 12.16 1.60 .20 .84 2.96 3.60 .41 .08 ND 99.51 MATRIX GLASS, N=10
   b. 74.76 11.52 2.24 .49 .89 2.96 3.84 .35 .07 .00 97.12 MELT INCLUSIONS, N=15

9. MARCH 19 1982 PUMICE. USNM115543-115.
   (60 wt% crystals; temperature = 935 C)
   a. 78.53 12.04 1.50 .15 .79 3.10 3.29 .33 ND .07 99.80 MATRIX GLASS, N=4
   b. 73.48 12.20 2.59 .72 1.11 2.76 4.33 .29 .00 .00 97.48 MELT INCLUSIONS, N=11

10. MARCH 19 1982 DOME FRAGMENT. USNM115543-125.
    (64 wt % crystals; temperature = ND)
    a. 77.79 11.16 1.35 .06 .57 3.50 3.88 .36 ND .03 98.69 MATRIX GLASS, N=7
    b. 75.30 9.52 2.90 .68 1.17 3.21 4.07 .55 .00 .07 97.46 MELT INCLUSIONS, N=13

TABLE 4

Crystal contents, inferred temperatures, and analyses of matrix and melt-inclusion glasses in the 1980-81 Mt. St. Helens eruptive sequence. Crystal contents are calculated from bulk K_2O and K_2O of matrix glasses, assuming most K_2O is in the matrix glass. Temperatures inferred using method of Lindsley (1976). Glass analyses are electron microprobe "moving-beam" analyses. First analysis is average composition of matrix glass; second analysis is average composition of melt inclusions in plagioclase. Although the moving-beam technique minimizes alkali loss, small amounts of Na and K may have been lost. Iron-titanium oxide temperatures have a precision of ± 30 degrees C. Oxidation during emplacement of some dome rocks precludes use of oxide geothermometry. ND = Not Determined. AF = Air-fall PF = Pyroclastic flow

decreases in these ratios; steady state supply or episodic resupply of magma and volatiles will cause these ratios to remain unchanged or to fluctuate, respectively) (Casadevall and others, 1983),
(3) changes in volatile content in glass inclusions of phenocrysts from one eruption to the next (an inherently post-eruption technique, with projections to the next eruption), and
(4) mineralogical indicators of water pressure, such as evidence of hornblende stability or instability, or anorthite contents of newly crystallized plagioclase.

Seismicity and (or) ground deformation might indicate exsolution of volatiles in excess of their saturation levels, but further research on this topic is needed to distinguish seismicity and deformation induced by volatile exsolution from that induced by other processes.

Viscosity appears to exert a major control on the explosive potential of water-poor magmas. Viscosity can be monitored indirectly, in most cases after eruptions, by determining:
(1) magma temperature;
(2) crystal content;
(3) composition and hence viscosity of matrix glass; and
(4) geophysical evidence for rate of magma movement.

The viscosity of a glass can be estimated from its composition (Bottinga and Weill, 1972; Shaw, 1972); the effective viscosity of a crystal-bearing magma, however, can only be estimated with additional information about its crystal content.

CONCLUDING REMARKS

New approaches to estimating volatile content and magma viscosity are needed if petrologic studies are to be a useful monitoring technique during volcanic crises. Domes and their associated explosions offer an opportunity to test these new approaches. We hope that readers will test the hypotheses presented here, using samples from the domes listed in Table 1 and associated pumice or scoria.

ACKNOWLEDGEMENTS

L. Siebert kindly supplied a list of examples in which domes were destroyed by explosive eruptions. Some of the Mount St. Helens samples were kindly provided by C. R. Kienle, D. A. Swanson, C. C. Heliker and K. V. Cashman. T. J. Casadevall, K. V. Cashman, D. A. Swanson, L. Siebert, N. S. MacLeod, R. L. Christiansen and L. J. P. Muffler made helpful suggestions on this manuscript. We also thank Michael Sheridan and Franco Barberi for organizing the workshop to which this volume is devoted.

REFERENCES

Anderson, A.T. Jr., 1974. Before-eruption H_2O content of some high-alumina magmas: Bull. Volcanol., 37: 530-552.

Bogoyavlenskaya, G.E. and Kirsanov, I.T., 1981. Twenty-five years of activity of Bezymianny Volcano: Vulkanologiya i Seismologiya, no. 2, pp. 3-13 (translated to English by D.B. Vitaliano).

Bottinga, Y. and Weill, D.F., 1972. The viscosity of magmatic silicate liquids: a model for calculation: Am. Jour. Sci., 272: 438-475.

Casadevall, T., Rose, W. Jr., Gerlach, T., Greenland, L.P., Ewert, J., and Wunderman, R., 1983. Gas emissions and the 1981-82 eruptions of Mount St. Helens, Washington: Science, in press.

Eggler, D.H., and Burnham, C.W., 1973. Crystallization and fractionation trends in the system andesite-H_2O-CO_2-O_2 at pressures to 10 kilobars. Geol. Soc. of Amer. Bull., 84: 2517-2532.

Hartmann, M.A., 1935. Die Ausbruche des G. Merapi (Mittel Java) bis zum jahre 1883: Neues Jahrbuch fur Mineralogie, Geologie, und Paleontologie: 75 (B): 127-162.

International Association of Volcanology and Chemistry of the Earth's Interior (IAVCEI), 1951-present, Catalogue of the Active Volcanoes of the World, including Solfatara Fields: Rome, IAVCEI, 22 volumes to date.

Johnston, D.A., 1978. Volatiles, magma mixing and the mechanism of eruptions of Augustine Volcano, Alaska: Univ. Washington, unpubl. Ph.D. dissertation, 177 pp.

Lindsley, D.H., 1976. Experimental studies of oxide minerals, in Oxide minerals: Mineralogical Society of America Short Course Notes, 3: 61-84.

Luhr, J.F. and Carmichael, I.S.E., 1980. The Colima volcanic complex, Mexico, Part 1. Post-caldera andesites from Volcan Colima: Contrib. Mineral. Petrol., 71: 343-372.

Marsh, B.D., 1981. On the crystallinity, probability of occurrence and rheology of lava and magma: Contrib. Mineral. Petrol., 78: 85-98.

Melson, W.G., 1982a. The cyclical nature of explosive volcanism: a petrological approach to volcano forecasting, in Martin, R.C. and Davis, J.F. (Editors), Status of volcanic prediction and emergency response capabilities in the volcanic hazard zones of California: Calif. Div. Mines and Geology, Special Publ. 63, Pt. II, pp. 111-133.

Melson, W.G., 1982b. Alternations between acidic and basic magmas in major explosive eruptions of Arenal Volcano, Costa Rica. in press. Costa Rican Boletin de Instituto de Vulcanologia.

Newhall, C.G. and Self, S., 1982. The Volcanic Explosivity Index (VEI): An estimate of explosive magnitude for historical volcanism: Jour. Geophys. Res., 87(C2): 1231-1238.

Niida, K., Katsui, Y., Suzuki, T. and Kondo, Y., 1980. The 1977-1978 eruption of Usu Volcano: Jour. Fac. Sci. Hokkaido Univ., ser. 4, 17: 357-394.

Powers, S., 1916. Volcanic domes in the Pacific: Am. Jour. Sci., 42: 261-274.

Rose, W.I., Jr., 1973. Patterns and mechanism of volcanic activity at the Santiaguito Volcanic Dome: Bull. Volcanol., 37: 73-94.

Shaw, H., 1972. Viscosities of magmatic silicate liquids: an empirical method of prediction: Amer. J. Sci., 272: 870-893.

Simkin, T., Siebert, L., McClelland, L., Bridge, D., Newhall, C. and Latter, J.H., 1981. Volcanoes of the World: Stroudsburg, PA, Hutchinson Ross, 232 pp.

Swanson, D.A., Chadwick, W.W., Iwatsubo, E.Y., and Heliker, C.C., 1982. Endogenous growth of the Mount St. Helens dacite dome (abstr.): EOS, 63: 1140.

Taylor, G.A.M., 1958. The 1951 eruption of Mount Lamington, Papua: Australian Bur. Min. Resources Geol. Geophys. Bull. 38: 117 pp.

Volcanological Society of Japan, 1961-present. Bulletin of Volcanic Eruptions: Tokyo, Volcanological Soc. Japan, numbers 1-18 to date.

Walker, G.P.L., 1973. Lengths of lava flows: Philos. Trans. Roy. Soc. London, A, 274: 107-118.

Williams, H., 1932. The history and character of volcanic domes, Univ. Calif. Publ. Geol. Sci., 21(5): 51-146.

Williams, H. and McBirney, A.R., 1979. Volcanology: San Francisco, Freeman Cooper and Co., 397 pp.

Note added in proof:

The volumes and rates of growth listed in Table 1 for Bezymianny and Santiaguito are in error. As of 1983, the Bezymianny dome has a volume of approximately 1.6×10^9 m^3, and its average rate of growth since 1956 has been 160,000 m^3/day. Santiaguito dome now has a volume of approximately 0.85×10^9 m^3, and its average rate of growth since 1922 has been 38,000 m^3/day. Corresponding corrections must be made in Figure 2. We apologize for these errors, and request that readers inform the first author of any other erroneous or out-of-date Table 1.

A VOLCANOLOGIST'S REVIEW OF ATMOSPHERIC HAZARDS OF VOLCANIC ACTIVITY: FUEGO AND MOUNT ST. HELENS

WILLIAM I. ROSE, RICHARD L. WUNDERMAN, MARY F. HOFFMAN, LISA GALE

Michigan Technological University, Houghton, MI 49931 (U.S.A.)

(Received August 8, 1982; revised and accepted November 15, 1982)

ABSTRACT

Rose, W.I., Wunderman, R.L., Hoffman, M.F., and Gale, L., 1983. A volcanologist's review of atmospheric hazards of volcanic activity: Fuego and Mount St. Helens. In: M.F. Sheridan and F. Barberi (Editors), Explosive Volcanism. J. Volcanol. Geotherm. Res., 17: 133-157.

The large amount of scientific data collected on the Mount St. Helens eruption has resulted in significant changes in thinking about the atmospheric hazards caused by explosive volcanic activity. The hazard posed by fine silicate ash with long residence time in the atmosphere is probably much less serious than previously thought. The Mount St. Helens eruption released much fine ash in the upper atmosphere. These silicates were removed very rapidly due to a process of particle aggregation (Sorem, 1982; Carey and Sigurdsson, 1982; Rose and Hoffman, 1982). There is some evidence to suggest that particle aggregation is particularly successful in removing glass shards with high surface areas/mass ratios. The primary atmospheric hazard of explosive eruptions is volcanic sulfur, which is converted to sulfuric acid and sulfate crystals. Although the Mount St. Helens dacite magma had a very low sulfur content before eruption, the eruptions did make a significant contribution to the stratospheric sulfate layer (Newell, 1982). Evidence based on measurements of S and Cl in erupted rocks, glass inclusions, gas samples, and atmospheric samples collected for both Mount St. Helens and Fuego volcanoes, suggests that both volcanoes released substantial contributions of S from intrusive (non-eruptive) magma. The amount of sulfur contributed to the atmosphere by an explosive eruption thus depends not only on the volume of magma erupted and its sulfur content, but also on the degree of near-surface non-eruptive magma.

The data collected to assess atmospheric hazard and to evaluate the processes and mechanisms of explosive volcanic eruptions have helped illuminate our understanding of: (1) the dispersion and atmospheric fractionation of volcanic ash and (2) the determination of the size and degassing energetics of shallow magma bodies beneath volcanoes.

INTRODUCTION

The 1980 activity of Mount St. Helens has provided the opportunity for a better evaluation of hazards to the atmosphere by volcanic activity. This paper is meant to summarize the results of this evaluation. A great many of the results have significant volcanological applications, as well. This paper focuses on volcanological topics, rather than atmospheric ones, and attempts to show which volcanological observations and measurements will further help illuminate this important subject.

Two main types of hazards to the atmosphere are perceived by most investigators. The first is hazard from fine silicate ash particles which can be injected into the upper atmosphere from explosive eruptions. The second is hazard that arises from the volcanic release to the atmosphere of large amounts of reactive chemical components (SO_2, H_2S, HCl, etc.) which can affect atmospheric chemistry. Much of the recent work on atmospheric hazard has tended to shift our focus from the first type of hazard to the second. This paper will deal with both types of hazard separately.

FINE VOLCANIC ASH

The volcanic ash which is of primary interest as an atmospheric hazard is that fraction which is less than a few micrometers in diameter, because larger particles of ash are removed quite readily by gravitational settling through the atmosphere in a short period of time. Yet, it is precisely this fraction of fine ash which we know least about, largely because it is difficult to sample. To evaluate atmospheric hazard we need to be able to estimate how much of this fine-grained ash is produced in an eruption and to obtain knowledge of the composition (mineral species and glass) of the ash.

AMOUNT OF FINE-GRAINED ASH FROM MISSING-VOLUME CALCULATIONS

One approach to this problem comes from the distribution of fallen ash. This data, presented as an isopach or isomass map can be used to estimate a "missing volume" of ash which fell outside of the last measured isopach. Clearly, this type of estimate is necessary to clearly evaluate the total volume of a pyroclastic eruption. Rose et al. (1973) described a method of integration to evaluate the total volume of ash-fall deposits, and this has since been applied to many other eruptions (see also Rose et al., 1978; Walker, 1980; Rose et al., 1981; Williams and Self, 1982). The integration is performed on an area-thickness (or area-mass) data set (from an isopach map) to which a function is fit. The integration can then be expanded beyond the last isopach to evaluate the "missing volume". Obviously, the best results come from the best, most complete isopach maps.

The Mount St. Helens eruption of May 18, 1980 was a critical example because the dispersal axis of the ash blanket was eastward directly over inhabited continental areas (too frequently these blankets are harder to study because of dispersal over water). Excellent maps of the ash fallout were prepared (Sarna-Wojcicki et al., 1981) and form the basis for a "missing volume" estimate. Fig. 1 is the data from isopach maps, plotted as isopach thickness versus area (the area enclosed by a given isopach). Table 1 shows a series of solutions of the $V = \int Adt$ function for a variety of intervals of t, starting with 0 to ∞ and ending with 0 to a very small value. The results indicate that 1/3 of the volume of the ash fell outside of the thinnest well-controlled isopach (0.25 cm). If corrected to a density for dacite, again using data from Sarna-Wojcicki et al. (1980), the total volume of the May 18, 1980 ash blanket is 0.26 km^3 (Table 2) of which 31% was estimated outside of the last measured isopach. The smaller estimated percentage of distal ash after density correction occurs because the density of distal ash is very low (0.2 g-cm^{-3}).

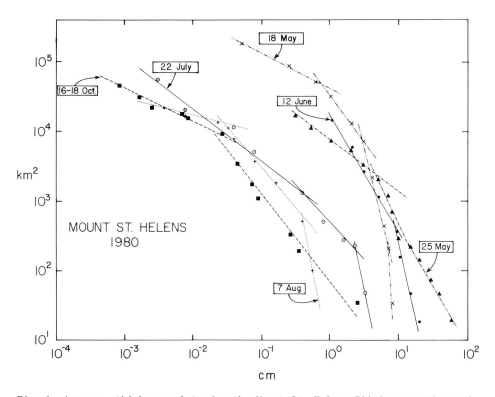

Fig. 1. Area vs. thickness plots for the Mount St. Helens Plinian eruptions of 1980. Integration of these curves provides the volume estimates of Table 2.

In other eruptions where similar integration of isopach data has been made, the proportion of "missing volume" has been higher (see Rose et al., 1973). We suspect that the main reason the Mount St. Helens example has a lower missing volume is that the isopach map is very complete. Table 2 shows the results of volume estimates for the other ash-fall blankets of Mount St. Helens. In these integrations, the proportion of "missing volumes" is similar to those of the May 18, 1980 example. Such data lead us to suggest that the potential amount of very fine and dispersed ash for a Mount St. Helens eruption is equal to about 5-35% of the total for a given eruption. The amount estimated to be of long term atmospheric significance must be smaller still.

AMOUNT OF FINE VOLCANIC ASH FROM GRAIN-SIZE STUDIES

Another approach to estimation of the amount of fine ash in an eruption comes from grain-size study of fallen ash. A number of such studies have been made of the Mount St. Helens materials (Rose and Hoffman, 1980; 1982; Fructer et al., 1980; Carey and Sigurdsson, 1982). A surprising feature of the ash samples from the distal areas is that they exhibit multimodal size distributions (Fig. 2). This seems unlikely because gravitational settling through the atmosphere could be expected to produce a very well-sorted fall deposit with size decreasing downwind. The multimodal size distributions (Fig. 2) imply that the finer particles have fallen out much faster than their grain size (and atmospheric gravitational sorting) would suggest. Table 3 shows calculated terminal velocities for ashes of various (constant density of 2.0 g-cm^{-3}) sizes and the

TABLE 1
Mount St. Helens May 18 ash blanket. Results of volume integration of curves plotted in Fig. 1. Method after Rose et al., 1973.

Isopach, cm	Volume outside, km^3
7	1.26
6	1.25
5	1.24
4	1.23
3	1.20
2	1.10
1	0.81
0.5	0.61
0.25	0.45
0.1	0.30
0.05	0.22
0.005	0.08
0.0005	0.03

TABLE 2
Volumes of Mount St. Helens fall deposits of 1980. Data obtained by integration of curves in Fig. 1.

Date	Volume, km^3				
	1	2	3	4	5
18 May 1980	0.9	0.4	1.3	0.26	.31
25 May 1980	0.03	0.017	0.049	0.025	.35
12 June 1980	0.02	?	(0.04)	(0.02)	?
22 July 1980	0.0025	0.00045	0.0028	0.0014	.16
7 August 1980	0.0010	0.00011	0.0013	0.0006	.08
16-18 Oct. 1980	0.0007	0.00005	0.0010	0.0005	.05

1. Volume integration only within known thickness, area limits (minimum volume).
2. "Far-flung" volume. Volume estimated to occur outside of last measured isopach or isomass line.
3. Total volume, including 1, 2 and additional near source volume.
4. Total volume, corrected to dense rock equivalent.
5. Proportion of ash which was beyond last measured isopach or isomass ($2 \sim 3$).

TABLE 3
Estimated terminal fall velocities of ash particles of various diameters.

Diameter, μm	Terminal Velocity cm-sec^{-1}	Time to Fall 5-20 km, hrs
90	40	3 - 10
25	3	45 - 190
10	.6	230 - 900

Data based on table of Lapple (1961) and assumes spheres with a bulk density of 2.0 g-cm^{-3} (Stokes-Cunningham factor, included) falling in air. Shape influence factors will decrease terminal velocities for the 90 μm ashes by factors of up to 5, based on Wilson and Huang (1979).

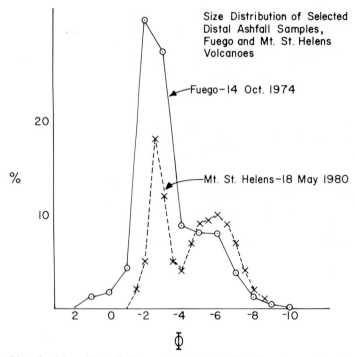

Fig. 2. Size distribution of selected distal ash samples from the Oct. 14, 1974 eruption of Fuego volcano (Murrow et al., 1980) and the May 18, 1980 eruption of Mount St. Helens (Carey and Sigurdsson, 1982).

times required for them to fall out as individual grains. For the May 18th Mount St. Helens eruption direct observations show that the ash fell out at times which are consistent with the size of the largest particle size peaks in the size distribution of the ash. The occurrence of a fine particle size peak indicates that small particles fell at velocities similar to those of larger grains, apparently because they were carried as aggregates, clinging to each other and to larger particles (Sorem, 1982; Rose and Hoffman, 1982; Carey and Sigurdsson, 1982). In fact, judging from the known fall-out times and the grain size distribution for the May 18 Mount St. Helens eruption, the majority of the mass of ash which fell out in Washington and Montana must have fallen as aggregates, rather than individual particles. This aggregation may be a general process which arises from moisture or electrical forces and therefore is an important control on the amount of fine-grained ash which has long-term atmospheric residence. Such aggregation has escaped recognition previously, probably because the aggregates are generally disrupted on impact and because most grain-size study techniques applied to ash have poorly sampled and described the size distribution of particles smaller than 50 μm (the size cut-off for hand seiving) thereby ignoring that fraction of the ash most likely to aggregate. It seems that the aggregation mechanism of ash, still not really understood, affects chiefly particles smaller than 50 µm.

The mean grain size of the Mount St. Helens ash is quite small when compared to that calculated for other eruptions. For example, the mean grain size of ash erupted in 1974 by Fuego volcano in an eruption of similar magnitude to the May 18th Mount St. Helens eruption was about 1 mm (Murrow et al., 1980) while the mean for the May 18, 1980 ash was about 40 µm (Carey and Sigurdsson, 1982). Comparison of the Mount St. Helens ash to world-wide standards of tephra fragmentation (fragmentation index of Walker, 1973) shows that it was highly fragmented. The cause of this high degree of fragmentation is not understood, but it has been suggested to be partly due to a phreatomagmatic mechanism (Rose et al., 1982). The overall fine grain-size of the Mount St. Helens tephra is important, because the effects of aggregation of ash during fall out (which probably effects only small ash particles) are generally enhanced by a fine grain-size distribution. Fig. 2 shows that there are secondary peaks in the size distribution of ash from the Fuego ash blankets as well, but the peaks are subdued because the ash as a whole is much coarser.

If the size distribution of ash samples from all parts of the ash blanket are determined, it is possible to calculate a "total grain-size distribution" for an eruption. Fig. 3 shows the results of such calculations for Fuego's Oct. 14, 1974 eruption and for May 18, 1980 Mount St. Helens eruption. This comparison leads to a fundamental question: Did the Mount St. Helens eruptions really produce a bimodal grain size distribution, or is the bimodality of the size distribution merely an artifact of fallout and aggregation processes? We will return to this question after considering the components of the ash and how they are affected by atmospheric fractionation.

COMPONENTS OF VOLCANIC ASH AND AGGREGATION

Volcanic ash is typically made up of several components: glass and several crystalline phases derived from fragmentation of the magma, and lithic fragments which represent accidental rock from the volcanic vent that were incorporated in the eruption. Coarse particles are often composites of glass and crystals, but finer ash particles are often single phases. The glass and the several crystalline phases have distinct densities and shapes which lead to fractionation in the atmosphere during fall out. In many eruptions lithic fragments tend to be less finely fractured than associated juvenile glass and and dense, equidimensional minerals. Therefore they fall out relatively near the vent causing over-representation in the coarse size-fractions of ashfall. The nature of atmospheric fractionation of ash is nicely demonstrated by the Mount St. Helens tephra (Table 4). The proportion of lithic fragments decreases rapidly with distance from the volcano. In the Mount St. Helens tephra, crystals and glass pyroclasts are also clearly fractionated in the atmosphere. Glass is consistently over-represented in the finest size splits and in most distal samples since its low density and platy, shard-like shape favor longer

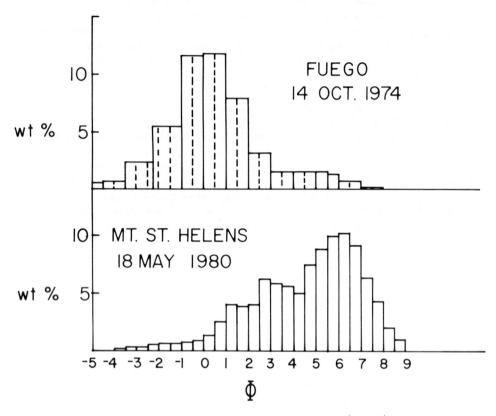

Fig. 3. Total grain-size distributions for the October 14, 1974 eruption of Fuego (Murrow et al., 1980) and the May 18, 1980 Mount St. Helens eruption (Carey and Sigurdsson, 1982).

atmospheric residency. The higher surface area/mass ratio of glass may also lead to more efficient accumulation than is true for other phases. This fact is dramatized by samples of ash which fell at 200-600 km distances (Richland, Spokane, Missoula) from Mount St. Helens after the May 18, 1980 eruption. The fine size population which fell much too fast to have been simple particles was consistently greatly enriched in glass (Table 4). One is tempted to conclude that the glass enrichment in fine-grained portions and in distal samples is balanced by enrichment of crystals near the vent, so that the ash of the whole eruption has the same proportion of crystals and glass as large pumices. However, observations on ash in the upper atmosphere, and of ash which fell at very great distances from the volcano suggest otherwise.

Farlow et al., (1980) report on silicate particles from Mount St. Helens in the size range of 0.5 to 2 µm which were collected in the stratosphere and which yielded glass and crystal proportions which ranged from approximately modal to crystal-enriched. Also, Fig. 4 shows ash particles which fell at Denver, Colorado on the morning of May 19, 1980. The time of fall is consistent with their descent as simple particles. Again, these ashes are relatively rich in crystalline fragments. From these sparse data we infer that non-aggregated silicate ash particles in the dispersed eruption cloud are probably enriched in crystals (i.e., that aggregation preferentially affects the glass shards) and that the ash with the longest atmospheric residence time is probably crystal-enriched when compared to the mode of the magma and certainly crystal-enriched compared to many of the ashes from distal locations. This conclusion is also consistent with observations by Rose et al. (1980) on ashes collected

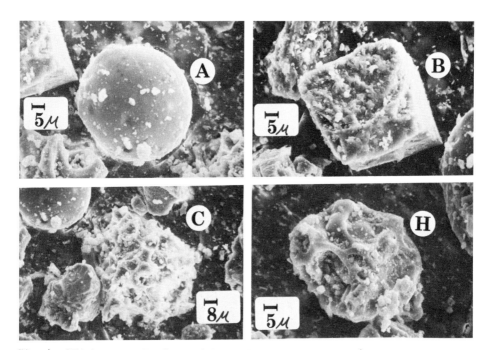

Fig. 4. SEM imagery of ash particles which fell May 19, 1980 at Bloomfield, Colorado. This ash was rather well sorted with mostly 30 μm particles. A is a silicate particle, with less K and more Ti than St. Helens glass. We suspect it is fly ash contaminant. B is a plagioclase fragment while C and H are vesicular Mount St. Helens glass (Rose and Hoffman, 1982).

TABLE 4

X-ray fluorescence analyses of selected bulk ash samples and various size splits representing grain-size peaks (Fig. 2). Below are the estimated proportions of lithic and dacitic populations, and selected modal values.

	RICHLAND				SPOKANE		
	Bulk	4ϕ	5.5ϕ	Pan	Bulk	4ϕ	Pan
SiO_2	58.5	54.4	60.3	62.1	64.3	57.1	66.3
Al_2O_3	18.3	18.0	17.4	17.4	17.2	18.5	16.8
Fe_2O_3*	5.8	7.9	4.5	5.1	3.9	6.6	3.3
MgO	2.6	3.3	2.4	2.7	1.8	2.5	1.8
CaO	5.4	6.1	4.8	4.3	3.6	5.6	3.0
Na_2O	4.6	4.3	4.2	4.6	4.8	4.6	4.9
K_2O	1.3	0.97	1.4	1.7	1.7	1.1	1.0
P_2O_5	0.16	0.17	0.17	0.18	0.12	0.15	0.11
TiO_2	0.83	1.11	0.66	0.65	0.59	0.97	0.49
	97.5	96.3	95.8	98.7	98.0	97.1	98.7
PROPORTIONS**							
LITHICS	67	100	42	40	16	80	0
DACITE	33	0	57	60	84	20	100

*Total Fe as Fe_2O_3

within small eruption clouds by instrumented aircraft. It is suggested strongly that the apparent bimodality of the total grain size distribution for the May 18th Mount St. Helens ash-fall deposits and the pattern of increasing proportions of glass in samples 200-800 km from the vent (Carey and Sigurdsson, 1982) is an artifact of aggregation/fallout processes which preferentially extract glassy 5-30 μm pyroclasts.

AN IMPORTANT DIGRESSION: CO-IGNIMBRITE ASH AT MOUNT ST. HELENS

Walker (1971) pointed out that silicic ignimbrites must scatter very large volumes of fine-grained glass-enriched ash during their eruption. This is true, according to Walker, because the fine-grained matrix of pyroclastic flows is almost invariably greatly enriched in crystals, compared to the magmatic mode which can be measured in large pumice blocks found in the ignimbrites. Often, the volume of fine-grained elutriated ash (co-ignimbrite ash fall) exceeds the total volume of the ignimbrite deposit. Because the co-ignimbrite fall can be thin, fine-grained and well-dispersed, it is poorly or not-at-all preserved for most eruptions (Walker, 1981). The recent Mount St. Helens activity provided an excellent opportunity to directly examine co-ignimbrite deposits and the ignimbrites with which they were associated. The problem is relevant to

Table 4 (continued)

	MISSOULA			Microprobe Glass	Bulk Pumice
	Bulk	5.5∅	Pan		
SiO_2	67.8	66.5	67.0	71.5	64.1
Al_2O_3	16.1	16.3	16.1	15.0	18.0
Fe_2O_3*	4.5	4.5	4.6	2.5	4.6
MgO	1.5	1.3	1.6	0.5	2.0
CaO	4.1	4.5	4.0	2.3	4.8
Na_2O	4.6	4.4	5.5	4.7	4.7
K_2O	1.7	1.6	1.8	2.0	1.45
P_2O_5	0.13	0.13	0.13	---	---
TiO_2	0.64	0.65	0.63	0.37	0.65
	101.0	99.9	100.9	99.0	99.9
PROPORTIONS**					
LITHICS	0	0	0		
DACITE	100	100	100		
GLASS	82			100	64
PLAG	11			0	26
HBD + PX	8			0	10
MAGN	1.7			0	2.2

**Mixing calculations using glass and mineral compositions. *Total Fe as Fe_2O_3.
Sources of Data: Glass: C. Meyer, BWEG, Menlo Park
          Pumice: Wozniak et al., 1980
          Ashes: Rose and Hoffman, 1982

atmospheric hazard because it is likely that during major ignimbrite eruptions the co-ignimbrite ash represents the main release of silicate ash to the atmosphere.

Table 5 gives a summary of relevant observations on the July 22, 1982 pyroclastic flow and its associated co-ignimbrite ash. The data show that the co-ignimbrite ash in fact does represent a volume which is at least as large as the pyroclastic flow itself. Table 6 gives a summary of volumes of pyroclastic flows at Mount St. Helens. The total amount of ignimbrite (not including co-ignimbrite fall) erupted in the 1980 activity is probably at least 0.07 km^3 which represents 23% of the total volume of magma erupted in 1980 at Mount St. Helens. In larger silicic eruptions the ignimbrite proportion would be much larger, perhaps as high as 90-95%. In this case, the co-ignimbrite process would be the main control of fine ash. For the Mount St. Helens deposits, the co-ignimbrite fall was not well dispersed. The majority of it was deposited within a few km of the margin of the pyroclastic flow itself. Especially thick deposits accumulated at the base of the "stair-steps", a steep channel north of the Mount St. Helens crater where the ignimbrites descended.

TABLE 5

Modal analysis, grain-size data and volume estimate of 22 July 1980 pyroclastic flows of Mount St. Helens.

I. MODAL DATA

Sample type (no.)	Glass	Plag.	Hbd + Px	Opaq.
Pumice (6)	61	28	10	1
Matrix (3)	52	33	14	2
Co-ignimbrite (3)	69	21	9	2

II. GRAIN SIZE DATA FOR CO-IGNIMBRITE

Size μm	62.5	44	24	20	16	12	10	8	6	5	2	1
%Coarser	37.4	47.6	67.7	73.5	78.1	82.6	86.6	89.7	91.9	93.8	97.2	98.0

Coulter Counter Analysis of July 22 co-ignimbrite

III. VOLUME ESTIMATE OF PYROCLASTIC FLOW:
1. Volume of flow deposit, sensu strictu = 0.003 km^3 DRE (Moore, 1981)
2. Volume ratio of co-ignimbrite/matrix determined from I. = 1:1
3. Total volume (flow deposit + co-ignimbrite) = 0.006 km^3 DRE

TABLE 6

Volume estimates of pyroclastic flows (dense rock equivalent), Mount St. Helens, 1980. After Rowley et al., 1981.

Date	Vol., km^3
18 May	0.06
25 May	<0.001
12 June	0.005
22 July	0.003
7 August	0.002
16 October	<0.001
Total	0.07 km^3

Note:
Does not include: (1) co-ignimbrite ash (see Table 5)
(2) possibly thick flow deposits surrounding the vent

PHREATOMAGMATIC OR HYDROMAGMATIC ERUPTION MECHANISMS

Recent work by many investigators (e.g., Self et al., 1980; Sheridan et al., 1981; Sheridan and Wohletz, this volume) has greatly increased the awareness of volcanologists to the role that phreatomagmatic processes play in eruptions.

This role has been expanded to include many eruptions which heretofore had been considered mainly magmatic.

The phreatomagmatic role in the recent Mount St. Helens eruptions is the subject of some controversy. Early eruptions in the March and April 1980 period were mainly phreatic, and demonstrated that a geothermal system existed within the volcano. To some investigators the overall fine grain-size of the ash, which is similar to that of phreatoplinian deposits (Self and Sparks, 1978), is suggestive of a phreatomagmatic mechanism in the May 18, 1980 activity. The early minutes of the May 18, 1980 activity were marked by a "lateral blast" (Hoblitt et al., 1981), "pyroclastic surge" (Moore, 1982) or "low aspect ratio ignimbrite" (Walker, 1981) eruption. The differences in nomenclature reflect the lack of consensus about the nature of this part of the eruption, but this early, very violent part of the eruption is associated with a phreatomagmatic mechanism, perhaps analogous to fuel-coolant interactions (Colgate and Sigurgeirsson, 1973; Wohletz, this volume). Accretionary lapilli were conspicuous features of some of the proximal fall deposits, and these are characteristic of phreatomagmatic activity. The angular fractured nature of juvenile glass blocks in the blast deposit may represent the hydrofractured carapace of the magma body. Given the likelihood of a significant phreatomagmatic component in the Mount St. Helens activity, we are left with some questions: (1) How frequently are phreatomagmatic mechanisms the stimulant for explosive eruptions? and, (2) What are the consequences of a phreatomagmatic mechanism with regard to the atmosphere? Although we feel that fragmentation of ash (and attendant generation of very small ash particles) will be enhanced, we should also suspect that because of increased moisture, aggregation of fine ash will be enhanced as well.

SULFUR AND CHLORINE EMISSIONS

Besides silicate ash, most eruptions also release significant amounts of volatile components which may themselves form particles or react with other components in the atmosphere. Large amounts of H_2O and CO_2 are released, but smaller releases of S and Cl species seem to have the most relevance to the atmosphere, which already contains high levels of H_2O and CO_2. Studies of the 1974 Fuego eruption have estimated that the mass of S contributed to the atmosphere was 1 to 2 orders of magnitude higher than the mass of silicates contributed by the same volcanic eruption (Murrow et al., 1980). This mass of S eventually contributed to the formation of a sulfate layer in the stratosphere (Cadle et al., 1976) which persists for several years following major eruptions. The impact of Cl release in the atmosphere may be most important in the ozone layer (Lazrus, et al., 1979), where Cl atoms liberated after HCl reaction with OH radicals could catalyze ozone decomposition.

MAGMATIC Cl AND S DEGASSING PATTERNS

This has been a difficult topic to directly examine, because of the uncontrolled escape of volatiles from magma at volcanoes. Direct sampling of gases is a poor way to gain information because direct sampling is feasible only when there is low-level activity. Furthermore, such samples are not representative of gases released during explosive activity. Direct sampling of magma before degassing is generally not possible either. We are left with indirect methods to piece together the S and Cl degassing patterns of shallow magma bodies. In spite of their deficiencies, such methods define important constraints that are difficult to establish in any other way. Rose et al. (1982) described how indirect observations can be combined to construct a S and Cl budget for Fuego Volcano. They concluded: (1) Most S, but only a small fraction of the Cl is released in explosive activity. Cl is mostly degassed in later, low level activity. (2) There is much S and Cl released from shallow magma bodies that is not erupted at all. This means that the erupted magma mass multiplied by the original magmatic concentration greatly underestimates the erupted S and Cl masses.

Table 7 is a summary of data on the S and Cl abundance of volcanic materials associated with the 1974 Fuego and 1980 Mount St. Helens eruptions. The table compares the pre-eruption magmatic concentrations of S and Cl (based on glass inclusions within phenocrysts) with the concentrations determined in fresh lava samples after eruption. Data for the concentrations of scavenged acid and salt from ash-fall tephra are given as well. The table documents the following: (1) There are large differences in the pre-eruption S content of magmas. This has also been shown by other studies (e.g. Anderson, 1975) but we still know too little about the general relationship of magma composition and S or Cl abundance

TABLE 7
Summary of S and Cl contents of magmas before and after (in parentheses) eruptions.

Silicate Concentration	1974, Fuego basalt[1]	1980 Mount St. Helens dacite[2,3]
S ppm	2800 (75)	300 (<20)
Cl ppm	800 (220)	1100 (400)

soluble concentration average (range) ppm
soluble Cl^- on ash	140 (60-350)[5]	700 (100-1500)[4]
soluble S as $SO_4^=$ on ash	530 (100-950)[5]	500 (50-1300)[4]

Data from: (1) Rose et al., 1982; (2) Melson et al., 1980; (3) Table 11, this paper; (4) Taylor and Lichte, 1980, Fructer et al., 1980; Rose and Hoffman, 1982, Stoiber et al., 1981; (5) Rose, 1977.

to generalize. (2) Nearly all of the S in magmas is lost during subaerial eruption. Even when the lava is quenched rapidly, less than 10% of the original S is left. (3) A significant fraction of the Cl is retained in quenched materials of the eruption. (4) Very large masses of both S and Cl are found on the ash-fall tephra, too large to have been derived only from erupted magma. It is significant that the latter point is also true for Mount St. Helens, because it represents a silicic, much more viscous magma than Fuego.

Table 8 lists estimates of the masses of S and Cl in the Fuego and Mount St. Helens eruptions, obtained by various methods. It is necessary to combine diverse measurements in order to establish the magma budget. These measurements include those in Tables 6 and 7 supplemented with data on actual measurements of S and Cl fluxes and ratios obtained between major eruptions. Fig. 5 gives the generalized pattern of SO_2 release from Mount St. Helens, measured (1 to 7 times a week) by correlation spectrometry (Casadevall et al., 1981). Table 9 gives data on the ratios of S and Cl in various types of volcanic emissions from both Mount St. Helens and Fuego. The data suggest that the S/Cl ratio is higher during explosive activity, a conclusion reached by numerous previous studies (Noguchi and Kamiya, 1963; Stoiber and Rose, 1970; 1973; 1974; Rose, 1977). Cl is evidently much more soluble in magmas and is one of the last volatile constituents purged from the melt; S on the other hand, appears to be one of the first components lost.

TABLE 8

Estimates of masses of S (in grams) associated with volcanic activity of Fuego (Oct., 1974) and Mount St. Helens (May 18, 1980).

	Fuego	St. Helens
1. sulfur detected in upper atmosphere	>5×10^{11}(1)	2×10^{11}(2)
chlorine detected in upper atmosphere	?	?
2. sulfur originally within magma mass erupted	6.2×10^{11}(3)	2×10^{11}(4)
chlorine originally within magma mass erupted	2×10^{11}(3)	7×10^{11}(4)
3. sulfur scavenged from atmosphere by ash	6×10^{11}(3)	3×10^{11}(5)
chlorine scavenged from atmosphere by ash	3×10^{10}(3)	5×10^{11}(5)
4. minimum total sulfur erupted	1.6×10^{12}(3)	5×10^{11}(6)
minimum total chlorine erupted	>3×10^{10}(3)	5×10^{11}(6)

Sources: (1) Lazrus et al., 1979
(2) Newell, 1982
(3) Rose et al., 1982
(4) Mass of magma erupted on May 18 (Table 2) times magmatic S concentration (Table 7)
(5) Mass of magma erupted on May 18 (Table 2) times soluble concentration of fresh ash fall
(6) Total of 1 + 3.

TABLE 9
S/Cl ratios in volcanic emanations of Fuego and Mount St. Helens

S/Cl ratios	Fuego	St. Helens
I. Ash leachates, representing eruptive periods	3.[1]	0.1-1.2[2]
II. Fumarolic sampling	0.5[1]	.06-0.5[3]
III. Airborne plume measurements during quiescence	0.5-3.0[4]	----
IV. Magmatic	3.5[5]	0.3[5]

Sources: (1) Rose, 1977
(2) Stoiber et al., 1981
(3) Zoller, W.H., pers. comm., 1980
(4) Lazrus et al., 1979
(5) Table 7

The difference in solubility of S and Cl in magmas can be demonstrated by data on residual S and Cl contents of quenched magma. Tables 7 and 9 and Fig. 6 present data for Mount St. Helens and Fuego. Table 10 reports a set of measurements on the 1968-82 andesitic lavas of Arenal Volcano, Costa Rica. Examination of these data reveals several important conclusions: (1) S is not retained in significant proportions in any of the samples. By the time these samples, erupted by highly variable mechanisms, are chilled nearly all of the magmatic sulfur has escaped. (2) High concentrations of Cl are retained in all of the samples. (3) The relationship of Cl content to time is systematic for both the Mount St. Helens and Arenal sample sets. Initially, samples show an increase in Cl, following the onset of activity. This is followed by a steady decline in Cl concentrations.

SITE OF S AND Cl IN VOLCANIC ROCKS

Sulfide blebs have formed in both the Fuego and Mount St. Helens magmas. These are small (1-60 μm) generally round bodies, which also contain major Fe and Cu. They occur in the groundmass and within many of the phenocrysts, but are especially common inside Fe-Ti oxides and amphiboles. Overall they account for a concentration of about 20 ppm S in the Mount St. Helens rocks, which is the concentration typically found in erupted and quenched dacite. Before eruption most of the S is found as a trace constituent in the silicate liquid. The concentration of sulfur in the silicate liquid is limited before eruption by the partitioning of sulfur into blebs. Thus in the Mount St. Helens calc-alkalic magmas, sulfur is saturated in the melt, and its concentration is prevented from rising. During eruptions the degassing sulfur comes mainly from the silicate glass, not the sulfide belbs, which are pristine after eruption.

TABLE 10
S and Cl concentrations in lavas of Arenal volcano Costa Rica. Determinations by X-ray fluorescence.

Date of Eruption	S, ppm	Cl, ppm
1968*	<10	426
1969	10	496
1971	19	485
1973	<10	458
1977	<10	412
1980	<10	346
1982	<10	264

Sequence of lavas erupted 1969-1982 from Arenal volcano have a composition nearly constant at 53.5 ± 1.0% SiO_2.

*1968 sample is a bomb from the initial explosive activity of July 1968.

Thus the solubility of sulfur in Fuego's basaltic magma is significantly greater than for the Mount St. Helens dacite. Because of the relevance of sulfur to atmospheric impact, we need to know more about sulfur solubilities and concentrations in different magmas.

Cl appears to be largely in the silicate glass in both the Fuego and Mount St. Helens magmas. This is shown by the mass balance of Cl concentrations in glass (20-40% higher) and whole rock concentrations.

AMOUNTS OF S AND Cl RELEASED INTO THE ATMOSPHERE BY ERUPTIONS

Table 8 summarizes the amounts of S released by the Fuego and Mount St. Helens eruptions. The S releases for Fuego are greater, even though the Fuego eruption was substantially smaller. This is mainly because of the high sulfur content of the Fuego basalt (nearly an order of magnitude greater than Mount St. Helens). Reports of the El Chichon (Mexico) eruption of March-April 1982 (Varekamp et al., 1982; Luhr et al., 1982) indicate that it is similar in magma volume to the May 18, 1980 Mount St. Helens eruption, yet the amounts of S in the stratosphere are much greater (P. McCormick, pers. comm., 1982). Luhr et al., (1982) report that the El Chichon magma is an alkalic trachyandesite. The large sulfur release is most likely to be due to a higher original sulfur content, or to a contribution of sulfur from evaporites below the surface of the volcano. The existance of evaporites under El Chichon is known from oil exploration records and is an unusual environmental characteristic for that volcano.

The impact of Cl on the atmosphere from volcanic activity has typically been less severe than for S. Lazrus et al. (1979) reported that little if any Cl reached the stratosphere from the Fuego activity, and inferred that it was efficiently removed by moisture in the troposphere. Cl was not conspicuous in the stratosphere following the Mount St. Helens activity either. Early reports of stratospheric sampling of the El Chichon cloud, however, indicate that substantial Cl is present (R. L. Chuan, pers. comm., 1982). We speculate that the alkalic magma composition of El Chichon or the evaporite contribution mentioned above may explain the Cl abundance.

INTERPRETATIONS AND DISCUSSION

In this section we will interpret the various measurements of S and Cl, and point out conclusions with volcanological and atmospheric significance. Table 11 gives an interpretation of S and Cl measurements based on the simple assumption of a single high-level magma body for the recent activity of Fuego, Mount St. Helens, and Arenal volcanoes. A minimum initial size of each shallow magma body is estimated by considering the total emission of S and Cl during an eruption cycle and multiplying this total by the magmatic concentrations of S and Cl. For each of these volcanoes the mass of the magma body is several times the mass of erupted magma. This is especially important because it means that estimates of S and Cl release to the atmosphere based on the volume of magma erupted and the magmatic S (and Cl) values (Kellogg et al., 1972) will be far too low. This also demonstrates that, in general, volcanoes erupt only a small fraction of the magma that comes close enough to the surface to substantially degas. This has been recognized for basaltic volcanoes, especially lava lakes, but the Mount St. Helens example shows that even silicic eruptions are marked by sustantial additional gas contributions from shallow level, non-eruptive magma. The contribution of gas to the atmosphere from intrusive magma at Mount St. Helens is graphically shown in Fig. 5. Here, the mass of S lost from May 20 to Dec. 31, 1980 was several times the sulfur supplied by the mass of magma erupted in that period. In the May 18 eruption, the mass of sulfur released was at

TABLE 11
Estimates of shallow magma bodies and degassing at Fuego and Mount St. Helens.

	Fuego[1]	Mount St. Helens[2]
Total size of shallow magma body	0.5 km^3	1.0 km^3
Total amount of magma erupted	0.1 km^3	0.3 km^3
Prop. of total S lost in initial eruption	80%	50%
Prop. of total Cl lost in initial eruption	10%	?

(1) Rose et al., 1982
(2) Casadevall et al., 1981; this paper

least 2-3 times the mass which could have been degassed from the mass of erupted magma (Table 11). Thus, a magma body of at least 1 km³ (about 3 times the amount of dense-rock-equivalent magma erupted) is necessary to account for the volatile release pattern seen. A similar series of arguments applies to Fuego and was explained by Rose et al., 1982.

It is very important to volcanic forecasting to be able to decide when the shallow magma body, which is the source of eruptive energy, is effectively degassed. The regular flux measurements of SO₂ (Fig. 5) appear to be one approach. During the several month period of increasing and rapidly decreasing SO₂ flux at Mount St. Helens, Plinian eruptions occurred periodically. The

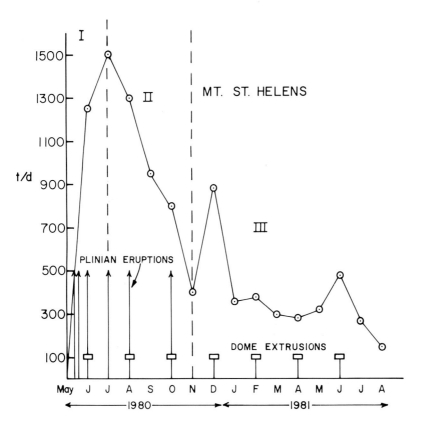

Fig. 5. Monthly mean SO₂ emission, based on measurements obtained 2 to 7 times a week by correlation spectrometry (Casadevall et al., 1981) at Mount St. Helens, 1980. Time of Plinian eruptions and dome extrusions are also indicated schematically. Regions labeled with I, II and III represent stages of activity. "Bl" refers to the May 18 blast dacite, "PF1", "PF2" and "PF3" to the three pyroclastic-flow deposits of May 18, from oldest to youngest.

decline in SO_2 following July 1980 and the continually smaller scale of Plinian activity are consistent with a degassing of a single shallow magma body. The transition in activity to dome extrusion without large explosions which occurred after October 1980 is also consistent with a mostly-degassed shallow magma source. The residual Cl concentrations of erupted lavas shown in Fig. 6 also seem to offer a way to monitor the degassing of the shallow magma bodies. It appears that Cl concentrations in these rocks may mimic the trend of magmatic concentrations, and provide an index of the Cl remaining in the shallow magma body. Thus, decreasing Cl concentrations in the Mount St. Helens and Arenal lavas may reflect degassing of the shallow magma body. Since Cl is a very soluble component in the magma, it is more likely to approximate quenched magmatic concentrations. Also, when Cl concentrations in the shallow magma body become a small fraction of the original magmatic concentration, it may be safely assumed that other volatiles are also of low concentrations. This surely means the danger of explosive activity from this magma body is less. We propose that monitoring of Cl in erupted lavas is a very useful method of monitoring the degassing of shallow magma bodies. It is simple and inexpensive to analyse rocks for Cl, and especially in the cases of frequent eruptive activity, when magma is repeatedly brought up from the shallow magma body, a very clear definition (compare Figs. 5 and 6) of degassing may be possible.

CONCLUSIONS

(1) Aggregation of volcanic ash within an eruption cloud during fallout is a widespread, probably general process which primarily affects ash particles smaller than 50 µm and causes premature fallout of ash. Some ash falls are made of mostly aggregated particles which would either not fall at all or descend at much slower rates. Since aggregation is likely to exert a major control on the amount of fine-grained ash left in the atmosphere, we need to evaluate which geological variables influence it.

(2) Partly because of aggregation, the mass of small silicates which reach and remain in the stratosphere can be small compared to the mass of sulfur. This is true, even for the highly explosive (and therefore well-fragmented) eruption of sulfur-poor magma (Mount St. Helens).

(3) Sulfur seems to be the main atmospheric hazard from eruptions, because it converts to small acid and salt particles in the stratosphere which can influence surface temperatures on earth.

(4) Many times more sulfur is erupted in explosive eruptions than would be indicated by the mass of erupted lava. Shallow magma bodies which are the source of eruptions apparently contribute the remainder of the sulfur.

(5) Mass flux data of SO_2 and other gases and the residual Cl contents of erupted lavas apparently are an index of the remaining gas content (energy) of shallow magma bodies below volcanoes.

(6) The hazard to the atmosphere from Cl released by eruptions is less serious than that posed by S, but is less well measured or understood.

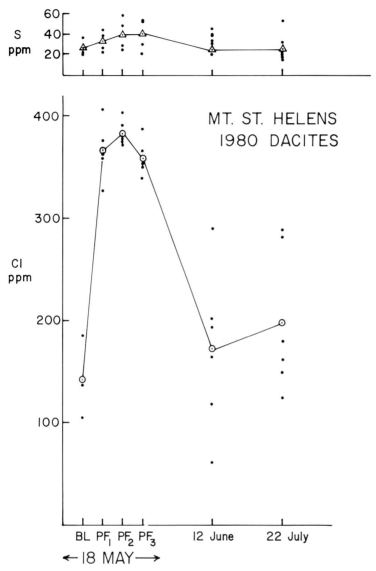

Fig. 6. Residual S and Cl concentrations in dacitic pumices of Mount St. Helens. Samples were collected of the blast dacite, (initial magma of May 18), three pyroclastic-flow units of May 18, and pyroclastic flows of 12 June and 22 July. Multiple samples of each unit were analyzed for major and 28 minor elements including S and Cl. Variations in other elements were small and consistent with data collected by others. In the plots, individual sample values and the group average are both plotted.

ACKNOWLEDGEMENTS

Financial support came from NASA (grant NAGW-84), NSF (grant DES-7801180), and the United States Geological Survey. Tom Casadevall and Robert Christiansen were enthusiastic and critical sounding boards for much of this work. Steve Self participated in some field sampling. Mark Bergion painstakingly described sulfide blebs in the Mount St. Helens dacites. Dan Sanville performed the Cl analysis of Arenal andesites. Discussions with L. Wilson, G. P. L. Walker, T. Bornhorst and A. T. Anderson led to some of the strategy used. Winton Cornell and Juergen Kienle reviewed the manuscript.

REFERENCES

Anderson, A.T., 1975. Some basaltic and andesitic gases. Rev. Geophys. Space Phys., 13: 37-55.
Cadle, R.D., C.S. Kiang and J.F. Louis, 1976. The global dispersion of the eruption clouds from major volcanic eruptions. J. Geophys. Res., 81: 3125-3132.
Carey, S.N. and Sigurdsson, H., 1982. Transport and deposition of distal tephra from the 18 May 1980 eruption of Mount St. Helens. J. Geophys. Res., 87: 7061-7072.
Casadevall, T., Johnston, D.A., Harris, D.H., Rose, W.I., Malinconico, L.L., Stoiber, R.E., Bornhorst, T.J., Williams, S.N., Woodruff, L., and Thompson, J.M., 1981. SO_2 emission rates at Mount St. Helens from May 29 through December 1980. In: P.W. Lipman and D.R. Mullineaux (Editors), The 1980 Eruptions of Mount St. Helens, Washington. U.S. Geol. Survey Prof. Paper 1250: 193-200.
Colgate, S.A. and Sigurgeirsson, T., 1973. Dynamic mixing of water and lava. Nature, 244: 522-555.
Farlow, N.H., Oberbeck, V.R., Snetsinger, K.G., Ferry, G.V., Polkowski, G., and Hayes, D.M., 1981. Size distributions and mineralogy of ash particles in the stratosphere from eruptions of Mount St. Helens. Science, 211: 832-834.
Fructer, J.S., et al., 1980. Mount St. Helens ash from the 18 May 1980 eruption: chemical, physical, mineralogical and biological properties. Science, 209: 1116-1125.
Harris, D.M., 1979. Geobarometry and geothermometry of individual crystals using H_2O, CO_2, S and major element concentrations in silicate melt inclusions. Geol. Soc. Amer. Abst. w. Prog., 11: 439.
Hoblitt, R.P., Miller, C.D. and Vallance, J.W., 1981. Origin and stratigraphy of the deposit produced by the May 18 directed blast. In: P.W. Lipman and D.R. Mullineaux (Editors), The 1980 Eruptions of Mount St. Helens, Washington. U.S. Geol. Survey Prof. Paper, 1250: 401-420.
Kellogg, W.W., Cadle, R.D., Allen, E.R., Lazrus, A.L. and Martell, E.A., 1972. The sulfur cycle. Science, 175: 587-596.

Kienle, J. and Shaw, G.E., 1979. Plume dynamics, thermal energy and long-distance transport of vulcanian eruption clouds from Augustine Volcano, Alaska. J. Volcanol. Geoth. Res., 6: 139-164.

Lapple, C.E., 1961. Characteristics of particles and particle dispersoids. Stanford Res. Inst. Journ., 5: 94.

Lazrus, A.L., Cadle, R.D., Gandrud, B.W., Greenberg, J.P., Huebert, B.J., and Rose, W.I., 1979. Sulfur and halogen chemistry of the stratosphere and of volcanic eruption plumes. J. Geophys. Res., 84: 7869-7875.

Luhr, J., Varekamp, J.C. and Prestegaard, K., 1982. The 1982 eruption of El Chichon Volcano, Chipas, Mexico, Part II. Mineralogy and petrology of the ejecta. Geol. Soc. Amer. Abst. W. Prog., 14: 551.

Melson, W.G., Hopson, C.A. and Kienle, C.F., 1980. Petrology of tephra from the 1980 eruption of Mount St. Helens. Geol. Soc. Amer. Abst. W. Prog., 12: 416.

Moore, J.G. and Sisson, T.W., 1981. Deposits and effects of the May 18 pyroclastic surge. In: P.W. Lipman and D.R. Mullineaux (Editors), The 1980 Eruptions of Mount St. Helens, Washington. U.S. Geol. Survey Prof. Paper, 1250: 421-438.

Murrow, P., Rose, W.I., and Self, S., 1980. Determination of the total grain size distribution in a Vulcanian eruption column and its implications to stratospheric aerosol perturbation. Geophys. Res. Lett., 7: 893-896.

Newell, R., 1982. Mount St. Helens eruption of 1980: Atmospheric effects and potential climatic impact. NASA SP 458.

Noguchi, K., Kamiya, H., 1963. Prediction of volcanic eruption by measuring the chemical composition and amount of gases. Bull. Volcanol., 26: 367-368.

Rose, W.I., 1977. Scavenging of volcanic aerosol by ash: atmospheric and volcanologic implications. Geology, 5: 621-624.

Rose, W.I., Bonis, S., Stoiber, R.E., Keller, H., and Bickford, T., 1973. Studies of volcanic ash from two recent Central American eruptions. Bull. Volcanol., 37: 338-364.

Rose, W.I., Anderson, A.T., Woodruff, L.G., and Bonis, S., 1978. The October 1974 basaltic tephra from Fuego volcano: description and history of the magma body. J. Volcanol. Geoth. Res., 4: 3-53.

Rose, W.I. and Hoffman, M.F., 1980. Distal ashes of the May 18, 1980 eruption of Mount St. Helens. Trans. Am. Geophys. Union, EOS, 61: 1137.

Rose, W.I., Harris, D M., Heiken, G., Sarna-Wojcicki, A., and Self, S., 1982. Volcanological description of the 18 May 1980 eruption of Mount St. Helens. NASA SP-458: 1-36.

Rose, W.I. and Hoffman, M.F., 1982. The May 18, 1980 eruption of Mount St. Helens: the nature of the eruption with an atmospheric perspective. NASA CP 2240: in press.

Rose, W.I., Stoiber, R.E., and Malinconico, L.L., 1982. Eruptive gas compositions and fluxes of explosive volcanoes: budget of S and Cl emitted from Fuego Volcano, Guatemala, In: R.S. Thorpe (Editor), Andesites and Related Rocks, J. Wiley & Sons, N.Y., pp. 669-676.

Rowley, P.D., Kuntz, M.A., and MacLeod, N.S., 1981. Pyroclastic flow deposits. In: P.W. Lipman and D.R. Mullineaux (Editors), The 1980 Eruptions of Mount St. Helens, Washington. U.S. Geol. Survey Prof. Paper 1250: 489-512.

Sarna-Wojcicki, A., Shipley, S., Waitt, R.B., Dzurisin, D., and Wood, S.M., 1981. Areal distribution thickness, mass, volume, and grain size of airfall ash from the 1980 eruptions of Mount St. Helens. In: P.W. Lipman and D.R. Mullineaux (Editors), The 1980 Eruptions of Mount St. Helens, Washington. U.S. Geol. Survey Prof. Paper, 1250: 577-600.

Self, S. and Sparks, R.S.J., 1978. Characteristics of widespread pyroclastic deposits formed by the interaction of silicic magma and water. Bull. Volcanol., 41: 196-212.

Self, S., Wilson, L. and Nairn, I.A., 1979. Vulcanian eruption mechanisms. Nature, 277: 440-443.

Sheridan, M.F., Barberi, F., Rose, M., and Santacroce, R., 1981. A model for Plinian eruptions of Vesuvius. Nature, 286: 281-282.

Sheridan, M.F. and Wohletz, K.H., 1983. Explosive hydrovolcanism: basic considerations. In: M.F. Sheridan and F. Barberi (Editors), Explosive Volcanism. J. Volcanol. Geotherm. Res., 17: (this volume).

Sorem, R.K., 1982. Volcanic ash clusters: tephra rafts and scavengers. J. Volcanol. Geoth. Res., 13: 36-41.

Stoiber, R.E. and Rose, W.I., 1970. Geochemistry of Central America volcanic gas condensates. Geol. Soc. Amer. Bull., 81: 2891-2912.

Stoiber, R.E. and Rose, W.I., 1973. Cl, F and SO_2 in Central America volcanic gases. Bull. Volcanol., 37: 454-460.

Stoiber, R.E. and Rose, W.I., 1974. Fumarole incrustations at active Central American volcanoes. Geoch. Cosmoch. Acta., 38: 495-516.

Stoiber, R.E., Williams, S.N., Malinconico, L.L., Johnston, D.A., and Casadevall, T.J., 1981. Mount St. Helens: evidence of increased magmatic gas component. J. Volcanol. Geoth. Res., 11: 203-212.

Taylor, H.E. and Lichte, F.E., 1980. Chemical composition of Mount St. Helens volcanic ash. Geophys. Res. Lett., 7: 949-952.

Varekamp, J.C., Luhr, J., and Prestegaard, K., 1982. The 1982 eruption of El Chichon volcano, Chiapas, Mexico. Part I. Stratigraphy, volume, and volatile element characteristics of the ash-fall deposits. Geol. Soc. Amer. Abst. W. Prog., 14: 637.

Walker, G.P L., 1971. Grain size characteristics of pyroclastic deposits. J. Geol., 79: 696-714.

Walker, G.P.L., 1972. Crystal concentration in ignimbrites. Contr. Miner. Petrol., 36: 135-146.

Walker, G.P.L., 1980. The Taupo pumice: product of the most powerful known (ultra-Plinian) eruption? J. Volcanol. Geoth. Res., 8: 69-94.

Walker, G.P.L., 1981. Generation and dispersal of fine ash and dust by volcanic eruptions. J. Volcanol. Geoth. Res., 11: 81-92.

Walker, G.P.L., Heming, R.F., and Wilson, C.J.N., 1980. Low aspect ratio ignimbrites. Nature, 283: 286-287.

Williams, S.N. and Self, S., in press. Grain size, distribution, and volume of the 1902 Plinian fall deposits of Santa Maria volcano, Guatemala. J. Volcanol. Geoth. Res.

Wilson, L. and Huang, T.C., 1979. The influence of shape on the atmospheric settling velocity of volcanic particles. Earth Planet. Sci. Lett., 44: 311-324.

Wohletz, K.H., 1983. Mechanisms of hydrovolcanic pyroclast formation: grain-size, scanning electron microscopy, and experimental data. In: M.F. Sheridan and F. Barberi (Editors), Explosive Volcanism. J. Volcanol. Geotherm. Res., 17: (this volume).

Wozniak, K.C., Hughes, S.S., and Taylor, E.M., 1980. Unpubl. Chemical Analyses, Ore. St. Univ.

THE PAST 5,000 YEARS OF VOLCANIC ACTIVITY AT MT. PELÉE MARTINIQUE (F.W.I):
IMPLICATIONS FOR ASSESSMENT OF VOLCANIC HAZARDS

D. WESTERCAMP[1] and H. TRAINEAU[2]

[1] Service géologique régional des Antiles et de la Guyane (B.R.G.M), B.P. 394, 97204 Fort de France, (MARTINIQUE, F.W.I.).
[2] Département Géothermie, B.R.G.M., B.P. 6009, 45060 Orleans CEDEX (FRANCE).

(Received July 8, 1982; revised and accepted November 24, 1982)

ABSTRACT

Westercamp, D. and Traineau, H., 1983. The past 5,000 years of volcanic activity at Mt. Pelée Martinique (F.W.I.): Implications for assessment of volcanic hazards. In: M.F. Sheridan and F. Barberi (Editors), Explosive Volcanism. J. Volcanol. Geotherm. Res., 17: 159-185.

The history of Mt. Pelée, Martinique, was subdivided into three stages based on field geology and ^{14}C data. The two first stages constructed an ancient Mt. Pelée and an intermediate cone between 0.4 m.y and 19,500 y.b.p. The third (or present) stage started 13,500 years ago, after a repose of 6,000 years. This paper focuses on the activity of Mt. Pelée during the past 5,000 years as a means to assess and zone volcanic hazards of the 23 magmatic eruptions during the past 5,000 years. The ages of 21 eruptions of this period are based on 75 new ^{14}C dates. The types of phenomena and distribution of pyroclasts relate to four main types of activity:
- The first type consists of pumice-and-ash flows that are not preceded by a Plinian fall. Two eruptions (named P6 and P4) illustrate this type, for which the mixture of gas, ash, and pumice simply overflow the vent and flood several valleys.
- The second type differs from the first by the occurrence of a preliminary moderate Plinian-fall stage. Four eruptions (P5, $P3_1$, P2 and P1) illustrate this type. Two eruptions ($P3_2$ and $P3_3$) experienced cataclysmic Plinian explosions and pumiceous surges.
- The third type is related to dome growth with the rise of viscous spines and the production of related block-and-ash flows. Five eruptions (1929, Sept. 1902-1904, NPM, NAB_2 and NMP) illustrate this type.
- The fourth type is characterized by violent ejection of more-or-less heterogeneous nuées ardentes. The direction of the blast, dictated by the morphology of the crater, has been towards the south several times at Mt.

0377-0273/83/$03.00   © 1983 Elsevier Science Publishers B.V.

Pelée. Four eruptions (May 1902, NAB1, NRP2 and NRP3) belong to this type.
Future magmatic eruptions at Mt. Pelée will very likely belong to one of these four types.

Assessment of hazards at Mt. Pelée is based upon the behavior of the volcano during the past 5,000 years because: (1) recognition of past magmatic eruptions is quite complete and well-dated, and (2) no structural change has occurred in the volcano. A probabilistic and statistical approach has been tentatively followed. Stochastic models of Wickman indicate: (a) the chance that contemporary abnormal seismic and/or phreatic activity at the volcano would signal an impending magmatic eruption is 20 percent, (b) Mt. Pelée exhibits a complex loading-time behavior that may be related to the buffering effect of a magma chamber (time is needed following a single or several closely spaced eruptions to establish conditions to initiate a new eruption), and (c) if the 1929 event can be regarded as the last episode of the major 1902 eruption, the volcano could remain in a dormant stage with respect to magmatic events for another century.

INTRODUCTION

The 1976 seismo-phreatic crisis of la Soufrière in Guadeloupe prompted detailed geological studies in the French West Indies in order to assess future volcanic hazards and to propose zoning maps. Field work carried out by the French Geological Survey at Mt. Pelée during the past three years has focused on detailed geological mapping (1: 20,000 scale) of the entire volcano, a survey of measured sections, and ^{14}C dating (about 180 new dates were obtained on charcoal and wood).

Mt. Pelée occupies the northern part of Martinique, one of the main islands in the central Lesser Antillean volcanic arc (Fig. 1). This composite explosive volcano rises about 1,400 meters above sea level and its deposits cover about 100 km^2. Since the famous descriptions of the 1902-1904 eruption by Lacroix (1904) and the 1929-1932 eruption by Perret (1935), studies have focused on:
- the general geological setting of the volcano (Grunevald, 1965; Westercamp, 1974 and 1980).
- chemistry of the volcanic products (Cheminee, 1973; Gunn et al., 1974; Smith and Roobol, 1976; Gourgaud, 1982; Traineau, 1982).
- surveys of stratigraphic sections, absolute age determination by ^{14}C, recognition of different volcanic phenomena (Roobol and Smith, 1975, 1976, and 1980; Fisher et al., 1980; Traineau, 1982).
- detailed geological mapping.
- preliminary appraisal and zonation of volcanic hazards (Stieltjes and Westercamp, 1978).

The purpose of this paper is: (1) to summarize new information on the volcanological history of Mt. Pelée that has been gathered by the French Geological Survey (BRGM) during the past three years, and (2) to interpret this

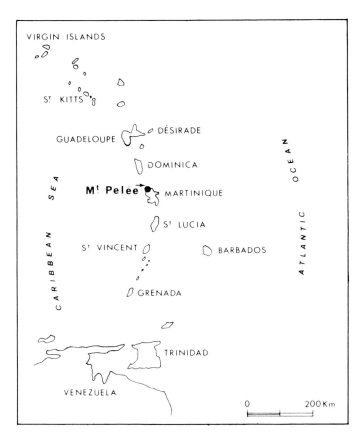

Fig. 1 Map of the Lesser Antilles showing the location of Martinique and Mt. Pelée.

information in terms of volcanic hazards.

VOLCANOLOGICAL HISTORY OF MT. PELÉE

The growth of Mt. Pelée (Fig. 2) can be subdivided into three stages on the basis of field geology and ^{14}C data. The first two stages are briefly summarized, while the third (present stage) is described in detail because of its importance in assessing volcanic hazards.

First stage: the ancient Pelée

Activity began around 0.4 m.y. ago (Bellon et al., 1974) between the Piton

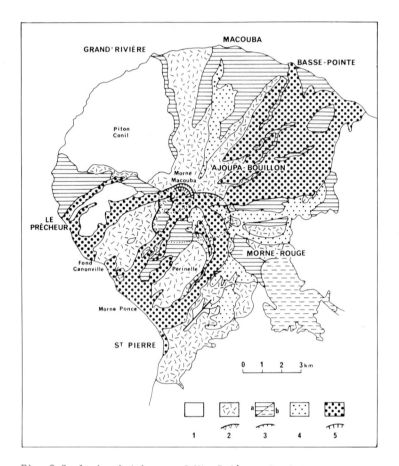

Fig. 2 Geologic sketch map of Mt. Pelée emphasizing the distribution of coarse pyroclastic flows emplaced during the past 5,000 years.
(1) Substratum. (2) Ancient Mt. Pelée. (400,000 y.b.p. > age > 200,000 y.b.p. (?) and caldera rim. (3) The intermediate cone (100,000 y.b.p. (?) > age > 19,500 y.b.p.) and crater rim. (3a) Pyroclastic flows. (3b) Lake deposits. (4) First stage of present cone (13,500 y.b.p. > age > 5,000 y.b.p.). (5) Second stage of present cone (age < 5,000 y.b.p.) and crater rim.

Mt. Conil located at the northern edge of the island and the volcanic complex of Pitons du Carbet and Morne Jacob which crop out, respectively, to the south and east. The early volcano consisted of more-or-less welded, coarse pyroclastic flows and thick lava flows of andesitic composition. Pyroclastic flows are well exposed on the west flank of the present-day volcano (e.g., Le Tombeau des Caraibes), whereas lava flows cap some ridges that trace the rim of an old crater or caldera. Judged by the amount of erosion on the flanks and the crater of ancient Pelée, activity probably ended 0.2 to 0.3 m.y. ago. This estimated age has not yet been confirmed by radiometric age determinations.

Second stage: the intermediate cone

A second stage of growth began near the northern rim of the old crater. The age of this activity is more than 40,000 y.b.p. (limit of ^{14}C method) but probably less than 0.1 m.y. ago, judged by the obvious long repose that separates the first stage from the second. The second stage was characterized by explosive eruptions that produced alternating pumice falls, pumice and/or scoria flows, and block-and-ash flows. St. Vincent-type eruptions of large volume occurred respectively at more than 40,000; 25,700 ± 1,200 and 22,300 ± 1,300 y.b.p. (Traineau et al., 1982). A two-kilometer-wide crater resulted from this volcanic activity, the northern rim of which is still recognizable (the Morne Macouba). It is noteworthy that this stage of activity ended around 19,500 y.b.p. with the most acid Plinian-type eruption ever experienced at Mt. Pelée.

Third stage: the present cone

Volcanic activity began again at Mt. Pelée around 13,500 y.b.p., after a 6,000-year period of quiescence. During the past 13,500 years of activity at least 34 magmatic eruptions have occurred. Among these, 10 occurred before 5,100 y.b.p. and 24 after. Probably other magmatic events occurred between 13,500 and 5,100 y.b.p., but have not been recognized due to lack of exposure. For example in the period from 13,500 to 8,000 y.b.p. only one eruption per millenium has been recognized. This earlier period is not described in detail. Assessment and zonation of volcanic hazards at Mt. Pelée is mainly based on the behavior of the volcano during the past 5,000 years of activity for which better data exist.

The period 13,500 to 5,100 y.b.p.

The first important eruptions produced heterogeneous nuées ardentes that flowed southward (13,500 ± 200 y.b.p.) and eastward (11,300 ± 420 y.b.p.). Typical sections consist of basic scoria flows at the bottom, overlain by grey and black, block-and-ash flows. Eruptions which followed alternated between Plinian pumice-flow and nuées ardentes eruptions. Distribution of coarse breccias suggests that the vents were located slightly to the south of the present volcano summit. This is particularly apparent with the nuée ardente, which blanketed the whole area of Morne Rouge 5,100 ± 60 y.b.p.

The period 5,000 y.b.p. to the present

This period of activity is well documented by field geology and radiocarbon ages. Fig. 3 reconstructs the eruptive history of Mt. Pelée based upon interpretation of stratigraphic sections such as those presented in Fig. 4. Pumiceous eruptions are designated by the letter P followed by the number of eruptions of this category. Nuées ardentes eruptions are named after the place where they were first recognized or after the area of best outcrop.

The age of the (P6) pumice deposit (Fig. 5a) is 4,610 ± 50 y.b.p. based on the average of eight ^{14}C dates. The first stage of this important pumiceous eruption was characterized by the eastward emission of scoreaceous basaltic-andesite flows. The nature of the eruption then changed and heterogeneous pyroclastic flows (mixtures of andesitic blocks, pumice, banded scoria, breadcrust bombs, and ash) were later expulsed. These deposits underlie white pumice flows that filled at least three valleys on the flank of the volcano. The last stage of the eruption was characterized by flows of banded pumice that are locally welded and columnar-jointed. The black bands of the pumice are slightly more basic than the white bands, both being more silicic than previous magmatic products. No related Plinian-fall deposits underlie the pyroclastic flows. On the other hand, up to 3 meters of very fine grained pumiceous ash-cloud deposits blanket the western flank of the volcano. The nuées ardentes of Pointe la Mare (NPM; Fig. 5b) are dated at 4,410 ± 120 y.b.p. based on the average of four ^{14}C dates.

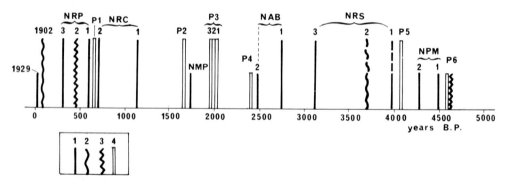

Fig. 3 Reconstructed eruptive history of Mt. Pelée for the past 5,000 years based on 75 radiocarbon-dated deposits.
(1) Homogeneous block-and-ash flows. (2) Slightly heterogeneous block-and-ash flows. (3) Heterogeneous scoria flows and/or block-and-ash flows (short line = eruption of the dome-collapse type; long line = eruption of the directed-blast type or eruption initiated with directed blast and then changing to the dome-collapse type). (4) Pumice-and-ash falls and/or flows (short line = eruption of the overflowing flow type; long line = Plinian eruption with subsequent pumice and ash flows. Names of eruptions explained in the text.)

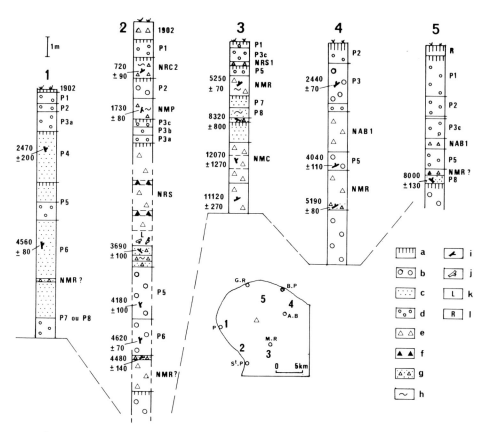

Fig. 4 Selected stratigraphic sections from Mt. Pelée. The eruptive sequences are based on paleosoils and ^{14}C data. (1) Upper part of Case-Viala section. (2) Morne Ponce. (3) Balisier-Calave. (4) Upper part of Riviere la Falaise. (5) Savane Petit. The dashed line joining the bottom of the columns corresponds to the 6,000 year repose period that separates deposits of the intermediate cone from those of present edifice. (a) Paleosoil. (b) Pumice-and-ash flows. (c) Pumiceous ash-cloud deposits. (d) Plinian falls. (e) Block-and-ash flows. (f) Basic scoria flows. (g) Lithic ash-cloud deposits. (h) Surges. (i) Charcoal. (j) Casts of old trees. (k) Lahar. (l) Reworked layer.

A series of more-or-less reworked block-and-ash flows crops out between Pointe la Mare and el Prêcheur. These slightly heterogeneous deposits are composed of both grey and dark-grey lithic or vesiculated andesitic blocks up to 8 meters in diameter. It is not yet established whether these resulted from one or two distinct eruptions. In the later case ^{14}C data suggest the older one to

166

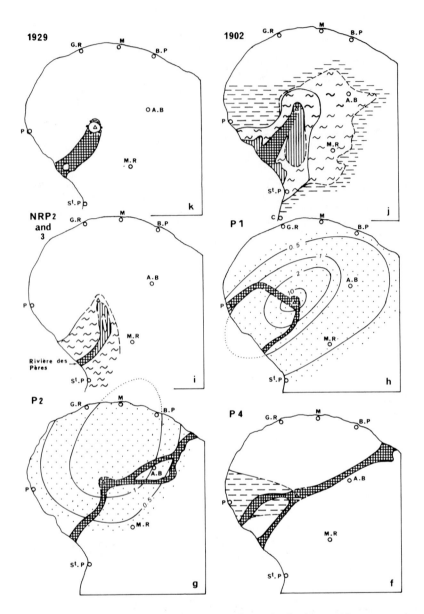

Fig. 5 Volcanological sketch map of the principal eruptions experienced at Mt. Pelée during the past 5,000 years (simplified from Westercamp and Traineau, in prep.)
1. pyroclastic flow - 2. coarse pyroclastic surges - 3. fine-grained pyroclastic surges and blast effects - 4. pumiceous ash cloud deposit more than 1 metre in

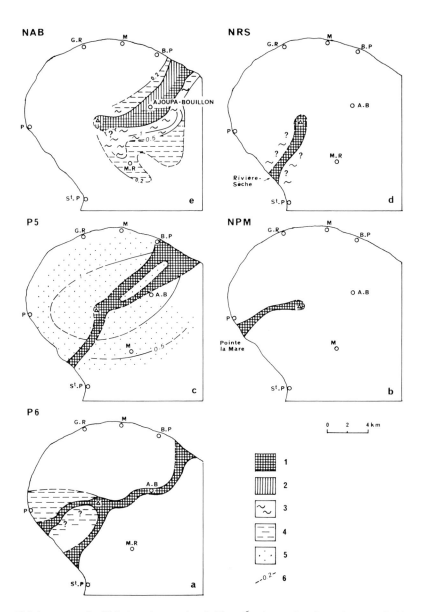

thickness — 5. Plinian-type air-fall — 6. isopachs in meters. Letters designate the principal towns around the volcano; B.P.: Basse-Pointe; M: Macouba; G.R.: Grand Rivière; P.: Prêcheur

have occurred around 4,500 ± 75 y.b.p. and the younger one around 4,320 ± 30 y.b.p. The restriction of deposits to the present Grande Savane area and the presence of huge blocks in the coarse deposits suggest that this event was related to the growth and collapse of a summit dome.

The age of the (P5) pumice deposit (Fig. 5c) is 4,060 ± 90 y.b.p. based on the average of three ^{14}C dates. This important pumiceous eruption started with a Plinian explosion. The resulting ash and lapilli fall is regularly distributed on the flanks of the volcano, reaching 1 m in thickness at a distance of 4-6 km from the summit. Plinian deposits on the windward side of the volcano are deeply altered to a "brown clay with yellow blobs". Deposits on the leeward side are fresh and characteristically exhibit a concentration of grey lithic lapilli of andesitic composition near the bottom of the layer. Coarse-grained pumice flows south of Basse Pointe spread widely over the Plinian deposits after filling pre-existing valleys. Thick pumice flows related to this eruption also are found in the Morne Ponce quarry.

The distribution of the nuées ardentes of Riviere Seche (NRS) is shown in Fig. 5d. Several distinctive eruptions between 4,060 y.b.p. (P5) and 3,000 y.b.p. are recorded on the western flank of Mt. Pelée. Field work and ^{14}C data distinguish three clusters of activity.

The first group (NRS1) has an age of 3,980 ± y.b.p. based on the average of three ^{14}C dates. This block-and-ash flow episode seems to be restricted to the present Vallée Blanche. The only known underflow* deposit crops out along the Riviere Claire (Roobol and Smith, 1976). Related (?) ash-cloud surge beds which make up the bottom layers of the Morne Ponce quarry are composed of thin and fine-grained lithic ash.

The second group (NRS2) has an age of 3,710 ± 30 y.b.p. based on the average of two ^{14}C dates. Pyroclastic flows belonging to this episode constitute the quarry wall at Morne Ponce. These coarse-grained lithic and slightly vesiculated block-and-ash flows are of andesitic composition. Thin lenses of darker and more mafic scoria and ash are interbedded within these units, suggesting laminar flow during emplacement.

The third group (NRS3) has an age of 3,130 ± 20 y.b.p. based on the average of two ^{14}C dates. This fine-grained, block-and-ash surge deposit has been encountered by Roobol and Smith, and Walker in a few places around the volcano (cf. Radiocarbon Journal). ^{14}C ages suggest that these deposits reflect a single event that terminated volcanic activity on the western flank of the volcano for a few centuries. The distribution of known outcrops suggests that a vertical explosion was responsible.

The distribution of the nuées ardentes of Ajoupa-Bouillion (NAB) is shown in Fig. 5e. Roobol and Smith (1976) recognized this important block-and-ash flow deposit on the eastern flank of the volcano, especially in the vicinity of

---

*Nuées ardentes can be separated into a basal avalance or underflow and an overriding expanding ash cloud containing fragments elutriated from the top of the avalanche (Sheridan, 1979; Smith and Roobol, 1980a).

Ajoupa-Bouillion. Underflows and ash-cloud deposits are up to 10 centimeters thick. Some stratigraphic sections indicate that two periods of magmatic activity were separated by a soil horizon. The available ^{14}C dates agree with this interpretation.

The first pyroclastic flows of this group (NAB1), have an age of 2,750 ± 50 y.b.p. based on the average of three ^{14}C dates. These heterogeneous block-and-ash flows up to 2 m thick filled a 2 km wide gully that descends the flank of the volcano toward the northeast. Deposits are a mixture of lithic ash enclosing more or less vitric and vesiculated andesitic blocks. Some deposits are normally graded and exhibit a concentration of the most pumiceous clasts in the upper part of the breccia. The related ash cloud blanketed a wider area down to Morne Rouge.

The second pyroclastic flows of this group (NAB2) have an age of 2,490 ± 10 y.b.p. based on the average of two ^{14}C dates. The coarse block-and-ash flow deposits that overlie NAB1 deposits at Ajoupa-Bouillion village are more homogeneous, mainly composed of dark hyaloporphyric andesitic blocks up to 3 m in diameter. They filled a narrow gully that intersects the NAB1 underflows. A coarse-grained, ash-cloud deposit that developed toward the south may be related to the NAB2 eruption based on a single ^{14}C age determination. It is thus suggested that some nuées ardentes developed in this direction but their underflows have not yet been recognized. Distribution, homogeneity, and size of products, suggest that the NAB2 eruption resulted from the activity of a growing dome. Conversely, the products of the NAB1 eruption are similar to deposits of the explosions that occurred on May 8 and 20 of 1902.

The P4 pumiceous deposits (Fig. 5f) have an age of 2,440 ± 50 y.b.p. based on the average of five ^{14}C dates. Coarse-grained pumice flows crop out at Grande Savane, Morne Ponce, and Vivé. These deposits are interpreted to represent a single event based on ^{14}C ages. Important features of this unit are the absence of a preliminary Plinian fall and the westward concentration of fine-grained, ash-flow deposits which are up to 2.5 m thick. These characteristics, similar to P6, suggest that no extensive vertical column developed over the vent during either the P4 and P6 eruptions. A mixture of pumice, ash, and gas gently overflowed from the vent. The fine-grained ash cloud that rose above the ground-hugging pyroclastic flow remained within the trade wind zone (below 5 km in altitude) and consequently was blown mainly westward.

The (P3) pumiceous deposits (Fig. 6) have an age of about 2,020 ± 140 y.b.p. based on 6 precise ^{14}C dates. This is the most powerful group of eruptions experienced at Mt. Pelée during the past 5,000 years. Three stages, all pumiceous in nature, have been recognized. The absence of a sharp boundary and paleosoil between the successive deposits strongly suggests that only a short time interval (less than 20 years) elapsed between the different stages. The radiocarbon ages from twelve charcoal samples recovered from the ash and pumice flows support this point of view. Within age uncertainties, all ^{14}C dates are identical regardless of stratigraphic position of the sample.

The first event, ($P3_1$) is a typical Plinian eruption that resembles the P5,

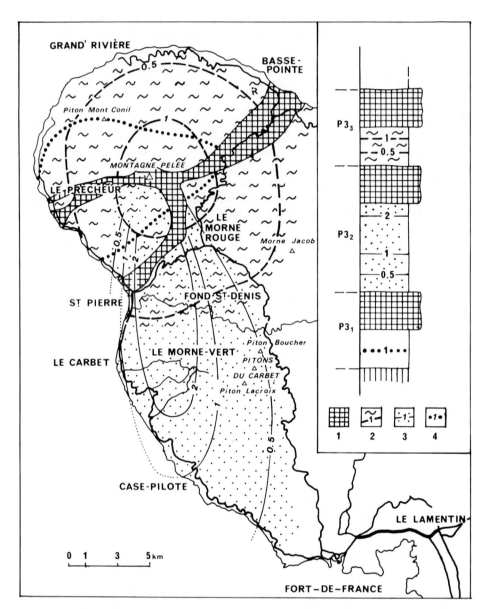

Fig. 6. Geologic sketch map of P3, the strongest eruption of the past 5,000 years at Mt. Pelée. All isopachs are in meters. (1) Pumice-and-ash flows. (2) Pumice- and crystal-rich ground surges ($P3_3$). (3) Plinian fall underlying the ash flow ($P3_2$). (4) Area buried by more than one meter of Plinian-fall deposit.

P2, and P1 events. The west flank of the volcano is buried beneath a one meter thick mixture of coarse-grained pumiceous ash, pumice, and oxidized lithic xenoliths. Overlying pumice flows occur in at least the Grande Savane valley.

The striking southward development of the second eruptive stage ($P3_2$) of P3 was first recognized by Roobol and Smith (1980, their horizon (a) on Fig. 3), who measured at least two meters of ash-flow deposits in the vicinity of Case-Pilote. Our own observations suggest a less voluminous eruption, but it is clear that a similar event today would be very destructive to life and property in Martinique.

The last stage ($P3_3$) of the P3 eruption began with a ground surge which buried nearly the entire cone of the volcano beneath more than half a meter of pumice and crystal-rich, pumiceous ash. Surge structures, such as elongated lenses of well-sorted pumice, are visible up to 6 km from the vent. If the limit of destruction more or less corresponds to the deposition of a 0.1-meter-thick ash layer, as suspected for lithic blasts, a serious hazard likely exists up to 10-12 km from the vent. The pumice-and-ash flows that followed are thought to have mainly flowed to the east.

The nuées ardentes of Morne Ponce (NMP) have an age of 1,730 ± 80 y.b.p. based on one ^{14}C date. Deposits of this block-and-ash flow episode have only been encountered in the northern part of the quarry of Morne Ponce. Ash layers belonging to P3 underlie homogeneous, slightly vesiculated ash-and-lapilli surges composed of a grey andesite and which exhibit flat, cross-bedded, and planar structures. These surges, owing to their homogeneity and restricted distribution, are thought to have resulted from a secondary lateral expansion (Fisher, 1979) from a Merapi-type nuée ardente (collapse of a homogeneous dome), rather than the result of a directed primary blast starting from the vent.

The P2 pumiceous deposit (Fig. 5g) has an age of 1,670 ± 40 y.b.p. based on the average of six ^{14}C dates. The lack of even a thin paleosoil at the top of the NMP surge suggests that the following pumiceous P2 eruption occurred shortly after the surge. The initial Plinian explosion blanketed the northern slopes of the cone, whereas the following pumice-and-ash flows flooded several valleys to the east and southwest. The juvenile material is a white homogeneous vesiculated andesite that is relatively acid in composition. In some places these pumice are banded.

The nuées ardentes of Riviere Claire (NRC) have an age of 1,140 ± 20 y.b.p. based on the average of three ^{14}C dates. Deposits of the first eruption that belong to this activity are poorly known. They have only been recognized by Roobol, Smith and Walker (cf. Radiocarbon Journal) in different places on the west flank of the volcano as lithic ash surges enclosing charcoal.

The second stage of nuées (NRC2) have an age of 690 ± 20 y.b.p. based on the average of three ^{14}C dates. They consist of a mixture of heterogeneous blocks up to 0.4 m in diameter. White pumice and dark glassy lapilli coexist in this deposit with a dominant, slightly-vesiculated, grey andesite. Similarities to younger, better-known heterogeneous nuées ardentes suggest that the subsequent

NRS2 deposits could have resulted from a southwest directed blast from the vent.

The (P1) pumiceous deposits (Fig. 5h) have an age of 650 ± 50 y.b.p. based on the average of two ^{14}C dates. As with P5, $P3_1$ and P2, this Plinian event was followed by pumice-and-ash flows that filled at least two valleys on the western flank of the volcano. These Plinian deposits are a mixture of white andesitic pumice, similar coarse crystal-rich ash, and oxidized lithic xenoliths. Some pumice at the top of the sequence are banded. The pumice-and-ash flow that flooded the Riviere du Prêcheur exhibits characteristic fines-depleted, coarse pumice lenses. The pumice flow in the Riviere Sèche that encloses banded pumice could be related to the final stage of the eruption.

Distribution of the nuées ardentes of the Riviere des Peres (NRP) is shown in Fig. 5j. Several scattered small heterogeneous block-and-ash flow deposits crop out along the Riviere des Peres and the upper Riviere Seche. A typical deposit is composed of rare black basaltic-andesite scoria and abundant white pumice enclosed within a block-and-ash mixture of lithic grey andesite. Breadcrust bombs and charcoal are also present. According to radiocarbon age determinations on charcoal (eight dates) and the state of the volcano in 1635 (date of the first Eruopean settlement), three distinctive eruptions have been considered.

The first deposit (NRP1), with an age of 590 ± 50 y.b.p. based on the average of two ^{14}C dates, is known through the studies of Smith and Roobol (1980b). The second deposit (NRP2), which has an age of 490 ± 20 y.b.p. based on the average of five ^{14}C dates, is illustrated by deposits along both the Riviere Seche and the Riviere des Peres. The third deposit (NRP3) has an age of 290 ± 35 y.b.p. based on the average of three ^{14}C dates. However, no magmatic event has occurred at Mt. Pelée between 1635, the date of the first Eurpoean settlement, and 1902. Reports by the first European inhabitants describe the mountain as vegetation free and grey in color. Thus the mountain was named Pelée (French term meaning hairless). Therefore it is likely that an explosive eruption which destroyed the vegetation occurred a few years before 1635. An age of 320 y.b.p. is within the analytical error of the average of the three ^{14}C determinations. Distribution of the deposits is consistent with a southward directed blast. The heterogeneity and low volume of the erupted material support this hypothesis. No dome grew in the crater after these explosive events, as indicated by the pre-1902 descriptions of the summit of the volcano (Fig. 5j).

The deposits of 1902 were described by the classical work of Lacroix (1904), who regarded the outbursts on May 8 and 20 as southward directed pyroclastic surges starting from the crater. This interpretation has recently been challenged by Fisher et al. (1980) and Fisher and Heiken (in press). Fisher and his co-workers suggest that the May 9 and 20 events resulted from the collapse of a vertically directed magmatic column into the crater. Upon collapse a pyroclastic flow emerged through a notch in the southern part of the crater wall and then flowed down the Riviere Blanche. Turbulent ash surges that elutriated upward from the underflow were able to move independently. At the mouth of the valley the confined underflow also expanded laterally as secondary underflows with over-riding ash-cloud surges. According to Fisher and co-workers, these secondary surges contained sufficient energy to overwhelm St. Pierre.

It is beyond the scope of this paper to discuss in detail the course of the 1902 events and the conflicting opinions above mentioned. However, we can point out the following remarks:
(1) Most of the dead bodies and remaining walls were aligned north-south (Lacroix, 1904).
(2) Several witnesses asserted that the top of the volcano was visible during the propogation of the blast.
(3) Coarse-grained breccias, up to 1.2 m thick, which belong to the 1902 eruption (proven by ^{14}C dates) have been encountered in the mid-slope part of the Riviere Seche and high on the slopes of Périnelle, south of the crater. Deposits within the Riviere Seche are fines-depleted pyroclastic flows consisting of dark grey and pale grey vesiculated blocks up to 0.8 m in diameter.
(4) Studies of prehistoric deposits suggest that southward directed outbursts have been a common feature during the recent activity at Mt. Pelée (cf. $P3_2$, NRP_2 and NRP_3).

These observations do not support the destruction of St. Pierre by secondary ash-cloud surges that originate from underflows that reached the mouth of the Riviere Seche. Conversely, they are compatible with a coarse pyroclastic surge that followed a more direct route between the notch and St. Pierre. This surge could have resulted from a lateral explosion (blast) directed southward to south-westward, as suggested by Lacroix (1904).

Secondary ash-cloud surges, as indicated by Fisher and co-workers, are likely to have occurred. Alone they might have been responsible for the destruction of the town, but it was already destroyed by the directed blast. Eye-witness accounts and geological surveys indicate that the outburst on May 20 was a repetition of the event of May 8. The further paroxysms on May 25, 26, June 6, July 9 and August 30 produced pyroclastic flows by collapse of vertical eruptive columns (Lacroix, 1904). During the later stages of the eruption, the presence of a dome averaging 300 m in height and the rise of a viscous spines had a great influence on the genesis of glowing clouds, both by explosion along the spines and by collapse of parts of the growing dome.

The eruption of 1929 (Fig. 5k) was described by Perret (1935), who stated that most glowing clouds formed during the course of this eruption resembled those of 1902. However, the area damaged was less than the climactic outbursts of 1902.

ASSESSMENT OF HAZARDS

Assessing future volcanic hazards at a given volcano based upon its past activity (historic and prehistoric) has gained worldwide acceptance since the pioneering work in the Cascade Range of Crandell and Mullineaux (1967, 1975, 1976). This method seems most applicable to the French Lesser Antillean volcanoes - Mt. Pelée and Soufriere de Guadeloupe (cf. Westercamp, 1980b; 1981).

Basis for discussions of Mt. Pelée

In order to establish a record against which possible future eruptive activity may be assumed, one must study a geological period that includes a sampling of eruptions that are representative of a volcano. Moreover, the studied period of time should exclude major structural changes at the volcano in order to include only comparable data. The studied deposits must be amenable to age dating through field studies and ^{14}C determinations. Establishing a detailed chronology is very important for a probabilistic approach in assessing hazards and in developing conceptual models of volcanoes.

Assessment and zonation of hazards at Mt. Pelée are based upon the behavior of the volcano during the past 5,000 years because: (1) Recognition of the past magmatic eruptions is thought to be complete or nearly so as a result of field study and excellent ^{14}C control (Fig. 7). (2) The vent has probably remained fixed after a slight northward shift after the NMR eruption (nuées ardentes of Morene Rouge) dated at 5,100 ± 60 y.b.p. During the past 5,000 y.b.p., 23 magmatic eruptions have been recognized, including the historical 1902 and 1929 events.

Phreatic events such as those of 1792 and 1851, generally do not leave a permanent geologic record due to the small volume of their products. The same argument applies to small magmatic eruptions that closely follow a large volume eruption. Thus, only the large volume magmatic eruptions form a record from which future events may be forecast.

FUTURE PHENOMENA AND TYPE OF ERUPTIONS

All types of volcanic phenomena that occurred during the past 5,000 years are likely to recur in the future. Table 1 lists these phenomena with their past frequency, based on the volcanological history summarized above. Roobol and Smith (1976) recognized an alternating pattern of nuées ardentes eruptions (lithic or poorly-vesiculated, juvenile block-and-ash flows) and pumice eruptions (highly-vesiculated, juvenile block-and-ash flows and falls). Another contrast is the occurrence of violent initial explosive phenomena among some eruptions while other eruptions lack initial outbursts. Four main types of magmatic eruptions seem to characterize the recent behavior of the volcano.
(1) The first type consists of pumice-and-ash flow eruptions which are preceded by a Plinian fall as illustrated by P6 and P4. Such eruptions result

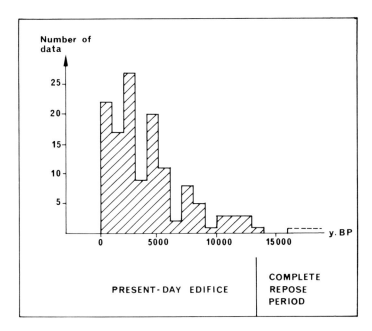

Fig. 7 Histogram showing the frequency distribution of radiocarbon ages by 1,000 year periods. Data are from Roobol and Smith (1976; 1980: cf. Radiocarbon Journ.), Walker (in Radiocarbon Journ.), and Grunevald (1965).

from the collapse of a very low vertical column of magmatic pumice. In fact, the mixture of gas, ash, and pumice may simply flow over the vent.

(2) The second type differs from the first by the occurrence of a preliminary sub-Plinian fall stage, as illustrated by P5, $P3_1$, P2 and P1. The Plinian phase results from the contact between gaseous magma en route to the surface and meteoric water trapped at the base of the volcanic edifice. After the meteoric water is completely consumed by the sub-Plinian phase, the eruption drops to the overflowing type. Cataclysmic events, such as $P3_2$ and $P3_3$, illustrate the peculiar scenario of this type of pumice-rich Plinian event.

(3) The third type involves the growth of a dome, the rise of viscous spines, and subsequent collapse that feeds glowing avalanches or heterogeneous, juvenile block-and-ash flows. The 1929 eruption, NPM, NAB2, and possibly NMP illustrate this type. Such magmatic activity has been called the Merapi-type by McDonald (1972) or the dome collapse-type by Wright et al. (1980).

(4) The fourth type is characterized by the violent ejection of heterogeneous mixtures of gas, lithic juvenile ash, lapilli, blocks, oxidized xenoliths, white pumice, and less commonly, black scoria and breadcrust bombs. NRP2, NRP3 illustrate this type. The direction of the blast is dictated by the morphology

TABLE 1

Eruptive phenomena during the past 5,000 years in order of decreasing frequency at Mt. Pelée.

Phenomena	Past - frequency
	100 / 75 / 50 / 25 %
(1) - phreatic explosion	▓▓▓▓ / ▓▓▓▓ / ▓▓▓▓ / ▓▓▓▓
(2) - precursory phreato-magmatic explosion	▓▓ / ▓▓ / ▓▓ / ▓▓▓▓ / ▓▓▓▓
- homogeneous lithic block and ash flow (related to a dome)	/ ▓▓ / ▓▓ / ▓▓▓▓
- homogeneous and slightly heterogeneous directed blasts (flow, surge, and ash cloud deposit) (unrelated to a dome)	/ / / ▓▓▓▓
- pumice and ash flow and related fine-grained ash cloud deposit	/ / / ▓▓▓▓
- pumice and ash fall (Plinian explosion)	/ / / ▓▓
- very heterogeneous flow and surge (unrelated to a dome)	/ / / ▓
- scoria and ash flow	/ / / ▓
- pumice crystal ground surge	/ / / ▓

1. Highest frequency of all volcanic phenomena. Preceeds all magmatic eruptions and also occurs alone.
2. Preceeds almost all magmatic eruptions.

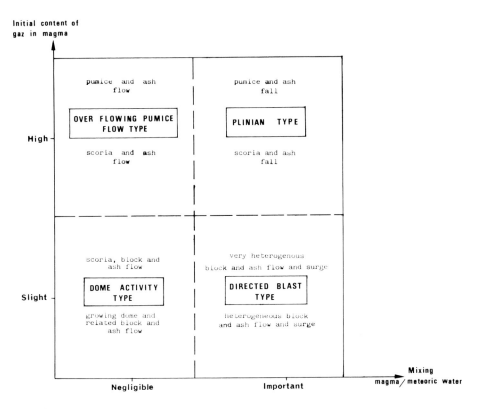

Fig. 8 Working classification of eruption types and pyroclastic deposits at Mt. Pelée. Explosive dynamics result from a combination of magmatic degassing and mixing of meteroric water with magma.

of the vent. The blast is normally vertical, but at Mt. Pelée southwest-directed blasts have occurred several times. Such blasts are initiated by phreatomagmatic explosions similar to those associated with Plinian eruptions. The difference is mainly due to the high content of magmatic gas for the Plinian type and low content of magmatic gas for the directed-blast type. Once the meteoric water is consumed, the eruption may revert to the dome collapse type. The course of the 1902-1904 eruption illustrate this situation. Probably some prehistoric eruptions as NAB1 also belong to this intermediate type, but this cannot be proven.

In conclusion, each eruptive scenario at Mt. Pelée depends principally upon: (1) the initial magmatic gas content which is illustrated by the vesicularity of lavas, (2) the degree of interaction between meteoric water and magma, and (3) the depth at which the magma and water contact occurs. Fig. 8 is a schematic classification of eruption dynamics at Mt. Pelée.

Probabilistic approach

Mt. Pelée, which has existed for at least 50,000 years (not taking into account the ancient Pelée), has been in a pseudo-stationary state for the last 5,000 years. In such conditions eruptions can be regarded as time independant stochastic events. Furthermore, the exact time of an eruption depends on a variety of factors, some of them unpredictable and/or only indirectly related with the volcano itself (e.g., a regional earthquake). Thus, stochastic models such as those introduced by Wickman (1966) may be applied. The simplest possible model can be elaborated through a straight forward statistical analysis of the recorded eruptive and repose periods. At Mt. Pelée all eruptions that occurred during the past 5,000 years are considered. Most repose periods are deduced from radiocarbon data (Table 2).

Wickman (1966) introduced the "survivor function":

$$F(t) = N.d(t)/No. \qquad (1)$$

where No is the total number of reposes and N.d(t) are the reposes which lasted more than time, t. The "survivor function" gives the probability that a repose period has not ended after the time (t) that has elapsed since the last eruption. Volcanoes characterized by random repose time intervals exhibit a straight survivor function line. Their probability of eruption is constant. Popocatepetl in Mexico illustrates this case (Fig. 9). Volcanoes characterized by a non-random repose time interval have a more-or-less complicated survivor function line. This is the case of Mt. Pelée (Fig. 10), the pattern of which resembles that of volcano Bromo, Java (Wickman, 1966b, Fig. 1) or volcano Tarumai, Japan (Wickman,, 1966c, Fig. 12). Until one century has elapsed following the last magmatic eruption, the probability of Mt. Pelée to produce another eruption is constant and quite high. After this period the probability drops and remains low up to a repose time of two centuries, after which probability increases.

A Wickman diagram for Mt. Pelée with only the repose time for intervals following a nuée ardente type eruption (our present situation) particularly emphasizes on this pattern (Fig. 11). According to this diagram, the probability of a new eruption would remain constant and quite high during the first 125 years after the last event (i.e., until 2,055 A.D.; 1930 + 125). A Wickman diagram for reposes following non-nuée eruptions describes a survivor function line shape that is characteristic of volcanos with a "loading time", such as Vesuvius or Hekla (Fig. 12; see Wickman, 1966d). Examination of the eruptive history of Mt. Pelée for the past 5,000 years (cf. Fig. 3) suggests that some eruptions have been linked to each other in time. This is especially apparent for the three P3 eruptions. NRS1 could have been triggered by a

TABLE 2

Magmatic eruptions and major clusters of activity for the past 5,000 years at Mt. Pelée.

length of following repose	age B.P.	major clusters of activity	identification of eruption	age B.P.	length of following repose
	50	C/1902	1929 1902	30 50	20
270	320	NRP3	NRP3	320	270
170	490	NRP2	NRP2	490 ± 10	170
150	660	C/P1	NRP1 P1 NRC2	590 ± 50 650 ± 50 690 ± 20	100 60 40
480	1140	NRC1	NRC1	1140 ± 80	450
530	1670	C/P2	P2 NMP	1670 ± 40 1730 ± 80	530 60
340	2010	P3	$P3_3$ $P3_2$ $P3_1$	$(1980)^1$ 2010 ± 140 $(2040)^1$	250 30 30
460	2470	C/P4	P4 NAB2	2440 ± 50 2490 ± 10	400 50
380	2750	NAB1	NAB1	2750 ± 50	260
280	3130	NRS3	NRS3	3130 ± 20	380
580	3710	NRS2	NRS2	3710 ± 30	580
310	4020	C/P5	NRS1 P5	3980 ± 30 4060 ± 90	270 80
390	4410	NPM	NPM2 NPM1	4300 ± 30 4515 ± 60	240 215
200	4610	P6	P6	4610 ± 50	95
490	5100	(NMR)	(NMR)	(5100 ± 60)	490

1. The time intervals between $P3_3$, $P3_2$, and $P3_1$ are arbitrary, fixed at 30 years.

"relaxation" of the magma chamber consequent to P5. The same kind of phenomenon could account for the 1929 eruption consequent to 1902. NAB2 and NRC2 look like precursory events, respectively, to the more important P4 and P1 eruptions. We tentatively define the main clusters of activity at the volcano in Table 2.

The Wickman diagram drawn from these clusters of events, each considered as a major eruption (Fig. 13), indicates a very low likelihood of a new eruption within 150 years of the preceding event. The probability increases suddenly

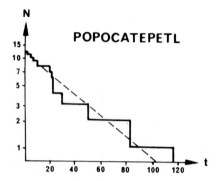

Fig. 9 Wickman diagram for Popocatepetl volcano, Mexico. N. = number of repose periods; t = length of repose in years.

after 150 years and remains moderate until 450 years. After this the probability increases once more and either remains constant at a high level or increases steadily with time. This pattern of loading time may be related to the "buffering" effect of the underlying magma chamber. Once partly emptied by an eruption, time is needed to refill the chamber to a new pre-eruptive condition. This time depends on the volume of magma extracted by eruption and the rate of deep magma supply. Following this hypothesis, and assuming that the 1929 eruption ended the last cluster of magmatic activity at Mt. Pelée, the next magmatic eruption is not likely before 2,080 A.D.

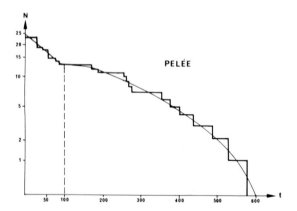

Fig. 10 Wickman diagram for the past 5,000 years of activity at Mt. Pelée based upon data of Table 2.

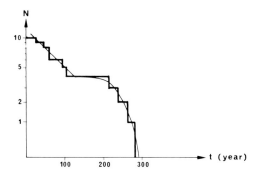

Fig. 11 Wickman diagram of Mt. Pelée considering only the repose periods following a nuée ardente type eruption (present situation). The number of reposes (ten) is at the lower limit for construction of a Wickman diagram (Wickman, 1966a).

STATISTICAL APPROACH

Probabilistic discussions are valuable only if each of the different types of eruption are time-independent. This is difficult to prove because of the large number of eruption types (four) in relation to the total number of eruptions. Another method of assessing volcanic hazards is to consider the cumulative frequency of definite repose time intervals (Fig. 14). Here the chance of a new eruption steadily increases with time because an inferred event should occur before the elapse of the longest past known repose time. Considering the present repose length (50 years), the chance that today some abnormal seismic activity beneath the volcano or appearance of fumaroles at the summit would

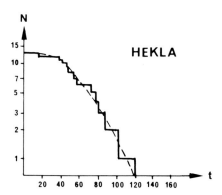

Fig. 12 Wickman diagram for Hekla, Iceland. This shows a typical loading-time survival number function.

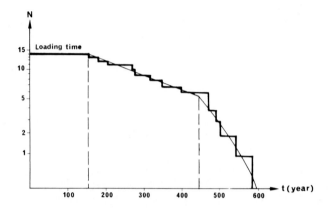

Fig. 13 Wickman diagram for the principal eruptions or clusters of eruptions at Mt. Pelée during the past 5,000 years based on data of Table 2. Compare the shape of the survival number function with that for Hekla.

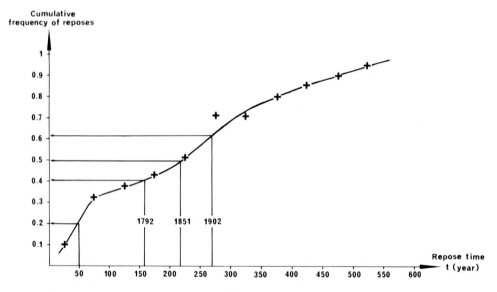

Fig. 14 Cumulative frequency for repose time at Mt. Pelée during the past 5,000 years per class of 50 years, based on data in Table 2. The length of the present-day repose, as well as the time intervals separating the historical eruptions of 1792, 1851, and 1902 from their respective preceding magmatic eruption, are outlined.

forecast an impending magmatic eruption is 20 percent. Prior to the 1902 eruption, the 1792 phreatic crisis had 40 percent chance of being a precursor to a new magmatic eruption; nothing occurred. At the time of the phreatic explosion in 1851 the chances had increased to 50 percent; but once more nothing occurred. In April 1902 the phreatic and abnormal seismic activity had more than a 60 percent chance of leading to an impending magmatic eruption; this time an eruption ensued.

CONCLUSIONS

The past 5,000 years of volcanologic history at Mt. Pelée have been tentatively reconstructed in detail. All significant magmatic eruptions have been recognized and dated. Consequent measurement of their repose-time interval permits a probabilistic and statistical approach to hazard assessment. However, a precise prediction of the time and type of a future magmatic eruption requires a conceptual model of the behavior of the volcano based on its past record. Recognition of the magmatic and dynamic patterns of the volcano is the only way to reach this goal. Such recognition requires detailed petrological and geochemical studies.

ACKNOWLEDGEMENTS

The author is indebted to W. A. Duffield, W. I. Rose, M. F. Sheridan and K. H. Wohletz for critical reading of the manuscript and helpful reviews.

REFERENCES

Bellon, H., Pelletier, B., Westercamp, D., 1974. Données géochronométriques relatives au volcanisme martiniquais, Antilles Françaises. C.R. Acad. Sc. Paris, 279: 457-460.

Cheminee, J.L., 1973. Contribution à l'étude des comportements du potassium, de l'uranium et du thorium dans l'évolution des magmas. Thèse Doct. Ett., C.N.R.S., 395 pp.

Crandell, D.R. and Mullineaux, D.R., 1967. Volcanic hazards at Mount Rainier, Washington. Geol. Survey Bull. 1238.

Crandell, D.R. and Mullineaux, D.R., 1975. Technique and rationale of volcanic hazards appraisals in the Cascade Range, N.W. U.S.A. Environmental Geol., 1: 23-32.

Crandell, D.R. and Mullineaux, D.R., 1976. Potential hazards from future eruptions of Mount St. Helens volcano, Washington. U.S. Geol. Survey open-file rept., 76-491, 25 pp.

Fisher, R.V., 1979. Models for pyroclastic surges and pyroclastic flows. J. Volcanol. Geoth. Res., 6: 305-318.

Fisher, R.V., Smith, A.L., and Roobol, M.J., 1980. Destruction of St. Pierre, Martinique by ash-cloud surges, May 8 and 20, 1902. Geology, 8: 472-476.

Fisher, R.V. and Heiken, G., 1982. Mt. Pelée, Martinique: May 8 and 20, 1902 pyroclastic flows and surges. J. Volcanol. Geoth. Res., 13: 339-372.

Gourgaud, A., 1982. Sur le déclenchement des éruptions historiques de la Montagne Pelée (Martinique) par injection d'un magma basique dans une chambre magmatique dacitique. 9ème R.A.S.T., Paris.

Grunevald, H., 1965. Géologie de la Martinique - Mémoires pour servir à l'explication de la carte géologique détaillée de la France. Imprimerie Nationale, Paris, 144 pp.

Gunn, B.H., Roobol, M.J., Smith, A.L., 1974. Petrochemistry of the Peléean-type volcanoes of Martinique. Geol. Soc. Amer. Bull., 85: 1023-1030.

Lacroix, A., 1904. La montagne Pelée et ses éruptions. Masson et Cie, Paris, 622 pp.

Perret, F.A., 1935. The eruption of Mt. Pelée. Carn. Inst. Washington, 458, 126 pp.

Romer, M.J. and Smith, A.L., 1975. A comparison of the recent eruptions of Mt. Pelée, Martinique and Soufrière, St. Vincent. Bull. Volc., 39: 1-27.

Roobol, M.J. and Smith, A.L., 1976. Mount Pelée, Martinique. A pattern of alternating eruptive styles. Geology, 4: 521-524.

Roobol, M.J. and Smith, A.L., 1980. Pumice eruptions of the Lesser Antilles. Bull. Volc. 43: 277-286.

Smith, A.L., and Roobol, M.J., 1976. Petrologic studies of Mount Pelée, Martinique. Bull. BRGM, Sect. IV, 4: 305-310.

Smith, A.L. and Roobol, M.J., 1980a. Andesitic pyroclastic flows. In: R.S. Thorpe (Editor), Orogenic andesites and related rocks. John Wiley and Sons, New York, N.Y.

Smith, A.L. and Roobol, M.J., 1980b. Late prehistoric and historic activity of Mt. Pelée, Martinique. 9th Carib. Geol. Conf. Dominican Republic, Abstr.

Stieltjes, L. and Westercamp, D., 1978. Première ébauche de zonation des risques volcaniques à la montagne Pelée. Unpubl. report, 78.ANT.08, BRGM.

Traineau, H., 1982. Contribution à l'étude géologique de la montagne Pelée, Martinique évolution de l'activité éruptive au cours de la période récente. Thèse IIIe cycle, Paris-Sud, ORSAY, 209 pp.

Traineau, H., Westercamp, D. and Coulon, C., 1982. Mélanges magmatiques à la Montagne Pelée (Martinique): origine des éruptions de type St. Vincent. 9ème R.A.S.T., Paris.

Westercamp, D., 1974. La montagne Pelée et la destruction de St. Pierre. In: Livre et guide d'excursions dans les Antilles françaises. 7th Carib. Geol. Conf. Pointe à Pitre, 149-160 pp.

Westercamp, D., 1980a. La montagne Pelée. In: Masson, (Editor), Martinique-Guadeloupe, Guides géologiques régionaux. 33-48 pp.

Westercamp, D., 1980b. Une méthode d'évaluation et de zonation des risques volcaniques à la Soufrière de Guadeloupe. Bull. Volc., 43: 431-52.

Westercamp, D., 1981. Assessment of volcanic hazards at Soufrière de Guadeloupe, FWI. Bull. BRGM, 2ième série, 2: 187-192.

Wickman, F.E., 1966a. Repose-period patterns of volcanoes I. Volcanic eruptions regarded as random phenomena. Arkiv für Mineralogi och Geologi, 4: 291-302.

Wickman, F.E., 1966b. Repose-period patterns of volcanoes, II: Eruption histories of some East Indian volcanoes. Arkiv für Mineralogi och Geologi, 4: 303-318.

Wickman, F.E., 1966c. Repose-period patterns of volcanoes, III: Eruption histories of some Japanese volcanoes. Arkiv für Mineralogi och Geologi, 4: 319-336.

Wickman, F.E., 1966d: Repose-period patterns of volcanoes, IV: Eruption histories of some selected volcanoes, Arkiv für Mineralogi och Geologi, 4: 337-352.

Wickman, F.E., Repose-period patterns of volcanoes, V: General discussion and a tentative stochastic model. Arkiv für Mineralogi och Geologi, 4: 353-366.

Wright, J.V., Smith, A.L., and Self, S., 1980. A working terminology of pyroclastic deposits. J. Volcanol. Geotherm. Res., 8: 315-336.

APPLICATION OF COMPUTER-ASSISTED MAPPING TO VOLCANIC HAZARD EVALUATION OF SURGE ERUPTIONS: VULCANO, LIPARI, AND VESUVIUS

MICHAEL F. SHERIDAN AND MICHAEL C. MALIN

Department of Geology, Arizona State University, Tempe, AZ 85287 (U.S.A.)

(Received October 21, 1982; revised and accepted March 18, 1983)

ABSTRACT

Sheridan, M.F. and Malin, M.C., 1983. Application of computer-assisted mapping to volcanic hazard evaluation of surge eruptions: Vulcano, Lipari, and Vesuvius. In: M.F. Sheridan and F. Barberi (Editors), Explosive Volcanism. J. Volcanol. Geotherm. Res., 17: 187—202.

A previously developed computer-assisted model has been applied to several pyroclastic-surge eruptions at three active volcanoes in Italy. Model hazard maps created for various vent locations, eruption types, and mass production rates reasonably reproduced pyroclastic-surge deposits from several recent eruptions on Vulcano, Lipari, and Vesuvius. Small-scale phreatic eruptions on the island of Vulcano (e.g. the 1727 explosion of Forgia Vecchia) pose a limited but serious threat to the village of Porto. The most dangerous zone affected by this type of eruption follows a NNW fissure system between Fossa and Vulcanello. Moderate-sized eruptions on Vulcano, such as those associated with the present Fossa Crater are a much more serious threat to Porto as well as the entire area within the caldera surrounding the cone. The less frequent surge eruptions on Lipari have been even more violent. The extreme mobility of surges like those produced from Monte Guardia (approx. 20,000 y.b.p.) and Monte Pilato would not only threaten the entire island of Lipari, but also the northern part of neighboring Vulcano. Eruptions at Vesuvius with energy and efficiency similar to that of the May 18, 1980 blast of Mount St. Helens would be still more destructive because of the great initial elevation of the summit vent. In addition, surge eruptions at Vesuvius are generally part of more complex eruption cycles that involve several other types of volcanic phenomena including Plinian fall and pyroclastic flows.

INTRODUCTION

Recognition of areas in Italy that are likely to be affected by various types of volcanic activity is of increasing importance because of population growth in the vicinity of many potentially active volcanoes. Pyroclastic surges are possibly the most dangerous of common eruption phenomena because of the large areas subjected to extreme blast velocities and high temperatures. Complete destruction of life and property may occur even in places where subsequent deposit thickness is measured only in centimeters (see Moore and Sisson, 1981).

Hydrovolcanic eruptions (Sheridan and Wohletz, 1981; this volume) commonly produce pyroclastic surges. However, deposits from such eruptions are not always easy to recognize, especially in their distal reaches. Detailed geologic mapping to determine the limits of surge emplacement is time-consuming and not always successful. Because the boundary for surge destruction sets an outer limit of the danger zone for many volcanic areas, an appropriate method for estimating the distribution of products from such eruptions would be extremely valuable.

We have developed a technique that uses a simple model of the energy relationships between a violent eruption, mobility of its pyroclastic flows, and the existing topography to create a nomograph that permits extremely rapid spatial and temporal analyses of its eventual deposits. A nomograph is a graphical representation of a function that permits rapid calculations of a desired result from specific observations. We have extended the technique into multi-dimensions by using interactive computer graphics applied to an initially simple relationship. Employing a model of linear energy dissipation with distance from the vent consequent on existing topography, our first-order application (Malin and Sheridan, 1982) produced a three-dimensional nomograph that we call an "energy cone model" or ECM. The "cone", seen as a nomograph, has no physical or real manifestation; rather it is simply a graphical representation of an analytical result. Because these models can be constructed in very short times they could prove useful during actual volcanic eruptions.

The accuracy of such computer models depends on the quality of field information, topographic data, and theoretical assumptions used to construct the nomograms. Even with little or no prior geologic data, the energy cone could produce a preliminary hazard map, given reasonable estimates of the theoretical energy relationships and models of the types of eruptive phenomena expected. The "energy cone" model, simple though it may be, has yielded a configuration for surge emplacement that closely mimics both the limits of destruction and resulting deposit thickness in several test cases.

Those who interpret simple ECM nomographs should keep in mind that the distribution of actual surge deposits may not conform in all directions to the perimeter and isopachs created by the model. This is because ground-hugging surges accelerate down steep gradients and have a variable tendency to remain within channels bounded by topographic relief lower than Δh. Depending on the relationship of the energy cone and the topographic surface, a surge may flow around a hill or ridge even though it has the potential to pass over it. This effect is similar to the movement of flood or lahar runoff after a heavy rain. An example of such preferred directional movement is the blast of May 18th at Mount St. Helens which spread to the north, northeast, and northwest due to the influence of the north-facing slope of the cone, the valley of Spirit Lake, and the drainage of the North Fork of the Toutle River.

Low amplitude topographic elements in the preferred sectors of flowage do not appreciably change the velocity or termination of the surge predicted by the model. However, topography has a profound effect on the total mass distribution of the products. Because of the importance of topography in restricting surge distribution, future refinements of the ECM for hazard evaluation should incorporate this effect. But for the present, the gross distribution of products within the total potential field must be qualitatively estimated by the operator.

Example hazard maps have been prepared for Vulcano, Lipari, and Vesuvius using our simple model. The diverse variety of pyroclastic-surge eruptions from these centers reflect a wide range in eruptive energy, several levels of surge emplacement efficiencies, and variable degrees of constraints by pre-eruption topography. Variations in venting and emplacement of surge deposits can be rapidly estimated using maps developed by these techniques. This method can also be used as an interactive field tool. A geologist can thus generate eruption models at any stage of the field work in order to optimize the efficiency of locating and interpreting critical exposures.

THE MODEL

Our energy cone model (ECM) uses a simple three-dimensional relationship based on the energy line concept of Heim (1882; 1932). An ECM of the May 18, 1980 "blast" deposits of Mount St. Helens (Malin and Sheridan, 1982) has qualitatively reproduced the sinuous outline of the devastated area and deposit thickness in the region north of the summit.

The principal assumption of the ECM is that the erupted material behaves as a cohesionless, gravity-driven suspension of particles and hot gas. The dispersal of material is assumed to have distributional aspects similar to large landslides as modeled by Heim (1882; 1932) and Hsü (1975; 1978). The potential energy of material erupted to some height above the volcano is converted to kinetic energy of mass flowage as the material moves laterally away from the vent. The efficiency of the system can be modeled as an energy loss that defines an energy line in two dimensions. The slope of the energy line may be empirically determined using the zero-vertical-velocity height of the particulate mass and the longitudinal distance of runout. Alternatively, theoretical values based on surge rheology could be chosen.

Although the models presented in this paper deal only with pyroclastic surge events, other types of eruptions could be similarly examined. The method can be adapted to model any type of gravity-driven flow, provided that appropriate empirical constraints are chosen. Our ECM is presently being expanded to include Plinian- and Strombolian-fall phenomena as well as confined flows such as lavas and lahars. Although the hazard from these latter phenomena is low on Vulcano and Lipari, they all could present a serious threat to areas surrounding Vesuvius (Rosi et al., 1980-81; Barberi et al., 1982).

Input for our implimentation of the ECM includes: the cone apex location, the cone apex elevation, the cone central depression angle, and a topographic data set. The computer then calculates Δh for each point in the data and displays the results on a television monitor. All operations are performed on image arrays.

The apex (vent) location is critical because on the flank of volcanic cones like Vesuvius and Mount St. Helens it can have the apparent effect of "directing" the resultant surge down the slope of greatest gradient. In many cases the cone apex location is adequately given by an active crater, as for the Fossa vent of Vulcano. For others, structural or geophysical data constrain the apex location. For past eruptions, the geometry of the deposits can be used to find the apex location by backfitting of computer maps to fit isopach lines.

The central angle of depression of the cone is assumed to describe an appropriate energy line for the eruption in question. The angles used in this paper were based on constraints of field data. The distal limits of the deposit give one or more fixed points for the establishment of the energy line slope. The limit of highest deposit on medial mountain ranges that were not overtopped provides more data points. The height of the eruption plume (either observed or theoretical) provides yet another fix on the slope.

The elevation of the energy cone apex above the vent is a function of the energy of the surge eruption. For historic eruptions, like Mount St. Helens, this value may be determined from actual observations. For older eruptions the apex height and slope of the energy cone may be fit to the geologic data by trial and error. The cone apex height above ground level is in the range of 350 m to 850 m for surge eruptions that we have modeled for Italian volcanoes. The Mount St. Helens surge had a height 1065 m (Malin and Sheridan, 1982), although the actual ground level changed drastically during this eruption. Phreatic eruptions may have a much lower apex height of about 100 m, as was the case at the 1976 eruption of La Soufriere of Guadeloupe (Sheridan, 1980).

Information from observed historic eruptions provides the best empirical data for the establishment of ranges of cone slopes for various classes of eruptions. At the present very few data of this type has been compiled. For our computations successive models with various apex elevations and slope angles were run to determine the best visual fit of the calculated deposit to the known outcrops. Values for cone slopes of surge eruptions range from 11° on Vulcano to 4° on Lipari. Phreatic explosions have a much steeper slope of 20° on Vulcano to 27° on La Soufriere (Sheridan, 1980).

Presumably, theoretical arguments or comparison with similar examples could also be used in the definition of the energy cone if an insufficient number of actual sections had been observed. For example, dry surges with abundant superheated steam are more mobile (i.e. have a lower energy slope) than wet surges in which the grains become cohesive as water condenses on their surfaces. It is also possible that the distributional efficiency of a surge could change with runout distance causing a corresponding modification in the slope of the energy line. In such cases the energy line could be modeled as a variable, non-linear function.

In backfitting the model to the geologic data, its sensitivity appears to be about 2° in slope and about 200m in height. However, even with these errors, eruption parameters appear to fall into distinct groups (Fig.1). Phreatic eruptions have low heights and steep slopes reflecting their weak eruptive energy and low mobility. Surge eruptions have moderate slopes and moderate heights, with the exception of highly mobile (Lipari, Monte Guardia) or highly energetic (Mount St. Helens) cases. An apparently wider range in both eruptive energies and slope angles for pyroclastic flows presented by Sheridan (1979) may reflect either a significant difference in energy relationships, or more likely, the more incomplete nature of the data used to calculate those energies and angles.

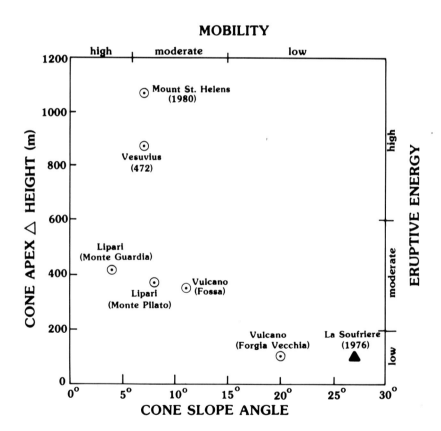

Fig. 1 - Plot of cone angle and apex height for modeled surge eruptions. Error estimate of 25% on slope angle and 25% on height is based on visual backfitting of eruption models to fit the studied deposits.

The quality of available data sets present the main constraints on the usefulness of the model. For carefully observed eruptions, such as the 1980 "blast" of Mount St. Helens, the boundary of the deposit was recorded on aerial photographs and by field mapping immediately after the event (Kieffer, 1981; Hoblett et al. 1981). The thickness was determined in a sufficient number of locations to construct an accurate isopach map (Moore and Sisson, 1981; Waitt, 1981). Thickness calculations have been subsequently determined for a large number of additional sections. Therefore the ECM for this eruption is tightly constrained and mimics the distribution of products quite well (Malin and Sheridan, 1982).

For older eruptions, such as those modeled in this paper, the field data are not as abundant and easily interpreted. The actual limits of the devastation can never be known for many of these eruptions. In general, the deposits can only be traced laterally until their thickness becomes less than some limit of perception, usually a few centimeters to tens of centimeters. Other factors that reduce the quality of available geologic data include: local removal by erosion, reworking by wind or water, hydrothermal alteration, cover by younger deposits, and limited outcrop exposure due to vegetation or other factors. A reasonable amount of field work might uncover 30 to 60 useful stratigraphic sections for deposits that are hundreds to thousands of years old. However, even this rather limited data base allows an approximate reconstruction of the deposit using a trial and error method to fit an energy cone model to the observations.

Topographic data for older events is likewise not as easily obtained. For example, new volcanic constructs, tectonic faults, or caldera collapses may have drastically altered the topography that constrained the deposit in question. Corrections may thus be needed to reconstruct the original topography at the time of the eruption. Therefore the most recent deposits provide the best test for this model.

Various kinematic properties, including acceleration, velocity, and distance at various times, can be calculated given the configuration of the topographic surface and the energy cone (Hsü, 1978; Malin and Sheridan, 1982). The most critical factor in the model for pyroclastic surges is the vertical distance (Δh) between the ground surface and the energy cone. This factor limits the maximum velocity (v_{max}) according to eq. 1.

$$v_{max} = (2*g*\Delta h)^{1/2} \tag{1}$$

The ultimate limit of destruction, as estimated from the thickness of past deposits, is the principal concern for volcanic risk evaluation. The limit of destruction in the ECM is given by the intersection of the energy cone with the topographic surface (i.e. where $\Delta h=0$). This boundary would be a circle for the trivial case of a perfectly flat surface. However, the outline of the area of destruction would have an irregular shape in regions of significant topographic relief. Areas with little or no surge impact could be located adjacent to areas of great devestation, depending on the elevation of each with respect to the energy cone surface (Δh).

An attractive feature of the ECM for volcanic risk evaluation is its real-time, interactive implimentation. Provided that digital topographic data is stored within the computer, an entirely new model can be generated in a few minutes at most. Thus, at the time of volcanic crisis on an active volcano, several eruption types can be generated for one or more sites using available or changing geophysical, geochemical, and geologic data. If the potential hazard zones were to suddenly move due to the migration of a seismic swarm between the summit and the flank, new models could be rapidly produced to indicate the changing areas of most probable hazard.

THE VOLCANOES

The volcanic eruptions chosen for this study come from a spectrum of activity types that probably represent the complete range of features to be expected from hydrovolcanic explosions at the active volcanoes of Italy. They include phreatic explosions and small surge eruptions at Vulcano, moderate hydrovolcanic surge blasts at Lipari, and cataclysmic surge eruptions associated with Vesuvius.

Vulcano belongs to the Aeolian Island Arc located in the Tyrrhenian Sea, north of Sicily. The general geology of Vulcano, described by Keller (1980), consists of four volcanic centers: Old Vulcano, Lentia, Fossa, and Vulcanello (Fig. 2). The two active vents, Vulcanello and Fossa, are quite different in character. Vulcanello, which appeared above sea level in 183 B.C., has not produced widespread pyroclastic deposits. Its explosive products are limited to a group of three small tuff cones near the central vent. The Fossa of Vulcano, which began to construct its cone between 11,400 and 8,300 y.b.p., has erupted a variety of products that represent a wide spectrum of hydrovolcanic phenomena (Frazzetta et al., this volume). The principal hazard to the village of Porto appears to be related to either phreatic eruptions along the tectonic line extending between Fossa and Vulcanello, or major surge eruptions from a vent on Fossa.

"Dry" surge explosions from the active Fossa crater are a common phenomena on Vulcano. Products of this type of event form a thick blanket within the caldera surrounding the Fossa, but they decrease in thickness on the cone as one approaches the vent (Frazzetta et al., this volume). Cross-stratified deposits of this type are plastered onto the caldera wall at several localities. Pyroclasts from these beds exhibit surface textures typical of base-surge grains (Sheridan and Marshall, 1981). A computer-generated map (Fig. 3A) fits the field distribution of products from this type of eruption quite well. Such a surge eruption would devastate the present town of Porto beneath its maximum deposit thickness. However, villas on the Vulcanello cone and farms on the high terrace of Piano would be relatively safe.

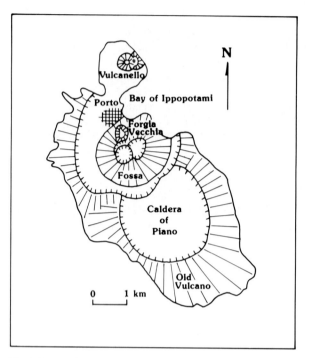

Fig. 2 - Main geologic features of the island of Vulcano (modified from Keller, 1980).

The second type of volcanic event that presents a threat to Porto is a phreatic eruption along the tectonic line extending between Fossa and Vulcanello. Eruptions of this type produced the Forgia Vecchia craters in the active period between 1731 and 1739 (Frazzetta et al., this volume). Deposits of Forgia Vecchia type were modeled from a vent at the location of the present crater on the northern flank of Fossa (Fig. 3B). This model produced a fan-shaped distribution pattern directed toward the town of Porto because of the topographic effect of the Fossa cone. The cone apex located 100 m above ground level and a 20° slope angle best simulates a deposit that corresponds to the actual beds.

Eruptions of the above type could occur anywhere along the tectonic zone between Fossa and Vulcanello. A second possible vent location was chosen at the bay of the Ippopotomi north of Porto, because this is the site of strong fumarolic activity and thermal springs. Eruption parameters identical to those at Forgia Vecchia produce a more limited distribution of products at this site because of the flat topography rather than the strong influence of a sloping surface (Fig. 3B). Phreatic eruptions present a significant threat to Porto, but such events on the flank of the Fossa cone are more dangerous than similar

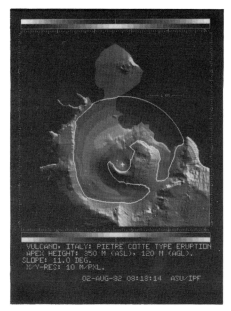 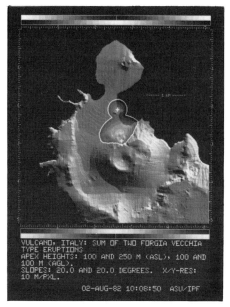

Fig. 3 - A. Computer-generated map of a typical surge eruption from the Fossa crater of Vulcano. B. Computer-generated map of identical phreatic eruptions centered at Forgia Vecchia on the Fossa cone and at Bay of the Ippopotami, Vulcano.

eruptions near sea level.

Lipari, located about 15 km to the north of Vulcano, is likewise a member of the Aeolian Arc. The general geology of this island, described by Pichler (1980), consists of an early stage of andesite cone-building followed by a late stage of rhyolitic activity (Fig. 4). Although the most recent eruption on Lipari was in the 6th century A.D. (Keller, 1970; Bigazzi and Bonadonna, 1973), at least two strong surge eruptions have occurred in the past 20,000 years (Crisci et al., 1981; 1982). Powerful eruptions such as these could be expected if this volcano becomes active in the future.

Although eruptions on Lipari are infrequent, they have been much more powerful that those on the neighboring Vulcano. The model constructed for the Monte Guardia eruption (approx. 20,000 y.b.p.) was designed to correspond to the data of Crisci et al. (1982). This eruption produced a thick, sub-plinian, pumice-fall deposit and widespread base-surge beds. The model that best fits the distribution of products has a modest cone apex height (420 m above ground level) and a very shallow energy line slope of 4° (Fig. 5A). This eruption devastated most of the island with only the peaks of Monte St. Angelo and Monte Chirica projecting above the blast. Even more surprising is the prediction that deposits from this eruption should be present on the island of Vulcano, a few kilometers to the south. In fact, surge deposits of an unknown source that could correlate with this event on Lipari are present on high surfaces surrounding the caldera walls of the Fossa on Volcano.

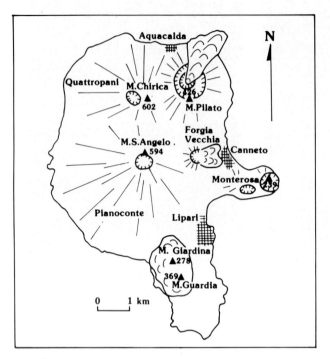

Fig. 4 - Main geologic features of the island of Lipari (modified from Pichler, 1980).

A second large base-surge event on Lipari was modeled assuming a vent in the vicinity of Monte Pilato. The source and distribution of this deposit are poorly known, at present but current research on these products should further constrain the model. A computer-drawn map (Fig. 5B) was constructed assuming that the vent was Monte Pilato and that the deposits are found in the swale between Monte St. Angelo and Monte Chirica, but not on these peaks. This model has a cone apex of moderate elevation (375 m above ground level) and a slope of 8°. Because surges from this eruption were less mobile than those from the Monte Guardia event, they would spare the southern part of Lipari and not effect Vulcano.

Vesuvius is a large stratovolcano located east of the Bay of Naples in south-central Italy. The oldest Plinian deposit of Vesuvius, dated at 25,000 years (Delibrias et al., 1979), post-date the Campanian ignimbrite, dated at 35,000 years (Barberi et al., 1979). The type of volcanism ranged from persistent activity that built scoria cones and lava flows to cataclysmic eruptions that produced Plinian falls, pyroclastic flows, pyroclastic surges, and lahars (Sheridan et al., 1980). Although all of these phenomena are real hazards for the inhabitants of the surrounding cities, the pyroclastic surges modeled by the present study present the greatest danger to life and property.

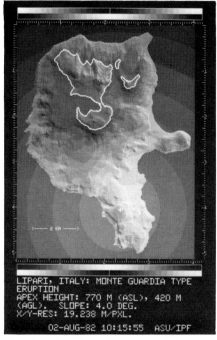 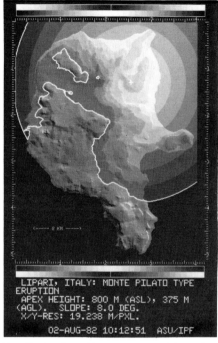

Fig. 5 - A. Computer-generated map of the Monte Guardia surge eruption at Lipari. B. Computer-generated map of the Monte Pilato surge eruption at Lipari.

Somma-Vesuvius has produced several cataclysmic events during the past 25,000 years (Delibrias et al., 1979). The most recent of these that involved widespread pyroclastic surges was the A.D. 472 Pollena event (Rosi and Santacroce, this volume). Although the Pollena deposits (Fig. 6) are covered by younger lava flows that spilled over the southern rim of the caldera, their thickness is fairly well known to the north. These deposits reach a maximum thickness near the towns of Pollena and San Sebastiano and isopachs apparently wrap around the Somma cone to the north.

The Pollena deposit lacks several key features that we have found useful to constrain the ECM for other volcanoes. (1) The boundary of devastation is not known, a nearly universal feature of prehistoric eruptions. (2) The topographic relief below the main Somma-Vesuvius cone is minimal so that no hills or deep valleys provided information on the slope of the energy surface. (3) Finally, a large portion of the original area of this deposit is covered by later lava flows so that the entire distribution is unknown. Despite these disadvantages, the distinctive pattern of deposit thickness (Fig. 6) and the recent archeological discovery of Pollena-age ash above volcanically destroyed buildings in an archelogical site in Naples provided sufficient controls to backfit an ECM to this eruption.

A computer-generated map was created to provide a best fit to the deposits

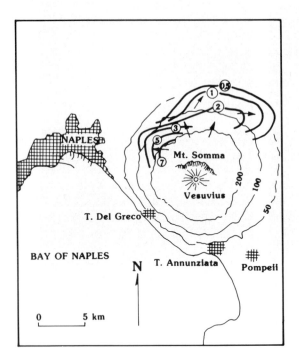

Fig. 6 - Location and isopach map for the A.D. 472 Pollena deposits at Vesuvius (modified from Rosi and Santacroce, this volume).

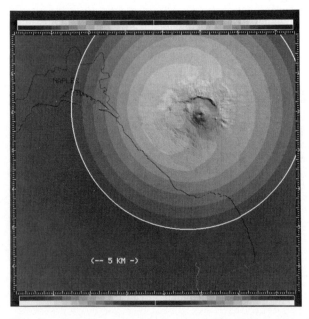

Fig. 7 - Computer-generated map of the Pollena eruption at Vesuvius.

assuming the limit of devastation to be in Naples and the vent to be at the western rim of the present caldera (Fig. 7). Density slices of the model for Δh were chosen to best fit the known distribution of the products. The model assumes a cone apex at 850 m above ground level and an energy line slope of 7.5°, thus the cone parameters are similar to those obtained for the 18 May, 1980 Mount St. Helens eruption.

Although this eruption was the smallest known Plinian event at Vesuvius, the accompanying surge was sufficiently strong to devastate the area occupied by most of the current towns at the base of the volcano. Even Naples was partially destroyed by this eruption. Pyroclastic surges probably accompany the smaller, but more frequent, eruptions that follow long repose periods, such as the eruption of 1631. The effect of these smaller surge events is difficult to predict. However, renewed activity with major surge blasts would certainly present a serious threat to the approximately 3,000,000 people who live in the Neapolitan urban area.

DISCUSSION

Published volcanic hazard maps for surge or surge-like eruptions display risk zones as circles centered at the principal vent (Walker, 1974; Westercamp, 1980; Barberi et al., 1982). Although the method is rapid and based on geological data, this technique can be quite inaccurate, as for the case of the May 18, 1980 eruption of Mount St. Helens. The method is purely geometrical and lacks a physical model as its basis. An obvious disadvantage of this type of representation is its inability to evaluate the effect of topography in restricting or augmenting surge advance. For example, compare the distribution of products from two identical phreatic eruptions at Vulcano, the Forgia Vecchia vent on the flank of the Fossa cone and the other on the plain near Porto (Fig. 3B). Likewise note that the considerably longer runout of the Mount St. Helens blast (Hoblitt et al., 1981; Kieffer, 1981; Moore and Sisson, 1981), compared with the A.D. 472 Pollena deposits of Vesuvius (Rosi and Santacroce, this volume), is due to its elevation on a higher stratovolcano.

The ECM overcomes the above problems and has several additional advantages. The deposit thickness and limit of extent are both shown. The effect of topography is an intrinsic part of the model and is easily represented. The model is based on a physical concept of surge emplacement so that physical parameters such as flow velocity could easily become model input or output. The display is on a realistic shaded relief base. So far the ECM has identified several families of common surge eruption types that could be easily applied to hazard maps of other volcanoes.

In addition to defining zones of potential risk preceeding an eruption, the ECM is a useful real-time predictive tool. After a vent has been located, either by geophysical evidence or the eruption of material, zones of possible blast effects can be predicted based on assumptions of eruptive strength and surge mobility. The effects of changes in vent location, efficiency of topographic bariers, and channeling by major valley systems are easily evaluated.

ACKNOWLEDGEMENTS

This paper has resulted from a cooperative research effort between researchers at the University of Pisa and Arizona State University. Many people have supplied geologic and topographic data as well as provided logistic support and cooperation in the development of the hazard maps. Franco Barberi has given his support and encouragement for the entire project and was essential to its success. Lilo Villari provided data and logistic support for field work of Tom Moyer and Ken Wohletz on the Aeolian Islands. Roberto Santacroce and Mauro Rosi generously donated their data on Vesuvius. Luigi La Volpe and Giovanni Frazzetta provided geological and topographic data for the Fossa cone of Vulcano and commented on early models. Rosanna de Rosa and Gian Zuffa helped to assemble the geologic data for Lipari. The computer graphic work was carried out at the Arizona State University Image Processing Facility, supported by the National Aeronautics and Space Administration. This work was partially supported by the Instituto Internazionale Volcanologia of Catani, NASA grant NAGW-245, and NFS grant INT 7823984.

REFERENCES

Barberi, F., Innocenti, F., Ferrara, G., Keller, J., and Villari, L., 1974. Evolution of Aeolian Arc volcanism (southern Tyrrhenian Sea). Earth Planet Sci. Lett., 21: 269-276.
Barberi, F., Rosi, M., Santacroce, R., and Sheridan, M.F., 1982. Volcanic hazard zonation: Mt. Vesuvius. In: H. Tazieff (Editor), Volcanic Hazards. Elsevier Sci. Publications, Amsterdam.
Bigazzi, G. and Bonadonna, F., 1973. Fission track dating of the obsidian of Lipari Island (Italy). Nature, 242: 322-323.
Crisci, G.M., Delibrias, G., De Rosa, R., Lanzafame, G., Mazzuoli, R., Sheridan, M.F., and Zuffa, G.G., 1981. Pyroclastic deposits of the Monte Guardia eruption cycle on Lipari (Aeolian arc, Italy). Int. Assoc. Sediment., Bologna.
Crisci, G.M., De Rosa, R., Lanzafame, G., Mazzuoli, R., Sheridan, M.F., and Zuffa, G.G., 1982. Monte Guardia deposits: a Late-Pleistocene eruptive cycle on Lipari (Italy). Bull. Volcanol., 44(3): 241-255.
Delibrias, G., Di Paolo, G.M., Rosi, M., and Santacroce, R., 1979. La storia eruttiva del complesso vulcanico Somma-Vesuvio ricostruita dalle successioni del Monte Somma. Rend. Soc. It. Min. Petr., 35 (1): 411-438.
Frazzetta, G., La Volpe, L., and Sheridan, M.F., 1983. Evolution of the Fossa cone, Vulcano. In: M.F. Sheridan and F. Barberi (Editors), Explosive Volcanism. J. Volcanol. Geotherm. Res., 17: (this volume).

Hoblitt, R.P., Miller, C.D., and Vallance, J.W., 1981. Origin and stratigraphy of the deposit produced by the May 18 directed blast. In: P.W. Lipman and D.W. Mullineaux (Editors), The 1980 eruptions of Mount St. Helens, Washington. U.S. Geol. Survey Prof. Paper 1250: 401-419.

Keller, J., 1970. Die historischen Eruptionen von Vulcano und Lipari. Deutsch. Zeitsch. Geol. Geoph., 121: 179-185.

Keller, J., 1980. The island of Vulcano. In: L. Villari (Editor), The Aeolian Islands. Ist. Inter. Vulcanol., Catania, pp. 29-75.

Kieffer, S.W., 1981. Fluid dynamics of the May 18 blast at Mount St. Helens. In: P.W. Lipman and D.W. Mullineaux (Editors), the 1980 eruptions of Mount St. Helens, Washington. U.S. Geol. Survey Prof. Paper 1250: 379-400.

Heim, A., 1932. Der Bergsturz von Elm. Z. Dtsch. Geol. Ges. 34: 74-115.

Heim, A., 1932. Bergstruz und Menschenleben. Fritz und Wasmuth Zurich, 218 pp.

Hsü, K.J., 1975. On sturzstroms - catastrophic debris streams generated by rockfalls. Geol. Soc. America Bull. 86: 129-140.

Hsü, K.J. 1978. In: Rockslides and avalanches, vol. 1, Natural Phenomena, B. Voight (Editor). Elsevier, N.Y., pp. 71-93.

Malin, M.C. and Sheridan, M.F., 1982. Computer assisted mapping of pyroclastic surges. Science, 217: 637-639.

Moore, J.G. and Sisson, T.W., 1981. Deposits and effects of the May 18 pyroclastic surge. In: P.W. Lipman and D.W. Mullineaux (Editors), The 1980 eruptions of Mount St. Helens, Washington. U.S. Geol. Survey Prof. Paper 1250: 421-438.

Pichler, H., 1980. The island of Lipari. In: L. Villari (Editor), 1st. Inter. Vulcanol., Catania, pp. 75-101.

Rosi, M., Santacroce, R., and Sheridan, M., 1980-81. Volcanic hazards of Vesuvius (Italy). B.R.G.M. Bull. (France), 4(2): 169-179.

Rosi, M. and Santacroce, R., 1983. The A.D. 472 "Pollena" eruption: Volcanologic and petrologic data for this poorly-known, Plinian-type event at Vesuvius. In: M.F. Sheridan and F. Barberi (Editors), Explosive Volcanism. J. Volcanol. Geotherm. Res., 17: (this volume).

Sheridan, M.F., 1979. Emplacement of Pyroclastic flows: a review. In: C.E. Chapin and W.E. Elston (Editors), Ash-flow tuffs. Geol. Soc. America Spec. Pap., 180: 125-136.

Sheridan, M.F., 1980. Pyroclastic block flow from the September, 1976 eruption of La Soufriere volcano, Guadeloupe. Bull. Volcanol., 43: 397-402.

Sheridan, M.F. and Marshall, J.R., 1981. SEM textural analysis of volcanic ejecta: an interpretation of base-surge behavior on Vulcano. Int. Assoc. Sedimentologists, 2nd Eur. Mtg., Bologna, p. 185-188.

Sheridan, M.F. and Wohletz, K.H., 1981. Hydromagmatic explosions: the systematics of water-pyroclast equilibration. Science, 212: 1387-1389.

Sheridan, M.F. and Wohletz, K.H., 1983. Explosive hydrovolcanism: basic considerations and review. In: M.F. Sheridan and F. Barberi (Editors), Explosive Volcanism. J. Volcanol. Geotherm. Res., 17: (this volume).

Sheridan, M.F., Barberi, F., Santacroce, R., and Rosi, M., 1981. A model for the Plinian eruptions of Vesuvius. Nature, 289: 282-285.
Waitt, R.B., 1981. Devastating pyroclastic density flow and attendant air fall of May 18 - stratigraphy and sedimentology of the deposits. In: P.W. Lipman and D.R. Mullineaux (Editors), The 1980 eruptions of Mount St. Helens, Washington. U.S. Geol. Survey Prof Paper, 1250: 439-458.
Walker, G.P.L., 1974. Volcanic hazards and the prediction of volcanic eruptions. Geol. Soc. London, Misc., 3:23-41.
Westercamp, F., 1980. Une methode d'evaluation et de zonation des risques volcaniques a la Soufriere de Guadeloupe. Bull. Volcanol., 43: 431-452.
Wohletz, K.H. and Sheridan, M.F., 1979. A model for pyroclastic surge. In: C.E. Chapin and W.E. Elston (Editors), Ash Flow Tuffs. Geol. Soc. America Spec. Paper, 180: 177-194.

GEOLOGIC, VOLCANOLOGIC, AND TECTONIC SETTING OF THE VICO-CIMINO AREA, ITALY

FEDERICO SOLLEVANTI

Agip S.p.A., Geothermal Exploration Department, S. Donato Milanese, Milano (Italy)

(Received September 16, 1982; revised and accepted March 8, 1983)

ABSTRACT

Sollevanti, F., 1983. Geologic, volcanologic, and tectonic setting of the Vico-Cimino area, Italy. In: M.F. Sheridan and F. Barberi (Editors), Explosive Volcanism. J. Volcanol. Geotherm. Res., 17: 203-217.

Although the Cimino and Vico volcanic complexes are located close to each other in a structural low, they belong to two different provinces: Cimino is connected to the Tuscan-Roman anatectic magmatic province and Vico to the Roman-Campanian potassic alkaline province (Barberi et al., 1971; Marinelli, 1975). The two volcanic centers lie on a basement formed by clays and sands of a Miocene-Pliocene aged sedimentary cycle.

The Mount Cimino complex forms a central volcano composed of a series of rhyodacitic domes. It was emplaced along a NW-SE trending fracture zone which can be followed for about 16 km. Cimino volcano was active from 1.4 m.y. to 0.95 m.y. ago. Eruption of rhyodacitic ignimbrites were preceded and followed by the emplacement of domes of the same composition. Fluid lavas of latitic and olivine-latitic composition were emitted from the central volcano during the final phase of activity.

Vico is a central volcano, mainly composed of lava, with a composite summit caldera. It was emplaced along a NE-SW trending fracture system. The Vico stratovolcano is formed by lavas of various compositions including phonolitic tephrites, tephritic phonolites and slightly undersaturated trachytes. The construction of the stratovolcano, the most ancient exposed products of which have an age of 0.4 m.y., was followed by the eruption of four pyroclastic flows totalling several cubic kilometers, which are dated at 0.3-0.15 m.y. The collapse of the caldera is most likely connected to the eruption of these pyroclastic flows. The ring fractures of the caldera are associated with the emplacement of a pyroclastic unit, most of which erupted along the inner northeast and north rim of the caldera. Post-caldera activity is characterized by the emission of tephritic-phonolitic and tephritic products that form the intra-caldera stratovolcano named Mount Venere.

0377-0273/83/$03.00   © 1983 Elsevier Science Publishers B.V.

The Vico-Cimino area (northern Latium), located between the Apennine chain and the Tyrrhenian sea, lies along the circum-Tyrrhenian belt (that extends from northern Tuscany to Campania). This zone has experienced post-orogenic tensional movements. Since the Middle Miocene, the teconic evolution of the Tyrrhenian basin is closely associated with acid anatectic and potassic alkaline volcanism along the Tyrrhenian margin of the Apennines. The age, composition, and geographic position of the volcanoes are controlled by different magnitudes of tensional tectonics and crustal thinning processes of the western margin of the Apennine chain.

GEOLOGIC SETTING

Morphology

The Vico-Cimino area (northern Latium) surrounds Lake Vico and geographically corresponds to the southward continuation of the Siena-Radicofani and Paglia-Val di Chiana-Tevere grabens (Locardi and Molin, 1975; Locardi et al., 1977), which are filled by a considerable thickness of Pliocene sediments (Fig. 1). Pre-Pliocene flysch sediments crop out in the western and southern sectors forming gentle mountain slopes which gradually link to those formed by volcanic products. Small outcrops of flysch in the extreme north of the area are abruptly separated from the volcanic products by two rivers that follow tectonic features (Vezza and Acqua Rossa rivers). Large zones with recent to current travertine accumulations that bound the volcanic area (Fig. 1), correspond to structural highs (Viterbo plain and Orte high).

Pliocene-Quaternary sediments, mainly consisting of clays, sands and gravels, crop out at the edges of the area along the Tiber and Vezza Rivers. Within the area are small outcrops of Pliocene clays mostly associated with some of the main Cimino domes.

Clay and flysch exposures mostly occur at the edges of a large ash-flow plateau formed by products of the Cimino and Vico volcanic complexes. Lava flows characterize the main positive volcanic landforms. Only some lavas reach the ash-flow plateau to modify its morphology. The induration of the "Cimino ignimbrite", which covers older Pliocene and Quaternary sediments, protects the substratum against erosion. Severe erosion mainly occurs along the valleys of the Tiber River and Vezza stream.

Most streams inside the area dissect Vico volcanic products which cover the large plateau of the Cimino ignimbrite and abut the southern flank of the Cimino volcano. Cimino and Vico volcanoes affect drainage directions in different ways. Vico volcano is characterized by a radial drainage network, except for the northeastern flank where the distribution and orientation of the streams are partly controlled by Cimino volcano. The Cimino drainage system has a radial trend with reference to the central volcano as well as modifications due to the predominant trend of the domes. Two tectonic trends (NW-SE and NE-SW) affect the drainage system along the eastern and southeastern flanks of Vico volcano as well as Cimino.

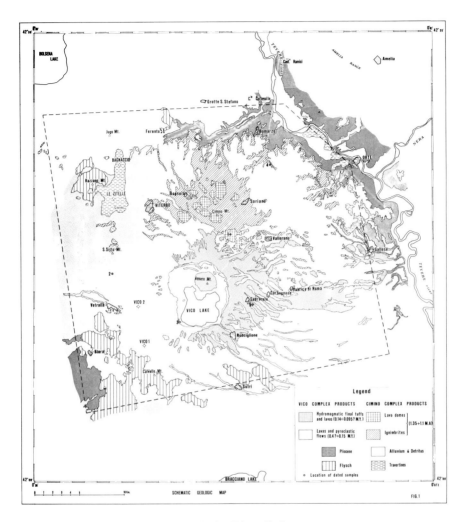

Fig. 1 Schematic geologic map of the Vico-Cimino area.

Geology

The area occupied by Vico and Cimino volcanoes corresponds to a graben zone approximately 25 km in width. The western edge of the graben is defined by the structural high of Mt. Razzano (Fig. 1) which consists of Eocene-Oligocene flysch sediments. The Mt. Razzano high can be followed using both surface and gravimetric data, for several tens of km down to the western edge of the Sabatini area (see De Rita et al., this volume). To the east, the graben is delimited by the Apennine chain (Amelia Range) and the Orte structural high, which is now buried by Pliocene-Quaternary sediments and is the continuation of the positive structure of Mt. Soratte. Flysch terrain also crops out northward (Ferento) and southward (Sutri) of Vico Lake. Outcrops of flysch at Sutri separate the southern products of Vico from those of Sabatini volcano. Generally the flysch terrain represents Cretaceous-Oligocene allochthonous nappes which lie on Mesozoic-Cenozoic Tuscan carbonate formations (Conforto, 1956; Fazzini et al., 1972).

The metamorphic basement beneath the transgressive Tuscan carbonate formations, consists of phyllitic and micaceous quartz schists. The closest outcrops are approximately 40 km NW of Cimino where the base of the Tuscan sequence consists of dolomitic limestones and anhydrites of the Upper Triassic (Calcare Cavernoso of Dessau et al., 1972).

Geophysical and geological data, and data obtained from bore holes drilled NW of Cimino and Vico area (Latera caldera area) reveal that the carbonate formations have undergone large translations. These movements began with the emplacement of flysch nappes in the Oligocene and Miocene and continued until the Middle-Upper Miocene (Beldacci et al., 1967; Fazzini et al., 1972). The resultant complex tectonic wedge has an eastern convergence that affected both the carbonate series and the flysch terrain. The period of rigid-style collapse and the formation of lake basins that were subsequently affected by marine transgressions, began in Messinian time. In the Cimino and Vico areas, sediments of the neo-authochthonous sedimentary cycle were deposited in a graben which is bordered on the west by the Mt. Razzano ridge and on the east by the Apennine chain.

The Upper Miocene transgression began with the deposition of a basal conglomerate formed by pebbles of flysch origin. In the Lower and Middle Pliocene the sedimentation continued with deposition of clays and sandstones (Beldi et al., 1974).

Middle Pliocene sediments composed of blue clays grading upward to silty clays and sands occur in the Bagnaia area on the northwestern flank of Mt. Cimino and in the vicinity of Fabrica di Roma. The top of these sediments is at anomalous altitudes, 60-250 m higher than that of similar isochronous sediments along the Tiber River in the western-most part of the Cimino area. Uplift of these sediments, which partly began on a regional scale at the end of the Lower Pliocene (Brandi et al., 1970; Micheluccini et al., 1971), is locally attributed to the emplacement of anatectic intrusions and domes.

VOLCANIC PRODUCTS

Cimino volcanic complex

Cimino volcanism began approximately 1.35 m.y. ago with the emplacement of rhyodacitic ignimbrites. Two ignimbrite units are separated by a fluvial-lacustrine deposit with ignimbrite clasts. Ignimbrites were followed by the emplacement of endogenous domes of similar composition (Bertini et al., 1971; Micheluccini et al., 1971; Puxeddu, 1971). The domes were emplaced along a

NW-SE trending strip about 16 km in length and up to 7 km in width. The available absolute age determinations of ignimbrites and domes range between 1.35 and 1.1 m.y. (Nicoletti, 1969). Most pyroclastic flows preceded emplacement of the domes. Locally, the ignimbritic plateau was uplifted by the domes. However, in the Bagnaia area an undisturbed ignimbrite overlies the domes testifying that some ignimbritic eruptions followed emplacement of the domes. This is confirmed by xenoliths of dome material within the ignimbrites. The Cimino ignimbrites form a plateau covering a surface of approximately 350 km^2. They were presumably erupted from a central volcano (Sabatini, 1912; Mittempergher and Tedesco, 1963; Veritriglia, 1963), as evidenced by their distribution which can be roughly encompassed in a circle centered at the Cimino volcano and having a radius of 15-20 km.

About 1 m.y. ago Cimino volcanism underwent an important change; lava flows with a different chemical composition were erupted from different vents on the central volcano. The first lavas were very viscous and had a latitic composition. Cimino volcanic activity ended about 0.95 m.y. ago with the eruption from the central volcano of very fluid lavas of olivine-latitic composition which spread as far as 6 km from the volcano.

Xenoliths within Cimino ignimbrites are mostly represented by thermally metamorphosed rocks showing two main parageneses. One type is characterized by skarn facies with ibschite (hydrogarnet), vesuvianite and diopside. This paragenesis is attributed to thermal metamorphism of marly rocks which could have come from the allochthonous flysch sequence or the top of the underlying Tuscan series. Ibschite is a good indicator of the crystallization conditions since it is stable in the range 250-420°C at 2 kb (Hsu, 1980).

Facies with more intense metamorphism are represented by mineralogical associations such as sanidine, spinel, biotite and plagioclase with the presence of accessory cordierite, melilite, corundum, and apatite. These facies not only occur in the Cimino ignimbrites, but also in the domes. Their foliation is due to orientation of sanidine and biotite crystalloblasts. Clay xenoliths, belonging to the poorly recrystallized Pliocene series, with some epidote and garnet, also occur in the Cimino ignimbrites.

Vico volcanic complex

The Vico volcanic complex consists of a central lava body composed of alternating undersaturated trachytes, phonolites, tephritic phonolites, tephrites, and subordinate tuffs. The basal units of the exposed sequence are leucititic and trachytic lavas with thin intercalations of leucite tuffs composed of ash, lapilli, and pumice. These tuffs extend for a distance of approximately 20 km. In the eastern sector of the volcano they cover a paleosoil above the Cimino ignimbrites. New K/Ar age determinations were made on samples from Vico collected in stratigraphic succession (Table 1). An age determination on a trachytic lava (Petrignano trachyte of Mattias and Ventriglia, 1970) exposed in the western sector of the volcano, which is one of the oldest stratigraphic units, yielded an age of 0.4 ± 0.008 m.y. An age

TABLE 1

K-Ar dating of Vico and Cimino samples made on sinidine. (Vv = Vico volcano; Cv = Cimino volcano)*
(Age determinations were performed by Laboratorio di Geogronologia e Geochimica Isotopica del CNR, Pisa.)
(Definitive data received September 22, 1982)

Sample Number	Rock type	K%	$^{40}Ar_{rad}$ ml STP g^{-1}	$\frac{^{40}Ar_{rad}}{^{40}Ar_{tot}}$	AGE (m.y.)
5	"Ignimbrite"D-Vv	11.23	$6.198 \cdot 10^{-8}$	.26	0.139 ± 0.016
4	"Tufo rosso a scorie nere"-Vv	11.37	$6.772 \cdot 10^{-8}$	.31	0.150 ± 0.007
3	Trachyte (SW base of caldera)-Vv	10.59	$1.283 \cdot 10^{-7}$	.14	0.305 ± 0.008
2	Trachyte (base of volcano)-Vv	9.98	$1.587 \cdot 10^{-7}$	.22	0.400 ± 0.008
1	Rhyodacite dome-Cv	10.27	$5.371 \cdot 10^{-7}$	.65	1.31 ± 0.02

*For location of samples, see Fig. 1.

determination on another trachytic lava which crops out at the base of the lava series inside the southwestern sector of the caldera, gave an age of 0.305 ± 0.008 m.y. This age is in good agreement with published data for a lava flow which is probably in the same stratigraphic position on the southern side of the caldera. this lava gave an age of 0.30 ± 0.07 m.y. (Nicoletti, 1969). Two other age determinations of 0.7 ± 0.21 m.y. and 0.82 ± 0.18 m.y. (Nicoletti, 1969), are not in agreement with the newly acquired data. Another age date (Everden and Curtis, 1965) yields an age of 0.095 m.y. for a leucitic lava of uncertain stratigraphic position near Vetralla.

The growth of the stratovolcano was followed by the eruption of three pyroclastic flows and a base-surge unit. The eruption of these tuffs, referred to as "Ignimbrites A, B, C and D" (Locardi, 1965), is presumably connected to the formation of the Vico caldera.

Pyroclastic unit A. This deposit, with a phonolitic tephritic composition, has a volume of approximately 1 km^3 of erupted magma. The typical facies is represented by a deposit of non-welded pumice and ash that varies from grey to purple in color. It contains scattered leucite crystals and large black pumices which can reach 30-40 cm in length. No evident structures are generally visible in these units, but flow units and many lateral variations locally can be distinguished. The outcrops of this unit extend as far as 12 km from the Vico volcano and the thickness reaches 50 m in paleovalleys. Xenoliths include leucititic lavas, sedimentary clasts, thermo-metamorphic rocks, and other rocks of the non-metamorphosed flysch series.

Pyroclastic unit B. This trachytic tuff extends as far as 12 km from the caldera rim. Its volume is perhaps less than half of the volume of Unit A. Deposits that range from semi-coherent to welded are dark grey or black in color. Black pumice rich in sanidine crystals are scattered in an ash matrix. In some sites there are vitrophyric facies with welded fiamme. The maximum thickness is tens of meters within paleovalleys cut into Unit A.

Pyroclastic unit C. This tephritic phonolitic tuff, 0.155 ± 0.01 m.y. old (Table 1) is the most important unit among the Vico pyroclastic products. It extends as far as 25 km from the volcano and covers an area of 1,200 km^2 with a volume of erupted magma equal to 3-5 km^3. Its proximal facies is characterized by a coarse basal breccia, a few meters thick, that reaches up to 6-7 km from the caldera. The breccia contains 50% lithic clasts up to 1 m in diameter ina matrix of black leucititic scoria, sanidine phenocrysts, and non-welded, light-colored pumices. Although this unit has a variable appearance, the typical facies consists of reddish-yellow pumice with large black scoria which can reach 50-60 cm in diameter. Lithification is mainly related to alteration. The structural characteristics of the deposit do not vary vertically. Xenoliths are evenly distributed, even where it reaches a thickness of 50-60 m. Inclusions consist of leucite-bearing rocks of various composition, trachytes, and more-or-less thermally altered sedimentary rocks, which belong to both the Pliocene and flysch series. The most frequent skarn minerals are vesuvianite and ibschite, which are similar to the xenoliths within the Cimino products. Great amounts of hauyne-bearing phonolitic lavas, which are especially common in the agglomeratic basal facies, are also contained in this unit. Some blocks of these lavas reach 1 m^3 in size and occur up to 5-6 km from the volcano. These lava inclusions are 0.250 ± 0.05 m.y. old (unpublished Agip data). These ages, together with those of the overlying unit D (0.144 ± 0.02 m.y. in Table 1), distinguish unit C from a similar unit exposed in the Sabatini area, (see De Rita et al., this volume). Age determinations range between approximately 0.52 and 0.42 m.y. (Everden and Curtis, 1965; Nicoletti, 1969; Ambrosetti et al., 1972) for the Satabini "red tuff with black scoria".

Pyroclastic unit D. Unit D almost exclusively occurs in the eastern sector of the volcano where it stretches as far as 20 km from the caldera. A few outcrops in the southwestern sector show this tuff separated from unit C by lacustrine deposits. In the eastern sector, however, a paleosoil is present between units C and D. This trachytic tuff consists of a light-colored vitric matrix which generally accretes around pumice. Vesicular grey pumice and obsidian blocks, both containing leucite and sanidine phenocrysts are uniformly distributed throughout the deposit. This coarsely-bedded tuff is fairly well lithified and shows flow structures, laminations, and cross-bedding. Because of these features, this unit is considered to be a pyroclastic surge (Mattson and Alvarez, 1973). Xenoliths include vesuvianite, ibschite, and diopside skarns formed by contact metamorphism of flysch sediments and by thermal metamorphism of limestones. Some xenoliths produce impact structures, especially in the upper part of the unit. This unit has an average thickness of some meters, except in paleodepressions where it can reach a thickness of 50-60 m near the towns of Caprarola and Carbognano.

Final products. The final erupted products from Vico rest on a paleosoil above pyroclastic unit D. These materials were erupted from vents that coincide with collapse fractures along the inner N and NE rim of the caldera. The position of these products suggests a post-caldera eruption because they unconformably overlie the inner walls of the caldera. The vent areas of these products are locally marked by a coarse volcanic agglomerate containing clasts which can reach 80 cm in diameter. These base-surge products form a complex which consists of beds with grey pumice containing pyroxene and biotite, levels of pumiceous vesicular tuffs, and lithified sandy tuffs (Mattson and Alvarez, 1973). This corresponds to the massive and sandwave facies of Wohletz and Sheridan (1979). The structures are typical of a base-surge deposit with dunes and antidunes approximately 5-10 m in wave length in the proximal zones. The chemical composition of the above described tuffs is phonolitic tephrite. Xenolith types include thermal metamorphic and volcanic rocks which belong to the Vico sequence as well as trachybasaltic lavas which were dated 0.99 m.y.

(unpublished Agip data). It is peculiar that xenoliths within the Vico products never include rocks belonging to Cimino volcano, although buried domes of this complex were identified by magnetic surveys on the eastern side of the Vico volcanic complex (Nannini et al., 1981).

The final volcanic episode of the Vico complex is the formation of the intracaldera stratovolcano of Mt. Venere. This cone consists of phonolitic tephrite and tephrite products. No age data is available for this feature at present.

TECTONIC SETTING

The main tectonic features of the Cimino and Vico areas, (Figs. 2 and 3) include the Mt. Razzano high, the uplifted Pliocene terrain associated with Cimino volcanism, the Tiber graben, and the Orte structural high, which is interrupted by an important NE-SW trending dislocation that stretches from Vico Lake to the Amelia Range (Borghetti et al., 1981). The uplifted Pliocene terrain is bordered on the north by a fault system along which the Vezza River cuts its river banks. The area of Vezza River has active NW-SE and NE-SE striking faults associated with gas exhalations. This system of faults has been active since the deposition of ignimbrites from Cimino volcano. These tectonic features limited the northward spread of the Cimino ignimbrites.

The location of basins associated with this fault system is evidenced by the distribution and thicknesses of the Vico "red tuff with black scoria" which fills the Vezza stream paleovalley and lies above the Cimino ignimbrite. After the deposition of this distinctive ignimbrite from Vico at 0.15 m.y. ago, erosion continued along faults which are still active. The difference in stratigraphic section on either side of the Vezza River valley is significant. Pliocene sediments, along the southern bank that corresponds to a structural high (Agip, 1980; Corda et al., 1980), are covered by the Cimino ignimbrite. But along the northern bank, the top of the clays lies about 50 m lower and these sediments are covered by travertine, fluvial, and lacustrine sediments with epivolcanic beds. This area corresponded to a structural high at the time of the deposition of the Cimino ignimbrites, as it restricted the distribution of these tuffs.

Old travertine deposits, dislocated and inclined as much as 35°, are present at the bottom of the sedimentary basin and thalweg of the Vezza River, thus confirming their source from fractures. Sediments of the volcano-sedimentary basin are affected by undulations, normal faults, and local reverse faults in the area between Ferento and Grotte Santo. The sedimentary basin in the Ferento area is bordered to the west by the structural high formed by flysch along a NW-SE trending fault (Figs. 1 and 2). Considerable hydrothermal and gas exhalations are present along these fractures. South of the Vezza River near the Cimino domes the altitude of the top of the Pliocene sediments gradually increases. Uplift of 200 or 300 m above the mean altitude of the beds at the margins of Cimino area are common. Small lenses of Pliocene sediments also

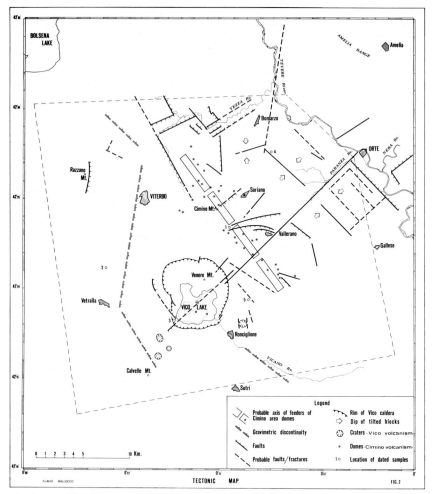

Fig. 2 Generalized tectonic map of Vico-Cimino area based on geophysical data.

occur at altitudes of approximately 650 m above the domes of Mt. Montalto and Mt. San Valentino. They are probably outliers of clays which were ripped off during the dome uprise. South of the Cimino area there is a slight 100 m uplift of Pliocene clays.

The domes of Mt. Cimino, which lie along the graben axis controlled by a NW-SE trending fault system do not show appreciable dislocations. The most evident dislocation is a NE-SW fault that cuts the Mt. Cimino volcano (Fig. 2) offsetting the Cimino ignimbrites for some tens of meters. Ring fractures surrounding the central volcano are possibly defined by the arrangement of domesaround the central volcano. A system of arc-shaped fractures with an approximate ENE-WNW direction is also evident south of Cimino volcano (Fig. 2). These fractures displace the Cimino ignimbrite by a southward subsidence and

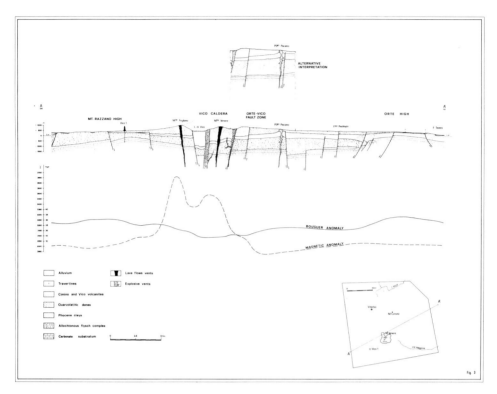

Fig. 3 Schematic section Vico-Cimino area. Geology based on geophysical surveys intergrated with surface and drilling data.

tilt of faulted blocks. The upwelling of rhyodacitic magmas also occurred along one of these faults, indicated by the emplacement of a ENE-WNW elongated dome. Furthermore, in the eastern part of the area, a NW-SE trending fracture zone in a position midway between the Cimino domes, the Tiber valley, and the Orte structural high, this 15 km long trend corresponds to the axis of two inward dipping blocks. Morphologic, geologic, and geophysical data also confirm this structure. The paleosurface of the Cimino ignimbrite, excluding the cover of the Vico pyroclastic flows which partly filled the paleodepression between the tilted blocks, accentuate this tilting. Similar tilted blocks also occur in the grabens of the Bomarzo area (Fig. 2).

The tectonism, for at least the eastern and northeastern zones of the area, was partly pre-volcanic, causing the first tilting of the Pliocene sedimentary substratum. Subsequent to the emplacement the Cimino ignimbrite, further tilting of the blocks occurred. One tectonic element which considerably affects the Cimino volcanic area is the Orte-Vico fault system to which the activity of

Vico volcano is presumably connected. This fault system, with a probable left transcurrent component, interrupts the structural high of Orte and extends from Vico Lake to the Amelia Range. Although this fault has not moved after the deposition of the Vico "red tuff with black scoria", it is associated with drainage anomalies, gas exhalations, and hydrothermal activity.

Geophysical data reveal a horst, approximately 2-3 km in width, which could be associated with a high of the carbonate basement along this fracture zone for a few kilometers east of Vico Lake (Fig. 3). However, this interpretation does not agree with the geological model. The most plausible explanation could be that the geophysical anomaly corresponds to the effect of high-density hydrothermally-altered rocks. Vico volcano lies near the western edge of the graben along the Orte-Vico fault system.

The most important structure which displaces rocks of the Vico volcano is the summit caldera. It is a composite, eccentric caldera with a diameter of approximately 7 km that consists of two main collapse structures: one forms the Vico valley, the other one hosts the lake. Collapse of the caldera followed a multiphase eruption of large pyroclastic units at about 0.15 m.y. ago. The shape of the caldera rim is fairly regular, being defined by two faults in the southwestern sector. The caldera rim is complicated on the SW side where it intersects the Orte-Vico fault system, and on the NE side where a hydrothermal alteration zone is present. Further evidence of the Orte-Vico fault system is given by the bathymetry of the lake. Isobaths abruptly deepen to the maximum depth of 50 m along a scar that corresponds to the northern shoreline. This morphology is consistent with recent activity of the Orte-Vico fault system (Fig. 2).

The structural framework of the area which includes Cimino and Vico volcanic complexes displays the general pattern of other post-orogenic grabens situated between the Tyrrhenian Sea and the Apennine chain. These grabens, elongated roughly parallel to the axial chain direction, stretch from northern Tuscany to Campania. Their origin is probably associated with tension related to the opening of the Tyrrhenian basin, which has been in progress since the Middle Miocene. These tensional features are related to the anti-clockwise rotation of the Italian peninsula (Alverez, 1972; Civetta et al., 1978; Scandone, 1979a, 1979b). The strip occupied by the grabens corresponds to a zone of general crustal thinning characterized by Bouguer anomaly values which are intermediate between the high values of the Tyrrhenian basin and the low values of the Apennine chain (Glese and Morelli, 1975). Both the anatectic Miocene-Pliocene-Quaternary volcanism of central-northern Tuscany and northern Latium, and the Quaternary potassic alkaline volcanism of Latium and Campania fall within this structural framework. The age, composition, and geographical position of volcanism appear to be controlled by the different amounts of extension along the western Apennine margin. The association between temporal migration of the anatectic acid volcanism from west to east and the more recent activation of the potassic alkaline volcanism probably lies in this context. The evolution of volcanism is consistent with the theories which consider an increase in the crustal thinning from NW (central northern Tuscany-Upper Tyrrhenian Sea) to SE (Latium, Campania, Lower Tyrrhenian Sea), following a model which postulates that the Tyrrhenian sea opened with increasing expansion from north to south (Scandone, 1979a).

ACKNOWLEDGEMENTS

I wish to express my thanks to AGIP S.p.A. for granting permission to publish this manuscript. I would also like to thank A. Sbrana for helpful discussions and suggestions during the writing of this paper, as well as G. Cimino and G. Borghetti for discussions that have helped to clarify some aspects of the manuscript and for their constructive criticism. Reviews of an early manuscript by G. Orsi and S. Self have improved the presentation. M. F. Sheridan critically reviewed the text. Finally, I would like to point out that the data concerning the tectonic setting are the result of a work performed previously by R. De Marinis, A. Sbrana, L. Zan and the author (in: Agip, 1980).

REFERENCES

Agip, 1980. Studio vulcanotettonico del permesso "Monti Cimini" Internal paper.
Alvarez, W., 1972. Rotation of the Corsica-Sardinia microplate. Nature, 235: 103-105.
Ambrosetti, P., Azzaroli, A., Bonadonna, F.P., Follieri, M., 1972. A scheme of Pleistocene chronology for the Tyrrhenian side of Central Italy. Boll. Soc. Geol. Ital., 91: 169-184.
Baldacci, F., Elter, P., Giannini, E., Giglia, G., Lazzarotto, A., Nardi, R., Tongiorgi, M., 1967. Nuove osservazioni sul problema della Falda Toscana e interpretazione dei Flysch arenacei tipo "Macigno" dell'Appennino Settentrionale. Mean. Soc. Geol. Ital., 6: 213-244.
Baldi, P., Decandia, F.A., Lazzarotto, A., Calamai, A., 1974. Studio geologico del substrato della copertura vulcanica laziale nella zona dei laghi di Bolsena, Vico e Bracciano. Mem. Soc. Geol. Ital., 13: 575-606.
Barberi, F., Innocenti, F., Ricci, C.A., 1971. Il magmatismo. In: M. La Bertini, C. D'Amico, M. Deriu, O. Girotti, S. Tagliavini, and L. Vernia (Editors), Note illustrative della Carta Geologica d'Italia. Foglio 143 Bracciano. Servizio Geologico d'Italia, 77 pp.
Bertini, M., D'Amico, C., Deriu, M., Girotti, O., Tagliavini, S., Vernia, L., 1971. Note illustrative della Carta Geologica d'Italia. Foglio 137 Viterbo. Servizio Geologico d'Italia, 109 pp.
Borghetti, G., Sbrana, A., Sollevanti, F., 1981. Vulcano tettonica dell'area dei M. Cimini e rapporti cronologici tra vulcanismo cimino e vicano. Mem. Soc. Geol. Ital. Note Brevi. In press.
Brandi, G.P., Cerrina Ferroni, A., Decandia, F.A., Giannelli, L., Monteforti, B., Salvatorini, G., 1970. Il Pliocene del bacino del Tevere fra Celleno (Terni) e Civita Castellana (Viterbo). Stratigrafia ed evoluzione tettonica. Atti Soc. Tosc. Sci. Nat., 77: 308-326.

Civetta, L., Orsi, G., Scandone, P., and Pece, R., 1978. Eastwards migration of the Tuscan anatectic magmatism due to anticlockwise rotation of the Apennines. Nature, 276: 603-604.

Conforto, B., 1956. Contributo ala conoscenza delle formazioni "fliscioidi" del Lazio. Boll. Soc. Geol. Ital., 75: 95-106.

Corda, L., DeRita, D., Sposato, A., 1980. Dati preliminari sulla neotettonica del Foglio 137 (VT). Publ. n°356 del Progetto Finalizzato Geodinamica, C.N.R.: 23-29.

DeRita, D., Funiciello, R., Rossi, U., and Sposato, A., 1983. Structure and evolution of the Sacrofano-Baccano caldera, Sabatini volcanic complex, Rome. In: M.F. Sheridan and F. Barberi (Editors), Explosive Volcanism. J. Volcanol. Geotherm. Res., 17: (this volume).

Dessau, G., Duchi, G., Stea, B., 1972. Geologia e depositi minerari della zona dei Monti Romani-Montete (Comuni di Capalbio (Grosseto) ed Ischia di Castro (Viterbo)). Mem. Soc. Goel. Ital., 11: 217-260.

Everden, J.F., Curtis, G.H., 1965. The potassium-argon dating of late Cenozoic rocks in East Africa and Italy. Curr. Anthrop., 6: 343-385.

Fazzini, P., Gelmini, Mantovani, M.P., Pellegrini, M., 1972. Geologia dei Monti della Tolfa (Lazio Settentrionale: prov. di Viterbo e Roma). Mem. Soc. Geol. Ital., 11: 65-144.

Giese, P. and Morelli, C., 1975. Main Features of crustal structures in Italy. In: C.H. Squyres (Editor), Geology of Italy. Petr. Expl. Soc. Libya, Tripoli, pp. 221-243.

Hsü, L.C., 1980. Hydration and phase relations of grossular-spessartine garnets at $pH_2O = 2kb$. Contr. Miner. Petrol., 71: 407-415.

Locardi, E., 1965. Tipi di ignimbriti di magmi mediterranei: le ignimbriti del vulcano di Vico. Atti Soc. Tosc. Sci. Nat., 72: 55-174.

Locardi, E. and Molin, A., 1975. Ricerche per uranio nel Lazio settentrionale. C.N.E.N. Roma, 106 pp.

Locardi, E., Funiciello, R., Lombardi, G., Parotto, M., 1977. The main volcanic groups of Latium (Italy): relations between structural evolution and petrogenesis. Geologica Romana, 15: 279-300.

Marinelli, G., 1975. Magma evolution in Italy. In: C.H. Squyres (Editor), Geology of Italy. Petr. Expl. Soc. Libya, Tripoli, pp. 221-243.

Mattias, P.P. and Ventriglia, U., 1970. La regione vulcanica dei Monti Sabatini e Cimini. Mem. Soc. Geol. Ital., 9: 331-384.

Mattson, P.H. and Alvarez, A., 1973. Base-surge deposits in Pleistocene volcanic ash near Rome. Bull. Volcanol., 37: 553-572.

Micheluccini, M., Puxeddu, M., Toro, B., 1971. Rilevamento e studio geo-vulcanologico della regione del M. Cimino (Viterbo-Italia). Atti Soc. Tosc. Sci. Nat., 78: 301-337.

Mittempergher, M. and Tedesco, C., 1963. Some observations on the ignimbrites, lava domes and lava flows of M. Cimino (Central Italy). Bull. Volcanol., 25: 1-16.

Nannini, R., La Torre, P., Sollevanti, F., 1980. Geothermal exploration in Central Italy: geophysical surveys in Cimini range area. Paper presented in 43rd Meeting of European Association of Exploration Geophysicists. Venezia, May 26-29.

Nicoletti, M., 1969. Datazione argon potassio di alcune vulcaniti delle Regioni vulcaniche Cimina e Vicana. Periodico di Mineralogia, 1: 1-20.

Puxeddu, M., 1971. Studio chimico petrografico delle vulcaniti del M. Cimino (Viterbo). Atti. Soc. Tosc. Sci. Nat., 78: 329-394.

Sabatini, V., 1912. I vulcani dell'Italia Centrale e i loro prodotti. Parte seconda: Vulcani Cimini. Mem. descr. della Carta Geol. d'It., 15: 617 pp.

Scandone, P., 1979a. Origin of the Tyrrhenian Sea and Calabrian Arc. Boll. Soc. Geol. Ital., 98: 27-34.

Scandone, P., 1979b. Inquadramento geologico del vulcanismo potassico. In: Tavola rotonda sul magmatismo potassico nell'area Tirrenica. Rend. Soc. Ital. Min. Petr., 35: 21-25.

Ventriglia, U., 1963. Il vulcano Cimino. Bull. Volcanol., 25: 181-199.

Wohletz, K.H. and Sheridan, M.F., 1979. A model of pyroclastic surge. In: C.E. Chapin and W.E. Elston (Editors), Ash-Flow Tuffs. Geol. Soc. Amer. Sp. Paper 180, pp. 177-194.

STRUCTURE AND EVOLUTION OF THE SACROFANO-BACCANO CALDERA, SABATINI VOLCANIC COMPLEX, ROME

D. DE RITA[1], R. FUNICIELLO[1], U. ROSSI[2], A. SPOSATO[3]

[1] Istituto di Geologia e Paleontologia, Università degli Studi, 00100 Roma (Italy)
[2] ENEL, Unità Nazionale Geotermica, P. za B. Sassoferrato 14, 56100 Pisa (Italy)
[3] CNEN, CSN Casaccia Laboratorio Geologico Ambientale, (Italy)

(Received September 17, 1982, revised and accepted December 23, 1982)

ABSTRACT

De Rita, D., Funiciello, R., Rossi, U., and Sposato, A., 1983. Structure and evolution of the Sacrofano-Baccano caldera, Sabatini volcanic complex, Rome. In: M.F. Sheridan and F. Barberi (Editors), Explosive Volcanism. J. Volcanol. Geotherm. Res., 17: 219-236.

Sacrofano eruptive center, in the eastern part of the Sabatini volcanic complex, was active between 0.5 and 0.09 m.y. ago. The Baccano geothermal area lies at the western edge of Sacrofano caldera. Sedimentary substrata that form the reservoir for this geothermal system were exposed at the surface at the beginning of volcanic activity. The volcanic history of the Sacrofano center can be divided into three stages:
(1) Construction of the Sacrofano pyroclastic edifice by predominantly Strombolian activity.
(2) Collapse of Sacrofano caldera following eruption of pyroclastic flows.
(3) Development of the Baccano explosive center at the western edge of Sacrofano caldera and collapse of Baccano caldera.

Hydrovolcanic activity that began at the end of the first stage of Sacrofano can be explained by a drop in the magma level in the conduit that relieved hydrostatic pressure in the aquifer. As intense fracturing associated with caldera collapse penetrated the carbonate reservoir, the entrapped water flowed towards the conduit to balance the pressure change. The close link between tectonism and volcanism in the Sacrofano center suggests that eruptions there may have been triggered by gravity faults related to regional tectonics.

INTRODUCTION

The Sabatini volcanic complex, located north of Rome (Fig. 1), lies within the Roman volcanic province. This region of highly potassic volcanism extends from the Alban Hills in the south to the Vulsini volcanic complex in the north. The evolution of volcanic centers in this province is closely related to extensional tectonism during the last million years. This paper concentrates on the volcanic activity of the Sacrofano eruptive center that formed in the eastern part of the Sabatini complex. This center is important not only because of its size, but also because of its long and varied activity.

Many authors have performed general studies in the Sacrofano region but they failed to consider the volcano as an entity. This study, based on 1:10,000 scale mapping and surface and subsurface stratigraphic data, has resulted in a reconstruction of the Sacrofano volcano. It has also led to a reinterpretation of the models previously proposed for the volcano-tectonic evolution of this area. The eruptive style of the Sabatini complex is governed by the regional geology, hydrology, and the position of the magma chamber within the crust.

Previous work

Moderni (1896) was the first geologist to associate the Sacrofano and Baccano volcanic centers with a larger and older E-W elongated elliptical structure. Moderni also described the close similarities between hydromagmatic products of Baccano (Corda et al., 1976) and those of the Albano volcanic center (Civitelli et al., 1975), south of Rome.

The stratigraphy of the eruptive products of Sacrofano was defined by Bertini et al. (1971) who recognized Sacrofano as a caldera but did not map its structure or deposits. The structure of Sacrofano caldera, along with additional stratigraphic data, were reported by Mattias and Ventriglia (1970). Correlation of the distribution of volcanic deposits with specific sources was prohibited because comprehensive facies maps or petrographic analyses were lacking.

Subsequent geological studies deal with specific problems. This work did not view Sacrofano as an entity and thus did not contribute to reconstruction of this volcanic edifice (Calamai et al., 1975; Cameli et al., 1976; Funiciello et al., 1976; Baldi et al., 1974; Funiciello et al., 1977; Funiciello et al., 1979). Other workers (Locardi and Sommavilla, 1974) proposed a model for the structural evolution of the entire Sabatini region which is no longer acceptable in view of the available geological data. The geological significance of the evolution of the Sacrofano volcanic structure is far from being completely understood. A monograph with complete descriptions of rock units and structures as well as a detailed map is now in press.

PRINCIPAL STRATIGRAPHIC UNITS

The Sabatini area was evaluated by ENEL for geothermal potential from 1975 to the present. The results of this program were examined together with data from a drill hole "Ladispoli" recently completed by AGIP at Sacrofano. This data allowed a three-dimensional reconstruction of the main volcanic and sedimentary

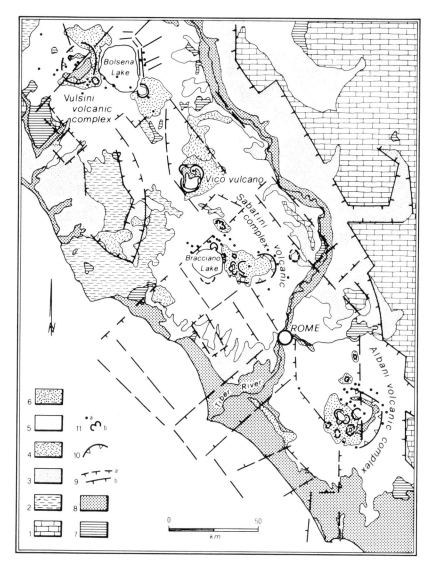

Fig. 1 Sketch map of the Latium area, Central Italy. (1) Latium-Abruzzi carbonate platform, Umbria-Sabina successions and Tuscan Nappe; (2) allochthonous complexes (Liguridi and Subliguridi); (3) Plio-Pleistocene marine sediments; (4) acidic volcanoes; (5) K-alkalic volcanic rocks; (6) hydromagmatic units; (7) travertine; (8) continental and marine sediments (Upper Pleistocene to Recent); (9a) normal faults; (9b) faults mapped by means of indirect investigations; (10) calderas; (11a) scoria cones; (11b) craters.

units of the area. The stratigraphic columns (Fig. 2) were based mainly on drill cuttings because only a limited number of core samples were made available by the companies working in the area. Consequently, doubt still remains regarding some stratigraphic correlations. Nevertheless, the data provide a tentative reconstruction of a few complete sections that cross this explosive volcanic complex. They also provide a relatively reliable picture of the underlying sedimentary units.

Details on the characteristics and structure of the deeper units of the "Tuscan Nappe" can be found in Funiciello et al. (1979). It is appropriate here to emphasize the role of the Soratte-Cornicolani Meso-Cenozoic structure as a recharge area for the Liassic carbonate reservoir that underlies the Sacrofano-Baccano caldera. The top of this reservoir appears to be lowered to the west by a throw of almost 2 km following a NNE-SSE trending line approximately along the Rignano Flaminio-Capena alignment. The top of the reservoir under the Sacrofano-Baccano area appears horizontal. An exception is the area near the tuff cone of Mte. Razzano where well RC1 encountered a thrust plane within carbonatic units that doubled the thickness of the series. Consequently, the structural top of the carbonate sequence in this area is uplifted by at least 500-600 m. It is in the vicinity of this structure that the carbonate rocks attain their greatest thickness in the entire Sabatini area.

This nappe structure must have had a considerable impact on hydrogeology. It may have contributed to the hydroexplosive activity of the Baccano caldera by favoring the inflow of major deep regional aquifers into this area. A significant portion of the late stage of explosive activity at Baccano-Sacrofano may have resulted from the circulation of geothermal fluids. In other words, circulation of hydrothermal fluids through karsted carbonate units favored hydroexplosive activity during its final stages.

The allochthonous units of the Sicilidi complexes form the near surface portion of the "Tuscan Nappe" (marly units of the Cretaceous and Eocenic "sacglia"). These Neogene and Quaternary marine units form a continuous impervious cover above the aquifer with an average thickness of 1500 m, except for the area near RC1 where they are considerably thinner.

The extent of the Plio-Pleistocene units was determined by drill cuttings in the easternmost area and their occurrence as xenoliths in the Mte. Razzano units. Their absence in wells C1, C2, C3, and C6 demonstrates that the carbonate structural high coincident with the Baccano caldera must have existed since the beginning of the volcanic activity in this area.

The multiple levels of lacustrine deposits encountered by drill holes occur in a variety of environments including:
(1) small basins in the most ancient volcanic units, interbedded in the pyroclastic units of Sacrofano;
(2) lakes within Sacrofano caldera which were subsequently drained due to collapses of the caldera;
(3) continuous, thick lacustrine deposits that are locally exposed within the Baccano caldera. These consist of several overlapping units, as can be seen

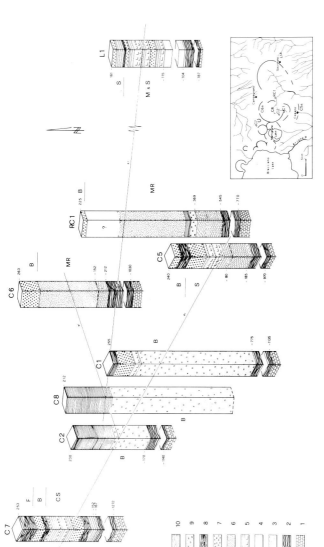

Fig. 2 Reconstruction of the stratigraphic succession from deep drilling in the Sacrofano–Baccano area. (1) Limestones and cherty limestones m = Lias n = Cretaceous. (2) allocthonous sedimentary complex; shales, marly, sandstones and limestones, Lower Cretaceous–Lower Miocene. (3) Clay, sand and conglomerates of post-orogenic sequence (U. Miocene–Quaternary). (4) Plio-Pleistocene sedimentary units. (5) Pyroclastic products and lava flows. (6) Mte. Razzano tuff cone units. (7) pyroclastic flow. (8) hydromagmatic products. (9) volcanic breccias. (10) lacustrine sediments. S: Sacrofano. B: Baccano. M: Morlupo-Castelnuovo di Porto. CS: Sabatinian complex products.

in well C8, in surface Holocene outcrops, and in outcrops south of Colle dell' Ellera where a peat horizon was dated at more than 40,000 years (Alessio et al., 1983).

The lacustrine sequence of well C2 represents early deposits within the Baccano caldera. Observations on the evolution of lacustrine basins suggest that they migrated from south to north. The total thickness of lacustrine deposits is greater in the northern part of the caldera, which also is the area of most recent explosive activity.

Volcanic sequences identified by drilling provide further data on the Baccano volcanic center. First, the most ancient pyroclastic units of the Sacrofano center are not present at Baccano. Moreover, these flows are concentrated to the south of the Sacrofano caldera rim, for example in well C5 as well as in outcrops. The sequence of lower pyroclastic flows of Sacrofano ("yellow tuff of Via Tiberina") and the Plio-Pleistocene sediments are approximately 100 m lower in well L1. The pyroclastic flow of Baccano, on the other hand, does not appear in well RC1 (for which the stratigraphy of surface units is doubtful). It has a minimum thickness in well C5 and maximum thickness in well C1, probably indicating an emission point to the north, which is supported by field data.

One of the most interesting and problematic aspects of the drilling at Cesano is the discovery inside the caldera of thick "breccias" related to hydromagmatic phenomena. These highly heterogeneous breccias contain volcanic, subvolcanic and sedimentary clasts within a hydrothermally altered matrix. In the deeper sections of the well the clasts, especially those of sedimentary origin, are highly inhomogeneous in vertical distribution.

Some brecciated zones were also identified entirely within limestone units at a depth of almost 3,000 m (well C3). Possible interpretations of the origin of these breccias include:
(1) fluidization,
(2) collapse, or
(3) near-surface explosions and emergence of rock fragments along fractured zones.

The detailed analysis of this problem will not be considered here. However, the following relationships are indicated:
(1) The breccias of well C8 are due to the most recent activity of the Baccano center.
(2) The breccias of well C1 are older than the others from which they are separated by intermediate deposits.
(3) The breccias of well C2 correlate with those of C8.

Hydrothermal alteration, which is concentrated in these breccias, cause them to exhibit an extremely low resistivity. As a result, the breccias, especially those related to the last explosions, can be located through detailed geoelectric studies (ENEL unpublished report, 1979).

EVOLUTION OF THE SACROFANO CENTER

Sacrofano is one of the most important volcanic structures of the Sabatini complex, not only because of its size but also because of its long and varied activity. It erupted various products from a differentiated magma chamber. The type of activity was related to structural and hydrologic settings that controlled magma/water interaction. Sacrofano volcano lies above a carbonate platform. These limestones, now buried by volcanic deposits, form the reservoir for the Baccano geothermal area. Sacrofano forms the eastern margin of the geothermal region. The history of the Sacrofano center can be subdivided into three stages:
(1) Construction of a pyroclastic edifice by predominantly Strombolian activity (Figs. 3b and 3c).
(2) Collapse of Sacrofano caldera (Fig. 3d).
(3) Development of the Baccano explosive center inside the old Sacrofano caldera and collapse of the Baccano caldera (Figs. 3e and 3f).

FIRST STAGE -- STROMBOLIAN ACTIVITY AT SACROFANO

The initial activity at Sacrofano is not well documented. A borehole drilled by AGIP (Ladispoli 1 see Fig. 2, section L1) SE of the present caldera reveals Plio-Pleistocene sediments overlain by pyroclastic beds and lava flows ascribed to the earliest products of this volcano. These volcanic materials rest on other pyroclastic deposits, most of which originate from another explosive center located to the east between the towns of Morlupo and Castelnuovo di Porto. The exact location of the other center is impossible to pinpoint because of extensive erosion. It has an evolution similar to that of Sacrofano, but its activity was more limited in time and space.

The first unit identified as a product of Sacrofano is a pyroclastic flow interbedded with deposits from Morlupo-Castelnuovo di Porto (Nappi et al., 1979; Mattias and Ventriglia, 1970). Most exposures of this deposit occur on the outskirts of the Sacrofano complex. It has been renamed the lower pyroclastic flow of Sacrofano (formerly called the "yellow tuff of Via Tiberina," Fig. 3b). The lower pyroclastic flow, although homogeneous in appearance, varies from tephritic-leucitic to phonolitic-leucitic composition (Scherillo, 1941). Locally the ash matrix contains accretionary lapilli and concentrations of lava or sedimentary lithic clasts. In distal sections it consists of two or more flow units (as defined by Smith, 1960). Individual flow units cannot be recognized near the vent, suggesting a rapid emplacement.

After the eruption of these pyroclastic flows Sacrofano entered a Strombolian stage. A very thick pyroclastic series (up to 100 m in proximal sections) with a primitive chemical composition comprise the main edifice. These pyroclastic-fall deposits are a part of the units called "varicoloured stratified tuffs of Sacrofano" and "varicoloured stratified tuffs of LaStorta"

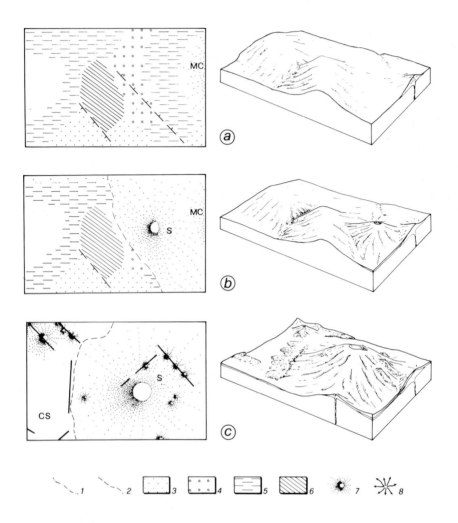

Fig. 3 Morphological and structural diagram of Sacrofano evolution. a: Diagram of sedimentary substrata reconstructed on the basis of deep drilling and analysis of inclusions of phreatomagmatic units of the final activity (from Funiciello et al., 1976). The eastern section shows the activity of the Morlupo-Castelnuovo di Porto center (MC). b: Beginning of Sacrofano activity (S) (first pyroclastic deposits and emplacement of lower pyroclastic flow -- 1st stage of activity of Sacrofano). c: Paroxysmal stage of Sacrofano (S) and of Sabatini volcanic complex (CS) (corresponding to emplacement of Sacrofano pyroclastic units and of lavas and cinder cones of Sabatini complex: 1st stage of activity of Sacrofano). d: Emplacement of upper pyroclastic flow of Sacrofano

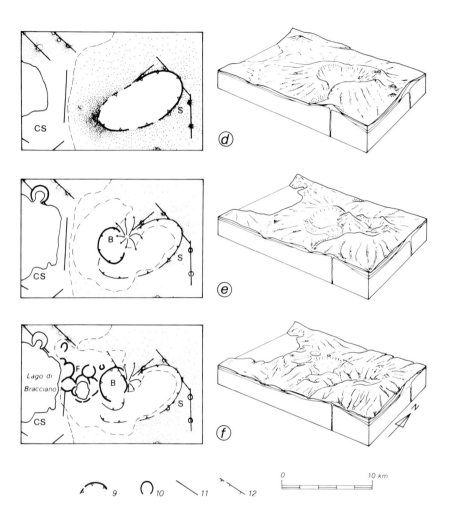

and formation of Sacrofano caldera (2nd stage of activity of Sacrofano). e: Building of the Mte. Razzano tuff cone and first stage of Baccano (B) activity (3rd stage of activity of Sacrofano). f: Terminal phreatomagmatic activity of Baccano and minor centers (F) (end of 3rd stage of activity of Sacrofano). Legend: (1) limit of products from Baccano and of terminal products; (2) limit of Sacrofano products; (3) Pleistocene sands and clays; (4) polygenic conglomerates and Upper Pliocene calcarenites; (5) Pliocene clays and sandy clays; (6) calcareous-clayey flysch (Sicilids); (7) main centers of activity; (8) tuff cone of Mte. Razzano; (9) calderas; (10) explosive craters; (11) main fractures feeding the lava and pyroclastic units; (12) main faults.

by previous authors. Small, mostly leucite-bearing lavas, flowed from scoria cones on the flanks and edges of this complex.

This stage of Sacrofano marks a period of paroxysmal volcanism for the entire Sabatini complex. The pyroclastic-fall products from Sacrofano are intercalated with scoria cones far to the west. Extensive leucititic and tephritic-phonolitic lava flows (lava di Vicarello, Vigna di Valle, La Casaccia, Canale Monterano, etc.) of this period are related, in part, to regional normal faults that controlled the formation of Lake Bracciano and the depression of the Cesano high. In the northern portion of the Sacrofano complex several scoria cones and local maars (e.g., Monterosi) produced alternating beds of phreatic, hydromagmatic (in the sense of Sheridan and Wohletz, 1981), and magmatic origin.

Sacrofano fall products also contain a interbedded "red tuff with black scoriae". The origin of this unit in the Sabatini area is a subject of much debate. Alvarez et al. (1975) hypothesized its origin from Vico volcano to the north, where a unit with this lithology has been dated at 0.4 m.y. (Everden and Curtis, 1965). The geologic map discriminates two units: a "red tuff with black scoriae" from Vico and an ignimbrite of phonolitic-tephritic to trachytic composition from Sabatini. Recent dating seems to corroborate the hypothesis of two units, assigning an age of 0.4-0.5 m.y. for the Sabatini "red tuff with black scoriae" while the Vico "red tuff with black scoriae" would be about 0.18 m.y. in age (Borghetti et al., 1981). Data from surface mapping and deep drilling indicate the absence of "red tuff with black scoriae" around the rim of the Sacrofano caldera. This implies that this unit originates from within the Sacrofano caldera, but an origin from fractures on the outskirts of the complex cannot be eliminated. The explosive activity of Sacrofano, and of the Sabatini complex in general, occurred over a long period. Episodes of reworking (Corda et al., 1978) filled small lacustrine and palustrine basins (Riano, Olgiata Cornazzano, etc.). Numerous erosion surfaces and thin levels of buried soils indicate brief pauses in activity (Fig. 3c).

SECOND STAGE -- HYDROEXPLOSIONS AND CALDERA COLLAPSE AT SACROFANO

The end of the stage-one magmatic eruptions marks the beginning of a critical change in behavior of Sacrofano and of the entire Sabatini complex. While volcanism waned to the north, small scoria cones continued to form along fractures on the rim of Sacrofano caldera. We speculate that this caused the magma level in the main conduit of Sacrofano to drop, destroying the hydrostatic equilibria and causing structural instability. The chamber walls, no longer supported by magmatic pressure, could collapse. As the fracturing of country rock from this process reached the reservoir of the regional aquifer, appreciable amounts of water could gain rapid access to magma in the conduit.

The ensuing hydroexplosion gave rise to a distinctive pyroclastic-flow deposit that crops out all around the northern and southwestern caldera rim. We call this deposit the "upper pyroclastic flow" of Sacrofano; previous authors termed it the "yellow tuff of Sacrofano." Local depositional and structural characteristics of this pyroclastic-flow deposit are comparable to the distinctive features of hydromagmatic units (sandwave beds, impact sags, accretionary lapilli, and sedimentary ejecta), suggesting water/magma interaction. However, the general depositional behavior and bedding characteristics are typical of pyroclastic flows, suggesting that the amount of water involved was smaller than that normally associated with hydromagmatic expolsions. SEM studies on surface features of crystals from different pyroclastic and hydromagmatic units as well as from the pyroclastic flow in question (De Rita et al., 1982) show a wide range in erupted form, size and number of impact fractures, and secondary mineral overgrowth reflecting the various levels of water/magma interaction that supports this hypothesis. The data suggests emplacement due to an eruptive cloud that collapsed to produce a pyroclastic flow.

A part of the Sacrofano caldera had probably already collapsed near the present area of Formello (Fosso della Mola) before this pyroclastic flow was emitted. Fosso della Mola must have been a wide valley that channelled most of the pyroclastic flow to the south. In its distal portion this unit has the appearance of a lahar. To the north, close to the semi-circular alignment of the scoria cones of Mte. Ficoreto, Mte. Regolo, Mte. Maggiore, Mte. Solforoso and Mte. Cucco, the caldera rim must have been only partially developed. These scoria cones, aligned on a circum-caldera fracture, apparently presented an obstacle to the northern transport of the pyroclastic flow. In this section the pyroclastic flow partially covers the scoria cones and extends back inside the caldera, where it contains a chaotic assemblage of large blocks (up to 50-60 cm). The final collapse of the caldera followed the violent explosion that produced this unit (Fig. 3d).

Formation of the caldera brought the magmatic history of the Sacrofano volcanic center to a close. Faults that controlled the caldera collapse probably also intersected the aquifer in the Baccano carbonate rocks. Because the thermal energy was not completely exhausted, water that flowed into the region of the magma produced violent hydromagmatic explosions. Deposits from these hydromagmatic to phreatic explosions constructed the Mte. Razzano tuff cone at the western edge of the caldera rim (Fig. 3e). This cone consists of a 100 m section of massive, yellow surge deposits that were emplaced in a sticky condition. The deposits are strongly lithified and rich in xenoliths (up to 6% of total volume). The Mte. Razzano edifice and a corresponding structure on the western rim of Baccano caldera, Mt. S. Angelo, display the typical features of tuff cones proposed by Wohletz and Sheridan (1979): appropriate morphology, wet pyroclastic-surge deposits, thin pyroclastic-fall beds, and dominant explosion breccias.

THIRD STAGE -- DEVELOPMENT OF BACCANO ERUPTIVE CENTER

The occurrence of non-recrystallized sedimentary lithics recognized in

Baccano caldera drillings from layers as deep as 3,000 m could be explained by means of an expanding fluid transport mechanism. The expansion of the fluid is related to extensional jointing along tectonic swarms and intrusion of a shallow magmatic body within a huge sedimentary reservoir. Geophysical data (Molina and Sonaglia, 1969) indicates magnetic anomalies with an E-W trend in the western sector of the Sabatini complex. These authors attribute the anomalies to flat apophyses of a hypothetical near-surface magmatic mass associated with a 2 km deep subvolcanic body. This latter interpretation is incompatible with new data collected by ENEL during drilling in the Baccano area. Drilling and surface data suggest that the anomalies could be due to small intrusions in the carbonate resevoir. The fragments of reservoir rock are less frequent than those from the shallower cover units. It is possible that the fluids sharply expand during their rise and their explosive products consequently involve a larger volume of near surface rocks.

The eruptive activity then moved westward into the area of the Baccano structural high which was by that time buried by volcanic materials. Baccano caldera was formed by consecutive collapses induced by numerous hydromagmatic explosions. Explosive style at Baccano differed from that of Sacrofano by the recurrence of hydromagmatic activity. This may have been due to a continuous recharge of the aquifer in the underlying carbonate units. Volcanism in the Baccano area may have been triggered by the contact of external water with the magma. Subsequent explosions then allowed magma to rise in the conduit, as evidenced by the emplacement of the Baccano pyroclastic flow that spread eastward towards the depression of the Sacrofano caldera. This pyroclastic flow lacks surge-like depositional features (sandwaves, inverse grading, stratification) common in the upper pyroclastic flow at Sacrofano. This difference suggests that, although genetic conditions for the two flows were similar, the triggering mechanism and the structural situation of the area at the time of activity were different. Eruption of the Sacrofano pyroclastic flow was triggered during a period of volcanic quiescence. We speculate that after the initial explosive paroxysm, the magma stagnated in the vent, dropping to lower levels, and giving rise to collapse of the wall and to movement along near surface caldera faults. Under these circumstances, the diminished hydrostatic pressure of the magma could allow aquifer water to enter the vent.

The eruption was presumably stronger than those that produced pyroclastic flows at Vesuvius as proposed by the model of Sheridan et al. (1981). Prior to emplacement of the Baccano pyroclastic flow the rise of magma in the vent exerted a pressure that sealed the aquifer. As a result, this pyroclastic flow was erupted without appreciable interaction of external water. After the emplacement of this unit the level of magma again fell in the conduit, causing the first caldera collapse. Volcanism then resumed with hydromagmatic characteristics, apparently derived from the reinstatement of conditions similar to those prior to collapse. A drop in magma level combined with hydrostatic pressure in the aquifer allows water to enter the conduit and to come into contact with magma.

Distribution of ejecta and morphological evidence show that the first
hydromagmatic explosions occurred in the southern section of the caldera, but
the final activity took place to the north. Concurrently with the last stage of
activity of Baccano, the Martignano center and other explosive vents north and
west of Baccano were active. These centers produced sporadic hydromagmatic and
phreatic explosions with only a few magmatic explosions (Fig. 3f). The magmatic
explosions may be linked with residual pockets of magma which were
intermittently put in contact with shallow aquifers, perhaps related to the
presence of Lake Bracciano, by gravity faults.

RELATIONSHIP OF VOLCANISM TO TECTONISM

The evolution of the Sacrofano caldera is closely related to regional
structures that reflect the tectonic style of the central Apennines (Fig. 1).
Sacrofano lies within a structural depression bounded on the west by the small
Cesano high that emerged early in the activity. The eastern boundary is the
carbonate buttress of the Soratte-Cornicolani mountains. The structural
configuration of the area influenced the location and the nature of the
volcanism as well as controlled aquifer recharge and the resultant variation in
water/magma interactions. The westward migration of volcanism within the
Sacrofano volcanic center is a response to active E-W extension in this area.
Hydromagmatic volcanism was concentrated on the Cesano structural high and was
also controlled by faults. The faults that controlled the gradual collapse of
the Cesano structure were a part of the distensive tectonics which involved the
central belt of the central-southern Apennine starting in Late Miocene time.
The carbonate rocks of Cesano form the reservoir for the regional aquifer.
Faulting could rapidly mobilize this supply of water leading to phreatic or
hydromagmatic activity.

These considerations allow a reasonable reconstruction of the
volcano-tectonic, hydrogeologic and climatologic characteristics of the Sabatini
area between 0.4 and 0.3 m.y. ago. It was during this period that most of the
large pyroclastic flows were emplaced. Climatologic data indicate a sudden drop
in temperature, indicated by isotopic analysis of oceanic sediments (Bowen,
1978). Contemporaneous sedimentary units in coastal outcrops north of Rome
(Conato et al., 1980) indicate a significant transgression ("S. Cosimato
formation"). Sedimentary units of this period in the vicinity of the volcano
rest on a surface due to the uplift and erosion which must have accompanied the
paroxysmal stage of Sacrofano.

A comparison of climatic conditions, structural evolution, and
characteristics of the sedimentary basement reveals a very significant
hydrogeological relationship that is relevant for understanding hydromagmatism
in the area. Recharge during this period is under maximum stress conditions
(Serva and Salvini, 1978). Their deep continuity is provided by thick (over
1,500 m) layers of Mesozoic carbonate units. Even the Upper Trias
(Norian-Rhaetian) is predominantly calcareous (Funiciello et al., 1979;

Mariotti, 1980) Considering the relationship between effective infiltration and lithology in various karst systems of central Italy (Boni and Bono, 1982), an almost "privileged" condition can be supposed for the Sacrofano area. During climatic conditions with inflows equal to or higher than the present and distensive tectonism at its peak, the thick limestones of the nappes allow deep circulation. Finally, circulation of active geothermal fluids in the main aquifers could have produced "hydrothermal karsts" which would be even more favorable for possible water/magma contact.

The close link between tectonism and volcanism suggest that eruptions in the Sabatini area may have been triggered by gravity faults that control the horsts and grabens of the more general regional tectonic pattern (Serva and Salvini, 1978). The volcanism in the eastern part of the Sabatini volcanic complex started along the edge of the main graben where tectonic movements were evidently more intense. These strongly fractured rocks provided easy routes for the rise of magma (Funiciello and Parotto, 1978). The principal zone of volcanism moved westward with time as a result of diminishing tension at the graben edges. During this stage, the faults converged, forming the "median ridge" of the Cesano high facilitating the rise of magma in this sector. The succession of hydromagmatic explosions in the Sabatini complex are connected with interaction between magma and regional aquifers. The main factors related to hydromagmatic activity include the depth, recharge capacity and hydrostatic pressures in volcanic conduits.

The first center of volcanic activity, prior to the formation of Sacrofano, was on the slope of the Mte. Soratte structure between the small towns of Morlupo and Castelnuovo di Porto. The initial volcanic products were chemically evolved (trachytic lavas of Morlupo). However, the chemical character of the volcanism soon changed to classic alkaline-potassic magmas. At the onset of volcanism in the Sabatini complex, structural conditions favored magma/water interaction. A volcanic edifice, the exact location of which remains uncertain, was essentially built of pyroclastic products with a silica unsaturated chemical character. Toward the end of the activity, falling magmatic pressures coupled with faulting that preceded the caldera collapse allowed the aquifer water to contact magma bringing about a series of hydromagmatic explosions and emplacement of surge deposits as the caldera finally collapsed.

Near-surface fracturing linked with the formation of the caldera probably caused the sudden isolation of the aquifers from the magma. Meanwhile, the structural evolution led to a new distribution of stresses concentrated at the edges of the "median ridge". Volcanism migrated westward along new routes of magma ascent starting the construction of the Sacrofano complex. The first stage of growth of Sacrofano was marked by the eruption of silica undersaturated magma which had high eruptive energy. Magmatic pressure was sufficient to keep the aquifer water away from the magma. The climax of activity at Sacrofano occurred at a time of intense volcanism throughout the Sabatini complex. This peak in eruption of silica unsaturated magmas corresponded to the peak of regional tensional movements in the area. It is not by chance that during the same period (0.4-0.3 m.y. ago), all of the main volcanic complexes of central Italy reached their paroxysmal stage. The Vulsini complex underwent volcano-tectonic movements that formed Lake Bolsena at approximately 0.4 m.y. ago. The intense volcanism associated with this tectonism included large pyroclastic flows and lavas with tephritic-leucitic composition. In the same period at Vico at least four pyroclastic flows were emitted, forming a caldera. Pyroclastic flows that cover an area of approximately 1,500 km^2 in the Latian Volcano were emplaced about 0.4 m.y. ago. A caldera was also formed there at the end of this stage. Almost all the volcanism of the Media Valle Latina developed in the same period and the Roccamonfina caldera was formed. The minor intro-Apennine volcanoes that developed in areas that had not previously experienced volcanism (S. Venanzo) may also belong to the same period.

Volcanic vents in the Sabatini area are predominantly concentrated along a NNW-SSE trending belt from Vico to Sacrofano. This zone is also probably associated with the most intense tectonic deformation. In the period between 0.4-0.3 m.y. ago, the depression of Lake Bracciano formed and the Cesano high further sank. Emptying of the Sacrofano chamber at the end of the paroxysmal stage caused the magma level to drop, relieving hydrostatic pressures. Intense fracturing associated with the caldera collapse penetrated the carbonate reservoir, mobilizing enough water to offset the magma pressure in the conduit. Gradual magma/water interactions began. A normal pyroclastic series with variable eruptive energy passed into sporadic hydromagmatic episodes and finally to the emission of a "wet" pyroclastic flow in a situation similar to that described for Vesuvius by Sheridan et al. (1981).

The violence of the latter explosion and the great amount of magma emitted led to the caldera collapse which brought volcanic activity of the Sacrofano center to a close. The collapse of Sacrofano must have been controlled by deep regional faults with N-S and E-W trends (ENEL report, in press). This fault system also influenced the simultaneous sinking of the Cesano high. Such faults brought about a sudden increase in water/magma ratios reflected in the erupted products. At this stage the Mte. Razzano tuff cone was constructed. The eruptive potential in this area was not yet exhausted, even though tectonic conditions favored a further westward migration of volcanism. The nature of volcanism changed slightly and the products of the new Baccano center, which developed on the western rim of the old Sacrofano caldera, produced a near-saturated magma composition. In this stage, water availability was probably greater owing to the fact that Baccano was located on the Cesano structural high, the only sizeable reservoir in the area.

The role played by the lake in the Bracciano basin remains doubtful. The lake must have had its present size at that time. Its shallow, but wide, aquifer could have interfaced with the volcanic activity. The deep aquifer within the carbonate units probably supplied the water for highly explosive and hydromagmatic volcanism of Baccano, while the small local centers at the edge of the lake were created as a result of minor contacts of magma with a shallow aquifer. In these centers, phreatic and/or hydromagmatic explosions alternated with definitely magmatic explosions (Trevignano-Villa di Valle).

The magmatic activity of these centers was probably connected to regional faults along which the Lake Bracciano basin sank. Rising magma intermittently contacted the lacustrine aquifer when the hydrostatic conditions were favorable. This near-surface hydromagmatic volcanism was intermittent and generally marked by only one explosion. The collapse of Baccano caldera ended central eruptive activity and the westward migration of volcanism. Only local explosions occurred due to occasional contacts between magma and shallow aquifers. Phreatic and hydromagmatic explosions brought to a close the volcanism over all the Sabatinian volcanic complex.

ACKNOWLEDGEMENTS

The new data presented in this paper was collected during a systematic exploration of the Cesano geothermal area by ENEL and detailed field studies conducted by the authors and other colleagues as part of the Geolozio Project of the Faculty of Science of Rome University, the Progetto Finalizzato "Geodinamica" of CNR, and the ENEL-AGIP joint venture research undertaken by a group of researchers from IGP and CNR. Early versions of this manuscript were read by M. F. Sheridan, D. M. Ragan, and T. C. Moyer, whose comments have improved this paper; the final version benefited from the substantial help of M. F. Sheridan.

REFERENCES

Alvarez, W., Gordon, A., and Rashak, E., 1975. Eruptive source of the "tufo russo a scorie nere," a Pleistocene ignimbrite north of Rome. Geol. Rom., 14: 141-154.

Baldi, P., Decandia, F.A., Lazzarotto, A., and Calami, A., 1974. Studio geologico del substrato della copertura vulcanica laziale nella zona dei laghi di Bolsena, Vico e Bracciano. Mem. Soc. Geol. Ital., 13: 575-606.

Bigazzi, G., Bonadonna, F.P., Iaccarino, S., 1973. Geochronological hypothesis on Plio-Pleistocene boundary in Latium region (Italy). Bull. Soc. Geol. It. 92: 391-422.

Bonadonna, F.P., Bigazzi, G., 1969. Studi sul Pleistocene del Lazio. VII Eta di un livello tufaceo del bacino diatomitico di Riano stabilita con il metodo delle tracce di fissione. Bull. Soc. Geol. It. 88: 439-444.

Boni, C.F., Bono, P., 1982. Relation entre infiltration efficace et lithologie dans dix systemes karstiques de l'Italie centrale. Bull. III Bureau Geol. et Miner. (in press).

Borghetti, G., Sbrana, A., Sollevanti, F., 1981. Vulcano tettonica dell'area dei M. Cimini e rapporti cronologici tra vulcanismo crimino e vicano. Soc. Geol. It. seduta scientifica "Vulcanismo e tettonica" 18-19 giugno 1981. Gargnano.

Bowen, D.O., 1978. Quaternary Geology. 221 pp.

Calamai, A., Cataldi, R., Dall'Aglio, M., Ferrara, G.C., 1975. Preliminary report on the Cesano hot brine deposit (northern Latium, Italy). U.N. Symp. on Geoth. Ener. S. Francisco, CA., USA 1: 305-313.

Cameli, G. M., Mouton, J., Toro, B., 1976. Contribution of geophysical surveying in the discovery of Cesano geothermal field (northern Latium, central Italy). Int. Congr. on thermal waters, Geoth. Ener. and Volc. of the Mediterranean area, Athens Proc. Geoth. Ener. 1: 130-143.

Civitelli, G., Funiciello, R., Parotto, M., 1975. Caratteri deposizionali dei prodotti del vulcanismo freatico nei Colli Albani. Geol. Rom., 14: 1-39.

Conato, C.F., Esu, D., Malatesta, A., Zarlenga, F., 1980. New data on the Pleistocene of Rome. Quarternaria XXII.

Corda, L., DeRita, D., Tecce, F., 1976. Caratteri granulometrici delle vulcaniti freatomagmatiche nell'area laziale. Bull. Soc. Geol. It., 95: 1235-1252.

Corda, L., De Rita, D., Tecce, F., Sposato, A., 1978. Le piroclastiti del sistema vulcanico sabatino: il complesso dei tufi stratificati varicolori dè la Storta. Bull. Soc. Geol. It., 97: 353-366.

De Rita, D., Sheridan, M., Marshall, J., 1982. SEM surface textural analysis of phenocrysts from pyroclastic deposits in the Sabatini volcanic field, Latium, Italy. Scanning Electron Microscopy in Geology, Geoabstracts Norwich England.

Evernden, J.F., Curtis, G.H., 1965. The potassium argon dating of late Cenozoic in East Africa and Italy. Cum. Antr., 6: 363-364.

Funiciello, R., Locardi, E., Lombardi, G., Parotto, M., 1976. The sedimentary ejecta from phreatomagmatic activity and their use for location of potential geothermal areas. Int. Cong. Thermal waters Geoth. Ener. and Volc. of the Mediterranean area Athens.

Funiciello, R., Locardi, E., Parotto, M., 1976. Lineamenti geologici dell'area satatina orientale. Bull. Soc. Geol. It., 95: 831-849.

Funiciello, R., Parotto, M., Salvini, F., Locardi, E., Wise, D.U., 1977. Correlazione tra lineazioni rilevate con il metodo shadow e assetto tettonico nell'area vulcanica del Lazio. anno XXXVI Bull. Geod. e Sc. affini n°4.

Funiciello, R., Parotto, M., 1978. Il substrato sedimentario nell'area dei Colli Albani (vulcano laziale). Geol Rom. 17: 23-287.

Funiciello, R., Mariotti, G., Parotto, M., Preite, Martinez M., Tecce, F., Toneatti, R., turi, B., 1979. Geology, mineralogy and stable isotope geochemistry of the Cesano geothermal field (Sabatini Mts. volcanic system, northern latium, Italy). Geothermics 8: 55-73.

Locardi, E., Sommavilla, E., 1974. I vulcani sabatini nell'evoluzione della struttura regionale. Mem. Soc. Geol. It. sup. 2, 13: 455-468.

Mariotti, G., 1980. Il Lazio settentrionale nella paleogeografia triassica: dati preliminari. Rend. Soc. Geol. It., 3: 25.

Mattias, P.P., Ventriglia, U., 1970. La regione vulcanica dei Monti Sabatini e Cimini. Mem. Soc. Geol. It., 9(3): 281-449.

Mattson, P.M., and Alvarez, W., 1973. Base surge deposits in Pleistocene volcanic ash near Rome. Bull. Volcanol., 37: 553-572.

Moderni, P., 1896. Le bocche eruttive dei vulcani sabatini. Bull. Re. Com. Geol. It., 27(1/4).

Molina, F., Sonaglia, A., 1969. Rilevamento geomagnetico degli apparati vulcani vicano e sabazio. Ann. Geoth. 22(2): 147-162.

Nappi, G., De Casa, G., Volponi, E., 1979. Geologia e caratteristiche del tufo giallo della Via Tiberina. Bull. Soc. Geol. It., 98: 431-445.

Serva, L., Salvini, F., 1978. Analisi statistica delle deformazionni meccaniche in alcune strutture dell-Appennino laziale. Bull. Soc. Geol. It., 95: 1213-1233.

Scherillo, A., 1941. Studi su alcuni tufi gialli della regione sabazia orientale. Per. Min., 12: 381-417.

Sheridan, M., Barberi, F., Santacroce, R., Rosi, M., 1981. A model for the Plinian eruptions of Vesuvius. Nature, 289: 282-285.

Sheridan, M.F., and Wohletz, K.H., 1981. Hydrovolcanic explosions: the systematics of water-pyroclast equilibration. Science, 212: 1387-1389.

Smith, R.L., 1960. Ash flows. Geol. Soc. Am. Bull. 71(6): 795-841.

Wohletz, K.H. and Sheridan, M.F., 1979. A model of pyroclastic surge. Geol. Soc. Am. Sp. Paper 180: 125-136.

A GENERAL MODEL FOR THE BEHAVIOR OF THE SOMMA-VESUVIUS VOLCANIC COMPLEX

ROBERTO SANTACROCE

Dipartimento di Scienze della Terra, University of Pisa, and Centro per la Geologia Strutturale e Dinamica dell'Appennino, CNR, Pisa (Italy).

(Received September 16, 1982; revised and accepted October 27, 1982)

ABSTRACT

Santacroce, R., 1983. A General Model for the Behavior of the Somma-Vesuvius Volcanic Complex. In: M.F. Sheridan and F. Barberi (Editors), Explosive Volcanism. J. Volcanol. Geotherm. Res, 17: 237-248.

Three classes of behavior can be distinguished in the eruptive pattern of the Somma-Vesuvius complex: 1) small-scale (volumes of magma equal about 0.01 km^3), mainly effusive activity, 2) intermediate-scale (V = 0.1 km^3), explosive or explosive-effusive events, and 3) large-scale (V = 1.0 km^3), explosive eruptions. The sequence within each major volcanic cycle (six occurred in the last 17,000 years) seems to be characterized by an inverse power distribution of energy release. The large-scale Plinian eruptions are the main perturbing events. Subsequent episodes tend to restore the pre-Plinian quasi-equlibrium conditions. The clear direct relation apparent between the length of the repose intervals, the volumes of erupted products and the degree of evolution of the erupted magmas, strongly suggest a continuous supply and storage of deep basic magma within a shallow magma chamber where magma is subject to crystal fractionation. A vertical compositional stratification forms within the reservoir because differentiation proceeds with continuous feeding of deep magma. The longer the repose-time and the larger the volume of stored magma, the more highly-evolved is the top of the stagnant magma. Triggering of the eruptions in closed conduit conditions essentially can be ascribed to the volatile pressure increase related to crystallization. The load of cover rocks is then a critical factor in determining the length of repose-time and the type of eruption.

INTRODUCTION

The activity of the Somma-Vesuvius volcanic complex can best be described by considering two levels of models: 1. specific models to explain the physical phenomena associated with each phase of activity and 2. a general model of the behaviour of the volcano through time. The specific models are strongly influenced by the degree of fractionation of magma and by the interaction with external water and country rocks. A model was proposed for the A.D. 79 event by Sheridan et al. (1981) and that can be extended to all major Plinian eruptions of Vesuvius (Rosi, Santacroce and Sheridan, in prep.). The most important factor in the general model appears to be the repose-time which controls the degree of fractionation within the magma chamber and the volume of magma erupted.

In this paper I will first discuss the types of eruptions that characterize the volcanological history of Vesuvius, followed by a qualitative, general predictive model of the volcano's activity based on its historical record.

TYPES OF ERUPTION

Vesuvius displays a complete spectrum of activity ranging from cinder cones and lava-flows to catastrophic Plinian eruptions. Its activity can conveniently be divided into three classes of behaviour:
1. Small-scale, mainly effusive (cinder cones and lava-flows);
2. Intermediate-scale, either solely explosive ("Pollena type") or explosive-effusive ("1631 type").
3. Large-scale, explosive ("Pompeii type").

The first type consists of small volumes, short repose-times and primitive magmas. The second type has moderate volumes, longer repose-times and intermediate plus primitive magmas. The third type has large volumes, long repose-times and evolved plus intermediate and, possibly, primitive magmas.

At least six major pumice eruptions have occurred at Vesuvius during the last 17,000 years (Delibrias et al., 1979; Rosi, Santacroce and Sheridan, 1981). The typical sequence of events during a Plinian pumice eruption displays a phreatomagmatic character (Sheridan et al., 1981). These highly explosive eruptions start with a paroxysmal ejection of large volumes of pumice (up to 4.0 km^3) with minor lithic blocks. Their characteristic pumice-fall beds have maximum thicknesses of about 3-4 m near the vent and extend in the dominant wind direction to cover several hundreds of km^2. Pyroclastic surges are common during the last part of the pumice-fall phase. The emplacement of pumice-flows, ash-flows and mud-flows generally concludes the eruption (Fig. 1). Ash-clouds, ground-surges and mud-hurricanes are generally associated with pyroclastic flows over limited distances. The duration of a single Plinian eruption cycle does not exceed a few days. Plinian events usually mark the beginning of a new eruptive cycle (Delibrias et al., 1979) and follow long periods of quiescence. This fact, coupled with the generally highly evolved (phonolitic to trachytic) chemical nature of the Plinian pumice, suggests that this kind of eruption requires the common basic magmas of Somma-Vesuvius (tephritic/basanitic to leucititic) to have a long residence in a shallow magma chamber in order to produce the residual liquids of the pumice by fractional crystallization (Barberi et al., 1982).

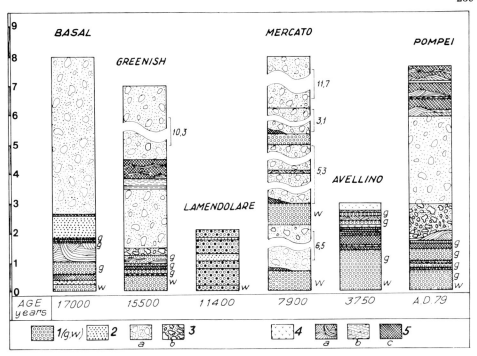

Fig. 1 Schematic type-sections from deposits of the six major Plinian eruptions of Vesuvius during the last 17,000 years. 1. pumice-fall deposits (g = dark pumice; w = light pumice); 2. black lapilli- and scoria-fall deposits; 3. pyroclastic-flow deposits (a = ash flows, b = pumice flows); 4. mud-flow and stream deposits; 5. pyroclastic surge deposits (a = sandwave beds, b = planar beds, c = massive beds).

Petrological data, on the depth of the magma chamber in the papers by Barberi and Leoni (1981) and Barberi et al.(1982) suggest on a P_{H_2O} of about 1.0 Kb for the fractionation of both Pompeii and Avellino pumice. The nature of the lithic ejecta provides further indications and suggests a possible lower pressure during the fractionation of the Lagno Amendolare pumice (absence of carbonatic clasts). The nature of the country rocks surrounding the magma chamber influences the modalities of the eruption: the hydromagmatic character is, in fact, significantly less marked in the Lagno Amendolare pumice eruption, possibly as a result of the minor importance of the aquifers within the volcanic sequence compared to those within carbonate rocks.

The strong darkening shown by the upper part of most pumice-fall levels ("Basal Pumice", "Greenish Pumice", "Avellino Pumice" and "Pompeii Pumice") or the interbedding of thin beds of dark and pale pumice ("Lagno Amendolare Pumice"), coupled with some evidence of mineralogical inhomogeneity, indicate that the upsurge of new basic magma is possibly responsible for triggering some

Plinian eruptions. Further verification of this hypothesis is now in progress.

Frequent large explosive eruptions that produced significant pumice-fall, surge, pyroclastic flow and mud flow deposits have occurred between the major Plinian episodes. These eruptions involve significantly smaller volume than their Plinian counterparts (0.1 to 0.5 km^3). Although the magma is less highly-evolved, the carbonate blocks and surge deposits suggest an important hydromagmatic component to these eruptions. The eruption cycle typically follows the pattern of pumice-fall, surge, pyroclastic-flow and mud-hurricane deposits. The Pollena eruption (Fig. 2), considered a typical intermediate eruptive sequence, has been studied in detail (Rosi and Santacroce, this volume).

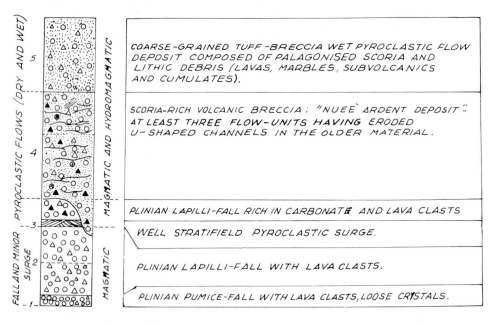

Fig. 2. Schematic section from deposits of the A.D. 472 "Pollena" intermediate-scale explosive eruption (after Rosi and Santacroce, this volume). Five eruptive phases have been recognized: 1. lapilli-fall tephra (light pumiceous lapilli and loose crystals); 2. greenish lapilli-fall with sparse lava clasts; 3. pyroclastic surge covered by greenish lapilli-fall with carbonate and lava clasts; 4. scoria-rich volcanic breccia (glowing cloud deposit) made up of at least three flow-units. It contains abundant carbonized wood; and lava and carbonate clasts; 5. coarse-grained tuff-breccia: lithified, wet pyroclastic-flow deposit containing large, black scoria and lithic clasts (lava, marble, subvolcanic, and cumulate rocks).

Volumes of products, comparable with those erupted during the Pollena-type episodes, characterize the severe eruption of A.D. 1631 that opened the recent period of Vesuvius activity. After more than a century of complete quiescence the eruption started at dawn on December 16th with strong explosions and emissions of large volumes of ash. During the night huge lava-flows that erupted from two subterminal vents rapidly reached the sea by means of several fingers between Torre del Greco and Torre Annunziata. After about a week of both strong effusive and explosive activity, the intensity of the eruption decreased and ended entirely on January 1, 1632. Several mud flows occurred throughout the eruption. The presence of pyroclastic flows is uncertain due to the difficulty of distinguishing the range of deposits pertaining to this eruption. Hydromagmatic phenomena, if present, were nevertheless much less conspicuous than those of the Pollena-type eruptions. The available data allows a tentative speculation on the temporal variation in the composition of the erupted lavas (phonolitic leucitites followed by tephritic leucitites).

The most recent period of the volcano history (A.D. 1694-1944) was characterized by semi-persistent, relatively mild activity (lava fountains, gases and vapour emission from the crater), frequently interrupted by short quiet periods that never exceeded seven years (Fig. 3). Rare minor eruptions that occurred during the semi-persistent, Strombolian-like activity of the volcano, mainly consisted of small lava-flows emitted from the summit crater. More important eruptions closed each of these short cycles. A schematic representation of these short cycles is shown in Fig. 4. Pasty fluid blebs, spherical bombs and ashes were violently ejected from the summit crater accompanying the effusion of moderately fluid lavas. Lavas were sometimes produced by lateral cinder cones on the flanks of the volcano. The emplacement of relatively small mud flows along stream valleys cutting the Monte Somma slopes was usual during these eruptions. This was especially true if the water gained access to the vent during a lateral eruption so that a large quantity of ash was produced. Likewise hot pyroclasts sometimes slumped from the oversteepened cone to form the so-called "hot-avalanches" (Perret, 1924). Such activity was not restricted to the last centuries but can also be recognized in the Monte Somma lava flows, the Pollena and Camaldoli cinder cones and possibly in other products ascribed to the time interval 3,800 y.b.p. to A.D. 79. Delibrias et al. (1979) suggest that periods of recurrent activity have taken place between most major pumice eruptions.

SOME FACTS

Available volcanological and petrochemical data lead to a greater understanding of the eruptive history of the volcano and contribute to modelling its future activity:
(1) The large-scale catastrophic Plinian eruptions follow long quiescent periods, probably exceeding ten centuries. Intermediate-scale events (both "Pollena" and "1631" types) have been preceeded by shorter pauses of the order

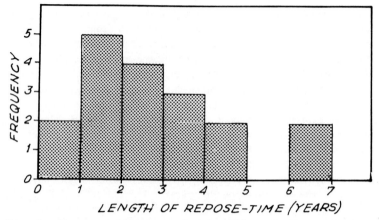

Fig. 3. Frequency histogram of repose-times lasting more than 1 year in the period A.D. 1694-1944.

of one or two centuries. Small-scale effusions occur after intervals between eruptions probably not exceeding ten (or a few tens of) years. A clear direct relation between the length of the repose intervals and the volumes of erupted products that is apparent from these qualitative data is tentatively illustrated in Fig. 5.

(2) The historic record suggests that the effusion rate is directly related to the volume of magma involved. Therefore, the duration of the repose intervals also controls the power of the eruption. This is qualitatively expressed by the volcanic explosivity index (VEI) of Newhall and Self (In Simkin et al., 1981).

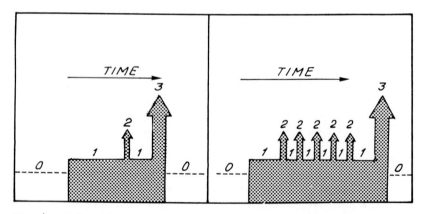

Fig. 4. Schematic representation of the short eruptive cycles of the recent (1694-1944) history of Vesuvius. 0 = repose; 1 = persistent activity; 2 = intermediate eruption; 3 = final eruption. (Redrawn after Principe, 1979).

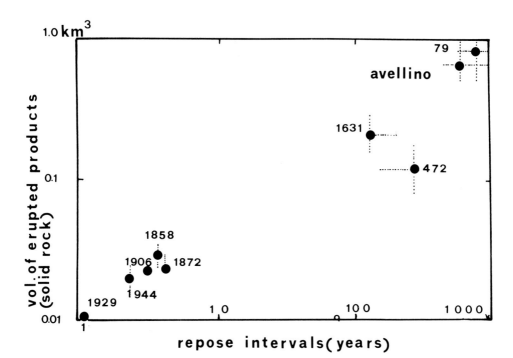

Fig. 5. Repose intervals vs. volume of erupted products (calculated as solid rock equivalent). Volume estimate after Lirer et al. (1973), Principe (1979) and unpublished data. Arrows suggest the range of possible errors.

(3) The long repose-time following the A.D. 1944 eruption and the emission of mafic cumulates in the last stages of this eruption strongly suggest the complete emptying of the shallow feeding system, the volume of which can be assumed to correspond to the volume of products erupted during such an eruption (0.023 km^3 according to Principe, 1979).

(4) The products of Somma-Vesuvius cover a large compositional spectrum including members of both the potassic and high-potassic series. The degree of evolution of the erupted magmas appears strongly related to the length of the repose intervals: salic (trachytic or phonolitic) magmas characterize Plinian episodes, while mafic ones (basanitic to leucititic) are typical of the small-scale to moderate eruptions. Significant compositional gradients (the degree of evolution of eruption products decreases with time) mainly appear in eruptive sequences of large- and intermediate-scale eruptions (Fig 6).

(5) Each cataclysmic Plinian episode (six in the last 17,000 years) marks a sharp change in the nature of the erupted products: the composition of the Plinian products proves distinct from post-Plinian products, but does show some affinities with pre-Plinian ones (Fig. 7).

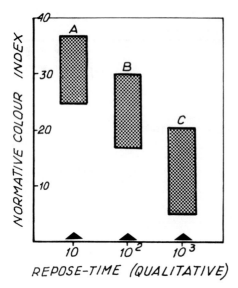

Fig. 6. Variation of the normative color index as a function of the order of magnitude of repose-time preceeding the eruption: A = 52 unpublished analyses of the A.D. 1694-1944 period; B = 7 analyses of the A.D. 472 Pollena eruption (Rosi and Santacroce, this volume); C = 12 analyses (mainly unpublished) of the Pompeii eruption (C.I. = 7.8-18.9); 8 analyses (mainly unpublished) of the Avellino eruption (C.I. 5.2-21.6); 6 analyses (unpublished) of the Mercato eruption (C.I. = 5.2-6.9).

THE MODEL

The activity of Somma-Vesuvius during the last 17,000 years can be divided into six major volcanological cycles, each opening with a catastrophic Plinian eruption and closing with a long repose-time lasting several centuries (Fig. 8). The reconstruction of events within each cycle is possible only for the youngest two. However, the reliability and detail of these reconstructions are quite different, due to the scarcity of significant complete sections older than A.D. 79. The sequence of event within each cycle seems to be characterized by an inverse power distribution of the energy release. The Plinian eruption is the main perturbing event and the following episodes tend to progressively restore pre-Plinian pseudo-equilibrium conditions.

The direct relation between repose-times and volumes of erupted products strongly suggests a continous supply of deep basic magma that is stored within a shallow magma chamber. Repeated rise of deep magma is revealed by the intermediate eruptions (Fig. 4) that occurred during short cycles of the recent activity. Some of these periods (e.g. in 1891-1894 and 1895-1899) produced larger volumes than that of the 1944 eruption. In these cases, the volume of

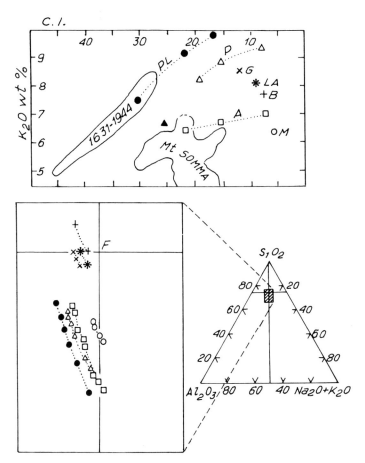

Fig. 7. Normative colour index vs. K_2O wt% and $Al_2O_3-SiO_2(Na_2O+K_2O)$ molecular diagrams of Somma-Vesuvius products. PL = A.D. 472 Pollena eruption; P = A.D. 79 Pompeii eruption; A = 3500 y.b.p. Avellino eruption; M = 7900 y.b.p. Mercato eruption; G = 15,500 y.b.p. "Greenish" pumice eruption; LA = 11,400 y.b.p. Lagno Amendolare eruption; B = 17,000 y.b.p. "Basal pumice" eruption. The fields of A.D. 1631-1944 and pre-17,000 y.b.p. lavas are also reported. F is the feldspar point.

products and the duration of the eruption are exclusively determined by the amount and rate of deep feeding. Within the chamber, the magma is subject to crystal fractionation. The two processes of differentiation and continous deep replenishment allow the development of a vertical compositional stratification. The longer the repose-time, the larger the volume of molten magma and the more evolved is the top of the stagnant magma body.

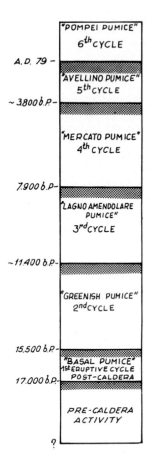

Fig. 8. The six main volcanic cycles in the last 17,000 years of the life of Somma-Vesuvius.

The inverse power distribution of energy release during each major volcanic cycle suggests, as a general rule, that there were no external causes triggering eruptions in closed conduit conditions. These events can be essentially ascribed to increase of volatile pressure related to degassing of crystallizing magma. The load of cover rocks is thus a critical factor in determining the length of the repose-time and hence the type of eruption. The cycle-opening Plinian eruption greatly lightens the system. At the same time it completely empties the chamber (Sheridan et al., 1981), provoking large gravitational collapse thus preventing the immediate rise of deep basic fresh magma to the surface. The energy needed to break the conduit plug progressively decreases through Pollena-type, intermediate-scale events until the point at which a 1631-type eruption leaves the chimney more or less wide open, establishing a regime of semi-persistent effusive activity Under these conditions the triggering of the intermediate eruptions of Fig. 4 should be ascribed to pulses of deep basic magmas, while the final eruptions could be provoked by moderate, discontinuous magma-water interactions, probably resulting from external causes (e.g. earthquakes). This small-scale activity tends to reinstate pre-Plinian conditions and possibly ceases when volumes of magma comparable with those dispersed during the eruptions have accumulated on the volcano slopes.

SOME SPECULATIONS

As a first approximation, the rate of magma replenishment can be considered to be constant over each magmatic cycle (terminating with a large-scale Plinian eruption). For the moment this can be approximated by the average eruption rate of undifferentiated basic magmas during the 1694-1944 period which is estimated at $1.6 \times 10^{-3} km^3$/year (after data of Cortini and Acandone, 1982). Thus, it would be possible to estimate the volume of magma involved for a Vesuvian eruption occurring today, 38 years after the last eruption in 1944. This calculated value of about $60 \times 10^{-3} km^3$ would place the eruption below the intermediate-scale class type, but would nevertheless, be the most violent since 1631. Such a value must be considered a maximum value because not all of the stored magma may be released.

The 38 year repose-time since the 1944 eruption distinguishes the present state of Vesuvius from the 1694-1944 period of its history (Carta et al., 1979) for which seven years was the maximum repose. The total volume of products erupted and accumulated on the volcano after the A.D. 79 event (roughly estimated as about 2.5 km^3) approximately equals the mass of products dispersed out of the volcanic system during A.D. 79 and subsequent explosive episodes. These facts could support the hypothesis that the volcano is presently in a pre-Plinian state that could be expected to continue for centuries longer.

The geochemical changes in the erupted magmas immediately following major Plinian eruptions cannot be fortuitous so that cause-effect relation must be investigated. It is unlikely that the load pressure decrease related to the eruption and dispersal of nearly one cubic kilometer of solid rock would be sufficient to bring about a significant change in the physical conditions for melting of the deep magmatic source. The pressure release connected to the Plinian event could, however, provoke an upward movement of volatiles and a consequent P_f decrease, possibly capable of explaining the observed chemical changes and probably reflecting a decrease with time of the degree of partial melting.

ACKNOWLEDGEMENTS

This paper was supported by CNR Italian Geodynamic Project, publication No. 496.

REFERENCES

Barberi, F., Bizouard, H., Clocchiatti, R., Metrich, N., Santacroce, R., Sbrana, A., 1982. The Somma-Vesuvius magma chamber: a petrological and volcanological approach and geothermal implications. Bull. Volcanol. (in press).

Barberi, F., Leoni, L., 1980. Metamorphic carbonate ejecta from Vesuvius plinian eruptions. Evidence of the occurrence of shallow magma chambers. Bull. Volcanol., 43 (1): 107-120.

Carta, S., Figari, R., Sartoris, G., Sassi, E., Scandone, R., 1981. A statistical model from Vesuvius and its volcanological implications. Bull. Volcanol., 44 (2): 129-151.

Cortina, M. and Scandone, R., 1982. The feeding system of Vesuvius between 1754 and 1944. J. Volcanol. Geotherm. Res., 12: 393-400.

Delibrias, G., Di Paola, G. M., Santacroce, R., 1979. La storia eruptiva del complesso vulcanico Somma-Vesuvio ricostruita dalle successio ni piroclastiche del Monte Somma. Rend. Soc. It. Min. Petr., 35 (1): 411-438.

Lirer, L., Pescatore, T., Booth, B., Walker, G.P.L., 1973. Two plinian pumice-fall deposits from Somma-Vesuvius, Italy. Geol. Soc. Am. Bull., 84: 759-772.

Perret, F. A., 1924. The Vesuvius eruption of 1906. Carnegie Ist. Wash. Publ., 339, 151 pp.

Principe, C., 1979. Le eruzioni storiche del Vesuvio: riesame critico, studio petrologico dei prodotti ed implicazioni vulcanologiche. Thesis, University of Pisa, Italy. Unpublished.

Rosi, M. and Santacroce R., 1983. The A.D. 472 "Pollena" eruption: a poorly-known Plinian event in the recent history of Vesuvius. In: M.F. Sheridan and F. Barberi (Editors), Explosive Volcanism. J. Volcanol. Geotherm. Res., 17: (this volume).

Rosi, M., Santacroce R., Sheridan, M. F., 1981. Volcanic hazards of Vesuvius (Italy). Bull. B.R.G.M., IV (2): 169-179.

Sheridan, M.F., Barberi, F., Rosi, M., Santacroce, R., 1981. A model for Plinian eruptions of Vesuvius. Nature, 289: 282-285.

Simkin, T., Siebert, L., McClelland, L., Bridge, D., Newhall, C., Latter, J.H., 1981. Volcanoes of the world. Hutchinson Ross, Stroudsburg, PA., 232 pp.

THE A.D. 472 "POLLENA" ERUPTION: VOLCANOLOGICAL AND PETROLOGICAL DATA FOR THIS POORLY-KNOWN, PLINIAN-TYPE EVENT AT VESUVIUS

MAURO ROSI AND ROBERTO SANTACROCE

Dipartimento di Scienze della Terra dell'Università di Pisa and Centro per la Geologia Strutturale e Dinamica dell'Appennino, CNR, Pisa (Italy)

(Received September 16, 1982; revised version accepted November 2, 1982)

ABSTRACT

Rosi, M. and Santacroce, R., 1983. The A.D. 472 "Pollena" eruption: Volcanological and petrological data for this poorly-known, Plinian-type event at Vesuvius. In: M.F. Sheridan and F. Barberi (Editors), Explosive Volcanism. J. Volcanol. Geotherm. Res., 17: 249-271.

The pyroclastic products of a poorly-known eruption of Vesuvius (ascribed by a combination of historic and radiocarbon data to A.D. 472) have been investigated from both volcanological and petrological points of view. The eruptive sequence starts with pumice-fall deposits (three units can be recognized) that darken upwards where there are sandwave interbeds. Surge deposits cover the pumice-fall bed and thick pyroclastic-flow deposits represent the uppermost levels of the deposit. Isopach maps of both the pumice-fall and pyroclastic-flow deposits led to an estimate of the total volume of tephra of about 0.32 km^3. The eruptive sequence and the distribution of lithic ejecta are similar to those of the major Plinian eruptions of Somma-Vesuvius (although the volume involved is significantly lower) and reflect an increase in the hydromagmatic character of the eruption with time. The products range in composition from phonolites (first-erupted) to phonolitic leucitites with gradual changes upwards. Whole rock chemistry and microprobe mineralogy indicate that the Pollena sequence represents a liquid line of descent towards the phonolitic minimum of petrogeny's residua system. Fractionation occurred within a shallow magma chamber (P_{H_2O} probably slightly higher than 1 kb) and was mainly controlled by leucite and clinopyroxene. The basic parental magma approached the composition of the recent period (A.D. 1631-1944), tephritic leucitites of Vesuvius. The phonolitic magma can be derived from a leucititic parent by fractionating about 50% solid phases. A two-stage fractionation model is suggested: the first stage occurred during the rise of magma from the deep source and the second within the shallow magma chamber. The rate of magma

0377-0273/83/$03.00 © 1983 Elsevier Science Publishers B.V.

introduction during the 150 to 200 year repose time preceeding the eruption probably averaged 1.2 to 1.7 x 10^{-3} km^3-yr^{-1}. These conditions were probably favorable for the occurrence of magma-mixing within the convecting zone of the magma chamber.

INTRODUCTION

The A.D. 79 "Pompeii" eruption marks the beginning of volcanology's written history thanks to the description of Plinius the Younger; it was the first historically documented eruption. Chroniclers and historians of the Late Roman Empire and Middle Ages unfortunately were not as interested in observing volcanic phenomena as Plinius. As a result, the data available on Vesuvius until the XVIIth century are rare, scattered, and incomplete, leaving unaccounted large periods of time.

This paper describes the pyroclastic deposits ascribed to the A.D. 472 eruption, which have so far been known only through the recorded ash fall in Costantinople and rare hagiographic documents (Marcellinus, Procopius, etc.). In spite of the relatively moderate volume of erupted products (if compared with the major catastrophic Plinian episodes recurring in the history of Somma-Vesuvius), the A.D. 472 eruption was extremely disruptive and must be considered the most violent and fatal during the last nineteen centuries. A similar eruption occurring today, considering the present urbanization of the area, would have apocalyptic consequences (see Sheridan and Malin, this volume.)

A set of volcanological and petrochemical data have been collected on the pyroclastic products ascribed to this eruption. They provide a complete picture of the event that allow it to be placed in the eruptive and petrogenetic framework of the Somma-Vesuvius volcanic complex.

STRATIGRAPHY, VOLUME, AND AGE

Products of this eruption have been recognized on the northern slopes of Monte Somma and extend northeastward as far as 30 km. Twenty-two measured sections provide the data-base for this study (Fig. 1). The complete pyroclastic sequence of the Pollena deposit, reconstructed by comparing sections 4, 5 and 12 (Fig. 1), is presented as a composite type section (Fig. 2). Lateral variations in stratigraphy are indicated in Fig. 3.

The first-erupted product is a greenish pumice fall with abundant loose crystals (mainly leucite, clinopyroxene and biotite). Lithic lava fragments are abundant and increase towards the top of the pumice-fall bed, where carbonate ejecta also occur. Three ash-fall units can generally be distinguished by the presence of thin interbedded tuff levels and/or sandwave beds. The pumice darkens upwards. The fall deposit reaches a maximum thickness of about 2.0 m in sections 1 and 5. At Avellino (28 km from the vent) it is only 26 cm thick. The available data on pumice-fall thickness indicate a distribution strongly influenced by NE trending winds. The lapilli-fall isopach map (Fig. 4) permits

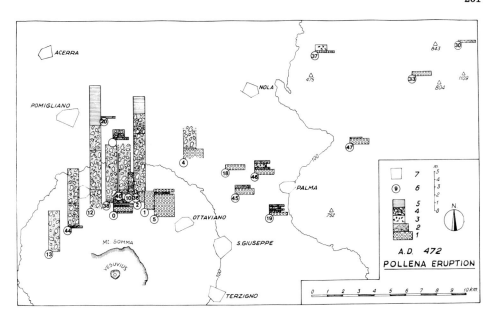

Fig. 1. Schematic stratigraphic sections of the Pollena deposit. 1. pumice-fall deposits; 2. surge deposits; 3. vesiculated tuffs; 4. pyroclastic-flow and glowing-cloud deposits; 5. coarse-grained tuff-breccia (wet pyroclastic-flow deposit?); 6. location and reference number of the measured sections; 7. major villages.

the volume of the fall deposit to be calculated at about 160×10^{-3} km^3.

Thin surge deposits showing lateral facies changes generally cover the pumice-fall bed. Sandwaves are mainly concentrated in outcrops near the vent (up to 5-6 km) whereas massive beds extend up to Palma Campana (12 km ENE) showing numerous transitions to vesiculated tuffs in the distal outcrops (mud hurricane deposits?).

Thick pyroclastic-flow deposits related to the last stages of the eruption have only been found in the NW sector of the volcano. They fill narrow paleovalleys and rapidly decrease in thickness where they run onto the plain surrounding the volcanic cone. The thickest observed sections occur at Pollena (more than 30 m, section 12) and Palmentello (about 12 m, section 36). Here four different flow units are recognizable. The first and the second are typical ash-flow tuffs containing large solid ejecta and abundant phaneritic blocks of leucite- and proxene-bearing scoria. The third flow unit exhibits typical features of glowing-cloud deposits. It is composed of a chaotic assemblage of dark grey, low-density, unwelded scoria, which also occur as scattered clasts in the underlying flows. The scoria are set in a fine sandy matrix containing blocks of old lavas and branches of carbonized wood with a random distribution. The uppermost unit is an ash-flow tuff that is thinner and

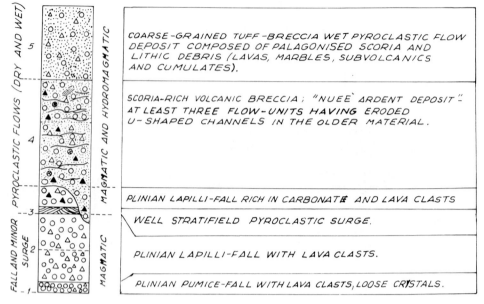

Fig. 2. Composite stratigraphic section of the Pollena deposit.

contains less solid ejecta than tuffs lower in the section. At Pollena (Sect. 12) the sequence ends with a coarse-grained tuff-breccia, probably resulting from the emplacement of a wet pyroclastic flow. It is composed of palagonized scoria and lithic debris (lava, marble, subvolcanics, and cumulates) scattered in a muddy, indurated matrix.

Isopachs were also drawn for the pyroclastic flow deposits. They depict a bi-lobate distribution due to the presence of an old topographic high (Fig. 4). Deposits thicker than 1.0 m occur as far as 8.0 km from the vent. According to the isopach map, the total volume of pyroclastic-flow deposits is estimated at 60×10^{-3} km^3 (the same as the lapilli-fall deposits). Flow isopachs allow a minimum volume estimate of these products. Flow channelization along paleovalleys has not been fully considered in the evaluation: reported measurements are mainly made on overbank deposits. The presence of pyroclastic-flow deposits buried under the younger products on the south side of the volcano cannot be excluded. Examples of the observed lateral variations, roughly radial to the vent, are shown in Fig. 3.

There is no evidence for lava flows associated with the eruption. However historical documents suggest that possible lavas could have characterized the last stages of the eruption. If so, these lavas would now be buried under the younger volcanic products that completely cover the southern half of the volcanic complex.

The directions of paleovalleys filled with pyroclastic-flow units are radial to the Vesuvius cone (see Fig. 4). The E-W trending channel at Pollena reflects

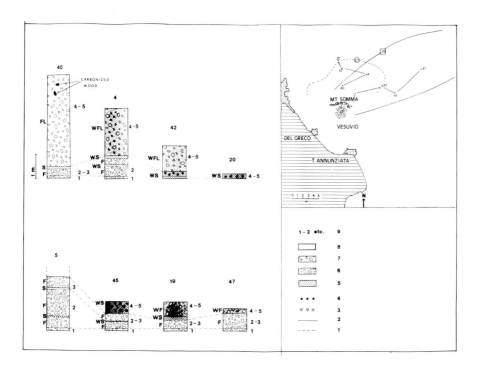

Fig. 3. Lateral variations of the Pollena pyroclastic sequence. 1. phase of the eruption according to Fig. 2; 2. carbonate ejecta; 3. magmatic ejecta; 4. carbonized plants; 5. scoria; 6. lapilli; 7. ash; 8. pisolites.

the present drainage which is controlled by an old topographic high. Surge directions, determined by cross-beds, carbonized tree orientation, and U-shaped erosional channels, seem poorly related to the Vesuvius central vent. This fact could possibly reflect the absence of true base surges in this eruption and indicates that these deposits may have resulted from ash-cloud surges or ground surges associated with pyroclastic flows. However, flow patterns of surges are commonly linked to topography (Malin and Sheridan, 1982; Sheridan and Malin, this volume).

According to changes in the nature of the erupted products, five different phases were recognized in this eruption (Fig. 2). The sequence is quite similar to that of the major Plinian eruptions (Sheridan et al., 1981). However, in this case the volume is significantly lower and reflects an increase in the hydromagmatic character of the eruption with time.

The deposits described here are dated at A.D. 472 on the basis of radiocarbon

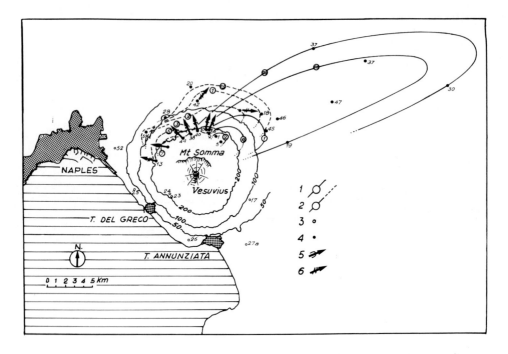

Fig. 4. Isopach map of the lapilli-fall and pyroclastic-flow deposits. 1. isopach lines (thickness in m) of pyroclastic-flow deposits, including surge deposits; 2. isopach lines of lapilli-fall deposits; 3. selection of measured sections in which Pollena pyroclastic products are lacking; 4. location of measured sections of the Pollena pyroclastic sequence; 5. directions of channels cut and filled by pyroclastic flows; 6. surge directions.

age measurements carried out on carbonized wood enclosed in both the pyroclastic-flow deposits and the paleosoil covered by the "Pollena" deposits. Even though the radiocarbon data (Table 1) do not exactly match the historical record, A.D. 472 is the best possible historical date for this eruption.

PETROCHEMISTRY

The pumice of the fall deposit is characterized by a remarkable color change, passing from pale at the base to greenish toward the top. Volcanic xenoliths (tephritic to leucititic in composition) are abundant mainly in the lower level while carbonate ejecta only occur in the upper levels. Pumice is porphyritic with a slight upward increase in phenocryst content. Leucite and clinopyroxene are the more abundant minerals. Other minor liquidus phases are sanidine, biotite, nepheline, brown amphibole and opaques. Other minerals, such as garnet, davyne, and fosteritic olivine, that sporadically occur are probably the

TABLE 1

Radiocarbon ages determined for the "Pollena" eruption

Sample	Location	Material	Age* y.b.p.	A.D.
83	Pollena quarry (12)	Carbonized wood within the glowing-cloud deposits	1600 ± 60	380 ± 60
86	Lagno Amendolare quarry (38)	Carbonized wood within the glowing-cloud deposits	1580 ± 60	400 ± 60
306	Case Trapolino quarry (2)	Paleosoil covering the Pollena pyroclastic sequence	1550 ± 60	430 ± 60
R.939	Arciprete quarry (20)	Paleosoil covered by the Pollena pyroclastic sequence	1630 ± 50	350 ± 50

*Ages of samples 83 and 86 from Delibrias et al. (1979); 306 unpublished (courtesy of G. Delibrias); R.939 from Alessio et al. (1974). Numbers in parentheses refer to location of sections in Fig. 1. Ages are uncorrected.

result of the interaction between the melt and calcareous country rocks.

The scoria from different outcrops of the glowing-cloud deposits are quite homogeneous, confirming that the unit was emplaced in a single episode. The mineralogy of the scoria is similar throughout the deposit and is represented by large phenocrysts of clinopyroxene and leucite with minor biotite. Small euhedral isotropic leucite, together with minor biotite plates, clinopyroxene needles and opaque grains, are set in the finely-vesiculated, optically fresh, glassy groundmass. Other minerals (the same as in the pumice) sporadically occur in variable quantities.

The main petrographical variation between the bottom pumice to the top scoria is the upward increase in clinopyroxene, accompanied by a roughly parallel decrease in leucite content. Modal estimates of the phenocryst content are reported in Table 2.

Seven rock samples were selected for major element chemical analyses, representing the pumice of the lapilli-fall deposits (phases 1, 2 and 3 of Fig. 2) and the scoria from different outcrops of the glowing-cloud deposits. The results are listed in Table 3, while Fig. 5 shows the chemical variations as a function of the normative color index (CI).

As with lavas of the recent (A.D. 1631-1944) Vesuvius activity, the pyroclastic products erupted during the "Pollena" event belong to the high potash series (Middlemost, 1971). They range in composition from phonolites to phonolitic leucitites. Within the fall sequence the composition becomes less

TABLE 2

Modal estimates (volume percent) of representative samples from the Pollena deposit.

Phase of Eruption	1	2	3	4
Sample	283	284	285	*
Leucite	8	12	8	5
Clinopyroxene	2	3	7	11
Biotite	-	-	<1	1
Garnet	<1	<1	<1	<1
Others (including nepheline and sanidine)	1	1	3	3
Xenoliths	9	<1	<1	<1
Glassy groundmass	80	84	82	80

Phases of eruption refer to Fig. 2.
*average of modal estimates of samples 142, 129, 139, and 152.
Point counting analysis: 3 sections for each sample, ~1000 points each section. Point distance chosen is about 1/10 the largest phenocryst diameter. Location of samples: 283, 284, and 285 pumice-fall sequence (base to top) at Trapolino quarry (5 in Fig. 1). 152, 142, 139, and 129 scoria from pyroclastic flows at Lagno di Pollena (11 in Fig. 1, sample 152), Cupa dell'Olivella (12 in Fig. 1, sample 142), Lagno Amendolare (38 in Fig. 1, sample 139), Palmentello (36 in Fig. 1, sample 129).

evolved from the bottom to the top, without sharp discontinuities. The magmatic material of the pyroclastic flows is, on the contrary, rather homogeneous and quite similar to the last erupted pumice fall. In Fig. 5 the fields of the pre- and post-A.D. 79 lavas ("Somma" and "Vesuvius" respectively) are also plotted together with the variation trends of both the "Pompeii" and "Avellino" Plinian eruptions. The chemical coherency of the Pollena products with the post-A.D. 79 family (mainly lavas of the 1631-1944 period) is quite evident.

MICROPROBE MINERALOGY

Three samples representing the eruptive phases 1, 3 and 4 were selected for microprobe analysis. The main results are summarized below. The quantitative chemical data were gathered by an AKL-SEMQ electron microprobe, equipped with crystal spectrometers, using 15KV acceleration voltage, 20μA specimen current, an appropriate 5μm diameter electron beam and standard Bence-Albee correction procedures.

TABLE 3

Representative chemical analyses and CIPW norms of the A.D. 472 pyroclastics

Phase of Eruption	1	2	3	4			
Sample	283	284	285	142	129	139	152
SiO_2	50.56	50.06	48.48	48.84	48.68	48.59	48.70
TiO_2	0.51	0.68	0.87	0.80	0.82	0.83	0.90
Al_2O_3	20.86	19.88	17.96	18.43	18.17	18.12	17.93
Fe_2O_3	2.87	2.93	5.77	3.76	3.65	3.82	4.18
FeO	1.88	2.92	2.45	2.68	3.02	2.89	3.03
MnO	0.15	0.15	0.16	0.14	0.14	0.14	0.14
MgO	0.81	1.64	3.22	3.21	3.51	3.52	3.59
CaO	5.19	6.49	9.02	8.47	8.67	8.81	9.04
Na_2O	5.35	4.52	3.41	3.79	3.98	3.98	3.26
K_2O	10.11	9.27	8.18	8.06	7.73	7.66	7.45
P_2O_5	0.14	0.28	0.48	0.44	0.46	0.45	0.49
l.o.i.	1.57	1.17	1.00	1.37	1.17	1.18	1.30
Sum	100.00	99.99	100.00	99.99	100.00	99.99	100.01
or	32.2	29.1	23.9	24.6	23.3	22.6	27.7
an	3.1	6.6	9.6	9.5	8.2	9.0	12.3
ne	24.5	20.7	15.6	17.4	18.2	18.2	14.9
lc	21.6	20.1	19.1	18.0	18.3	17.7	12.8
wo	6.3	3.4	4.1	2.5	1.6	2.2	1.1
di	5.3	12.8	19.3	18.7	21.6	20.9	20.9
mt	4.2	4.3	2.7	5.5	5.3	5.5	6.1
hm	-	-	3.9	-	-	-	-
il	1.0	1.3	1.7	1.5	1.6	1.6	1.7
ap	0.3	0.7	1.1	1.0	1.1	1.1	1.2

XRF analyses (analyst Dr. G. Crisci, University of Calabria); MgO by Atomic Absorption and FeO by titration (analyst G. Sbrana).
Analyzed samples represent a selection on petrographic bases of about 60 samples collected in different areas and are representative of the entire eruptive sequence.

Leucite is an ubiquitous phase in all three analyzed samples as both large phenocrysts and small grains scattered in the groundmass. Representative analyses of phenocrysts from the phonolitic pumice and the phonolitic-leucititic scoria are shown in Table 4. The higher apparent Na content of the leucite in pumice probably reflects either a low crystallization temperature or volatilization during the analysis.

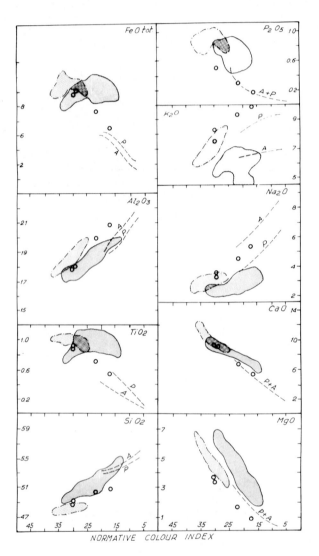

Fig. 5. Chemical variation diagrams as a function of the normative color index (CI). Circles for the Pollena rocks; P = A.D. 79 Pompeii Plinian eruption trend; A = Avellino Plinian eruption trend; dark field = pre-17,000 y.b.p. Mt. Somma lavas; light field = A.D. 1631-1944 Vesuvius historical lavas.

TABLE 4
Representative microprobe analyses of leucite, nepheline and davyne phenocrysts and microphenocrysts

	Leucite		Nepheline			Davyne
Sample	283	152	283	285	152	283
SiO_2	56.2	55.8	42.6	43.7	42.4	33.9
Al_2O_3	22.7	22.6	33.5	33.5	33.7	28.3
Fe_2O_3	0.38	0.29	0.44	0.49	0.39	0.32
CaO	-	-	2.2	2.2	2.1	10.2
Na_2O	1.1	0.11	15.2	15.2	17.1	12.2
K_2O	19.7	19.9	6.1	5.7	4.3	4.9
Sum	100.08	98.70	100.04	100.79	99.99	89.82
Number of ions on the base of	6 oxygens		32 oxygens			12(Si+Al+Fe^{+++})
Si	2.024	2.033	8.216	8.328	8.153	6.026
Al	0.964	0.971	7.617	7.526	7.640	5.931
Fe^{+++}	0.010	0.008	0.064	0.070	0.056	0.043
Ca	-	-	0.455	0.449	0.432	1.943
Na	0.077	0.008	5.683	5.616	6.376	4.205
K	0.905	0.925	1.502	1.386	1.056	1.111

* = All Fe as Fe^{+++}

Nepheline occurs as small crystals in all the analyzed samples. All nephelines contain more silicon and less aluminium than is represented by the formula $NaAlSiO_4$. Significant amounts of CaO are present, resulting in a normative anorthite content ranging around 10-12 wt%. An increase in kalsilite content is observed from scoria to pumice, probably reflecting different temperatures of crystallization (higher for K-poorer nepheline of less evolved host rocks). Representative analyses are presented in Table 4.

Davyne is a rare but ubiquitous phase in the analyzed samples. Weissemberg single crystal X-ray spectra revealed the pertinence of the mineral to the cancrinite-wishnevite series. The cell parameters and the optical sign ascribe the mineral to davyne. A representative analysis is reported in Table 4. Note its strong chemical similarity with the classical SO_3-Cl-rich microsommite of Mt. Somma (Rauff, 1878)

Alkali feldspar. Rare laths and microlites of this mineral were found in the pumice (samples 283 and 284). Microprobe analyses constantly show low totals (92-95) probably indicating large proportions of Ba and Sr components (not determined in our study). Available data from sample 285 show a narrow

TABLE 5

Representative microprobe analyses of clinopyroxene phenocrysts and microphenocrysts

Sample					152 Scoria			
	WPh	GPh	Wc	Gm	Gr	Wc	Wm	Wr
SiO_2	52.4	40.0	51.1	45.6	43.8	51.5	49.7	48.2
TiO_2	0.25	2.3	0.49	2.1	1.5	0.27	0.64	0.88
Al_2O_3	1.3	11.3	2.7	7.2	8.6	2.1	3.6	4.7
Cr_2O_3	0.54	0.55	-	-	-	-	-	-
FeO*	3.1	12.2	4.7	8.2	10.8	3.9	5.7	7.2
MnO	-	0.19	0.14	-	0.14	-	-	0.13
MgO	16.8	7.9	15.5	11.8	9.4	16.4	14.5	12.7
CaO	22.9	22.9	23.1	22.9	22.6	23.2	23.9	24.0
Na_2O	0.16	0.36	0.23	0.41	0.64	0.18	0.18	0.29
Sum	97.45	97.70	97.86	97.31	97.48	97.55	98.22	98.10
			Number of ions on the basis of 6 oxygens					
Si	1.959	1.589	1.920	1.765	1.717	1.932	1.876	1.840
Al^{IV}	0.041	0.411	0.080	0.235	0.283	0.068	0.124	0.160
Al^{VI}	0.017	0.118	0.038	0.090	0.113	0.023	0.035	0.051
Cr	0.016	0.018	-	-	-	-	-	-
Ti	0.008	0.069	0.014	0.036	0.043	0.007	0.018	0.025
Fe^{+++}	0.096	0.403	0.150	0.264	0.355	0.123	0.181	0.231
Mn	-	0.006	0.004	-	0.004	-	-	0.005
Mg	0.936	0.468	0.865	0.678	0.552	0.918	0.816	0.724
Ca	0.918	0.975	0.928	0.951	0.949	0.934	0.967	0.979
Na	0.012	0.028	0.015	0.031	0.049	0.013	0.014	0.021
WXY	2.01	2.08	2.01	2.05	2.06	2.02	2.03	2.04
Ca cat%	47.1	52.6	47.7	50.2	51.0	47.3	49.2	50.5
Mg "	48.0	25.3	44.5	35.8	29.7	46.5	41.5	37.4
Fe "	4.9	22.1	7.8	14.0	19.3	6.2	9.3	12.1

WPH or W = colorless phenocryst; GPH or G = greenish phenocryst; c = core; m = mantle; r = outer rim. - means not detected; FeO* = total iron as FeO.

	285 Greenish Pumice					283 Light Pumice		
Wc	Wr	Gc	Gm	Gr	Gc	Gm	Gr	
52.8	49.3	45.5	43.2	45.1	45.3	43.0	46.0	
0.35	0.76	1.5	1.9	1.5	1.3	1.8	1.3	
1.6	4.5	7.4	10.0	7.4	7.4	9.9	7.5	
–	0.23	–	–	–	–	0.12	0.08	
3.6	6.4	10.0	11.0	10.2	10.0	10.8	9.8	
–	0.12	0.16	0.13	0.13	0.12	0.10	0.08	
17.0	13.0	10.0	9.5	9.8	10.7	9.3	10.7	
24.4	22.7	23.0	23.5	22.3	23.4	23.1	23.5	
0.15	0.31	0.33	0.33	0.31	0.36	0.35	0.29	
99.90	97.32	97.89	99.56	96.75	98.58	98.47	99.25	
1.936	1.877	1.765	1.662	1.770	1.750	1.671	1.757	
0.064	0.123	0.235	0.338	0.230	0.250	0.329	0.243	
0.004	0.079	0.105	0.115	0.110	0.084	0.122	0.096	
–	0.007	–	–	–	–	0.004	0.002	
0.010	0.022	0.042	0.055	0.044	0.037	0.054	0.038	
0.111	0.203	0.324	0.354	0.334	0.321	0.348	0.311	
–	0.004	0.005	0.004	0.005	0.004	0.003	0.003	
0.935	0.744	0.581	0.546	0.574	0.619	0.544	0.612	
0.956	0.927	0.954	0.969	0.939	0.969	0.960	0.962	
0.011	0.023	0.024	0.024	0.024	0.027	0.026	0.022	
2.03	2.01	2.03	2.07	2.03	2.06	2.06	2.05	
47.8	49.4	51.1	51.7	50.7	50.7	51.8	51.0	
46.7	39.6	31.2	29.2	31.0	32.4	29.3	32.4	
3.5	11.0	17.7	19.1	18.3	16.9	18.9	16.6	

compositional range ($Or_{74}Ab_{23}An_3$-$Or_{77}Ab_{21}An_2$) which is slightly poorer in Or than historic Vesuvius lavas (Or_{80} from Baldridge et al., 1981).

Plagioclase is scarce in samples 152 and 285 and absent in 283. It has a very calcic composition in the range An_{88} - An_{77} and is consistently resorbed and circled by a glassy rim.

Clinopyroxene. Two zoned clinopyroxenes are optically recognizable: a yellow-green slightly pleochroic variety ($2V_\gamma$=50°-68°), and a colorless one ($2V_\gamma$=50°-68°), respectively fassaitic and salitic in composition (Table 5 and Fig. 6). The composition of two varieties overlaps roughly in the interval (Wo_{50}-En_{35}-Fs_{15}) - (Wo_{50}-En_{30}-Fs_{20}), but no phenocrysts have been observed to fully cover the compositional range. Oscillatory zoning is frequent mainly in pyroxenes from scoria 152 . The pumice 183 (the most salic) does not contain white salitic clinopyroxene. The usual positive correlation between Al and Fe/(Fe+Mg) is apparent from Fig. 7. Also Ti increases with the increase in Al. The Pollena clinopyroxenes overlap the variation trends of clinopyroxenes from the recent A.D. 1631-1944 lavas. They show, however, a wider Fe/(Fe+Mg) range as a consequence of the more evolved nature of the host rock. Clinopyroxenes from older products of Somma-Vesuvius activity depict, on the contrary, different trends, as schematically indicated in Figs. 6 and 7.

Biotite phenocrysts and microphenocrysts are rare components of scoria 152 and pumice 285 while they are totally absent in pumice 283. The mineral has a distinctly magnesiferous composition (Fe/Fe+Mg = 0.2-0.3), moderately rich in TiO_2, quite similar to biotites found in other volcanic products of the Somma-Vesuvius complex (Hermes, 1975; Baldridge et al., 1981; Barberi et al., 1982). The F, Cl and OH anions have not been determined; however, according to Baldridge et al. (1981) significant amounts of F should be expected to occur. In Table 6 the analyses have been recalculated on the basis of 24 OH and O simply by assuming OH = 4.00.

Amphibole is a relatively common phase in all three analyzed samples. It can be classified as a potassic ferropargasite distinctly poorer in K than in the Pompeii pumice (see Table 6). An increase in the Fe/(Fe+Mg) ratio from scoria to pumice is dubiously reported.

Olivine is a rare component of scoria 152. It is very magesian and similar (Table 7) to that reported from the contact metamorphosed limestone xenoliths (Hermes, 1977). Olivine has probably been inherited from the reaction skarn bordering the magma chamber. In Table 8 analyses of igneous olivines from basic lavas of Somma-Vesuvius are reported for comparison.

Garnet has a melanitic composition with andradite content in the range 50-85%, with significant Ti contents. The mineral appears to be irregularly zoned (Table 8). Barberi et al. (1982), found glassy magmatic inclusions within similar phenocrysts occurring in both Pompeii and Avellino pumice, demonstrating that the mineral crystallized from the magma, possibly slightly modified as a result of interaction with calcareous wall rocks, possibly slightly modified as a result of interaction with calcareous wall rocks.

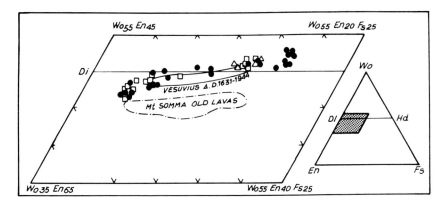

Fig. 6. Plot of the Pollena pyroxenes in the Wo-En-Fs triangle. Triangles refer to cpx from sample 183, squares from 185 and solid circles from 152. Variation fields of pyroxenes from both the recent A.D. 1631-1944 "Vesuvius" and old pre-17,000 y.b.p. "Somma" lavas are indicated.

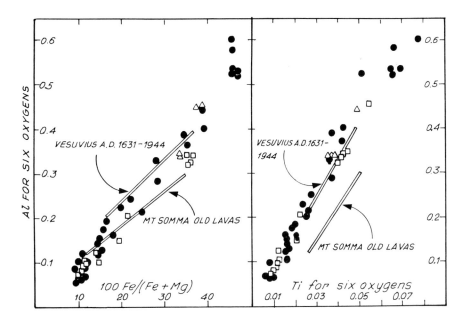

Fig. 7. Pollena clinopyroxenes; Al for six oxygens vs. 100Fe/(Fe+Mg) and Ti for six oxygens. Same symbols as in Fig. 6.

TABLE 6

Representative analyses of biotite and amphibole from various Somma-Vesuvius products

	Biotite			Amphibole		
Sample	285	94.14	N6	152	285	157W
SiO_2	26.9	26.9	37.3	35.7	–	34.5
TiO_2	3.2	3.2	3.1	2.4	–	2.2
Al_2O_3	14.7	12.2	15.7	13.4	–	14.7
FeO	9.6	12.3	7.6	22.6	23.8	25.2
MnO	0.06	0.14	0.08	–	–	0.73
MgO	17.6	17.0	20.0	5.4	3.8	3.6
CaO	0.02	0.07	0.01	11.2	9.2	11.1
BaO	–	3.0	0.67	–	–	–
Na_2O	0.30	0.7	0.32	2.2	–	1.4
K_2O	9.9	7.8	9.5	2.4	–	3.5
Cl	–	0.10	–	–	–	–
F	–	5.7	–	–	–	–
H_2O	–	(1.2)	–	–	–	–
Sum	92.28	100.3	94.28	95.30		96.93

Number of ions for 24 O, OH, F, Cl

	285	94.14	N6	152	285	157W
Si	5.578	5.680	5.474	5.811		5.631
Al^{IV}	2.422	2.213	2.256	2.189		2.369
Al^{VI}	0.198	–	0.190	0.382		0.456
Ti^{IV}	–	0.107	–	–		–
Ti^{VI}	0.364	0.263	0.342	0.294		0.273
Fe	1.214	1.583	0.933	3.076		3.440
Mn	0.007	0.018	0.010	–		0.101
Mg	3.965	3.901	4.374	1.310		0.877
Ca	0.003	0.011	0.002	1.953		1.946
Ba	–	0.180	0.039	–		–
Na	0.087	0.207	0.092	0.694		0.455
K	1.909	1.532	1.778	0.499		0.725
F	–	2.775	–	–		–
Cl	–	0.026	–	–		–
OH	4.000[1]	(1.232)	4.000[1]	2.000[1]		2.000[1]
Z	8.00	8.00	8.00	8.00		8.00
XY	5.75	5.78	5.85	5.06		5.15
W	2.00	1.93	1.91	3.15		3.13
(OH, F, Cl)	4.00[1]	4.03	4.00[1]	2.00[1]		2.00[1]
Fe(Fe+Mg)	0.235	0.291	0.177	0.701	0.779	0.801

285 = Pollena pumice; 94.14 lava from recent Vesuvius activity (after Baldridge et al. 1981); N6 lava from Mt. Somma pre-17,000 y.b.p. activity (Hermes, 1976). 157W = Pompeii eruption white pumice (after Barberi et al. 1982). 152 = Pollena scoria.

Numbers in parenthesis obtained by charge balance. (1) = postulated values in calculations of the number of ions. – means not detected.

TABLE 7
Microprobe analyses of olivine

Sample	152	N3	94.8	10
SiO_2	41.6	41.5	–	38.25
FeO	3.1	4.5	25.3	22.9
MnO	0.18	0.22	0.53	0.33
MgO	53.1	52.8	35.8	38.5
CaO	0.31	0.74	0.57	0.32
Sum	98.29	99.75		100.31

Number of ions for four oxygens

Si	1.005	0.997		0.996
Fe	0.062	0.090		0.499
Mn	0.004	0.004		0.007
Mg	1.912	1.891		1.490
Ca	0.008	0.022		0.009
Y	1.99	2.01		2.01
Fe/(Fe+Mg)	0.03	0.05	0.22	0.25

152 = Pollena scoria; N3 = Skarn rock (Hermes, 1976); 94.8 phonolitic leucite lava of the recent activity of Vesuvius (Baldridge et al., 1981); 10 = leucite tephritic lava of the old Mt. Somma activity (Barberi et al. 1982).

GENESIS OF THE POLLENA ROCK SEQUENCE

Despite their significant normative An content (up to 12.3% in sample 152), plagioclase does not appear as a stable phase in the Pollena pyroclastic rocks. The rare crystals are largely resorbed and show anorthitic to bytownitic compositions that are poorly compatible with the evolved nature of the host rocks. This fact and the salic nature of the Pollena rocks (D.I. between 55 and 80) allow the petrogeny's residua system (Fig. 8) to be utilized in a discussion on their chemical variation. Leucite and nepheline (phenocrysts and microphenocrysts) are present in the entire rock sequence, suggesting that the liquids lie on the nepheline-leucite boundary surface. The more salic rocks (samples 283 and 284) contain small amounts of sanidine phenocrysts, indicating that the ternary point nepheline-leucite-alkali-feldspar of the system is reached. The coherency of the observed mineralogical variations with the plot in the petrogeny's residua system suggests that the entire Pollena sequence could represent a liquid line of descent towards the phonolitic minimum. Fig. 8 suggests, moreover, that the Pollena products crystallized under a P_{H_2O}

TABLE 8

Garnet microprobe analyses

Sample	152		283				
	rim	core	rim	mantle	core	rim	core
SiO_2	33.9	32.4	35.3	36.7	36.4	35.7	32.7
TiO_2	4.7	5.2	3.7	2.5	3.2	2.9	8.0
Al_2O_3	10.1	8.2	7.7	10.8	9.9	7.9	5.9
Fe_2O_3	16.3	19.4	18.6	11.5	15.0	17.7	20.0
FeO	3.1	2.1	2.3	2.2	2.5	2.8	0.7
MnO	0.58	0.87	0.68	0.60	0.52	0.76	0.31
MgO	0.50	0.36	0.28	0.21	0.20	0.22	0.52
CaO	30.7	31.1	30.8	31.8	32.0	30.1	31.4
Sum	99.88	99.63	99.36	99.31	99.72	98.08	99.53
Number of ions for 24 oxygens							
Si	5.503	5.349	5.775	5.879	5.795	5.898	5.386
Al^{IV}	0.497	0.651	0.225	0.103	0.205	0.102	0.614
Al^{VI}	1.436	0.945	1.260	1.943	1.822	1.437	0.531
Ti	0.574	0.646	0.455	0.302	0.383	0.360	0.991
Fe^{+++}	1.991	2.410	2.290	1.753	1.797	2.201	2.479
Fe^{++}	0.420	0.290	0.315	0.295	0.333	0.386	0.096
Mn	0.080	0.122	0.094	0.082	0.070	0.106	0.044
Mg	0.121	0.088	0.068	0.050	0.048	0.055	0.128
Ca	5.340	5.502	5.399	5.475	5.458	5.328	5.542
Z	6.00	6.00	6.00	6.00	6.00	6.00	6.00
Y	4.00	4.00	4.00	4.00	4.00	4.00	4.00
X	5.96	6.00	5.88	5.90	5.91	5.88	5.81
Andr %	63.7	76.4	67.8	50.5	53.8	63.3	86.2
Gross %	25.8	15.3	24.0	42.4	38.6	27.4	9.3
Alm %	7.1	4.8	5.4	5.0	5.6	6.6	1.6
Pyr %	2.0	1.5	1.2	0.7	0.8	0.9	2.2
Spess %	1.4	2.0	1.6	1.4	1.2	1.8	0.7

The ratio Fe_2O_3/FeO was calibrated to make Y = 4.00.

slightly higher than 1 kb, similar to that controlling the fractionation of the Pompeii pumice (Barberi et al., 1982) erupted during the A.D. 79 Plinian event.

In the search for the basic parental magma of the Pollena sequence, the variations diagrams of Fig. 5 and some mineralogical data (mainly from

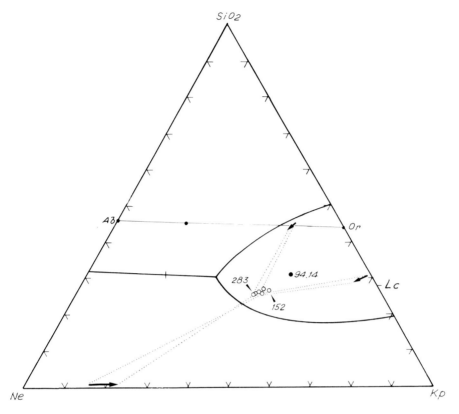

Fig. 8. Plot of Pollena rocks and minerals in petrogeny's residua system (Hamilton and McKenzie, 1965 at 1.0 kb H_2O). Sample 94.14 from Baldridge et al. (1981). Arrows indicate variations observed in composition of the minerals from basic to salic host-rocks.

clinopyroxene) demonstrate the strong affinity between these rocks and the recent lava products of the A.D. 1631-1944 cycle. One of these lavas (sample 94.14, Fig. 8) taken from Baldridge et al. (1981), can be used as a possible parental composition of the Pollena sequence. Plagioclase, a rare phenocryst phase in this lava, ranges in composition from An_{90} to An_{65}. The presence of plagioclase accounts for the divergence of this sample from the tie line of leucite-Pollena sequence (Fig. 8). According to Baldridge et al. (1981) plagioclase and salitic clinopyroxene crystallized first, followed by leucite. Olivine is confined to the groundmass, together with single-phase titaniferous magnetite.

On this basis it is now possible to estimate the degree of fractionation that will account for the transition from the inferred initial liquid (represented by the 94.14 tephritic-leucititic sample) to the final salic term of the Pollena suite (leucititic phonolite 283). Table 9 illustrates the transition we have considered, the solid phases utilized in the computation and the results obtained using a XLFRAC program (Stormer and Nicholls, 1978).

One problem arising from such a fractionation model is the disappearance of plagioclase at the level of the transition B-C of Table 9, a fact not in agreement with an isobaric fractionation within the phonolitic penthahedron, as schematically illustrated by Carmichael et al. (1974). After the liquid reached the surface of the leucite volume, allowing the contemporaneous crystallization of leucite and plagioclase, a pressure drop expanding the leucite volume can account for the unstable plagioclase in Pollena rocks. A two-stage fractionation model thus seems appropriate. During the first stage the deep leucititic magma slowly rises while fractionating plagioclase, clinopyroxene and olivine. When the surface of the leucite volume is reached, leucite joins plagioclase which possibly becomes rapidly unstable. During the second stage, mildly differentiated magma stagnates in the shallow magma chamber where fractionation occurs within the feldspathoidal volume of the phonolitic pentahedron. Leucite is a stable phase throughout the graded magma column. However, its denisty prevents a significant participation in the fractionation process which is mainly controlled by heavy mafic phases (cpx and bt). In Table 9 an alternative C-D transition suggest a possible, but extremely limited, participation of nepheline in the last stage of fractionation.

The proposed fractionation model is obviously a first order characterization of the problem to show a semiquantitative agreement of observed chemical and mineralogical variations with such a process. The actual process we imagine is somewhat more complicated involving repeated arrivals of deep basic magma within the chamber. The geochemical and mineralogical evolution of magma during fractional crystallization under these conditions (see O'Hara, 1977) needs a more complete and detailed study which is now in progress.

DISCUSSION

The pyroclastic products of the Pollena eruption (a volume of about 0.32 km^3) have an average density of about 1.4 g-cm^{-3}. The equivalent volume of solid rock (d=2.7) is consequently about 0.165 km^3. Half of this volume is represented by pyroclastic-flow deposits chemically corresponding to sample 152. The fall deposit is roughly made up of 20% pale pumice (sample 283), 60% greenish pumice (sample 284) and 20% darker pumice (sample 285 to scoria 152). This means that the Pollena products roughly consisted of
- 0.100 km^3 of solid rock similar to the composition of the sample 152;
- 0.050 km^3 of solid rock similar to the composition of the sample 284;
- 0.015 km^3 of solid rock similar to the composition of the sample 283;
By comparing these data with Table 10, the required volume of primary basic liquid (sample 94.14) can be calculated at about 0.24-0.25 km^3.

The repose time preceeding the Pollena event, according to available documentation, was between 150 and 200 years. The rate of magma production

TABLE 9

Fractionation model of the Pollena sequence

Liquids and Solids	A	B	C	D	1	2	3	4	5	6	7	8
Sample	94.14	152	284	283	ol	cpx1	plg	bt	cpx2	lc	ne	ox
SiO$_2$	49.44	49.87	51.03	51.67	36.6	50.8	50.5	40.0	46.6	56.5	42.4	–
TiO$_2$	1.01	0.92	0.69	0.52	–	0.78	–	3.5	1.5	–	–	13.2
Al$_2$O$_3$	15.60	18.36	20.27	21.32	–	5.1	30.8	15.9	7.6	22.9	33.7	4.5
FeO tot	7.39	6.95	5.67	4.56	33.5	6.8	0.89	10.4	10.2	0.26	0.35	81.8
MgO	6.51	3.68	1.67	0.83	29.3	13.2	–	19.1	10.2	–	–	0.53
CaO	11.67	9.26	6.62	5.30	0.50	23.0	14.4	0.02	23.5	–	2.10	–
Na$_2$O	2.29	3.34	4.61	5.47	–	0.30	3.2	0.33	0.34	0.11	17.1	–
K$_2$O	6.10	7.63	9.45	10.33	–	–	0.31	10.7	–	20.2	4.3	–

Transition	% Fraction Solid	R^2	Composition of fractionated solid (tot %) of mineral phases						
A – B	23.5	0.44	13.1	80.6	6.3				
B – C	23.6	0.35			10.9	18.8	70.3		
C – D	9.0	0.69					97.1	2.9	
(C – D)	(12.3)	0.16					(69.4)	(23.3)	(0.2)	(7.1)

Analyses recalculated to 100. Data for A, 1, 2, 3, 8 from Baldridge et al. (1981); 4 and 5 from sample 285 (= 152), 6 and 7 from sample 152.

(≈ feeding rate) consequently ranges between 1.7 and $1.2 \times 10^{-3} km^3$/year, in good agreement with the $1.6 \times 10^{-3} km^3$/year average eruption rate of the 1694-1944 period (practically continous eruptions of poorly-differentiated, basic magmas) obtained from the data of Cortini and Scandone (1982). The assumption by Santacroce (this volume) that the magma feeding rate was constant during the last (post A.D. 79) cycle of Vesuvius activity thus is confirmed by the Pollena eruption.

The direct relationship in the Somma-Vesuvius volcanic complex between repose time on the one hand and volumes and degree of differentiation of erupted products on the other, strongly suggests a continous supply and storage of deep basic magma within a shallow depth magma chamber (Santacroce, this volume.) The chemical variations shown by the pyroclastic deposits of Pollena and other Plinian eruptions derive from a compositionally stratified magma chamber.

The two co-existing processes (differentiation and gradation of the stagnant magma disturbed by more or less periodic arrivals of new basic magma) may split the magmatic body into a two-part system, characterized by a compositionally graded zone above a more basic convecting zone, with magma mixing phenomena occurring at each arrival of fresh magma (e.g., O'Hara, 1977). This corresponds to a model illustrated in Williams and McBirney (1979). Magma mixing phenomena are possibly evidenced by the large oscillatory zonation shown by the clinopyroxene of samples 152 and 185, which are assumed to be the expression of the convecting zone within the magma chamber. Arrival of fresh magma within the magma chamber is probably a pulsating phenomenon. Each pulse could result in the triggering of an eruption.

Finally we should like to point out that the proposed two-stage model can account for the slight, but significant, chemical differences observed in magmas erupted during the recent A.D.1694-1944 period of semi-persistent, Strombolian-like, activity of Vesuvius. They should, in fact, reflect different rising-times of different deep magma pulses.

ACKNOWLEDGEMENTS

We are grateful to R. Rinaldi and G. Vezzalini of the University of Modena for the helpful technical assistance in the microprobe mineralogical analyses. G. Crisci of the Universty of Calabria kindly provided XRF analyses. This research was supported by CNR, Italian Geodynamic Project, publication No. 497.

REFERENCES

Allessio, M., Bella, F. Improta, S., Belluomini, G., Calderoni, G., Cortesi G., Turi, F., 1974. Univerity of Rome Carbon 14 dates XII. Radiocarbon, 16 (3): 358-367.

Alfano, G.B. and Friedlander, I., 1929. La storia del Vesuvio illustrata da documenti coevi. Karl Hohn; Ulm.

Baldridge, W.S., Carmichael, I.S.E., and Albee, A.L., 1981. Crystallization path of leucite-bearing lavas: examples from Italy. Contrib. Mineral. Petrol., 76: 321-335.

Barberi, F., Bizouard, H. Clocchiatti, R., Metrich, N., Santacroce, R., Sbrana, A., 1982. The Somma-Vesuvius magma chamber: a petrological and volcanological approach and geothermal implications. Bull. Volcanol., 44(3): 295-315.

Carmichael, I.S.E., Turner, F.J., and Verhoogen., 1974. Igneous petrology. McGraw-Hill, New York, N.Y., 739 pp.

Cortini, M., and Scandone, R., 1982. The feeding system of Vesuvius between 1754 and 1944. J. Volcanol. Geotherm. Res., 12: 393-400.

Delibrias, G., Di Paola, G.M., Rosi, M., and Santacroce, R., 1979. La storia eruttiva del complesso vulcanico Somma-Vesuvio ricostruita dalle successioni del Monte Somma. Rend. Soc. It. Min. Petr., 35(1): 411-438.

Hamilton, D.L., and McKenzie W.S., 1965. Phase-equilibrium studies in the system $NaAlSiO_4$-$KAlSiO_4$-SiO_2-H_2O. Mineral. Mag., 34: 214-231.

Hermes, O.D., and Cornell, W.C., 1978. Petrochemical significance of xenolithic nodules associated with potash-rich lavas of Somma-Vesuvius volcano. NSF Tech. Rept., Rhode Island University. (Unpubl.)

Malin, M.C. and Sheridan, M.F., 1982. Computer-assisted mapping of pyroclastic surges. Science, 217: 637-369.

Middlemost, E.A.K., 1975. The basalt clan. Earth Sci. Rev., 11: 337-364.

O'Hara, M.J., 1977. Geochemical evolution during fractional crystallization of a periodically refilled magma chamber. Nature, 266: 503-507.

Principe, C., 1979. Le eruzioni storiche del Vesuvio: riesame critico, studio petrologico dei produtti ed implicazioni vulcanologiche. Thesis, Univ. Pisa, Italy (unpubl).

Rauff, H., 1878. Uber die chemische Zusammensetzung des Nephelins, Cancrinits und Microsommits. Zeit. Krist., 2: 445.

Santacroce, R., 1983. A general model for the behavior of the Somma-Vesuvius volcanic complex. In: M.F. Sheridan and F. Barberi (Editors), Explosive Volcanism. J. Volcanol. Geotherm. Res., 17: (this volume).

Sheridan, M.F., Barberi, F., Rosi, M., and Santacroce R., 1981. A model for plinian eruption of Vesuvius. Nature, 289: 282-285.

Sheridan, M.F. and Malin, M.C., 1983. Application of computer-assisted mapping to volcanic hazard evaluation of surge eruptions: Vulcano, Lipari, and Vesuvius. In: M.F. Sheridan and F. Barberi (Editors), Explosive Volcanism. J. Volcanol. Geotherm. Res., 17: (this volume).

Stormer, J.C. and Nicholls, J., 1978. XLFRAC: a program for the interactive testing of magmatic differentiation models. Comput. Geosci., 4(2): 143-159.

Williams, H. and McBirney, A.R., 1979. Volcanology. Freeman Cooper and Co., San Francisco, 397 pp.

THE PHLEGRAEAN FIELDS: STRUCTURAL EVOLUTION, VOLCANIC HISTORY AND ERUPTIVE MECHANISMS

M. ROSI[1], A. SBRANA[2], C. PRINCIPE[2]

[1] Dipartimento di Scienze della Terra, University of Pisa - Centro di Geologia Strutturale e Dinamica dell'Appennino - C.N.R.- Pisa (Italy)
[2] Agip S.p.A. - Servizio Esplorazione Geotermica - Milano (Italy)

(Received July 15, 1982; revised and accepted December 20, 1982)

ABSTRACT

Rosi, M., Sbrana, A., and Principe, C., 1983. The Phlegraean Fields: structural evolution, volcanic history, and eruptive mechanisms. In: M.F. Sheridan and F. Barberi (Editors), Explosive Volcanism. J. Volcanol. Geotherm. Res., 17: 273-288.

The main event within the volcanic history of the Phlegraean Fields was the eruption, about 35,000 years ago, of a huge alkali trachytic ignimbrite (80 km^3, dre) followed by caldera collapse. The pre-caldera activity (evidence from geothermal wells and surface outcrops) changed from submarine to subaerial shortly before the ignimbrite eruption. The caldera was subsequently invaded by the sea and became progressively filled with tuffites and subordinate submarine flows in its northern half. The last submarine explosive activity dates back to 12,000-10,500 yr. with the eruption of mostly "hydroplinian" yellow tuffs. Recent, mostly subaerial, activity consists of a series of explosive eruptions whose volume decreases with time, until the last eruption of Mt. Nuovo in 1538 A.D. Two peaks in recent activity occur at 10,000-8,000 and 4,700-3,000 years ago. During post-caldera activity, eruptive vents migrated from the caldera rim toward the center where there are two distinct zones of recent vents. Minor collapses occurred within the caldera after the main eruptions; often they are eccentric to the vent and displaced toward the caldera center. Phlegraean explosive activity is characterized by water/magma interaction; eruptive events without a hydromagmatic component are extremely rare.

In most cases, the evolved trachytic magmas interacted with surface water (either the sea or intracaldera lakes). However, there is evidence of interaction of magma with deep-seated aquifers for some Plinian events. Explosivity, i.e. the efficiency of the transformation of thermal into kinetic energy during magma/water interaction, is largely controlled by the primary fragmentation of the magmatic melt.

0377-0273/83/$03.00 © 1983 Elsevier Science Publishers B.V.

INTRODUCTION

The Phlegraean Fields, (Naples), a classic volcanic site, includes many explosive eruptive vents and lacks a well-defined central volcano. The geology was described in two classical papers (De Lorenzo, 1904; Rittmann, 1950a) but a comprehensive picture of the Phlegraean Fields volcanism based on sound stratigraphic correlations and detailed volcanological studies of the various vents has been lacking.

This paper summarizes the results of a detailed field survey during which more than 150 stratigraphic sections were measured; stratigraphic guide horizons were located and dated by the radiocarbon method, volcanic structures and tectonic lineaments were mapped. This surface study, coupled with deep drilling data, has permitted the reconstruction of a new history of the volcanic activity at the Phlegraean Fields.

THE SEQUENCE OF THE ERUPTIVE EVENTS

The main volcano-tectonic features of Phlegraean Fields are shown in Fig. 1. The main result of the field survey was the recognition of a large (12 km in diameter) caldera related to the eruption of the 35,000 year old Campanian Ignimbrite. This impressive pyroclastic-flow deposit covers 7000 km^2 (Barberi et al., 1978), with a volume equivalent to 80 km^3 of trachytic liquid (Thunell et al., 1978). Development of the caldera is certainly the main volcano-tectonic event of Phlegraean Fields. It is therefore convenient to describe the volcanic history of the area in reference to this major feature.

Pre-caldera activity

Pre-caldera activity includes a period of submarine volcanism, as determined from deep drilling data. Rocks from this period consist of trachytic and latitic lavas, tuffs, and tuffites interbedded with silty, arenaceous and marly sediments, of Quaternary or Plio-Quaternary age (Fig. 2). The submarine sequence is covered by subaerial products which can be related to the activity of several vents (lava and/or pyroclastic), some of which are still recognizable outside the caldera, in zones not affected by the collapse.

The oldest rocks date back to about 50,000 years ago. Fig. 3 illustrates the stratigraphy of these products on various points of the caldera rim. They include lavas, hydromagmatic tuffs, and pyroclastics derived from subaerial activity and form cones and domes. This older phase ended with a peculiar pyroclastic unit: the so called "Breccia Museo". As suggested by its name, the Breccia Museo is characterized by an abundance of lithic blocks of various kinds. Most of these lithic ejecta consist of lava blocks which cover the whole compositional spectrum of Phlegraean products. Hydrothermally altered or contact metamorphosed rocks, very similar to those found at depth in the geothermal wells, are also present. The occurrence of fumerolic pipes within the Breccia Museo indicates that this deposit originated as a pyroclastic flow.

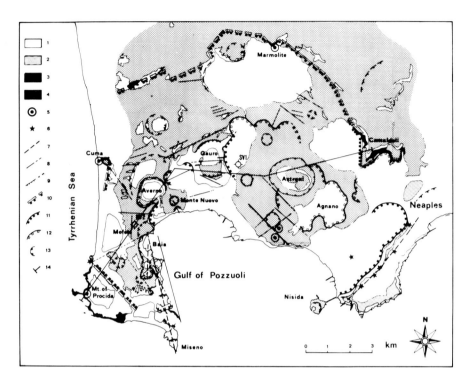

Fig. 1 - Geological sketch map of Phlegraean Fields; 1. urban areas and soil cover; 2. post-caldera deposits, mostly subaerial tephra and minor lavas, age 10,500 years b.p. to 1538 A.D.; 3. post-caldera mostly submarine activity, age comprised between 35,000 and 10,5000 years b.p., yellow chaotic tuffs and yellow layered tuffs; 4. pre-caldera activity, age older than 35,000 years b.p., lava domes, tephra and products of Campanian Ignimbrite eruptions; 5. lava domes; 6. inferred eruptive centers; 7. feeding fractures; 8. fractures; 9. faults; 10. caldera rim; 11-12. volcano-tectonic collapses related to post-caldera activity (full squares 35,000-10,500 yr.; open squares 10,500-present); 13. crater rim; 14. trace of the sections of Fig. 5.

Areal distribution, grain-size, petrographic, and chemical data of the Breccia Museo suggest that it may be the fines-depleted proximal facies of the Campanian Ignimbrite, possibly erupted through a ring fracture of a cone-sheet feeding system. This deposit is exposed on opposite sides of the caldera rim separated by a 12 km distance. Whereas the granulometric features and composition of the associated juvenile magma fragments are basically the same, the heterogeneity of the lithics suggests that different sequences of pre-existing solid rocks were sampled by the explosions.

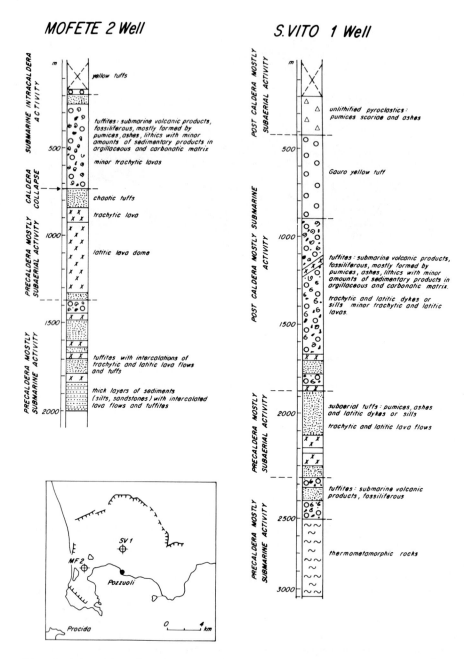

Fig. 2 Schematic subsurface stratigraphy of Mofete and S. Vito areas (from geothermal well data)

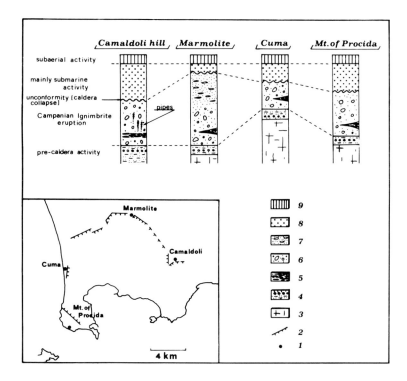

Fig. 3 Stratigraphic sections on the caldera rim of the Phlegraean Fields; 1. location; 2. caldera rim; 3. lava domes; 4. pyroclastic sequence composed of stratified yellow tuffs, Plinian pumice-fall layers and ashes interbedded with paleosoils; 5-6-7. products of Campanian Ignimbrite eruption, 35,000 years ago. - 5. lenticular levels of welded fiamme ("piperno"). 6. fines-depleted, lithic-rich pyroclastic flow (Breccia Museo); 7. welded ignimbrite facies; 8. post-caldera activity, 16,000 (?) to 10,500 years ago, yellow tuffs (Yellow Neapolitan tuff, yellow layered tuffs): mostly subaerial deposits erupted from submarine vents; 9. post-caldera subaerial activity, 10,500 to 3,700 years ago, tephra with interbedded paleosoils.

Post-caldera activity

After caldera collapse, volcanic activity resumed in the Phlegraean Fields. Radiocarbon age measurements and paleogeographic considerations allow the recognition of three main phases:
Phase A - Ancient submarine volcanism (35,000 - 10,500 y.b.p.).
Phase B - Early subaerial volcanism (10,500 - 8,000 y.b.p.).
Phase C - Recent subaerial volcanism (4,500 y.b.p. - A.D. 1538).

Phase A - After collapse, the caldera depression was invaded by the sea. Volcanic activity occurred within the caldera, which was progressively filled, during the time from 35,000 to 10,500 years ago. The products of this activity are tuffites with minor latitic to trachytic lava intercalations (observed only in the geothermal wells). All rocks are so deeply affected by hydrothermal metamorphism that radiometric dating is not possible. The chronology of this old submarine phase of the post-caldera activity is consequently unknown.

Phase A includes the transition from submarine to subaerial activity with the deposition of yellow lithic tuffs within the caldera depression and mantling its rim. One of the most famous products of Phlegraean Fields, known as the "Neapolitan yellow-tuff", has been attributed by some authors (Lirer and Munno, 1975) to a single eruption of this period.

In contrast, results confirm the view of Rittmann that these tuffs were erupted from several vents and consist of both chaotic (i.e. Camaldoli) and stratified (i.e. Mofete, Miseno) facies (Fig. 4). Eruptive vents were probably located under shallow marine water whereas the deposition of the yellow tuff occurred both in marine and subaerial environment. Feeding vents for the yellow tuff were mainly concentrated along the inner rim of the caldera (examples are the "volcanic arcs" of Miseno-Monticelli in the western sector and that of Posillipo in the east; Fig. 4). Gauro volcano, however, testifies to the presence of yellow tuff cones in the central part of caldera.

Some of the main eruptions of yellow tuff were followed by limited collapses of their volcanic edifices: the San Vito - Toiano, Bagnoli, and Lucrino plains must be related respectively to the partial sinking of the Gauro, Posillipo Hill, and Mofete-Gauro volcanoes. Such collapses affect the parts of the volcanoes which are nearer to the center of the caldera.

After deposition of the yellow-tuffs, the activity became predominantly subaerial and the chronology and the volcanic history can be reconstructed more precisely. As a whole, post-caldera subaerial activity was concentrated into two main periods at 10,500 to 8,000 years ago (Phase B) and 4,500 to 3,000 years ago (Phase C). The last eruption of the Phlegraean Fields (Mt. Nuovo) occurred in 1538.

Phase B - Radiocarbon dates indicate 11,000 years age for both the Agnano eruption and the underlying Vomero yellow tuff. This suggests that a long quiescent period did not occur between deposition of the yellow tuff and the following subaerial activity.

During Phase B vents were concentrated in two main areas within the caldera; Baia and Fondi di Baia in the western sector and Montagna Spaccata, S. Martino, Pisani and Agnano in the central-eastern sector (Fig. 4). The main volcanic event of this phase was the Plinian eruption of Agnano which was followed by the collapse of the eruptive center, forming the Agnano caldera (3 km across).

The eruptive vents migrated, during this phase, toward the central part of the caldera, around the Gulf of Pozzuoli. Exceptions are the trachybasaltic vent of Minopoli, lying on the caldera rim, and the latitic vents of Concola and Fondo Riccio, on the inner rim of the caldera. All of these "peripheral" vents

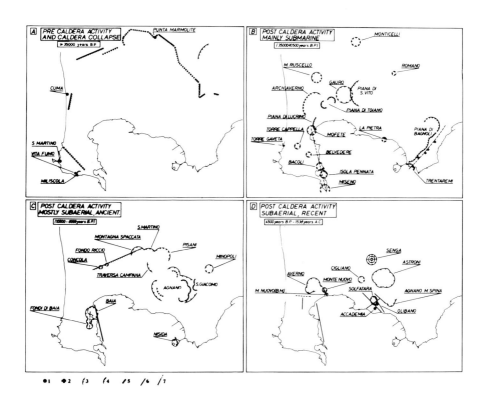

Fig. 4 Main phases of Phlegraean Fields evolution; 1. lava dome, 2. inferred eruptive centers; 3. crater rim; 4. volcano-tectonic collapse; 5. caldera rim; 6. fractures; 7. faults.

erupted basic products (trachybasalt or latite) whereas the products of vents near the caldera center are dominantly trachytic. Phase B was followed by a rather long repose period, as indicated by the occurrence of a very thick paleosoil cover, which is continuous throughout the whole Phlegraean area.

Phase C - The recent phase of post-caldera subaerial activity started approximately 4,500 years ago. It was preceded by the uplift of the northern sector of the Gulf of Pozzuoli, a local tectonic event. This uplift is represented by the Starza marine terrace. The uplift dates back 5,400 years and may be attributed to the injection of magma at shallow depth.

The most recent eruptions occurred through vents located on or near the uplifted area, in a brief interval barely 1,000 years after the probable injection of magma into the shallow basin. Again, two eruptive areas can be identified in this phase (Fig. 4); Averno and Mt. Nuovo in the west and Solfatara, Astroni, Agnano, and Senga in the east. All erupted products, both pyroclastic and lava, have a trachytic to trachyphonolitic composition. During this phase, a further migration of the eruptive vents toward the central part of the caldera clearly occurred (Fig. 4).

EXPLOSIVE ACTIVITY IN THE PHLEGRAEAN FIELDS

A great number of explosive eruptions occurred during the volcanic history of the Phlegraean Fields. Effusive activity is subordinate and only a few lava domes were formed. Pyroclastic deposits of the Phlegraean Fields volcanoes are represented by both tephra and tuffs in which pumice and ash are by far the most abundant constituents. The volume of the erupted products decreases with time from the 80 km^3 liquid equivalent of the Campanian Ignimbrite to the 0.03 km^3 of Mt. Nuovo.

The explosive subaerial eruptions of the last 10,000 years of the Phlegraean Fields were either of Plinian, sub-Plinian, or Strombolian types. The Plinian eruptions were the most voluminous and were associated with widespread deposits of pumice falls, and pyroclastic flows and surges. Sub-Plinian and Strombolian eruptions were characterized by the small-scale accumulation of products around eruptive vents.

Data from deep wells (Fig. 2) reveal that the sea invaded the caldera after the eruption of the Campanian Ignimbrite; consequently the first post-caldera explosive eruptions occurred below the sea giving rise to strong magma/water interactions.

These interactions played an important role in controlling the eruptive mechanism, the type of products emitted, and the volcanic structures formed thereafter. The wide variety of Phlegraean explosive eruptions can be considered the result of a combination of the primary magmatic explosivity and the eventual interaction of fragmented magma with surface, or phreatic waters.

Strombolian eruptions

Volcanoes formed by subaerial Strombolian-type activity include Senga, Mt.

Spaccata, Fondo Riccio and Mt. Nuovo. Their deposits consist of locally welded scoria containing a limited amount of shallow-origin lithics. The scoria cones commonly have an atypical shape with a high base to height ratio, suggesting a somewhat higher than usual explosive energy for this kind of activity. Some of the deposits contain layers of pyroclastic surge. The surge deposits of Senga are made of sandy ash beds with cross-laminations. Vesiculated tuffs with accretionary lapilli are common in the ash layers. The pyroclastic-surge eruptions can be interpreted as hydromagmatic phases caused by the inflow of limited quantities of phreatic water into the volcanic conduit, following the model proposed by Sheridan and Wohletz (1981).

Hydromagmatic activity also occurred during the eruption of Mt. Nuovo (1538 A.D.). According to several witnesses (Parascandola, 1946) this predominately Strombolian eruption was accompanied by heavy falls of mud and there were repeated horizontally-directed base-surge explosions which were presumably responsible for uprooting some huge trees in the town of Pozzuoli.

The Astroni and Averno volcanoes are typical of tuff rings in the Phlegraean Fields. Their deposits are represented by a thick series of pyroclastic surge deposits composed of pumice and ash and accompanied by a few pyroclastic-fall layers. The distribution of surge facies around the vent nicely matches the model of pyroclastic surge emplacement proposed by Wohletz and Sheridan (1979) in which sandwave deposits gradually pass to massive and subsequently to planar sequences with increasing distance from the vent. All the features classically related to hydrovolcanism, such as the abundance of ash and the presence of quenched glass fragments, accretionary lapilli and vesiculated tuffs, are common in these deposits. At the time of the eruptions of Astroni and Averno volcanoes, it seems probable that a shallow lake existed inside the caldera. The eruptive mechanism of these volcanoes was then controlled by Strombolian explosions occurring within this lake, the kinetic energy being greatly increased by magma/water interaction. The shallow origin of the lithics and shape of the crater, with its high diameter/depth ratio, suggest a shallow depth of the explosion. The degree of magma/water interaction is reflected in the ash to pumice ratio. The presence of a large mass of pumice in the deposits of the Phlegraean tuff rings could corroborate the hypothesis that subaerial explosions alternated with hydroexplosions.

The eruptions that took place at shallow depths in the sea created stratified, yellow-tuff edifices. This type of volcano is commonly referred to as a tuff cone (see: Sheridan and Wohletz, this volume). The erupted magma in the Phlegraean Fields is usually of trachytic composition, therefore diagenetic alteration products are zeolites rather than palagonite, which is common to basaltic tuff cones. There are several examples of trachytic tuff cones in the Phlegraean Fields, their products frequently contain pyroclastic surge beds accompanied by fall deposits. Accretionary lapilli layers and vesiculated tuffs are also a common feature near the craters, indicating rapid deposition of wet products. The tuff stratification found in the Phlegraean area typifies the intermittent nature of the explosive activity.

Plinian eruptions

During the last 10,000 years there have been several Plinian eruptions associated with volcanotectonic collapse in the Phlegraean Fields area. They are distinguished by correlating pumice-fall layers found mostly outside the caldera. One feature common to all deposits of these eruptions is that they are spread over very wide areas. Associated with the pumice falls (initial phase of the Plinian-type eruption) are pyroclastic flows and surges, the deposits of which occur within the caldera, especially in the vicinity of eruptive centers. The materials emitted during the Phlegraean Plinian eruptions were so widely dispersed that they created no significant volcanic morphology.

The youngest Plinian eruption in the Phlegraean Fields is that of Agnano Mt. Spina. This eruption took place 4,300 years ago within the Agnano caldera. During the initial phase of the eruption, pumice and lithic fragments were thrown to great heights; pumice were scattered by the wind to form a wide sheet to the ENE of the vent. Then pyroclastic flows formed, probably by gravitational collapse of the eruptive column, and some of them reached the location of the present town of Pozzuoli.

The Agnano eruption ended with base-surge activity which produced large quantities of ash. The sequence of eruptive events is identical to that described by Sheridan et al. (1981) for the Plinian eruptions of Vesuvius and a similar eruptive model can be applied. The final ash deposit thus represents the result of magma/water interaction caused by the inflow of deep phreatic water into the magma chamber.

The almost exclusively fine ash of the final surge deposits of the Plinian eruptions could be the result of hydromagmatic explosions at depth (see: Wohletz, this volume). The partly-closed environment of these explosions facilitate a maximum of interaction between the water and magma. Geothermal boreholes now being drilled in the area have confirmed that fractured rocks containing fluids exist in the subsurface. Identification of various Plinian eruptions during the last 10,000 years and reconstruction of the paleogeographic evolution of the Phlegraean Fields, lead us to the conclusion that a few of these Plinian eruptions (during the interval between 34,000 and 10,000 years ago) occurred in areas that were then covered by surface waters.

Self and Sparks (1978) have described Plinian eruptions which have occurred in lacustrine environments and used the term of "phreatoplinian" to define the deposits produced by these eruptions (see: Self, this volume). Although we would prefer the term "hydroplinian" -- because the water involved in the process is not an underground fluid -- these Phlegraean eruptions are very similar to those described by Self and Sparks (1978). Walker et al. (1980) and Self (this volume) recently presented more data on phreato-Plinian eruptions.

Some of the eruptions that produced the Neapolitan yellow tuff could have been phreato-Plinian. This yellow tuff appears is chaotic volcanic deposit made up of pumice of varying size and lithic fragments embedded in a predominantly ash matrix. Similar to the stratified yellow tuff of the tuff cones, it commonly contains accretionary lapilli and is vesciculated.

Campanian Ignimbrite eruption

One of the most important results of this study concerns the relationship between the Campanian Ignimbrite and the Phlegraean Fields volcanic area. The stratigraphic position of the "Breccia Museo" deposit of the Phlegraean Fields and its structural relationship with the caldera collapse suggest that it was formed during the Campanian Ignimbrite eruption. The main evidence for a flow emplacement of the Breccia Museo are the following.

The deposit tends to thicken in topographically low areas. The bulk of the deposit often shows laminations, lenses enriched in coarse lithics, and crude cross-stratification. The presence of fumarolic pipes within the breccia indicate that the fines were lost through the escape of gas from fluidized material at rest.

The granulometric composition of the Breccia Museo and its relatively high lithic content correspond to fines-depleted ignimbrites (see: Walker et al., 1980; Walker, this volume). The interpretation is also supported by the discovery at Marmolite, on the northern rim of the collapsed caldera, of a typically welded fines-rich Campanian Ignimbrite horizon overlying the Breccia Museo. A similar stratigraphic relationship is also described for the Taupo ignimbrite of New Zealand by Walker et al. (1980).

In all its outcrops, the Breccia Museo occurs as an unwelded, poorly-sorted breccia made up of large lithic fragments (maximum diameter 1 m) and minor pumice. Levels of welded scoria ("piperno") occur in the lower-middle part of the unit. The degree of welding and flattening of the "piperno" layers could be explained by the weight of the overlying breccia.

Near Lago di Patria (6 km from the caldera rim) there is a typically welded fines-rich Campanian Ignimbrite deposit covered by a thin breccia layer (less than 1 m thick). This breccia is almost entirely composed of lithic clasts possibly ejected by a phreatic phase. These lithic blocks consist mostly of hornfels and intrusive rock that comes from the deepest part of the pre-caldera sequence. The occurrence of this breccia at the end of the magmatic phase of the Campanian Ignimbrite eruption and the deep origin of most of its solid ejecta suggest that it originated when phreatic water came into contact with the hot wall-rocks of the magma chamber, after the magmatic gas pressure was mostly released and the roof of the emptied chamber probably had collapsed.

SUMMARY AND CONCLUSIONS

The Phlegraean Fields have long been regarded as a very complex volcanic zone characterized by the occurrence of a great number of mostly monogenetic pyroclastic vents. However, detailed field survey coupled with geophysical investigations (La Torre and Nannini, 1980) and deep geothermal drilling has permitted reconstruction of the volcanic evolution of the area. This shows the structural setting is less complex than previously believed.

In light of these new results, the Phlegraean Fields can be considered as the expression of activity within a large caldera located above a progressively cooling, shallow magma chamber. Reconstruction of the 12 km wide caldera, recognition of its lower limit by drilling data, and location of its rim by surface study are the keys to understanding the volcano-tectonic evolution of Phelgraean Fields (Fig. 5). The caldera formed 35,000 years ago as a consequence of the Campanian Ignimbrite eruption. The geometry of the collapsed area and the isopach reconstruction of the products related to this eruption suggest that nearly 80 km^3 of trachytic liquid was rapidly erupted before the caldera collapsed. The study of the products related to the Campanian Ignimbrite eruption cropping out on the caldera rim, in particular the fines-depleted pyroclastic flow (Breccia Museo), indicates that the eruption may have been fed by a ring fracture connected to the magma chamber through a cone-sheet structure. The volume of the chamber at that time was not less than 240 km^3, as suggested by the petrochemical study of the erupted products (Armienti et al., this volume).

Since caldera collapse, all volcanic activity of Phlegraean Fields has been confined to the depressed part. The few eruptive vents outside the caldera preceded the Campanian Ignimbrite eruption. Little is known about the volcanic activity between 35,000 and 10,500 years ago. Drilling evidence suggests that several submarine eruptions took place in this time span and that their products, together with marine sediments, progressively filled the caldera, particularly its northern half. When the sea within the caldera became shallow, a series of eruptions produced the yellow tuff deposits.

Since 10,500 years ago, volcanic activity has been mostly subaerial, with only a few, important sublacustrine eruptions. As illustrated in Fig. 4, there has clearly been a progressive migration of eruptive vents towards the caldera center. This time-space migration of the eruptive vents, coupled with the observations that: 1) minor collapses always affect the internal part of the associated edifice, that 2) there is a large predominance of highly evolved compositions (trachytic to peralkaline phonolitic trachyte) among the erupted products, and that 3) magma erupted from the more peripheral vents at any activity period is basic, strongly suggest that Phlegraean Fields have been fed, during the last 10,000 years by a progressively cooling magma chamber, as discussed in the companion paper of Armienti et al. (this volume).

Furthermore, examination of the spatial distribution of eruptive vents during the major phases of Phlegraean post-caldera activity, shows that vents tend to occur along "arcs" in distinct zones of the caldera floor. We think that this distribution results from two combined factors: 1) the possible persistence in time of a cone-sheet type of feeding system, that becomes progressively smaller in size as the magma in chamber cools down and 2) intrusions of magma pockets at shallow depth controlling preferential paths of ascent to the surface. An example of this is the inflation which occurred 5,300 years ago to produce the La Starza terrace, and to concentrate there most of the recent eruptions. Extrapolating the past eruptive history to the future, the most dangerous area of the Phlegraean Fields is in the center of the caldera, practically coincident with the town of Pozzuoli, where a major inflation has been under observation for a few years (Corrado et al., 1977).

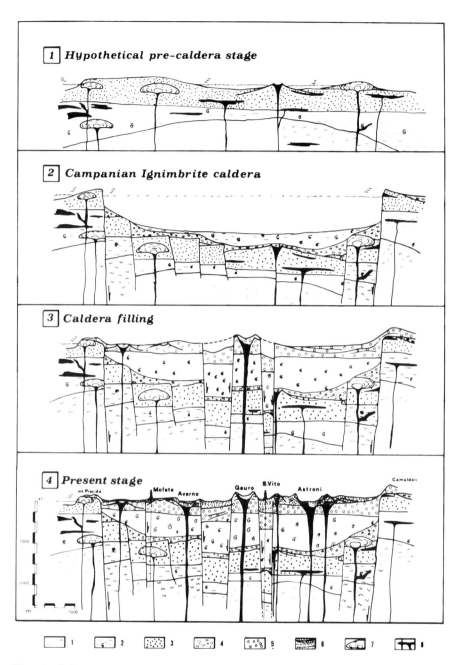

Fig. 5 Scheme of the main evolution stages of the Phlegraean Fields. 1. thermometamorphic rocks; 2. tuffites; 3. subaerial tuffs; 4. products of the Campanian Ignimbrite eruptions; 5. yellow tuffs; 6. tephra and tuffs of the recent, mainly subaerial activity; 7. lava domes; 8. lavas, subvolcanic rocks and volcanic conduits.

From a tectonic point of view we can conclude that the feeding fractures of the Phlegraean Fields seem to be controlled, at least in the last 10,000 years, by the stress field induced by the processes occurring within the magma chamber, rather than by the regional stress field. We believe, however, that the regional stress field was responsible for the initial emplacement of a huge magmatic body.

More work has to be done in order to more precisely reconstruct the eruptive mechanisms of all the Phlegraean eruptions. Some points that seem worthy of emphasis at this stage are:

- The study of the nature of the xenolithic ejecta has been a powerful means for investigating the depth at which magma and water interacted, and in particular the depth of the water.
- The degree of primary fragmentation of magma coming into contact with water seems to be a main factor in controlling the extent and efficiency of magma/water interaction, and the degree of transformation of thermal into kinetic energy in hydromagmatic explosions. Another parameter to express the extent of the water/melt contact surface seems, therefore, more appropriate than the bulk melt/water weight ratio proposed by Sheridan and Wohletz (1981).
- Because of the resulting large increase of the kinetic energy of the eruption, the processes of water/magma interaction is a factor of major importance in the assessment of volcanic hazards. The distribution of surface water as well as the underground hydrology in active volcanic areas deserve a much more attention than they have received in the past.

ACKNOWLEDGEMENTS

This research has been sponsored by an AGIP-ENEL joint-venture for geothermal exploration. The authors wish to thank AGIP and ENEL for permission to publish this work. Particular thanks to Franco Barberi for many helpful discussions and suggestions. The final paper has been improved through the useful comments of R. Greeley and R. V. Fisher who read an early version and provided many useful comments. CNR Italian Geodynamics Project, publication no. 498.

REFERENCES

Armienti, P., Barberi, F., Bizouard, H., Clocchiatti, R., Innocenti, F., Metrich, N., Rosi, M., Sbrana, A., 1982. The Phlegraean Fields: magma evolution within a shallow chamber. In: M.F. Sheridan and F. Barberi (Editors), Explosive Volcanism. J. Volcan. Geotherm. Res. 17: (this volume).

Alessio, M., Bella, F., Belluomini, G., Calderoni, G., Cortesi, C., Fornasieri, M., Franco, E., Improta, F., Scherillo, A., Turi, B., 1971. Datazioni con il metodo del C-14 di carboni e livelli humificati (paleosuoli) intercalati nelle formazioni piroclastiche dei Campi Flegrei (Napoli). Rend. Soc. Ital. Min. Petr. 27(II): 305-308.

Alessio, M., Bella, F., Improta, S., Belluomini, G., Cortesi, C., Turi, F., 1973. University of Rome C-14 dates IX. Radiocarbon, 13: 395-411.

Alessio, M., Bella, F., Improta, S., Belluomini, G., Calderoni, G., Cortesi, C., Turi, F., 1973. University of Rome C-14 dates X. Radiocarbon, 15: 165-178.

Alessio, M., Bella, F., Improta, S., Belluomini, G., Cortesi, C., 1974. University of Rome C-14 dates XII. Radiocarbon, 16: 358-367.

Barberi, F., Innocenti, F., Lirer, L., Munno, R., Pescatore, P., 1978. The Campanian Ignimbrite: a Major Prehistoric Eruption in the Neapolitan Area (Italy). Bull. Volcanol., 41: 1-22.

Bruni, P., Sbrana, A., Silvano, A., 1981. Risultatti geologici preliminari dell'esplorazione geotermica nell'area dei Campi Flegrei. Rend. Soc. Geol. Ital. (note brevi), in press.

Corrado, G., Guerra, I., Lo Bascio, A., Luongo, G. and Rampoldi, R., 1977. Inflation and microearthquake activity of Phlegraean Fields, Italy. Bull. Volcanol., 40-43: 1-20.

Delibrias, G., Kieffer, G. and Pelletier, H., 1969. Datation par la méthode du C-14 de l'Astroni, volcan des Champs Phlégreéns (Campanie). Comptes Rendus Academie des Sciences, Paris, 269: 2070-2071.

Delibrias, G., Di Paola, G.M., Rosi, M., e Santacroce, R., 1979. La storia eruttiva del complesso vulcanico Somma-Vesuvio ricostruita dalle successioni piroclastiche del M. Somma. Rend. Soc. Ital. Min. Petr. 35(1): 411-438.

De Lorenzo, G., 1904. The history of volcanic action in the Phlegraean Fields. Quart. Jour. Geol. Soc., vol LX.

Di Girolamo, P., Keller, J., 1972. Zur Stellung des grauen Campanichen Tuffs, innerhalb des quartären Vulkanismus Campaniens (Süd Italien). Ber. Nature. Ges. Freiburg., Br. 61/62, 85-92.

La Torre, P. and Nannini, R., 1980. Geothermal well location in southern Italy: the contribution of geophysical methods. Boll. Geof. Teor. Appl., 22: 201-209.

Lirer, L. and Munno, R., 1975. Il "tufo giallo napoletano", (Campi Flegrei). Per. Miner. Roma, 44,(1): 103-118.

Lucini, P., Tongiorgi, E., 1959. Determinazione con C-14 dell'eta di un legno fossile dei Campi Flegrei (Napoli). Studi e Ricerche della Divisione Geomineraria C.N.E.N., 2: 97-99.

Parascandola, A., 1946. Il Monte Nuovo ed il lago di Lucrino. Boll. Soc. Natur. Napoli, 55: 151-312.

Rittmann, A., 1950a. Sintesi geologica dei Campi Flegrei. Boll. Soc. Geol. Ital., 69: 117-128.

Rittmann, A., 1950b. Rilevamento geologico della Collina dei Camaldoli nei Campi Flegrei. Boll. Soc. Geol. Ital., 69: 129-177.

Rosi, M., Santacroce, R., and Sheridan, M.F., 1980. Volcanic hazards of Vesuvius (Italy). Bull. B.R.G.M., 4 (2): 169-178.

Self, S., 1983. Large-scale phreatomagmatic silicic volcanism I: a case study from New Zealand. In: M.F. Sheridan and F. Barberi (Editors), Explosive Volcanism. J. Volcanol. Geotherm. Res., 17: (this volume).

Self, S. and Sparks, R.S.J., 1978. Characteristics of pyroclastic deposits formed by the interaction of silicic magma and water. Bull. Volcanol., 41: 196-212.

Sheridan, M.F. and Wohletz, K.H., 1981. Hydrovolcanic explosions: the systematics of water-pyroclast equilibration. Science, 212: 1387-1389.

Sheridan, M.F. and Wohletz, K.H., 1983. Hydrovolcanism: basic considerations. In: M.F. Sheridan and F. Barberi (Editors), Explosive Volcanism. J. Volcanol. Geotherm. Res., 17: (this volume).

Sheridan, M.F., Barberi, F., Rosi, M., and Santacroce, R., 1981. A model for Plinian eruptions of Vesuvius. Nature, 289: 282-285.

Thunell, R., Federman, A., Sparks, S., Williams, D., 1978. The Age, Origin and Volcanological Significance of the Y-5 Ash Layer in the Mediterranean. Quaternary Res., 12: 241-253.

Walker, G.P.L., 1983. Ignimbrite types and Ignimbrite problems. In: M.F. Sheridan and F. Barberi (Editors), Explosive Volcanism. J. Volcanol. Geotherm. Res., 17: (this volume).

Walker, G.P.L., Wilson, C.J.N., Froggatt, P.C., 1980. Fines-depleted ignimbrite in New Zealand: The product of a turbulent pyroclastic flow. Geology, 8: 245-249.

Wohletz, K.H., 1983. Mechanisms of hydrovolcanic pyroclast formation: grain size, SEM, and experimental studies In: M.F. Sheridan and F. Barberi (Editors), Explosive Volcanism. J. Volcanol. Geotherm. Res., 17: (this volume).

Wohletz, K.H. and Sheridan, M.F., 1979. A model of pyroclastic surge. In: C.E. Chapin and W.E. Elston (Editors), Ash-flow tuffs. Geol. Soc. Am., Spec. Paper 180: 177-194.

THE PHLEGRAEAN FIELDS: MAGMA EVOLUTION WITHIN A SHALLOW CHAMBER

P. ARMIENTI[1], F. BARBERI[1], H. BIZOUARD[2], R. CLOCCHIATTI[2],
F. INNOCENTI[1], N. METRICH[2], M. ROSI[1], A. SBRANA[3].

[1] Dipartimento di Scienze della Terra, Università di Pisa - Centro di Geologia Strutturale e dinamica del'Appennino - CNR - PISA. (Italy).
[2] Laboratoire de Petrographie et Volcanologie, Université Paris XI, Orsay (France).
[3] Agip S.p.A., Servizio Esplorazione Geotermica, S. Donato Milanese, Milano (Italy).

(Received September 16, 1982; revised and accepted February 2, 1983)

ABSTRACT

Armienti, P., Barberi, F., Bizouard, H., Clocchiatti, R., Innocenti, F., Metrich, N., Rosi, M., and Sbrana, A., 1983. The Phlegraean Fields: magma evolution within a shallow chamber. In: M.F. Sheridan and F. Barberi (Editors), Explosive Volcanism. J. Volcanol. Geotherm. Res., 17: 289-311.

A systematic petrological and chemical study of the volcanic products of the Phlegraean Fields has been accomplished based on the new stratigraphy described by Rosi et al. (this volume). The majority of Phlegraean rocks belong to the "potassic" series of the Roman province. The compositional spectrum ranges from trachybasalts to latites, trachytes, alkali trachytes, and peralkaline phonolitic trachytes. Trachybasalts are extremely rare and there is a sharp compositional gap between them and the latites. The series between latites and the trachytic varieties is complete. Trachytic rocks are much more voluminous than latites. The order of appearance of the main solid phases is: olivine, clinopyroxene, plagioclase, alkali feldspar, biotite, and oxides. Mineral compositions obtained by microprobe analyses are compatible with the evolution of the rock chemistry. However, primitive compositions of plagioclase (An_{80}) and clinopyroxene (diopside) persist in the cores of phenocrysts, even at the trachytic and alkali trachytic stage. Fractional crystallization within a shallow magma chamber has been the dominant process for the generation of Phlegraean rock series. The volume of the magma chamber is estimated to have been at least 240 km^3 at the moment of the eruption of the Campanian Ignimbrite, nearly 35,000 yr. ago. This event was followed by a large caldera collapse.
  The depth of the chamber cannot be precisely evaluated. However, its top must have been very shallow, probably at 4-5 km, as suggested by contact metamorphic rocks obtained from deep geothermal wells within the caldera.

0377-0273/83/$03.00   © 1983 Elsevier Science Publishers B.V.

Volcanological and petrological data favor a model of upward migration of lighter liquids produced mostly by fractionation along the cool walls of the chamber, the deeper part of which is occupied by a convecting trachybasaltic magma. Progressive migration of eruptive vents toward the caldera center and the contemporaneous strong reduction in the volume of the erupted products, suggest that the chamber behaved as a closed system. The volume of magma was progressively reduced by both cooling and extraction to the surface.

INTRODUCTION

The study of processes within the shallow magma chambers that feed active volcanoes is one of the most promising and interesting problems in present-day volcanology. These studies are essential for understanding the causes of compositional variations observed in the erupted magmas and for elucidating the relationships between deep processes, magma ascent, and eruptive mechanisms. Furthermore, magma chambers represent a potential heat source for high-temperature geothermal systems. Therefore their study has important practical implications both for volcanic hazards assessment and energy development. The aim of this paper is to discuss a model for the magma chamber of the Phlegraean Fields (Naples) based on a systematic petrological and chemical study of the erupted products.

The Phlegraean Fields, although one of the most famous volcanic areas in the world, until now lacked a comprehensive, modern volcanological study. The detailed reconstruction of the volcanic stratigraphy and structural evolution made by Rosi et al. (this volume), together with the new information obtained through geophysical exploration (La Torre and Nannini, 1980) and several deep geothermal wells recently drilled in this area, form the basis for our study.

GEOLOGICAL AND VOLCANOLOGICAL OUTLINES OF THE PHLEGRAEAN FIELDS

The Phlegraean Fields are one of the classical areas of Quaternary potassic volcanism of the Neapolitan Region that also includes Vesuvius, the island of Ischia, and Roccamonfina. These volcanoes are located on the Tyrrhenian margin of the Apennine chain in a zone intensely affected by extensional tectonics with a conjugate NW and NE trending fault system during Quaternary (Fig. 1).

The main structure of the Phlegraean area is a large (nearly 12 km diameter) caldera formed about 35,000 years ago by collapse after the eruption of the Campanian Ignimbrite (Barberi et al., 1978). Half of the caldera is submerged beneath the sea in the Gulf of Pozzuoli (Fig. 2). The half on land is almost totally filled by volcanic and volcano-sedimentary products that cover most of the caldera rim. Pre-caldera rocks found only in local exposures along the rim also occur in deep geothermal wells within the caldera. Unfortunately these rocks are so strongly affected by hydrothermal metamorphism that no radiometric dating has been possible (Bruni et al., 1981).

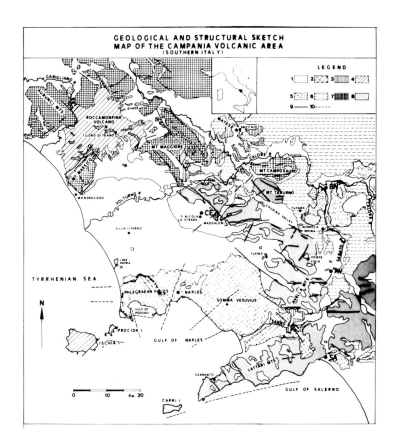

Fig. 1 Geological and structural sketch map of the Campania volcanic area (after Carberi et al., 1978). Quaternary: (1) Alluvium and lacustrine deposits; (2) Volcanic complexes mainly built up after the Campanian Ignimbrite eruption (Somma-Vesuvius and Phlegraean Fields); (3) Campanian Ignimbrite; (4) Volcanic complexes mainly built up before the Campanian Ignimbrite eruption (Roccamonfina, Ischia, Procida). Pleistocene-Pliocene; (5) Ariano unit: mostly clastic deposit. Miocene-Upper Trias; (6) Irpinia units (Middle Miocene): deep sea terrigenous deposits, Sicilid units: pelagic variegated shales, sandstones and limestones (Upper Cretaceous-Early Miocene); (7) Matese-Mt. Maggiore unit: dolomites, limestones and reef complexes (Upper Trias - Upper Cretaceous); calcarenites grading upwards to marly flysch (Langhian-Tortonian); (8) Alburno-Cervati unit: dolomites, limestones and reef complexes (Upper Trias-Upper Cretaceous); calcarenites and calcirudites grading upwards to marly-sandy flysch (Aquitanian-Langhian); (9) Faults; (10) Faults inferred from geophysical data; (11) Caldera rim. CE = Caserta, BN = Benevento, AV = Avellino, SA = Salerno.

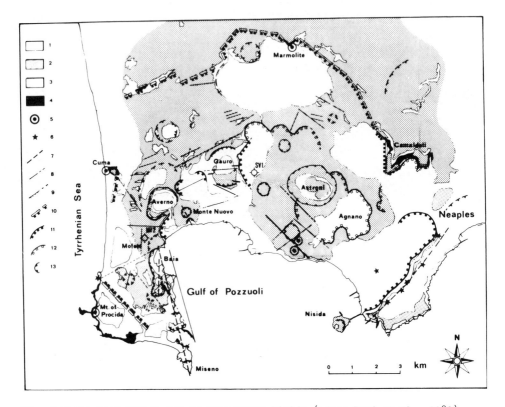

Fig. 2 Geological sketch map of Phlegraean Fields (after Rosi et al., 1982). (1) Urbanized areas and soil cover; (2) post-caldera mostly subaerial activity, tephra and minor lavas, age 10,500 y.b.p. to 1,538 A.D.; (3) post-caldera mostly submarine activity, age 35,000 (?) to 10,500 y.b.p., yellow chaotic tuffs and yellow layered tuffs; (4) pre-caldera activity, age 35,000 y.b.p., lava domes, tephra and products of Campanian Ignimbrite eruption; (5) lava domes; (6) eruptive centers, uncertain location; (7) feeding fractures; (8) fractures; (9) faults; (10) caldera rim; (11) volcano-tectonic collapses related to post-caldera activity (full squares 35,000 to 10,500 y.b.p.; open squares 10,500 to present); (13) crater rims.

Based on drilling evidence, the volume of the caldera is estimated to be about 80 km^3. After the collapse, the area was invaded by the sea and then progressively filled with products of submarine activity (chaotic tuffites, breccias, and minor lava flows). The submarine phase of activity culminated 14,000 to 10,500 years ago with the eruption of huge quantities of yellow hydromagmatic tuffs from several vents.

After this phase (traditionally called "second Phlegraean period" by De Lorenzo, 1904 and Rittmann, 1950) the activity was mostly subaerial. The intensity decreased with time and eruptive vents migrated towards the center of the caldera. Two additional eruptive phases occurred at 10,000 to 8,000 years ago and 4,500 to 3,000 years ago. The most recent eruption (Mt. Nuovo) dates back to 1538. The Phlegraean Fields are presently affected by vertical deformation, intense fumarolic emission and microseismic activity (Corrado et al., 1977).

VOLCANIC PRODUCTS OF THE PHLEGRAEAN FIELDS

Most of the volcanic products of the Phlegrean Fields are pyroclastic. Lava flows and domes are of small volume and mostly of pre-caldera age. In order to reconstruct the magmatic evolution, all exposed volcanic varieties have been sampled as well as the various types of lava xenoliths found as ejecta in volcanic breccias or in other pyroclastic deposits. In total, 116 rock samples have been collected and chemically analysed. Representative fresh samples are rare because most of the pyroclastic rocks are affected by alteration. Hydromagmatic products are usually strongly zeolitized (e.g. yellow tuffs). Primary feldspars are usually transformed into adularia. Secondary fluorite and, to a lesser extent, calcite, hematite, albite, epidote, and analcime are common. Glass is particularly affected by post-depositional alteration, mainly showing a modification of the alkali content and K_2O/Na_2O ratio.

A combination of petrographic, X-ray, and chemical studies allowed the selection of 68 fresh samples that represent the compositional spectrum of the entire eruptive sequence. Only the yellow tuffs of the post-caldera submarine phase of activity are poorly represented because of their deep alteration. Their trachytic nature is however well established.

The composition of the pre-caldera products has been derived mainly from lava xenoliths from a polygenetic breccia ("breccia museo") that crops out on the caldera rim. We interpret this unit as a fines-depleted facies of the Campanian Ignimbrite. With exception of a few deeply-transformed, leucite-phyric lavas found as ejecta in explosive breccias and crossed by some drill holes, all of the erupted products of the Phlegraean Fields belong to the "potassic" association of the Roman province (Civetta et al., 1981).

CHEMISTRY

Phlegraean products show a compositional spectrum that ranges from trachybasalts to peralkaline phonolitic trachytes. Trachytes and alkali trachytes represent by far the most abundant rocks. The less evolved members are scarcely represented. Trachybasalts are rare and a clear compositional gap separates them from the less evolved latites (Fig. 3). If lava xenoliths from the pre-caldera sequence are considered, exactly the same compositional pattern emerges (Armienti, 1981). Chemical analyses of selected samples that represent

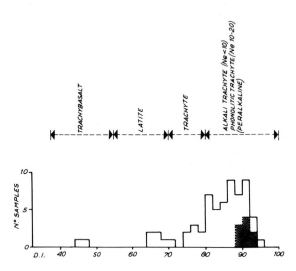

Fig. 3 Differentation Index (D.I.) frequency plot for fresh samples of Phlegraean rocks. D.I. values used for rock classification are indicated. All rocks but peralkaline phonolitic trachytes have normative Ne < 10%. Shaded area represents peralkaline rocks.

the whole compositional range are reported in Table 1.

Most of the Phlegraean rocks are undersaturated in silica (nepheline normative). The degree of silica undersaturation is very low from trachybasalts to trachytes but increases in alkali trachytes and especially in peralkaline phonolitic trachytes. Some of the latites are nearly silica saturated or even slightly oversaturated (sample PF22, Table 1). However, this last sample has a higher iron oxidation ratio and a lower K_2O content than the other analysed latites. The weak silica oversaturation in a few samples of trachytic or alkali trachytic pumice was probably produced by removal of alkalies by glass alteration.

The samples plotted on a variation diagram (Fig. 4) cover the whole activity span, including both outcropping eruptive sequences and pre-caldera lavas. A compositional gap separates trachybasalts from latites (D.I. 47-60). Trachybasalts are only represented by a few xenoliths from pre-caldera lavas and by products of the Minopoli vent that formed during the old post-caldera subaerial phase of activity (between 9,000 and 8,000 years ago). They exibit a variety of textures that range from porphyritic holocrystalline to vitrophyric. Chemical differences are mainly in the TiO_2, P_2O_5 and Al_2O_3 contents.

TABLE 1

Selected representative analyses and CIPW norms of Phlegraen Fields volcanics.

Sample	Trachybasalts		Latites		Trachytes		Alkali trachytes					Peralk. phon. Trachyte
	ZR 44 s. Minopoli	PF 20 l. (*)	PF 13 l. (*)	PF 22 l. (*)	ZR 15 p. Agnano top	ZR 14 p. Agnano bottom	PF 50 l. Astroni Rotond.	CFA 34 p. Astroni	CFA 31 p. Senga	PF 1 l. (*)	Campanian Ignimbrit. (**)	PF 17 s. M.Nuovo
SiO_2	51.47	49.15	56.96	55.61	54.59	57.08	59.07	59.55	58.58	62.36	62.30	59.99
TiO_2	0.90	1.35	0.77	1.14	0.75	0.58	0.48	0.50	0.52	0.43	0.46	0.43
Al_2O_3	16.44	18.90	19.04	18.32	17.44	18.00	19.87	18.51	18.42	18.21	18.98	19.17
Fe_2O_3	1.02	4.88	2.94	4.82	2.47	2.74	2.72	2.15	2.39	1.71	1.77	2.35
FeO	6.62	2.49	1.67	1.14	4.05	2.32	0.89	1.63	1.71	0.99	1.65	0.72
Mn O	0.14	0.11	0.08	0.10	0.14	0.14	0.10	0.12	0.13	0.18	0.25	0.22
Mg O	5.10	5.32	1.87	2.39	2.65	1.40	0.85	0.80	0.83	0.41	0.35	0.21
Ca O	9.78	9.86	5.45	6.62	6.21	4.03	2.89	2.98	3.14	1.63	1.62	1.71
Na_2O	3.09	2.87	3.28	3.07	3.05	3.56	3.90	3.91	4.11	5.88	5.51	7.47
K_2O	3.47	3.86	7.01	5.33	7.08	9.03	8.08	8.81	8.93	7.34	7.03	6.89
P_2O_5	0.43	0.81	0.44	0.44	0.29	0.18	0.15	0.14	0.15	0.05	0.08	0.02
H_2O	1.53	0.40	0.49	1.02	1.28	0.93	1.00	0.89	1.09	0.80	-	0.82
Total	99.99	100.00	100.00	100.00	100.00	99.99	100.00	99.99	100.00	99.99	-	100.00
Q				2.92								
or	20.50	22.81	41.42	31.49	41.83	53.35	47.74	52.05	52.76	43.37	41.70	40.71
ab	19.79	16.99	27.74	25.96	17.13	16.37	31.11	26.85	23.21	44.16	45.85	35.72
an	20.74	27.29	16.53	20.47	12.99	6.47	12.86	6.94	5.45	1.63	6.12	
ne	3.44	3.95			4.70	7.44	1.02	3.37	6.26	3.02	0.43	13.29
ac												2.59
wo								0.95	1.35			2.88
di	20.49	12.71	5.93	7.39	13.00	9.91	0 39	4.88	5.49	2.26	1.18	1.12
hy			0.23	2.52								
ol	9.31	5.15	1.17		3.40	0.01	1.36				1.10	
mt	1.48	4.47	3.41	0.70	3.58	3.97	1.80	3.12	3.47	2.48	2.55	1.79
hm		1.80	0.59	4.34			1.48					0.22
il	1.71	2.56	1.46	2.16	1.42	1.10	0.91	0.95	0.99	0.82	0.91	0.82
ap	1.02	1.92	1.04	1.04	0.69	0.43	0.36	0.33	0.36	0.12	0.19	0.05
D.I.	43.73	43.75	69.16	60.37	63.66	77.16	79.87	82.27	82.23	90.55	87.98	92.31

(*) Lithics in Breccia Museo; l = lava, s = scoria, p = pumice
(**) Inferred original composition, average of the three freshest pumice, from Barberi et al.(1978)

A fairly continuous series exists from latites to alkaline and peralkaline trachytes. Only Al_2O_3 shows a rather wide dispersion of values. A good negative correlation is observed for MgO, FeO tot., CaO, TiO_2 and P_2O_5, whereas Na_2O progressively increases. SiO_2 shows a regular increase; however, a dispersion is observed at high D.I. values (i.e., in peralkaline phonolitic trachytes), for which SiO_2 content decreases. A peculiar trend is shown by K_2O. It increases regularly up to a D.I. of nearly 80; then it sharply decreaes. The trend inversion occurs where sanidine become the dominant phenocrystic phase at the transition between trachytes and alkali trachytes. The rate of Na_2O sharply increases at the same D.I. value. Part of Al_2O_3 (and SiO_2) dispersion near this D.I. value could also be due to a similar inversion of trend. The variations in the trend of alkalies are also well displayed in the triangular K_2O - Na_2O - CaO diagram (Fig. 5).

MINERALOGY

The percentage of phenocrysts in the Phlegraean rocks decreases from trachybasalts and latites (phenocrysts content of 30% to 20%) to trachytes. Alkali trachytes are subaphyric; peralkaline phonolitic trachytes are practically aphyric. Lavas and pumice samples covering the whole

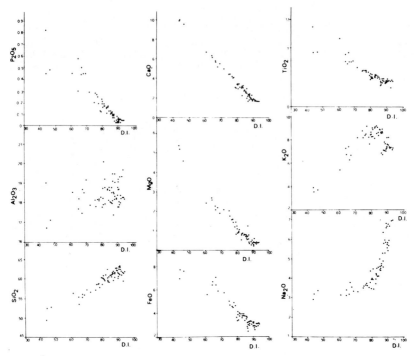

Fig. 4 Variation diagram of major oxides versus Differentation Index (D.I.) of Phlegraean volcanic rocks.

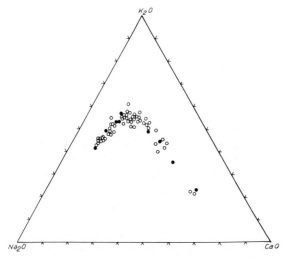

Fig. 5 CaO - K_2O - Na_2O plot for Phlegraean rocks. Dots are xenolithic ejecta of pre-caldera lavas. Open circles are samples of exposed volcanic rocks.

Compositional spectrum were selected fro microprobe study.

Clinopyroxene. The clinopyroxene composition (Table 2) varies from diopside to ferrosalite, with a general trend of Fe and Ca enrichment with rock evolution (Fig. 6). The abundance of clinopyroxene phenocrysts and microphenocrysts decreases from trachybasalts to alkali trachytes; only scarce microlites are observed in the peralkaline phonolitic trachytes. Colorless, homogeneous or weakly-zoned diopside ($Wo_{47.5}$ $Fs_{5.9-7.9}$; Table 2) is the dominant phenocryst of trachybasalts. The minimum crystallization temperature based on glass inclusions has been estimated at 1160 ± 10°C (Metrich, 1982). Diopside also occurs as cores of phenocryst in some latites and trachytes. Salite, present asmicrophenocrysts or as outer rims of phenocrysts in the holocrystalline

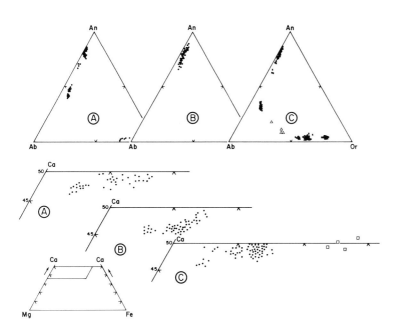

Fig. 6 Clinopyroxene and feldspar phenocryst compositions in molar Ab-Or-An and Ca-Mg-Fe proportions.
(A) Trachybasalts (PF20-ZR44) phenocrysts and microphenocrysts.
(B) Latites (PF22-PF13).
(C) Trachyte (PF50, crosses); alkali-trachyte (PF1, dots); phonolitic trachyte (PF17, triangles).
In the pyroxene plot (C), open squares are microlites from phonolitic trachyte.

TABLE 2

Selected representative microprobe analyses of pyroxenes.

	TRACHYBASALT				LATITE						
	ZR 44				PF 22				PF 13		
	Ph		MPh		Ph		MPh		Ph		MPh
	Core	Rim	Sector – zoning		core	rim					
SiO_2	53.14	51.49	48.86	46.24	51.54	50.62	50.60	51.42	48.96	52.09	51.25
TiO_2	0.38	0.69	1.23	1.59	0.64	0.75	0.72	0.63	0.73	0.52	0.53
Al_2O_3	2.55	3.59	6.22	9.04	3.51	3.35	3.08	2.32	3.57	3.22	3.73
FeO	3.73	4.93	7.22	8.39	4.76	7.26	7.60	8.06	8.43	8.22	8.17
MnO	0.20	0.11	0.10	0.08	0.13	0.22	0.21	0.13	0.34	0.24	0.23
MgO	16.55	15.61	13.17	12.53	16.27	14.35	14.48	15.68	12.96	13.47	13.32
CaO	23.51	23.13	22.79	22.50	22.32	22.17	22.08	20.34	22.59	23.22	23.34
Na_2O	0.12	0.11	0.17	0.18	0.18	0.33	0.29	0.18	0.42	0.27	0.14
Cr_2O_3	0.27	0.07	–	–	–	–	–	–	–	–	–
Si	1.932	1.893	1.814	1.706	1.895	1.889	1.889	1.922	1.958	1.914	1.887
Al IV	0.068	0.107	0.186	0.294	0.105	0.111	0.111	0.078	0.142	0.086	0.113
Al VI	0.041	0.049	0.085	0.098	0.047	0.036	0.025	0.044	0.018	0.054	0.049
Ti	0.010	0.019	0.034	0.044	0.018	0.021	0.020	0.018	0.021	0.014	0.015
Fe^{3+}	0.007	0.026	0.004	0.118	0.035	0.055	0.066	0.031	0.113	0.022	0.058
Fe^{2+}	0.106	0.126	0.180	0.140	0.111	0.170	0.171	0.221	0.154	0.230	0.193
Mn	0.006	0.003	0.003	0.003	0.004	0.007	0.007	0.004	0.011	0.007	0.007
Mg	0.897	0.856	0.754	0.693	0.892	0.798	0.806	0.874	0.733	0.738	0.731
Ca	0.916	0.911	0.907	0.889	0.879	0.886	0.883	0.815	0.919	0.914	0.921
Na	0.008	0.008	0.013	0.013	0.013	0.024	0.021	0.013	0.031	0.019	0.024
Cr	0.008	0.002	–	–	–	–	–	–	–	–	–
Wo	47.5	47.5	48.2	48.3	45.9	46.4	45.8	42.0	47.9	48.0	48.4
En	46.6	44.6	39.4	37.7	46.5	41.8	41.8	45.0	38.5	38.7	38.4
Fs	5.9	7.9	12.0	14.0	7.6	11.8	12.4	13.0	13.9	13.3	13.2

TABLE 2, continued.

	TRACHYTE	ALKALI TRACHYTE					PERALKALINE PHONOLITIC - TRACHYTE	
	PF 50	CFA 31		PF 1			PF 17	
	MPh	Ph core	Ph rim	MPh	MPh	MPh	G.M.	G.M.
SiO$_2$	49.66	52.28	49.05	49.92	51.20	52.27	44.37	49.06
TiO$_2$	0.80	0,35	0,90	0,65	0.75	0.66	2,40	1,20
Al$_2$O$_3$	3,91	2.08	4.19	3,20	2.92	2.05	5.98	3.21
FeO	8.60	3.71	8,22	7,97	9.33	9.19	14,79	14.31
MnO	0.30	0.09	0.24	0.17	0.74	0.86	0.98	1.26
MgO	12.93	16.71	12.89	13.20	12.56	13.12	7.02	8.45
CaO	23.12	23.71	25.40	23.76	22.43	22.50	22.05	22.18
Na$_2$O	0.41	0.11	0.38	0.41	0.75	0.66	1.04	0.84
Cr$_2$O$_3$	-	0.35	0.05	-	-	-	-	-
Si	1.853	1.919	1.837	1.867	1.899	1.926	1.731	1.873
Al IV	0.147	0.081	0.163	0.133	0.101	0.074	0.269	0.127
Al VI	0.025	0.009	0.022	0.008	0.027	0.015	0.007	0.017
Ti	0.022	0.010	0.025	0.018	0.021	0.018	0.070	0.034
Fe^{3+}	0.106	0.051	0.117	0.118	0.087	0.070	0.201	0.103
Fe^{2+}	0.162	0.063	0.140	0.132	0.202	0.213	0.272	0.354
Mn	0.009	0.003	0.008	0.005	0.023	0.027	0.032	0.041
Mg	0.720	0.914	0.720	0.736	0.694	0.721	0.408	0.481
Ca	0.925	0.932	0.939	0.952	0.891	0.888	0.922	0.907
Na	0.030	0.008	0.028	0.030	0.054	0.047	0.079	0.062
Cr	-	0.010	-	-	-	-	-	-
Wo	48.2	47.6	49.0	49.1	47.9	46.9	50.8	49.1
En	37.5	46.6	37.6	38.0	37.3	38.1	23.5	26.2
Fe	14.3	5.8	13.4	12.9	14.7	15.0	27.8	24.7

Ph : Phenocryst. - MPh: Microphenocryst. - G.M.: Groundmass.

trachybasalt (PF 20), becomes the dominant phenocryst phase in the more evolved rocks. Compositional variation in the latite-trachyte field is rather limited (Table 2, Fig. 6).

Only the pyroxene of the peralkaline phonolitic trachytes shows a marked iron enrichment. It has a ferrosalitic composition, being more calcic and slightly enriched in Al, Ti, and Na, in agreement with the peralkaline nature of the host rocks. The presence of small quantities of Al(IV) agrees with the undersaturated character of the rock suite. High Al_2O_3 variations (2-9%) are associated with sector zoning of trachybasalt microphenocrysts. The Al/Ti ratio usually remains high (>7) reflecting the low TiO_2 content of Phlegraean rocks.

Optical thermometry on glass inclusions indicates a similar minimum salite crystallization temperature of 1020-1040 ± 10°C for latites and trachytes (Metrich, 1982). Solid inclusions of apatite, Ti-magnetite and plagioclase are abundant especially in trachytes.

Feldspars. Plagioclase composition (Table 3) ranges from calcic bytownite to sodic andesine (An_{87-35}). As with clinopyroxene, plagioclase crystals (phenocrysts and microphenocrysts) of the holocrystalline trachybasalt (PF20) show nearly the same variations observed in plagioclase phenocrysts from the whole rock spectrum (Fig. 6). Calcic bytownite (An_{87-80}, Or_{2-4}) phenocryst cores in latites and trachytes are commonly mantled by sanidine. Plagioclase phenocrysts become more sodic (An_{42} Or_{9-10}) in alkali trachytes. There is a tendency toward a progressive increase of Na and K in plagioclase phenocrysts of the more evolved rocks.

Sanidine occurs as small laths ($Or_{73.6}$ $An_{3.4}$) in trachybasalts as an outer rim of some plagioclase phenocrysts in latites, and as the dominant phenocryst in trachytes. The highest Or content ($Or_{81.5}$, $An_{2.4}$) is observed in the CFA31 trachyte (D.I. 82) that shows the highest K_2O content of the Phlegraean series (Fig. 4). For higher D.I. values the sanidine becomes more sodic and calcic up to Or_{63-65} and An_3 in peralkaline phonolitic trachytes.

In some trachytes a corroded sodic plagioclase (An_{31-35}) is mantled by sodic sanidine (Or_{56-60}). This could be an evidence that plagioclase has reached the "critical composition" corresponding to the termination of the plagioclase - alkali feldspar surface (Rahman and Mackenzie, 1969). Heating stage experiments on K-feldspar glass inclusions (methodology in Clocchiatti, 1975) suggest the same minimum crystallization temperature as for trachytic pyroxenes (1050 ± 10°C).

Biotite. Biotite (Table 4) occurs as a late crystallization phase in the groundmass of trachybasalts. Its first appearance as a phenocryst is in latites, and it persists in trachytes as rare microphenocrysts, commonly with an opacite rim. The Fe/(Mg + Fe) ratio increases slightly from zoned phenocrysts in latite (0.21 - 0.30) to trachytes (up to 0.34).

Oxides. A single titanomagnetite phase occurs in the Phlegraean rocks (Table 5) as phenocrysts and microphenocrysts, mostly in latites, but also sporadically in trachybasalts and trachytes. The ulvospinel content tends to decrease from trachybasalts and latites (25.4 - 27.5%) to trachytes (19-20%). The Al_2O_3 and

TABLE 3

Selected representative microprobe analyses of feldspars.

	TRACHYBASALT			LATITE						
	ZR 44			PF 22				PF 13		
	MPh	MPh	MPh	Ph. core	Ph. rim	Ph. core	Ph. rim	Ph. core	Ph. rim	MPh core	MPh rim
SiO_2	47.56	48.78	63.95	45.74	46.88	45.88	63.11	46.93	49.02	49.38	63.67
Al_2O_3	32.31	31.52	19.27	33.38	32.48	32.94	19.67	33.53	31.82	31.65	19.61
Fe_2O_3	0.69	0.83	0.37	0.73	0.80	0.85	0.46	0.75	0.72	0.71	0.39
CaO	16.20	15.77	0.72	17.44	16.33	16.92	1.23	16.73	14.90	14.74	1.12
Na_2O	2.00	2.00	2.56	1.29	2.01	1.59	3.61	1.67	2.47	2.70	3.19
K_2O	0.37	0.75	12.56	0.21	0.26	0.22	10.62	0.34	0.58	0.56	11.25
Total	99.13	99.65	99.43	98.79	98.76	98.40	98.70	99.95	99.51	99.74	99.22
Ab	17.8	18.0	22.5	11.7	17.9	14.3	32.0	15.0	20.2	22.3	28.5
An	80.2	78.0	3.4	87.1	80.5	84.4	6.0	83.0	77.1	74.3	5.5
Or	2.0	4.0	73.6	1.2	1.6	1.3	62.0	2.0	2.7	3.4	66.0
cations	5.02	5.00	5.01	5.00	5.00	5.00	5.02	5.01	5.00	4.99	5.01

Table 3, continued

	TRACHYTE						ALKALI TRACHYTE				PERALKALINE PHONOLITIC-TRACHYTE	
	PF 50					CFA31		PF 1			PF 17	
	Ph. core	Ph. rim	Ph. core	Ph. rim	MPh	GM	Ph.	Ph. core	Ph. rim	MPh	MPh	Ph. core	Ph. rim
SiO_2	63.76	63.74	46.47	64.59	48.41	63.57	63.28	63.05	65.50	64.68	60.50	65.64	63.90
Al_2O_3	19.27	19.69	33.04	19.82	32.19	20.02	19.03	18.96	19.22	19.09	24.24	19.29	20.46
Fe_2O_3	0.19	0.44	0.74	0.48	0.62	0.42	0.22	0.26	0.46	0.40	0.54	0.18	0.39
CaO	0.54	1.05	16.74	1.08	15.41	1.28	0.47	0.62	0.81	0.91	6.42	0.51	1.84
Na_2O	1.97	3.69	1.66	4.01	2.34	4.55	1.80	3.69	4.74	3.97	6.79	3.83	5.36
K_2O	13.49	10.63	0.27	10.55	0.43	8.98	13.59	11.11	9.70	10.66	1.91	11.41	7.77
Total	99.22	99.24	98.92	100.53	99.40	98.82	98.39	99.67	100.43	99.71	100.40	100.86	99.72
Ab	17.6	32.7	15.0	34.7	21.0	40.7	16.1	32.5	40.2	34.5	58.5	32.0	46.5
An	2.7	5.1	83.4	5.2	76.4	6.3	2.4	3.0	3.9	4.4	30.6	2.4	9.0
Or	79.7	62.2	1.6	60.1	2.6	53.0	81.5	64.5	55.2	61.1	10.9	65.4	44.5
cations	5.03	5.00	5.00	5.02	5.00	5.00	5.01	5.01	5.01	5.00	5.01	5.03	5.02

Ph : Phenocryst. - MPh: Microphenocryst. - G.M.: Groundmass.

MgO contents are relatively high (respectively 2-5% and 1.5 - 2.9%) in agreement with the silica undersaturated nature of the host rocks.
Olivine. Olivine (Fo 82%, Table 5), present as phenocrysts only in trachybasalts is generally altered to iddingsite. The distribution of Fe and Mg suggests equilibrium between olivine and liquid (whole rock) (Kd = 0.30; Roeder and Ernslie, 1970).

TABLE 4
Selected representative microprobe analyses of biotites.

	TRACHYBASALT		LATITE					TRACHYTE			ALKALI TRACHYTE
	PF20	ZR44	PF22		PF13		PF50	CFA31			PF1
	Ph.	G.M	Ph. core	Ph. rim	MPh	Ph	MPh	MPh	Ph core	Ph rim	MPh
K_2O	9.25	9.13	9.14	9.01	9.10	9.53	9.40	9.31	9.25	8.99	9.30
Na_2O	0.59	0.36	0.55	0.53	0.51	0.37	0.36	0.40	0.42	0.41	0.67
FeO	12.33	12.15	11.23	12.17	12.56	12.91	13.96	13.45	12.93	13.20	14.09
MgO	14.62	15.91	17.18	16.19	15.69	15.28	15.08	15.36	15.66	14.75	15.00
TiO_2	4.26	5.11	3.57	4.62	6.09	5.99	5.47	5.57	5.63	5.68	5.23
MnO	0.20	0.13	0.09	0.07	-	0.13	0.14	0.18	0.20	0.18	-
Al_2O_3	14.39	15.26	14.49	15.09	14.25	14.61	14.60	14.65	14.10	14.41	13.65
SiO_2	37.01	35.55	37.17	36.38	36.25	37.31	37.38	36.25	36.10	35.37	37.71
Total	96.65	93.60	94.32	94.06	94.65	96.20	96.54	95.17	94.31	92.99	96.23
Fe/Mg+Fe	0.370	0.300	0.208	0.297	0.310	0.323	0.342	0.329	0.316	0.334	0.336

<u>Brown hornblende</u> is a rare microphenocryst in latites, trachytes and in alkali trachytes. Common accessory minerals are apatite, zircon and sphene (Table 5).

FRACTIONATION OF PHLEGRAEAN MAGMA

Chemical and mineralogical data are sufficiently coherent to suggest that fractional crystallization played a significant role in the genesis of Phlegraean volcanic rocks. The observed chemical trends of major elements are compatible with a fractionation process initially dominated by olivine, clinopyroxene, plagioclase, and magnetite. Alkali feldspar appears later and becomes the dominant phase in the final part of the series. The scatter of Al_2O_3 values can be due to the relative proportion of clinopyroxene and feldspar (both plagioclase and sanidine) in the sample, and also to the highly variable Al_2O_3 content of sector-zoned clinopyroxene phenocrysts.

A plot of the trachytic compositions in Petrogeny's Residua System (Fig. 7) shows an initial displacement parallel to the feldspar line and normal to the isotherms. Then a clear trend moves toward the phonolitic minimum. The most evolved compositions are peralkaline phonolitic trachytes. The transition into the peralkaline field with increasing silica undersaturation is observed in the molecular SiO_2 - Al_2O_3 - alkali plot of Fig. 8. This, together with the behavior of K_2O, Na_2O and SiO_2, confirms the dominant role of late-stage sanidine fractionation. Because plagioclase still occurs in the alkali

TABLE 5

Selected microprobe analyses of Fe-Ti oxides, olivine, amphibole and sphene.

	TITANOMAGNETITE						OLIVINE	HORNEBLENDE	SPHENE
	TRB	LAT.		TR			TRB	PH-TR	TRB
	PF 20	PF 22	PF 13	CFA 31	PF 1		ZR 44	P 17	ZR 44
	Ph	Ph	Ph	MPh	Ph		Ph	Ph	MPh
Fe_2O_3	50.30	47.67	52.38	52.29	51.81	SiO_2	40.52	40.43	30.70
FeO	36.03	35.11	33.60	33.00	35.33	TiO_2	-	3.32	37.90
MgO	1.55	2.89	2.49	2.55	1.67	Al_2O_3	-	11.39	2.18
TiO_2	8.56	9.05	6.73	6.60	8.57	FeO	16.22	17.31	0.53
Al_2O_3	2.12	4.31	4.17	3.74	2.10	MnO	0.31	0.82	0.10
MnO	0.76	0.48	0.56	0.48	1.72	MgO	43.64	9.03	0.04
SiO_2	0.03	0.10	0.10	0.11	0.07	CaO	0.35	12.11	28.60
Cr_2O_3	0.10	-	-	-	-	Na_2O	-	2.32	-
						K_2O	-	1.94	-
Total	99.45	99.61	100.03	98.77	100.73		100.86	98.67	100.05
% Usp	25.4	27.5	20.4	20.1	18.7	% Fo	82.2		

TRB trachybasalt; LAT: latite; TR : trachyte

PH-TR Peralkaline phonolitic trachyte

Usp = ulvospinel; Fo = forsterite;

trachytes, it can be inferred that the transition to the peralkaline compositions is due to the "plagioclase effect" (Bowen, 1945).

A least-squares mass balance program (Stormer and Nicholls, 1978) has been used to test the mechanism of crystal fractionation (Table 6). Calculations show a good fit in the various transitions obtained using the mineral compositions determined by microprobe. Results are also compatible with the order of appearance or disappearance of solid phases in the series.

The fit of the last transition (alkali trachyte to peralkaline phonolitic trachyte) is less good (Σr^2 = 2.2). This requires clinopyroxene, whereas no clinopyroxene was needed in the previous transition from trachyte to alkali trachyte, in agreement with the petrography. Such an adjustment might be due to the effect of accessory minerals, such as apatite and sphene, that have not been considered in the computation. But it might also reflect some other process superimposed on crystal fractionation. This could possibly also account for the limited chemical dispersion which is observed at the alkali trachytic stage.

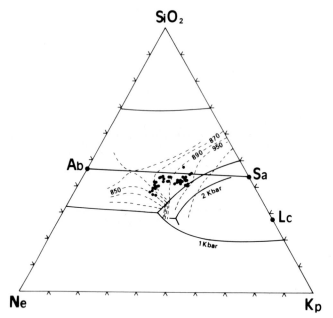

Fig. 7 Plot of the most evolved Phlegraean rocks (trachytes, alkali-trachytes and peralkaline phonolitic trachytes) in the Nepheline-Kalsilite-Quartz system. The boundary lines for P_{H_2O} of 1 and 2 Kb and the isotherms are taken from Gittings (1979).

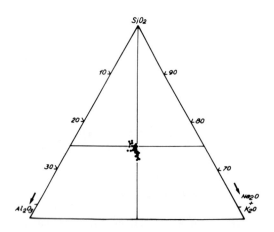

Fig. 8 Silica-alumina-alkalies molecular plot of the most evolved Phlegraean rocks. The horizontal line expresses silica saturation. The vertical line separates peralkaline from subalkaline compositions. The crossing point is the alkali feldspar composition.

TABLE 6

Fractionation of Phlegraean magmas by least-squares mass-balance computation.

Initial composition	Final composition	% Total Solid removed	Ol	Cpx	Pl	Mt	Bi	K-fd	r^2
ZR 44 (43.7)	PF 22 (60.4)	31.8	2.5	14.9	11.0	3.4	–	–	0.79
ZR 44	PF 13 (69.2)	35.0	2.3	17.4	11.4	3.9	–	–	1.55
PF 13	CFA 31 (82.2)	19.3	–	6.3	11.1	–	1.9	–	0.62
PF 13	PF 50 (79.9)	18.4	–	8.8	6.6	–	3.0	–	0.39
PF 50	CFA 31	16.5	–	–	5.2	–	–	11.3	2.05
PF 50	PF 51 (86)	14.6	–	–	6.9	–	3.1	4.6	0.72
PF 50	PF 1 (90.6)	41.3	–	–	11.1	–	8.7	21.5	1.87
PF 1	PF 17 (92.3)	42.4	–	2.7	–	–	–	39.7	2.20

D.I. - Values are reported in **brackets** - For the compositions see Tables 1 to 5

TIME-SPACE VARIATIONS OF PHLEGRAEAN VOLCANISM

It is only possible to reconstruct the time-space evolution of Phlegraean volcanism with some precision for the most recent phases of post-caldera activity (i.e., for the last 11,000 years). The pre-caldera phase had an unknown duration. Data from geothermal wells and lithic ejecta only show that the erupted products include a compositional range that exactly overlaps that of the more recent products. The compositional gap between trachybasalts and latites is also present in the older rocks and the volume of trachytic products dominate over the more basic ones. This phase ended nearly 35,000 years ago with the caldera collapse which followed the eruption of the Campanian Ignimbrite.

The geometry of the collapsed caldera inferred by deep wells is consistent with the isopach reconstruction of the co-ignimbrite ash-fall deposit (Thunnel et al., 1978), both indicating that a volume of about 80 km^3 of evolved liquid was ejected during this eruption. Barberi et al. (1978) have shown that this pyroclastic deposit is so deeply affected by post-depositional chemical changes that any original compositional variations that may have existed are now largely obscured. All of the freshest pumice fragments so far analysed have an alkali trachytic composition. Any primary variations, if present, must lie within the trachytic field of the Phlegraean volcanic series.

The situation is practically the same for the older products of post-caldera activity that filled the collapsed area during the submarine stage. A rough estimate the magma-equivalent volume of the products erupted in this phase, based on drill-hole data, is 23 km^3 (Fig. 9). The most basic products so far encountered in deep wells are strongly altered latites. The whole range of

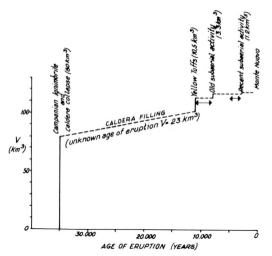

Fig. 9 Volume of erupted products (recalculated as liquid) versus age of eruptions. The volumes and age of eruptions prior to caldera collapse cannot be evaluated. During the long initial phase of caldera filling, a total volume of 23 km³ of liquid was erupted, but the chornology of the volcanic events is unknown. Arrows indicate the time buondaries for the two main phases of post-caldera activity.

composition seems somewhat more restricted than for pre-caldera rocks (Fig. 10). Activity was confined to the area of caldera collapse. The yellow tuffs that end the submarine phase of post-caldera activity were erupted nearly 11,000 years ago. Most of their vents are located near the caldera rim (Fig. 2). The largest eruptions were followed by collapses, the location of which generally migrated toward the caldera center. This has been interpreted as evidence for a cone-sheet feeding system from a magma body restricted to the caldera center (Rosi et al., this volume).

The magma-equivalent volume for this phase is estimated at about 10.5 km³. The composition seems dominantly trachytic, but alteration is so strong and widespread that primary compositional variations are hard to recognize. The less-evolved products (latites and trachytes with low D.I.) were erupted from the more peripheral vents (e.g. Monticelli and Torregaveta). Morphology and state of preservation of the rocks are much better in the subaerial products of post-caldera activity for which it is possible to reconstruct the time-space variations with some accuracy (see Rosi et al, this volume). The main elements for the final phase of activity are the following:

(1) There is a clear temporal migration of the eruptive vents toward the caldera center (Fig. 2) coupled with a progressive decrease of the volume of the erupted products (Fig. 9).

(2) Eruptions of basic products (e.g. the trachybasalt of Minopoli and the latite of Montagna Spaccata) occur on the external side of contemporary

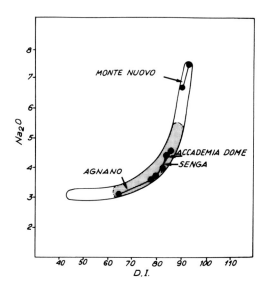

Fig. 10 Compositional variations observed during some single eruptions in the Phlegraean Fields. The enclosed area is the compositional range of the entire Phlegraean series. Dotted area is the variation field of the yellow tuffs.

trachytic vents.

(3) The erupted products cover the entire variation range of the Phlegraean series, with trachytic compositions dominating (Fig. 10).

(4) Compositional variations occur in the products of a single eruption (Fig. 10). Examples include: Agnano: from latite to trachyte (D.I.: 64.5 to 78); the lava dome of the Accademia: alkali trachytes with D.I. ranging from 80 to 86.5; Monte Nuovo: alkali trachytes with a 91 to 93 D.I. range. The widest variation occurs in the most voluminous eruption (Agnano). The temporal sequence is always from the more-evolved to the less-evolved composition. These "within-eruption" trends perfectly overlap the general trend of the Phlegraean series. Particularly significant in this respect is the opposite trends of Na_2O and K_2O in the eruptive sequence of Monte Nuovo, which is identical to that observed in the final part of the Phlegraean series.

CONCLUSIONS

Reconstruction of the volcanic history combined with the detailed chemical and mineralogical study allows the creation of a model for the magma feeding system of the Phlegraean Fields. The migration of eruptive vents toward the center of the caldera, the inward displacement of collapse zones with respect to the related eruptive vents (Rosi et al., 1982), and the progressive decrease in volume of erupted products with time, are all evidence that the Phlegraean feeding system consists of a shallow magma chamber which acted as a closed system, the volume of which progressively diminished by cooling and magma extrusion onto the surface. The center of this chamber is located beneath the town of Pozzuoli where the last eruption occurred. A concentration of microseismic activity and the center of a major zone of active inflation is located there (Corrado et al., 1977).

The petrochemical evolution of the erupted products helps in understanding magmatic processes within the chamber. We have shown that the Phlegraean products form an evolutionary series which is continuous from latites to peralkaline phonolitic trachytes. However, there is a gap between latites and trachybasalts. Petrology and chemistry suggest that crystal fractionation has been the dominant petrogenetic process. Another important indication comes from the compositional variations observed with some of the major post-caldera rock units. These indicate that the magma chamber was compositionally zoned with trachytic liquids occupying the highest part and grading downward towards latites. The trachybasalts vented at Minopoli suggests that at the moment of eruption trachytic liquids were confined to the shallow top of the chamber, whereas trachybasalt occupied the deeper outer zone.

All these data favor a magma chamber characterized by an upper part that is compositionally zoned from trachyte to latite, with liquids stratification probably controlled by density. The persistence of the same chemical characteristics in pre-caldera and post-caldera trachybasalts, their mineralogical homogeneity, and the compositional gap between trachybasalts and latites, further suggest the possible occurrence of a convecting trachybasalt layer in the deepest part of the magma chamber (Fig. 11).

Trachybasalts have been tapped from this layer only through deep peripheral fractures like the one that fed the Minopoli vent. We envisage that crystal fractionation, mainly near the cool walls, has been the major process that produced the compositional variation of the upper part of the magma chamber, as in the model of McBirney (1980). Other processes, like the interaction with wall rocks or upward fluid diffusion, may be superimposed on the main liquid-solid fractionation process.

The persistence of high-temperature mineral phases in some trachytes can be explained by imperfect removal of the phenocrysts during the ascent of the lighter evolved liquids along the chamber walls. The very low phenocryst content in trachytic products of the entire span of Phlegraean activity together with the experimental indication of a relatively high crystallization temperature, could be related to the existence of a relatively thick, high-temperature trachybasaltic zone underlying the graded zone. The perfect overlap of petrochemical trends in products erupted during the different periods of Phlegraean volcanic activity suggests that the magmatic process in the feeding system has been basically the same.

Eruption of the Campanian Ignimbrite tapped the maximum volume of trachytic liquid (~ 80 km^3). Using this value and the degree of fractionation according

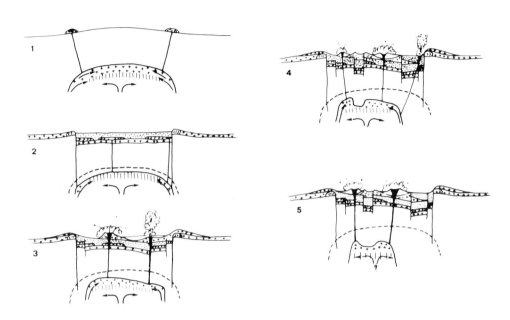

Fig. 11 Model for the evolution of the Phlegraean Fields.
(1) 35,000 y.b.p. (pre-caldera). Zoned magma chamber with 80 km³ of trachyte grading downward to latite produced by crystal fractionation along the chamber walls. A convective trachybasalt layer occupies the chamber bottom. (2) 35,000 to 15,000 y.b.p., Submarine phase of post-caldera activity. Submarine eruptions of latites and trachytes (23 km³ of liquid) and sedimentation progressively fills the caldera. Accumulation of lavas restores the lithostatic load; the system closes again producing upward re-accumulation of trachytic liquid. (3) 11,000 y.b.p. (yellow tuffs). 10 km³ of trachytes are erupted along fissures near the caldera rim, fed by a partially cooled magma chamber. (4) 1,000 to 8,000 y.b.p. (old subaerial phase). Magma chamber has futher cooled. Eruptive trachytic vents concentrate in a smaller internal zone. More voluminous eruptions show compositional changes from more evolved to less evolved liquids (e.g. trachyte-latite). Trachybasalt is erupted from a peripheral vent (Minopoli). (5) 4,500 y.b.p. to present (recent subaerial phase). By progressive cooling, liquids are now confined to a small central part of the consolidated ancient chamber. Small volume trachytes are erupted from vents located near the caldera center (Monte Nuovo).

to the calculation reported in Table 6, we estimate the minimum volume of trachybasalt in the pre-caldera magma chamber to have been 240 km^3. Since that time, volcanological evidence suggests a progressive reduction in the chamber volume. However, the magma chamber probably acted as a closed system only in the last 10,000 years. A partial refilling through ascent of a trachybasalt magma from depth cannot be excluded for the earliest times.

This chamber must be very shallow, its top probably lying at 4-5 km depth, as suggested by contact metamorphosed rocks and by the very high temperatures found in geothermal wells (Rosi et al., this volume). No precise information on the crystallization pressure can be obtained from petrological data. The lack of leucite crystals in undersaturated trachytic rocks indicates only that these liquids have crystallized under a P_{H_2O} higher than 1 kbar (see Fig. 7).

The Phlegraean Fields, as many other volcanic systems, are characterized by a large predominance of evolved products over intermediate and basic ones. We think this is an indication that the interaction of contrasting factors, such as lithostatic load and stress field, controls the availability of magma for eruptions. Only the less dense and volatile-rich trachytic melts usually rise to the surface. It is interesting to note that the main eruptions of latitic lavas occurred just after the caldera collapse, when the lithostatic load was greatly reduced. The accumulation of these latitic lavas together with sediments on the caldera floor, progressively restored the lithostatic load. The system, then being closed, reestablished the conditions that permit only the ascent of light trachytic magma that erupts onto the surface.

ACKNOWLEDGEMENTS

This work has been sponsored by the AGIP-ENEL joint venture of geothermal exploration and by the National Research Council (CNR) of Italy. Chemical analyses of rocks were done by M. Menichini (X-ray fluorescence) and A. Sbrana (MgO and FeO, wet method). The authors wish to thank Agip and Enel for permission to publish this work.

REFERENCES

Armienti, P., 1981. L'evoluzione della serie potassica dei Campi Flegrei: studio petrologico degli inclusi ignei e delle lave. Atti Soc. Tosc. Sc. Nat. Mem., 88: 83-116.
Barberi, F., Innocenti, F., Lirer, L., Munno, R., Pescatore, T. and Santacroce, R., 1978. The Campanian Ignimbrite: a major prehistoric eruption in the Neapolitan Area (Italy). Bull. Volcanol., 41: 10-31.
Bowen, N.L., 1945. Phase equilibria bearing on the origin and differentiation of alkaline rocks. Am. Jour. Sci., 243 A: 75-89.
Bruni, P., Sbrana, A., Silvano, A., 1981. Risultati geologici preliminari dell'esplorazione geotermica nell'area dei Campi Flegrei. Rend. Soc. Geol. Ital. (note brevi) in press.

Civetta, L., Innocenti, F., Manetti, P., Pecerillo, A. and Poli, G., 1981. Geochemical characteristics of potassic volcanism from Mts. Ernici (Southern Latium, Italy). Contrib. Mineral. Petrol., 78: 37-47.

Clocchiatti, R., 1975. Les inclusions vetreuses de cristaux de quartz. Mem. Soc. Geol. France, 122: 1-96.

Corrado, G., Guerra, I., Lo Bascio, A., Luongo, G. and Rampoldi, R., 1977. Inflation and microearthquake activity of Phlegraean Fields, Italy. Bull. Volcanol., 40-43: 1-20.

De Lorenzo, G., 1904. L'attività vulcanica nei Campi Flegri. Rend. Acc. Sc. Fis. Mat. Nat., 310.

Gittings, J., 1979. The feldspathoidal alkaline rocks. In: H.S. Yoder, Jr. (Editor), "Evolution of Igneous Rocks", Princeton Univ. Press, pp. 351-390.

La Torre, P. and Nannini, R., 1980. Geothermal well location in southern Italy: the contribution of geophysical methods. Boll. Geof. Teor. Appl., 22: 201-209.

McBirney, A.R., 1980. Mixing and unmixing of magmas. J. Volcanol. Geotherm. Res., 7: 357-371.

Metrich, N., 1982. Chemical variations of calcic clinopyroxenes as an indicator of petrological processes (Example: Pumices of Phlegraean Fields). Rap. P.I.R.P.S.E.V. (in press).

Rahman, S. and McKenzie, W.S., 1969. The crystallization of ternary feldspars: a study from natural rocks. Am. J. Sci., 267 A: 391-406.

Rittmann, A., 1950. Sintesi geologica dei Campi Flegrei. Boll. Soc. Geol. It., 69: 117-128.

Roeder, P.L. and Emslie, R.F., 1970. Olivine liquid equilibrium. Contrib. Mineral. Petrol., 29: 275-288.

Rosi, M., Sbrana, A. and Principe, C., 1983. The Phlegraean Fields: structural evolution, volcanic history, and eruptive mechanisms. In: M.F. Sheridan and F. Barberi (Editors), Explosive Volcanism. J. Volcanol. Geotherm. Res., 17: (this volume).

Stormer, J.C. Jr. and Nicholls, J., 1978. XLFRAC: a program for interactive testing of magmatic differentiation models. Computer and Geosci., 4: 143-159.

Thunell, R., Federman, A., Sparks, S., Williams, D., 1978. The age, origin, and volcanological significance of the Y-5 ash layer in the Mediterranean. Quaternary Res., 12: 241-253.

EVIDENCE FOR MAGMA MIXING IN THE SURGE DEPOSITS OF THE MONTE GUARDIA SEQUENCE, LIPARI

ROSANNA DE ROSA[1] and MICHAEL F. SHERIDAN[2]

[1] Dipartimento di Geologia, Università di Calabria, Cosenza (Italy)
[2] Department of Geology, Arizona State University, Tempe, AZ 85287 (U.S.A.)
(Received August 4, 1982; revised and accepted March 17, 1983)

ABSTRACT

De Rosa, R. and Sheridan, M.F., 1983. Evidence for magma mixing in the surge deposits of the Monte Guardia Sequence, Lipari. In: M.F. Sheridan and F. Barberi (Editors), Explosive Volcanism. J. Volcanol. Geotherm. Res., 17: 313-328.

The base-surge products of the Monte Guardia sequence on Lipari contain evidence suggesting that magma mixing accompanied the main eruption. Principal data include at least two essential glass populations with discrete compositions, dominance of crystals that are incompatible with the rhyolitic glass, and contrasting quench and superheating textures in the various types of glass pryoclasts. Four distinct types of non-phenocrystic essential pyroclasts are present: vesicular white glass, non-vesicular white glass, brown devitrified clasts, and black microcrystalline clasts. The first three clast types have a uniform rhyolitic composition that is quite distinct from the alkali trachytic composition of the fourth type. These latter black microcrystaline grains have a wide range of compositions that fall in the trachyte field. Discrete crystal grains also occur as essential pyroclasts. The dominant crystal phases are augite and bytownite, with minor amounts of orthopyroxene, olivine pseudomorphed by iddingsite, magnetite, ilmenite, and sanidine. Biotite and hornblende are trace constituents. The diverse mineral assemblage and composition of the most abundant crystals suggest that their origin is different from that of the rhyolitic clasts. Pyroclast textures indicate that these magmas of different composition have been intimately mixed during eruption. The cryptocrystalline textures of the trachyte pyroclasts is due to their chilling within the rhyolitic magma. Strongly vesiculated rhyolite and glass in contact with augite and bytownite crystals suggests that these crystals were able to locally cause exsolution of volatiles. These relationships suggest a separate source for the mafic crystals, and discrete chambers for the trachytic and rhyolitic glass. The eruption model thus involves the mixing of two magmas during the interaction of the erupting materials with near-surface (sea) water.

INTRODUCTION

The Aeolian Islands lie at the southern margin of the spreading Tyrrhenian basin near the suture of the African and European plates (Fig. 1). This archipelago is generally considered to be a volcanic island arc located above a northwestern dipping Benioff zone (Barberi et al., 1974; Keller, 1980). Lipari, the largest of the Aeolian Islands, has a surface area of approximately 38 km^2. This island, composed entirely of volcanic materials, rises approximately 1100 m above the Tyrrhenian Sea floor. Although eruptions on Lipari during the last millenia have been less frequent than those on the neighboring islands of Vulcano and Stromboli, the potential for renewed activity nevertheless exists. The most recent eruption on Lipari occurred in the sixth century A.D. (Keller, 1970).

The rocks of Lipari display a calc-alkaline trend (Pichler, 1980) that is distinct from the shoshonitic series comprising the latest products of Vulcano (Keller, 1980) and Stromboli (Rosi, 1980). Because no significant increase in the $^{87}Sr/^{86}Sr$ ratio is noted for the shoshonites, Barberi et al. (1974) consider both series to have a common origin. The abrupt change from calc-alkaline to high-K products in the Aeolian Islands during the last million years is interpreted by Barberi et al. (1974) to be due to the rapid sinking of the subducted slab beneath the islands.

Pichler (1980) has identified four periods of volcanic activity on Lipari.

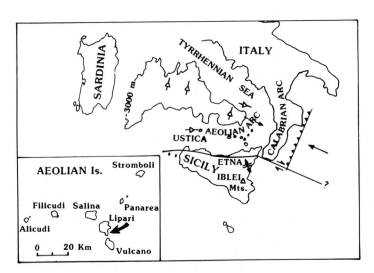

Fig. 1 Location map of Lipari showing its general tectonic setting relative to the Tyrrhenian Sea. Lines with open triangles show directions of seamount chains. The crosshatched line represents assumed contact of the sinking Ionian slab and the continental plate (Modified after Barberi et al., 1974).

The first and second periods are dominated by the construction of andesitic composite volcanoes. The third and fourth periods are characterized by rhyolitic domes and pyroclastic deposits which comprise the type "Liparites". No mafic to intermediate lavas have been previously reported to be associated with these silicic volcanic products. However, Crisci et al. (1981a) note black-and-white banded pumice from the basal explosion breccia of the Monte Guardia sequence which they speculated could represent mixed acid and basic magma.

The samples studied for this paper come from the Monte Guardia sequence (Crisci et al., 1981a), which belong in Pichler's third period. An eruption of short duration produced these deposits that lie between brown ash-flow tuffs with soil horizons dated at 22,600 ± 300 and 16,800 ± 200 years ago by ^{14}C (Crisci et al., 1981b). The initial deposit of the Monte Guardia sequence is a lithic-rich explosion breccia which is overlain by a series of sub-Plinian, lapilli-fall deposits. Intermittent surge beds increase in frequency until the end of the cycle, when a viscous lava dome plugged the vent. Crisci et al. (1981a) interpret the record of this pyroclastic sequence to indicate an increasing level of explosive water/melt interaction with time.

Our current examination of surge pyroclasts from the Monte Guardia sequence reveals essential clasts with a wide range in composition. The purpose of this paper is to present some new data on the composition of associated glass and crystal pyroclasts that have a strong bearing on the interpretation of the petrologic evolution and eruption of this volcano. These data support a new model for late-stage bimodal (rhyolite and trachyte) volcanism at Lipari.

VITRIC AND MICROCRYSTALLINE PYROCLASTS

Surge deposits contain various types of grains, including vitric, lithic, and crystal pyroclasts. The grain size of surge clasts typically fall in the sand (2.0 to 0.063 mm) range (Wohletz and Sheridan, 1979), although pumice lapilli greater than 2.0 mm are also common in the Monte Guardia beds. In addition to white pumice of rhyolitic composition, the Monte Guardia beds contain a variety of non-vesiculated vitric to aphanitic particles including white glass, brown, partly crystalline pyroclasts, and black microcrystalline clasts.

The white glass occurs either as pumice or as less common non-vesicular clasts. The pumice usually have long tubular vesicles (Figs. 2A and 2C), but some pumiceous clasts contain sub-spherical bubbles (Fig. 2B). In addition, there are blocky white glass particles without vesicles. Neither the pumice nor the blocky grains contain appreciable phenocrysts. Both types have a uniform rhyolitic composition.

The brown pyroclasts are generally devitrified (Figs. 3A and 3B). These non-vesiculated grains have large smooth fracture surfaces that result in a blocky morphology typical of hydrovolcanic clasts. Their homogeneous rhyolitic composition (Table 1) is nearly identical to that of the white glass, with the exception of slightly higher sodium and potassium.

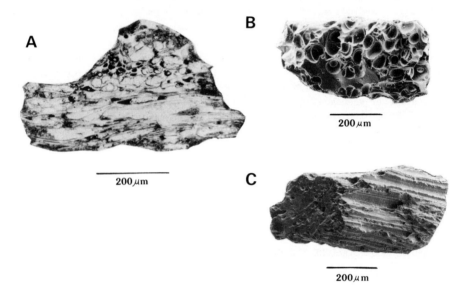

Fig. 2 Types of rhyolitic pumice. A. Pumice with both elongate and sub-spherical vesicles (petrographic micrograph). B. Pumice with multiple sub-spherical vesicles (SEM image). C. Pumice with tubular vesicles.

Black, microcrystalline clasts have a rough appearance with the SEM due to their irregular surface texture. Broken surfaces exposed within these grains may appear smooth and conchoidal. Microphenocrysts within these fragments include augite and plagioclase. This type of fragment lies in the broad compositional range of trachyte (Table 1) although they are also similar to quartz latites or rhyodacites.

One of the main petrologic problems related to this deposit is to explain the co-existence of several types of vitric and microcrystalline particles. Because of their fresh, unaltered surfaces all of these pyroclast types are considered to have come from liquid magma that fractured at the time of the eruption. The rhyolitic pumice resulted from normal magmatic degassing. The non-vesiculated clasts (Fig. 4) resemble typical hydrovolcanic grains that resulted from the

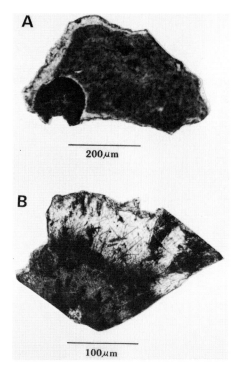

Fig. 3 Brown, partly crystalline rhyolitic pyroclasts. A. Completely devitrified clast in transmitted light (petrographic micrograph). B. Partially devitrified.

explosive mixture of external water and magma (Sheridan and Wohletz, this volume; Wohletz, this volume).

The white vitric pyroclasts have extremely smooth surfaces and delicate edges (Fig. 4A). This smooth surface morphology is a result of spalling after emplacement and does not reflect eruption or transport phenomena. The brown, partly crystalline pyroclasts have a similar blocky grain morphology, but a slightly rougher surface due to their microcrystalline texture (Fig. 4B). Finally, the black microcrystalline pyroclasts have a completely different appearance with a rough surface and strong edge modification (Fig. 4C). Only broken surfaces that expose grain interiors may appear smooth with small-scale fracture details (Fig. 4D). The difference in appearance of SEM images of these grains led to an investigation of their chemical composition.

SOURCE OF CLASTS

The rhyolite clasts (Table 1) are similar in composition to other rhyolites on Lipari (Pichler, 1980), as well as those on Lentia of Vulcano (Keller, 1980). The occurrence of pumice lapilli together with clasts of

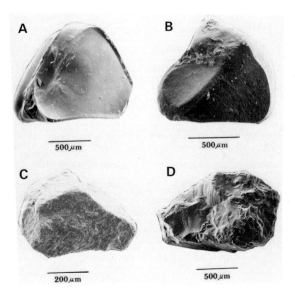

Fig. 4 SEM images of blocky glass fragments. A. White rhyolitic glass. B. Brown rhyolitic pyroclasts. C. Black trachytic pyroclast. D. Black high-K andesitic pyroclast. The surface of D appears smooth due to breakage during transport.

weakly vesiculated glass in the same bed is compatible with a near sea level sub-Plinian eruption during which sea water gained access to the vent (Crisci et al., 1981a). The interfingering of lapilli-fall beds with surge horizons supports this hypothesis.

The black, microcrystalline clasts (Table 1), on the other hand, are more difficult to explain. Their alkali trachyte composition does not correspond to any analysed lava from Lipari. They are, in fact, similar to shoshonitic series lavas from Vulcano (see Keller, 1980). Their unaltered appearance and fresh chemical composition are compatible with a juvenile rather than accidental origin. The lack of shoshonitic lavas on Lipari, the primary hydroclastic morphology of the microcrystalline grains, and the rare black-and-white banded pumice in the basal breccia of the Monte Guardia series are compatible with the hypothesis that these fragments are products of a juvenile magma alien to Lipari. This magma, together with the rhyolite, erupted under variable conditions of water/melt interaction, as reflected by the range in vesiculation and quenching of the glass and the primary depositional structures.

CRYSTAL PYROCLASTS

Large crystals comprise another important pyroclast type. Black clinopyroxene is the dominant crystal phase, although plagioclase is also

TABLE 1
Composition (water-free) of analysed clasts

	1.	2.	3.	4.	5.	6.	7.
SiO_2	77.30 ± .71	77.91 ± .94	76.65 ± .60	61.12 ± 2.17	59.00	59.30	58.4
TiO_2	.15 ± .05	.14 ± .04	.15 ± .04	.97 ± 0.25	.70	.60	.6
Al_2O_3	12.89 ± .29	12.86 ± .25	13.01 ± .25	16.32 ± 2.37	17.80	16.50	15.9
FeO*	1.33 ± .10	1.45 ± .10	1.39 ± .19	6.98 ± 1.80	5.14	6.37	6.1
MnO	.14 ± .06	.17 ± .04	.17 ± .03	.18 ± 0.05	0.10	0.10	0.1
MgO	.21 ± .10	.17 ± .04	.21 ± .13	1.82 ± 0.98	2.30	2.20	3.2
CaO	.66 ± .11	.67 ± .06	.66 ± .09	5.49 ± 1.42	6.40	5.30	5.2
Na_2O	2.79 ± .28	2.07 ± .43	2.84 ± .49	2.91 ± 0.59	2.90	2.00	3.8
K_2O	4.64 ± .31	4.89 ± .31	5.26 ± .37	3.68 ± 0.85	3.70	3.50	5.4

1. non-vesicular white glass (average of 8 grains).
2. white pumice (average of 7 grains).
3. brown, devitrified pyroclasts (average of 12 grains).
4. black, microcrystalline pyroclasts (average of 18 grains).
5. quartz latites on Lipari (average of 6 analyses from Pichler, 1980).
6. rhyodacites on Lipari (average of 8 analyses from Pichler, 1980).
7. trachytes on Vulcano (average of 2 analyses from Keller, 1980).

*Total iron calculated as FeO.

abundant. Minor amounts of orthopyroxene, magnetite, ilmenite, sanidine, and iddingsite pseudomorphs after olivine are also present. Only rare grains of hornblende and biotite were noted.

Some perfect crystals occur free of a glass coating (Fig. 5A). More common are glass-coated, broken crystals. Typically, the glass that is in contact with crystals is vesiculated (Figs. 5B, 6A, and 6B). Some pyroxene crystals appear to have been corroded along their contact with the white rhyolitic glass (Fig. 6C). Other pyroxene crystals are surrounded by an inner layer of brown, devitrified glass with an outer coating of white vesicular glass (Fig. 6D).

Chemical analyses were made of individual crystals using two different techniques. Preliminary data were obtained using a KEVEX microanalyser during SEM textural examination of grain surfaces. Although this method emphasizes chemical variations on grain surfaces due to hydration, recrystallization, or secondary mineral growth, the main compositional types are easily identified. Subsequent quantitative analyses were made on individual grains preselected using a stereo microscope. These single grains were mounted in epoxy and polished so that grain interiors could be analysed using a Cameca microprobe.

Chemical analyses made on single crystal grains show that clinopyroxene with a uniform diopsidic augite composition (Table 2) dominates. Less abundant

Fig. 5 SEM images of crystals. A. Clinopyroxene cluster free of adhering glass. B. Orthopyroxene crystal coated with vesiculated glass.

orthopyroxene (Fig. 7) of bronzite composition (Table 2) and plagioclase (An_{62-77}; Table 3) are also common. Analyses were also made of sanidine (Table 3), ilmenite (Table 4) and titanomagnetite (Table 4).

INTERPRETATION OF CHEMICAL AND TEXTURAL DATA

Two magma types appear to be represented in the Monte Guardia surge products: (1) aphyric rhyolitic magma (SiO_2 = 73 to 76%) and (2) trachyte magma (SiO_2 = 56 to 63%) with augite, hypersthene, and Ca-plagioclase phenocrysts. Glassy to aphanitic pyroclasts representing the liquid part of the two magma types analysed with the microprobe are given in Table 1.

Rhyolitic glass of homogeneous composition forms pyroclasts of the two textural types: white vitric glass and brown divitrified glass. Neither of

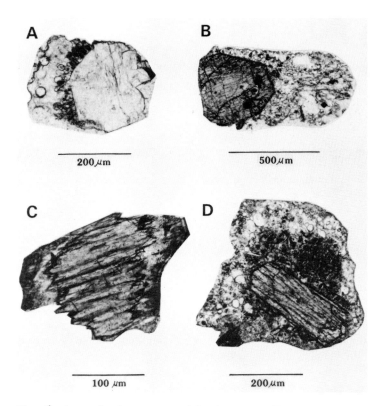

Fig. 6 Crystals in contact with glass. A. Plagioclase crystal surrounded by white transparent glass. Bubbles developed at the contact. B. Clinopyroxene crystal surrounded by white vesicular glass. C. Clinopyroxene in disequilibrium with white transparent glass. D. Clinopyroxene surrounded by black devitrified glass in turn surrounded by white vesiculated glass.

these pyroclast types commonly contains phenocrysts. The composition of individual rhyolitic pyroclasts (Fig. 8) is similar to that of rhyolitic lavas from the other Aeolian Islands as reported by Pichler (1980, Fig. 8) and Keller (1980, Fig. 16).

Microcrystalline trachytic products have a rough surface appearance on SEM images. These fragments generally contain microphenocrysts of augite and plagioclase. Analyses of these opaque black pyroclasts show a linear trend in composition ranging from 56 to 63 percent SiO_2 (Fig. 8). The variation of these clasts is similar to that of trachytes from Vulcano analysed by Keller (1980). The composition of these fragments is also similar to some of the glass on the 1888-90 breadcrust bombs of Vulcano (see Keller, 1980), to some high-K series trachytes from Stromboli (Rosi, 1980), and to the second period products of Lipari described by Pichler (1980).

TABLE 2
Analyses of individual pyroxene crystals

	CLINOPYROXENE								
	1	2	3	4	5	6	7	8	9
SiO_2	52.29	51.97	51.96	51.31	51.24	51.01	50.88	50.37	50.15
TiO_2	0.41	0.66	0.33	0.63	0.59	0.52	0.80	0.55	0.57
Al_2O_3	2.64	1.97	1.96	3.56	2.65	3.42	4.24	3.32	2.98
FeO	9.24	12.44	9.46	9.52	12.00	5.99	9.64	9.28	9.49
MnO	0.33	0.39	0.30	0.27	0.30	0.15	0.36	0.09	0.36
MgO	15.53	14.46	16.18	13.43	15.18	14.55	12.99	13.24	12.99
CaO	21.76	19.06	18.90	22.72	18.75	23.12	22.74	22.07	22.27
Na_2O	–	0.23	0.28	0.27	0.48	–	0.39	0.28	0.36
Total	102.25	101.18	99.37	101.71	101.19	98.76	102.04	99.20	99.17

Cations

O-basis	6	6	6	6	6	6	6	6	6
Si	1.91	1.93	1.94	1.89	1.90	1.91	1.87	1.90	1.90
Ti	.01	.02	.01	.02	.02	.02	.02	.02	.02
Al	.11	.09	.09	.15	.12	.15	.18	.15	.13
Fe	.28	.39	.30	.29	.37	.19	.30	.29	.30
Mn	.01	.01	.01	.01	.01	.04	.01	.00	.01
Mg	.85	.80	.90	.74	.84	.81	.71	.75	.73
Ca	.85	.76	.76	.90	.75	.93	.90	.89	.90
Na	–	.02	.02	.02	.03	–	.03	.02	.03

	CLINOPYROXENE						ORTHOPYROXENE		
	10	11	12	13	14	15	1	2	3
SiO_2	50.11	49.71	49.55	49.36	49.03	48.42	53.47	53.00	51.85
TiO_2	0.69	0.50	0.62	0.57	0.70	0.67	0.28	0.19	0.20
Al_2O_3	4.41	1.86	4.23	1.83	2.14	1.86	1.46	1.66	0.97
FeO	10.26	12.05	10.20	11.67	10.43	12.12	19.66	17.69	19.19
MnO	0.24	0.27	0.24	0.32	0.23	0.45	0.50	0.58	0.53
MgO	12.49	15.11	12.73	14.71	14.31	13.45	23.41	24.46	22.74
CaO	22.10	17.50	22.05	18.14	19.83	18.96	2.30	1.97	1.97
Na_2O	0.38	0.22	0.33	–	0.31	0.29	–	–	0.31
Total	100.68	97.22	99.95	96.60	96.98	96.22	101.08	99.55	97.76

Cations

O-basis	6	6	6	6	6	6	6	6	6
Si	1.87	1.92	1.87	1.92	1.90	1.91	1.95	1.95	1.96
Ti	.02	.01	.01	.02	.02	.02	.01	.01	.01
Al	.19	.08	.19	.08	.10	.09	.06	.07	.04
Fe	.32	.39	.32	.38	.34	.40	.60	.54	.61
Mn	.01	.01	.01	.01	.01	.01	.01	.02	.02
Mg	.70	.87	.72	.85	.83	.79	1.28	1.34	1.28
Ca	.88	.72	.89	.76	.82	.80	.09	.08	.08
Na	.03	.02	.02	–	.02	.02	–	–	.02

TABLE 3
Analyses of individual feldspar crystals

	PLAGIOCLASE					SANIDINE	
	1	2	3	4	5	1	2
SiO_2	53.79	52.47	52.09	50.26	49.26	65.87	64.94
TiO_2	0.17	0.26	0.20	-	0.18	-	0.09
Al_2O_3	26.74	24.44	28.44	29.70	30.22	18.65	18.68
FeO	1.97	2.55	0.79	0.46	0.98	-	-
MgO	0.53	1.00	-	0.11	0.14	-	0.12
CaO	10.89	11.29	12.96	14.18	14.86	-	0.13
Na_2O	3.96	3.00	3.50	3.26	2.44	3.33	3.41
K_2O	1.04	0.78	0.69	0.22	0.18	11.38	10.84
Total	99.09	95.79	98.67	98.19	98.26	99.23	98.21
Cations							
O-basis	8	8	8	8	8	8	8
Si	2.48	2.50	2.41	2.34	2.30	3.01	2.99
Ti	.01	.01	.01	-	.01	-	.00
Al	1.45	1.37	1.55	1.63	1.66	1.00	1.02
Fe	.07	.10	.03	.02	.04	-	-
Mg	.04	.07	-	.01	.01	-	.01
Ca	.54	.58	.64	.71	.74	-	.01
Na	.35	.28	.31	.29	.22	.29	.30
K	.06	.05	.04	.01	.01	.66	.64

TABLE 4
Analyses of individual iron-titanium oxide crystals

	TITANOMAGNETITE					ILMENITE	
	1	2	3	4	5	1	2
SiO_2	0.40	0.31	0.40	0.46	0.46	0.20	0.55
TiO_2	14.13	11.78	11.33	9.33	9.03	49.50	48.28
Al_2O_3	5.01	3.60	3.24	5.35	5.27	0.48	-
FeO	78.23	82.36	79.60	77.31	76.30	48.18	48.40
MnO	0.71	0.39	0.58	0.53	0.28	1.34	1.42
MgO	1.37	2.53	2.73	4.12	4.00	0.32	0.38
Total	99.85	100.97	97.88	97.10	95.34	100.02	99.03
Cations							
O-basis	4	4	4	4	4	3	3
Si	.01	.01	.01	.02	.02	.00	.01
Ti	.43	.36	.36	.29	.29	.95	.94
Al	.24	.17	.16	.26	.26	.01	-
Fe	2.64	2.82	2.81	2.70	2.72	1.03	1.05
Mn	.02	.01	.02	.02	.01	.03	.03
Mg	.08	.15	.17	.26	.25	.01	.01

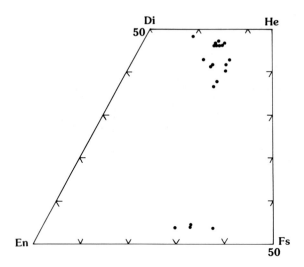

Fig. 7 Pyroxene composition plot.

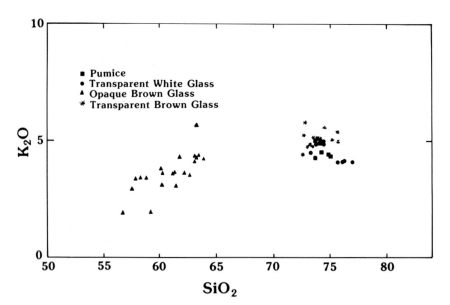

Fig. 8 Plot of alkalies vs. silica for essential glass grains in the Monte Guardia surge beds. All analyses are from a single sample.

Fig. 9 Plagioclase crystal with an inclusion of basaltic glass. Compositions are given in Table 5.

The main crystal phases occurring as separate grains in the Monte Guardia surge deposits (augite, bronzite, and Ca-plagioclase) are not compatible with the rhyolitic magma. KEVEX analysis of a glass inclusion observed in a SEM image of a plagioclase pyroclast (Fig. 9) indicates a mafic but high-K glass (Table 5) that is different from all other analysed clasts. Several grains analysed by the KEVEX on the SEM showed a similar mafic glass, but no grains of this composition were found during the microprobe analyses. The main crystals indicated above are generally surrounded by dark vesiculated glass which microprobe analyses show to be trachytic to high-K andesitic in composition (Table 5).

Most investigators of pyroclastic deposits choose large scoria or pumice for petrologic analyses because such fragments are considered to best represent the magma composition. Although present, such black-and-white banded pumice are rare in the Monte Guardia deposit. It was through the careful examination of the sand-sized (2.0 to 0.063 mm) clasts in the Monte Guardia surge beds that the diverse nature of the mixed magmas has been documented. It is possible that particles in this size range in other volcanic deposits may contain important petrologic data, especially where water mixing or magma mixing are significant factors.

THE MODEL

Textural evidence suggests that liquid magmas with different temperatures

TABLE 5
Analyses of glass associated with crystals

	Inclusion in plagioclase		Glass surrounding phenocrysts					
	Plag.	Glass	Plag.	Glass	Opx	Glass	Cpx	Glass
SiO_2	50.79	48.18	50.86	60.24	52.23	57.48	50.43	61.74
TiO_2	1.08	1.35	-	0.57	0.18	0.61	0.65	1.07
Al_2O_3	30.97	18.65	31.89	21.65	1.50	21.68	2.75	18.38
FeO	1.03	3.89	1.12	3.42	23.21	5.0	10.60	6.21
MnO	-	-	-	0.14	0.50	-	0.41	-
MgO	2.01	11.87	0.43	0.38	22.19	1.01	14.89	0.69
CaO	9.88	12.66	15.13	7.26	2.01	8.68	20.17	5.15
Na_2O	1.46	1.48	3.05	3.38	-	2.98	-	3.39
K_2O	2.06	1.23	0.35	3.16	0.04	2.93	0.06	4.29
Total	99.46	99.31	102.84	100.20	102.86	100.37	99.05	100.93

came into contact during the eruption of the Monte Guardia products. The heating effect is illustrated by vesiculation of the cooler magma in contact with hotter material. For example, rhyolitic glass surrounding trachytic glass is vesiculated whereas isolated rhyolitic fragments are not. Likewise, trachytic glass surrounding crystals is vesiculated, whereas isolated trachytic glass is without bubbles.

The effect of cooling was also observed. Hotter magmas in contact with cooler material is quenched forming brown or black microcrystalline grains with a blocky morphology. Such a microcrystalline texture produces the rough surface observed on SEM images. Similar textures due to supercooling are caused by contact of magma with water to produce the typical blocky morphology observed in surge pyroclasts (see Wohletz, this volume).

The rhyolitic grains do not show a range in composition. There are no grains with compositions between those of the rhyolites and trachytes. In addition, they lie off the evolutionary trend defined by the trachytes. This suggests that the rhyolite magma could have come from an isolated homogeneous magma chamber. The composition of the iron-titanium oxides from the same sample suggest a temperature between 660 and 710°C and the log fO_2 between -16.5 and -18.2 for this chamber. The sanidine plus the scarce hornblende and biotite also could have been stable phases in this magma.

Below the rhyolitic chamber we assume another resevoir that is compositonally zoned from trachyte to high-K andesite. This chamber, represented by the black microcrystalline clasts, could have been periodicaly fed by a more mafic magma. Some resorbed pyroxene and bytownite crystals as well as the altered olivine grains could have come from this body. The phenocrysts are homogeneous and similar in compositon to those of leucite tephrites of Vulcanello (Keller, 1980).

DISCUSSION

Most authors (Barberi et al., 1974; Rosi, 1980; Keller, 1980) refer to a regular temporal change in composition of volcanic rocks in the Aeolian Islands from early calc-alkaline basalts and andesites to present-day shoshonitic magmas. Evidence from the Monte Guardia beds indicates that magmas from two series (rhyolites and trachytes) were contemporaneous and intimately mixed during the eruption. This suggests that the volcanism here is complex. The composition of augite, bronzite, and bytownite phenocrysts is similar to those in the lavas erupted at Vulcanello during the last two millenia. The trachytic magma (microcrystalline clasts) is similar in composition to the trachytes on Vulcano. The magma chamber for this material, represented by clasts from the Monte Guardia beds, appears to have been compositionally zoned. It could possibly have been periodically fed by more mafic magma.

The rhyolitic magma is similar in composition to lavas on Lentia and Fossa of Vulcano (Pichler, 1980) and the Monte Pilato complex on Lipari (Keller, 1980). The compositional gap between these rhyolites and the trachytes suggests that the rhyolitic chambers were homogeneous but isolated from the trachytic chambers.

There is a strong similarity in composition and time of eruption (about 20,000 years ago) of rhyolites on Lipari and those of Lentia on Vulcano. In addition, Monte Guardia, Vulcanello, and Fossa of Vulcano all lie along a 6 km long segment of a prominent north-trending fracture system. Therefore, all of the rhyolitic vents probably belong to the same volcanic feeder system.

The simple subduction model with evolution from calk-alkaline magmas to shoshonites should be re-examined in light of complex volcanological relationships. It appears that a bimodal (rhyolite and trachyte) volcanic system existed on Lipari and Vulcano for the last 22,000 years. Frazzetta et al. (this volume) report magma mixing for the rhyolites of the Fossa of Vulcano. Tephrites with minor trachytes have erupted on Vulcanello since 183 A.D.. Rhyolite and trachyte lavas, similar in composition to those on Fossa but approximately contemporaneous to the Monte Guardia complex, are present on Lentia of Vulcano.

We can thus reconstruct the Lipari-Vulcano volcanic system as follows. If the mafic end-member rises without contacting a shallow magma chamber, leucite tephrites and trachyte result as at Vulcanello. If the rising mafic magma intersects a pocket of rhyolitic magma, it produces a mixed eruption such as the Monte Guardia products or the Pietre Cotte and Commenda lavas of Vulcano (Frazzetta et al., this volume). Because the rhyolitic magmas are generally contaminated by more mafic xenoliths, we can assume that the rhyolitic eruptions are triggered by magma mixing. The physical character of the eruption (domes, lavas, pumice falls, wet surges, or dry surges) depends on the level of water/melt interaction in the fragmenting magma interface (Crisci et al., 1981a).

ACKNOWLEDGEMENTS

Microprobe analyses were done at Arizona State University on an instrument funded by National Science Foundation. We would like to thank Sonia Esperanca for help in preparing samples and reducing the microprobe data. Kenneth Wohletz helped with the initial SEM images and the EDAX analyses at Los Alamos National Laboratory. The complete set of SEM images of Monte Guardia pyroclasts (900 images) were made in the laboratory of David Krinsley at Arizona State University. This work was partially supported by NSF Grant INT-7823984 and NASA Grant NAGW-245.

REFERENCES CITED

Barberi, F., Gasparini, P., Innocenti, F., and Villari, L., 1973. Volcanism of the southern Tyrrhenian Sea and its geodynamic implications. J. Geophys. Res., 78: 5221-5232.

Barberi, F., Innocenti, F., Ferrara, G., Keller, J., and Villari, L., 1974. Evolution of Eolian Arc volcanism (southern Tyrrhenian Sea). Earth Planet. Sci. Letters, 21: 269-276.

Crisci, G.M., De Rosa, R., Lanzafame, G., Mazzuoli, R., Sheridan, M.F., and Zuffa, G.G., 1981a. Monte Guardia deposits: a Late-Pleistocene eruptive cycle on Lipari (Italy). Bull. Volcanol., 44(3): 241-255.

Crisci, G.M., Delibrias, G., De Rosa, R., Lanzafame, G., Mazzuoli, R., Sheridan, M.F., and Zuffa, G.G., 1981b. Pyroclastic deposits of the Monte Guardia eruption cycle on Lipari (Aeolian Arc, Italy). Int. Assoc. Sediment., Bologna.

Keller, J., 1970. Die historischen Eruptionen von Vulcano und Liapri. Deutsch. Zeitsch. Geol. Geoph., 121: 179-185.

Keller, J., 1980. The island of Vulcano. In: L. Villari (Editor), The Aeolian Islands. Ist. Inter. Vulcanol., Catania, pp. 29-75.

Pichler, H., 1980. The island of Lipari. In: L. Villari (Editor), The Aeolian Islands. Ist. Inter. Vulcanol., Catania, pp. 75-101.

Rosi, M., 1980. The island of Stromboli. In: L. Villari (Editor), The Aeolian Islands. Ist. Inter. Vulcanol., Catania, pp. 5-29.

Wohletz, K.H., and Sheridan, M.F., 1979. A model for pyroclastic surge. In: C.E. Chapin and W.E. Elston (Editors), Ash-Flow Tuffs. Geol. Soc. Am., Spec. Pap., 180: 177-194.

Wohletz, K.H., 1983. Mechanisms of hydrovolcanic pyroclast formation: size, scanning electron microscopy and experimental studies. In: M.F. Sheridan and F. Barberi (Editors), Explosive Volcanism. J. Volcanol. Geotherm. Res., 17: (this volume).

EVOLUTION OF THE FOSSA CONE, VULCANO

GIOVANNI FRAZZETTA[1], LUIGI LA VOLPE[2] and MICHAEL F. SHERIDAN[3]

[1] Istituto Internationale di Vulcanologia, v.le Regina Margherita 6, Catania, (Italy)
[2] Istituto di Mineralogia e Petrografia, Universita, Piazza Umberto I°, No. 1, Bari, (Italy)
[3] Department of Geology, Arizona State University, Tempe, AZ, 85287 (U.S.A.)

(Received July 8, 1982; revised and accepted October 28, 1982)

ABSTRACT

Frazzetta, G., La Volpe, L., and Sheridan, M.F., 1983. Evolution of the Fossa cone, Vulcano. In: M.F. Sheridan and F. Barberi (Editors), Explosive Volcanism. J. Volcanol. Geotherm. Res., 17: 329-360.

In decreasing order of abundance, the principal emplacement mechanisms for deposits of the Fossa edifice of Vulcano include: dry-surge, wet-surge, pyroclastic-fall, lahar, lava-flow, pyroclastic-flow and epiclastic processes. Most samples from pyroclastic beds on Fossa contain two grain-size populations. The coarser mode at $0.5 - 1.5$ ϕ resulted from modified ballistic transport whereas the finer mode at $2.5 - 3.5$ ϕ is related to surge transport. Reconstruction of the entire stratigraphy of the cone permits the beds to be grouped into five principal eruptive cycles, all of which postdate the lower Pilato ash of Lipari (dated at 11,000 to 8,500 y.b.p): Pte. Nere cycle, Palizzi cycle, Commenda cycle, Pietre Cotte cycle, and the modern cycle. Additional cycles may be present, but exposures on the cone are insufficient for their adequate definition. All cycles follow a similar stochastic pattern starting with surge eruptions and ending with effusion of lava from the crater rim. The frequency of pyroclastic-fall events increases with time throughout each cycle. This progressive decrease in the efficiency of water/melt interaction through the duration of each cycle may be due to a gradual rise of each new magma column coupled with an increase in magmatic volatiles.

Tectonism and magma mixing both play a role in triggering eruptions on Fossa. The present crater lies at the intersection of two prominent tectonic trends. A NNW fracture system parallels the main eruptive centers and an ENE set aligns with the migration of minor vents. The silicic lavas of Fossa contain a mixture of two components. Xenocrysts of plagioclase, augite, and olivine, surrounded by a dark glass are intimately dispersed into an evolved rhyolitic matrix.

0377-0273/83/$03.00 © 1983 Elsevier Science Publishers B.V.

INTRODUCTION

In order of their decreasing length of repose periods, the active volcanic islands of the Aeolian archipelago include Lipari, Vulcano and Stromboli (Fig. 1). Numerous explosive eruptions have ocurred on Vulcano during historic times. The last eruptive episode on Vulcano (1888-1890) produced a 5 m thick pyroclastic deposit on the Fossa cone and large (~1 m) breadcrust bombs were ejected to distances of more than 1 km from the vent (Mercalli and Silvestri, 1891). The present activity is restricted to fumarolic emissions (T = 280°C) along the northern rim of the Gran Cratere of the Fossa edifice.

Because of the potential volcanic risk at Vulcano, it is important to evaluate the consequences of renewed explosive activity using historic observations of eruptive phenomena and the geologic record of pyroclastic deposits. To this end we have carefully examined 84 sections exposed on the Fossa edifice in order to reconstruct the eruptive phenomenology throughout its history. The volcanic record, constrained by stratigraphy and models for pyroclastic dispersal, forms the basis for volcanic hazard evaluation at this vent.

The tectonic map of Frazzetta et al. (1977) shows a strong interconnection between structural patterns on Vulcano and Lipari with an emphasis on two principal directions of lineaments (NNW and ENE). The main volcanotectonic trend (NNW) controls the location of the Fossa and old Vulcano structures. The other tectonic trend (ENE) controls the minor migration of vents on the Fossa and Vulcanello tuff cones. Recent ground deformation data (Villari, per. comm.) supports the above structural model. Other authors (Keller, 1970, 1980; Lo Giudice et al., 1975) emphasize a NS trending structural element by connecting the rhyolitic vents on Lipari and Vulcano.

The geology of Vulcano has been studied in greatest detail by Keller (1980) who shows the island to be composed of four eruptive centers. Old Vulcano is mainly composed of trachybasalt and trachyandesite lava flows and pyroclastic deposits. Its activity ended with a collapse caldera that is filled with pyroclastic beds and lava flows. The second center is the Lentia lava dome complex to the north composed of rhyolites, trachytes, and latites. The domal activity ended with a collapse caldera that is the site of the third center, the presently active Fossa vent. Silicic ash with minor lava of trachytic to rhyolitic composition are the dominant products of the Fossa tuff cone. The fourth volcanic center is Vulcanello at the northern tip of the island. This cone, dating from 183 B.C., is mainly composed of tephritic lavas and tuffs.

Our study concentrates on the evolution of the Fossa edifice, which is the most active center on the island of Vulcano. On the Fossa cone Keller (1980) noted three main pyroclastic units separated by lavas. His Fossa I pyroclastic deposit consist of dark gray tuffs with a basal date between 11,000 and 8,500

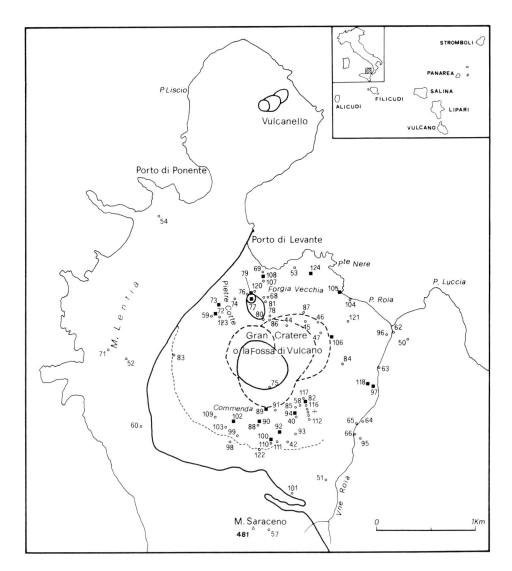

Fig. 1. Location map showing major vent complexes of Vulcano and geographic locations mentioned in the text. Measured stratigraphic sections are indicated by Arabic numerals. Squares represent selected sections of Fig. 2.

B.C. His middle tuff unit (Tufi Rossi) consists of fine-grained, lithified ash deposits of reddish to varicolored appearance. His uppermost unit is composed of dark gray tuffs that are very fresh appearing. Our detailed stratigraphic study has discovered several discrepancies in Keller's simple pattern. For example, there are actually red tuff layers which correlate to each of the several vent complexes on the Fossa edifice (with the exception of the Pte. Nere cycle), rather than a single stratigraphic level of such deposits. This new stratigraphic interpretation is presented later in this paper.

The first detailed pyroclastic stratigraphy on the Fossa of Vulcano concentrated on the recent volcanic products above the uppermost 'Tufi Rossi' bed (Sheridan et. al., 1981). The above work introduced the concept of volcanic cycles in order to group the various layers into coherent genetic units. The present paper reinterprets the stratigraphy of Fossa and extends the analysis of cycles to include the total products of this cone.

The methodology of the present paper has deep roots in several newly developed concepts of hydrovolcanism. For example, the distinction of wet-surge and dry-surge deposits (Sheridan and Wohletz, 1981) is useful for interpreting the activity of Vulcano; as is the relationship of water/melt ratio in the vent to energy partition and melt fragmentation (Wohletz, 1979, this volume; Wohletz and Sheridan, 1980, 1981). Finally, the sequential development of various textures in the deposit based on water/melt ratios stems from the discussion of tuff rings and tuff cones by Wohletz and Sheridan (in press).

HISTORICAL RECORD

Three types of information must be used to reconstruct the activity pattern of Fossa: geological relationships, early historical records, and modern scientific observations. Geologic data must be used for the period prior to the historical accounts of the Roman era. There are only a few events from the early historic period that can be directly correlated with deposits on Fossa. The upper Pilato ash, erupted from a vent on Lipari in the 6th century A.D. (Keller, 1970) forms a prominent marker bed on the Fossa cone. Only the last eruptive event, starting in 1888, provides extremely useful scientific information of high quality.

There is considerable discrepancy in the dates reported for the more recent historic deposits on Fossa. No certain date can be given for the large phreatic crater of Forgia Vecchia I. However, we assign a date of 1727 to the small phreatic crater of Forgia Vecchia II on the northwestern slope because deposits produced by this vent underlie the Pietre Cotte lava.

The age of the Pietre Cotte lava is also in question. Mercalli and Silvestri (1891) suggest a date of 1771 for this lava flow. De Fiore (1922) reports a visit of Le Duc who saw a recent lava flow on the northwestern flank of Fossa in 1757. According to De Fiore, this lava is the Pietre Cotte flow and it was extruded at the end of the 1731 to 1739 active period. We accept this latter suggestion.

Numerous volcanic explosions have been recorded between extrusion of the Pietre Cotte lava of 1739 (1771?) and the 1888-90 eruption. For example, Mercalli (1891) records strong explosive activity that produced juvenile material for periods exceeding one month in 1822-23, 1873, and 1886.

The most recent activity of Fossa (1888-1890) is the only eruption on Vulcano described in detail (Mercalli and Silvestri, 1891). Because this eruption is the type example of Vulcanian activity, it merits a brief review here. The activity started on August 2, 1888 after a relatively quiet period lasting about one century. The initial explosion emitted only clasts of old vent materials but later eruptions ejected juvenile blocks, bombs and ashes. The distinctive breadcrust bombs of the rim were produced during the middle part of the active period. The main phenomena consisted of intermittent explosions separated by quiet periods lasting from a few minutes to a few days. Explosive intensity covered a wide range but only the strongest explosions ejected blocks and bombs. Such strong eruptions were separated by long periods of quiescence or vapor emission. There were no effusive lavas or domes, although the large breadcrust bombs are similar in composition and texture to some of the earlier lavas. The activity stopped on 22 March, 1890 after 20 months of persistent explosions.

The only activity on Vulcano after 1890 has been fumarolic emission of gas. The fumarolic intensity reached two maxima in this period: 615°C in 1924 and 317° in 1979. The minimum fumarolic temperature between peaks was 110°C in 1962. The 1981 temperatures were about 280°C.

METHODOLOGY

Our approach to understanding the activity of Vulcano is basically stratigraphic. More than 80 sections were selected on the Fossa cone to include the majority of the best exposed outcrops (Fig. 1). Sections were measured on a bed-by-bed basis with data recorded to centimeter detail where warranted. For representation in this report the beds were grouped into several genetic types listed below in their approximate order of increasing water-melt interaction: lava flow, pyroclastic-fall, pyroclastic-flow, dry-surge, wet-surge, and lahar deposits. An additional distinctive type, which has alternating coarse-and-fine beds at the centimeter level, is considered to have an origin transitional between fall and surge.

Samples were taken from many of the sections for characterization of the deposit, correlations between units, and confirmation of origin. The techniques employed include mechanical size determination, petrographic characterization, and SEM analysis of grain texture and morphology.

Unfortunately, soil horizons and datable material are lacking. Although beds are discontinuous and lateral facies changes are common, key marker beds and unconformities are sufficient to allow correlation between sections on different parts of the structure. Most useful for unravelling the history of activity on Vulcano are larger groups of genetic units, here termed cycles (Sheridan et al.,

1981). A cycle represents the total output of materials from a single vent produced by a genetically related progression of activity that has a definite beginning and ending. Each cycle on Fossa marks the opening and development of a new vent and crater system of the main edifice that is related to the rise of a magma body with intermittent eruptions en route to the surface. A similar coherent eruptive pattern typifies every eruption cycle. This pattern can be directly applied to a predictive model for evaluation of probable volcanic risks associated with future activity.

STRATIGRAPHY

Critical to this study is an accurate evaluation of the stratigraphy of the Fossa edifice based on a genetic interpretation of depositional units. Our reconstruction of the cone starts above the tephritic lava flow of Pta. Roia. Although Keller (1970; 1980) included this lava as the first product of the Fossa vent, we can neither confirm nor contradict this interpretation. The deposits above this lava are the first to show the development of a cone, hence we prefer to start the Fossa stratigraphy at this contact.

The stratigraphic units in Figs. 2 and 3 are generalized into a few genetic types based on similar textural and structural characteristics. The principal types in decreasing order of abundance include: dry-surge, wet-surge, pyroclastic-fall, lahar, lava-flow, epiclastic and pyroclastic-flow deposits. Units range in thickness from a few centimeters to several meters, whereas cycles range from 2 to 33 m in maximum measured thickness (Fig. 3). Marker beds recognized in several sections include: trachytic pumice of Palizzi, explosion breccia, upper Pilato ash, and rhyolitic pumice.

The lava flows associated with the Fossa edifice range in composition from trachyte to rhyolite (Keller, 1980; Castellet y Ballara et al., 1981). All lavas emerge from a crater rim, but only the trachytes reach the base of the cone. Trachytic types have an aa surface but the rhyolites form thick stubby flows with an obsidian core surrounded by an autobrecciated carapace and traction carpet.

The rhyolite lavas, (Pietre Cotte and Commenda flows) as well as the 1888-90 breadcrust bombs, exhibit strong textural evidence for the mixing of two melts of different composition. The Pietre Cotte lava contains dark blebs with chill textures surrounding mafic xenocrysts. The aphyric rhyolite glass is cluttered with xenocrysts of plagioclase, clinopyroxene, and olivine. The Commenda lava contains 30% trachytic fragments mixed within a glassy rhyolitic matrix. Both trachyte and rhyolite were fluid as evidenced by vesicles, amoeboid outlines, and chill margins of the trachyte blobs. The bulk composition of this lava is transitional between trachyte and rhyolite because of the abundant xenocrysts (clinopyroxene, plagioclase, rare olivine, and sanidine). The trachytes have less obvious textural evidence for mixing. Perhaps the large xenocrysts of plagioclase and the glomeroporphyritic clots of plagioclase, clinopyroxene and minor olivine are relics of a more primitive magma.

Fig. 2. Selected sections measured on the Fossa cone of Vulcano. Section locations are shown in Fig. 1. (1) Lava flows. (2) Fall: a) non-vescicular angular fragments, b) pumice fragments. (3) Coarse-and-fine. (4) Dry surges. (5) Wet surges. (6) Pyroclastic flows: a) block-and-ash type, b) ash type. (7) Lahars. (8) Sheet wash and alluvium. (9) Accretionary lapilli. (10) Upper Pilato ash. Correlation lines represent marker beds and stratigraphic horizons used for construction of the schematic section of Fig. 3.
Note: pyroclastic-fall beds are shown by the maximum projection from the base line. Lava flows and pyroclastic-flow deposits have an intermediate projection. Surge and coarse-and-fine deposits have the least projection. Lahars have the maximum projection from the back side of the base line; sheet-wash and alluvial deposits have the least projection.

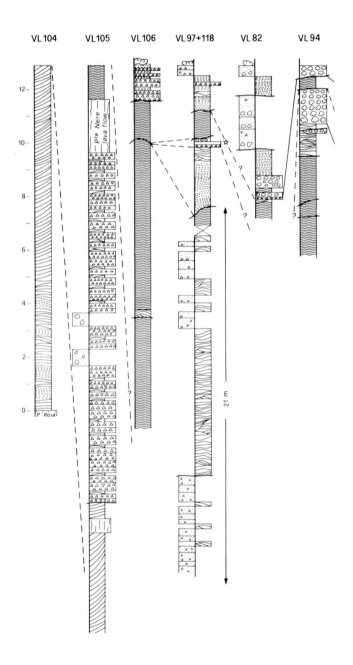

Fig. 2 CONTINUED

337

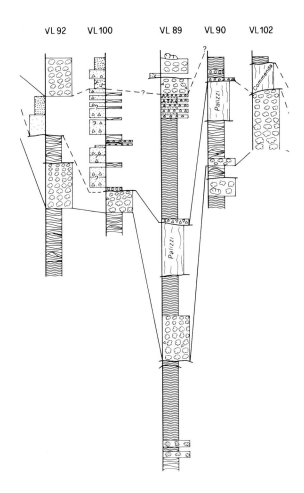

338

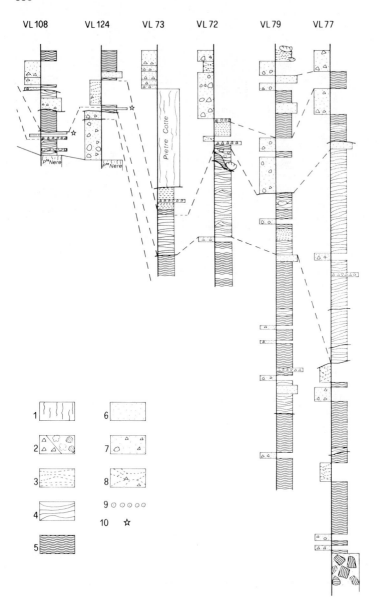

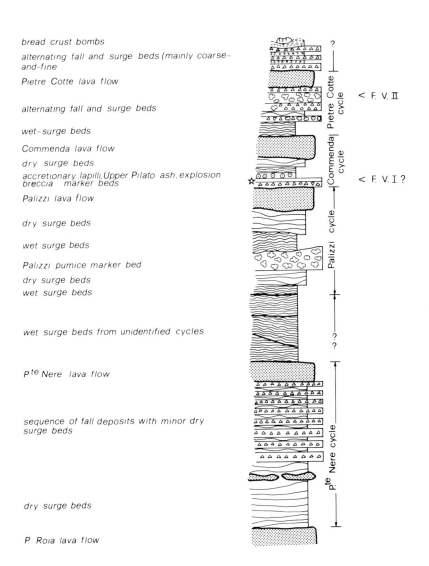

Fig. 3. Composite columnar section of the Fossa edifice.

Several types of pyroclastic-fall deposits are recognized on Fossa. Because pumice-fall deposits are so distinctive, the most widespread of these are included as key marker beds. Fall deposits, most common within the Pte. Nere cycle and the modern cycle, are generally composed of nonvesicular, angular fragments. Stratified beds, 2-3 dm thick generally have normal grading. Thin fall deposits mantle topography but locally a fall-and-roll mechanism has produced thickening and reverse grading. Thick fall deposits increase in dip to a maximum inclination of 30° where they form cones. Minor surge beds locally interfinger with fall deposits.

Coarse-and-fine stratified deposits of pale gray color mainly occur in the modern cycle, but some beds of this type are also in other cycles. The concentration of such beds in depressions near the crater rim and on the outer slopes of the cone indicates a strong control by topography. Individual beds are only a few mm thick (Fig. 4) and lack cross stratification. The clasts are generally sand size but some fragments of lapilli size occur. The grain-size characteristics, discussed in a later section, indicate a bimodal population of clasts, suggesting two separate emplacement mechanisms.

Dry-surge deposits (as defined by Sheridan and Wohletz, 1981) are common on the flanks of the cone and on the plain around the structure but they pinch out up-slope towards the rim. The effect of local topography on their deposition is strong in that units are thicker in gullies but very thin on over-bank surfaces. Bedding is also thicker in the gullies, but individual layers seldom exceed 2 mm. Dry surges commonly erode U-shaped channels through underlying deposits. These dark gray to blue-gray deposits generally contain sand-sized blocky clasts, but locally (e.g., near Pietre Cotte) Pele's hair is present.

Fig. 4. Coarse-and-fine deposits.

Large-scale sandwaves are typical bedforms (Fig. 5), but some massive beds also occur.

Wet-surge deposits (as defined by Sheridan and Wohletz, 1981) are common at the rim of the cone but decrease in abundance outward. They are generally red colored but may also be yellow, green, or gray. Their bed forms are generally planar and massive types and deposits are coherent due to secondary mineral growth as seen in SEM images. Because of grain cohension during emplacement these deposits stick to steep erosional surfaces (Fig. 6). Their common disturbed beds, vesicular textures (see Lorenz, 1974), and accretionary lapilli attest to an initial high content of interstitial water.

Widespread phreatic explosion breccias composed of large angular fragments of yellow fumarolized material are rare on Vulcano. The largest unit of this type is described in the section on marker beds. Another small phreatic breccia occurs at the base of the 1888-90 products.

There are two types of pyroclastic-flow deposits on Fossa. The first type is represented by thin greenish ash-flow tuffs on Forgia Vecchia that have a very fine grain size. Individual beds of these local flows reach a maximum thickness of 2.5 m. The second type is the block-and-ash flow related to the explosion breccia marker bed. Deposits of this unit reach 3 m in thickness in the Palizzi area and locally contain vertical gas pipes (Fig. 7).

Fig. 5. Dry-surge sandwave beds.

Fig. 6. Wet-surge deposits with interstratified pyroclastic flows. The three main units have quaquaversal dips and rest on a strong unconformity at a crater rim. Wet-surge beds adhere to the steep wall of a previous crater of the Pte. Nere cycle.

Water-worked units include lahar, sheetwash and alluvial deposits, that occur on the lower slopes of the cone where individual beds rarely exceed 1 m in thickness. Locally, as in the vicinity of Palizzi, units resembling lahars exceed 3 or 4 m. These deposits, which generally contain a sand-sized matrix with large dispersed clasts, may actually be pyroclastic flows. Some sheetwash lahars that formed during our field study contain vesicles formed by air bubbles even though they were emplaced cold. Thus, bubbles do not necessarily indicate a hot origin. One deposit, formed in a small intermittent body of water to the east of the Palizzi lava, is composed of thin, disc-like beds that have pumice layers at their surfaces.

Fig. 7. Block-and-ash flow deposit related to the phreatic breccia marker bed. Fumarolic pipes are common near the top of this 3-m-thick bed.

MARKER BEDS

Several local marker beds were useful in correlation between sections. The lowest marker horizon in our study (TP) is a trachytic pumice-fall bed which originated from a vent in the Palizzi area. This bed has a maximum thickness of 2.1 m but subsequent erosion has removed it from several sections. The bed consists of normal-graded pumice ranging from a few cm to 15 cm diameter. Large isolated pumice blocks reach 30 cm and scattered angular obsidian clasts up to 2-3 cm are common. The pumice contains phenocrysts of sanidine, plagioclase, clinopyroxene, and rare dark mica. Scarce dark amphibole is also present in some samples. None of the other units on Fossa contain hydrous phases.

Marker bed EB is an explosion breccia that possibly originated from a vent located near the center of the present crater. The breccia contains both fall and flow sub-units. The fall, consisting of nonstratified angular fragments of lava in a sparse matrix, has a maximum thickness of 40 cm. Clasts range from a few centimeters to 10-25 cm maximum size, but in places larger isolated blocks reaching 30 cm have produced impact sags at the base of the unit. The fragments have a yellow surface alteration due to fumarolic action within the vent prior to explosion. Blocks in both the explosion breccia and pyroclastic-flow deposits are similar, ranging in composition from trachyte to andesite. The pyroclastic-flow deposits reach a maximum thickness of 3 m in the Palizzi area. Although the lithic components of the pyroclastic flow are similar to those of the breccia fall, the flow has an abundant fine matrix and numerous fumarolic pipes (Fig. 7). Thin breccia-fall deposits are scattered around the crater, whereas the flow deposits only crop out near the base of the cone in the Palizzi area. Directly above the fall breccia and flow breccia lie the upper Pilato ash and a vesiculated tuff layer composed of very fine ash and accretionary lapilli.

The upper Pilato ash (UP) forms a prominent marker horizon on Fossa. This bed, dated at the 6th century (Keller, 1970; Bigazzi and Bonadonna, 1973), gives the only absolute age for older deposits on the Fossa cone. The source of this ash is Mte. Pilato on Lipari, 12 km to the north. This bed ranges in thickness from 2.5 cm on the NE flank of the cone to 0.5-1.5 cm on the southern side. Erosion has removed the thin upper Pilato marker bed from most areas so that it is identified in only a few sections. This ash is dominantly composed of pyramidal, aphyric glass fragments with minor amounts of vesiculated pumice.

The uppermost marker is a normal-graded to nongraded rhyolitic pumice horizon (RP) that occurs along the present crater rim and on the southern part of the cone. Near the rim it reaches a maximum thickness of 2.5 m with an additional upper 1.5 m of mixed pumice and lithic fragments. Pumice have a median size of 10 cm but their maximum diameter reaches 20 cm. The phenocryst assemblage within pumice include plagioclase with minor clinopyroxene and rare olivine. Near the vent this horizon divides into two pumice layers separated by a thin surge bed.

ERUPTIVE CYCLES

Stratigraphic units are grouped into five cycles that display a coherent pattern of activity. These identified cycles account for at least 75 percent of the Fossa deposits. Because Fossa produced numerous small eruptions of local distribution from several vents, measured sections through deposits of the same cycle differ with position on the cone. Most sections expose deposits of two or more cycles. A composite column for the Fossa cone is given in Fig. 3.

The Pte. Nere (1) cycle on the northern part of the cone between Pte. Nere and Pte. Roia (Fig. 8a) is the lowest cycle. This cycle begins with dry-surge deposits composed of a lower coarse-and-fine unit overlain by sandwave beds with massive beds in the upper part. The middle part of the cycle is composed of a fragmented lava flow of trachytic composition overlain by a thick block-fall (5-20 cm clasts) sequence. The morphology of the cone at this stage was controlled by the angle of repose of the clasts which show inverse size-grading and local cross stratification. The cycle is capped by the trachytic lava flow of Pte. Nere. Clasts in the fall deposit are similar to the capping lava flow, both having abundant sanidine phenocrysts with minor clinopyroxene, plagioclase and rare olivine.

The Palizzi cycle (2) begins with wet-surge deposits that mantle a collapsed or eroded crater rim (Fig. 8b). Conformably above these units are dry-surge

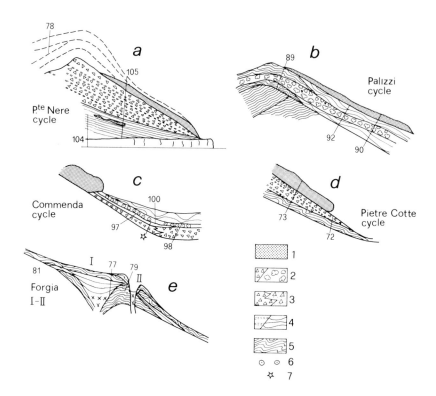

Fig. 8. Generalized sections of the principal cycles on Fossa. The numerals refer to measured sections incorporated into the general diagrams. A. Pte. Nere cycle. B. Palizzi cycle. C. Commenda cycle. D. Pietre Cotte cycle. E. Forgia Vecchia parasitic cone. (1) Lava flows. (2) Fall deposits including lapilli, blocks, and pumice. (3) Explosion breccia and associated block-and-ash pyroclastic flow. (4) Dry surge deposits. (5) Wet surge deposits. (6) Accretionary lapilli. (7) Upper Pilato ash.

beds and a trachytic pumice deposit. The pumice may have been locally retransported because they are rounded, reverse-graded on slopes, and increase in thickness downslope. The pumice beds are essentially of fall origin because they lack a matrix and are normally graded where they lie on level topography. Above the pumice horizon is a surge sequence that contains a proximal wet-surge facies and a more distal dry-surge sequence. The cycle closes with the extrusion of the Palizzi trachytic lava flow.

Above this series is the unique Commenda cycle (3) that begins with the explosion-breccia (EB) marker horizon (Fig. 8c). Locally, the breccia grades into a pyroclastic flow unit which varies in thickness from less than 1 m to more than 3 meters and it contains abundant fumarolic pipes. The pyroclastic-flow unit is locally overlain by the upper Pilato ash (UP) and a thin (8 cm) bed with accretionary lapilli that range from 0.5 to 4 mm diameter. Also above the explosion breccia is a dry surge sequence with a 6 cm-thick interbedded Pele's hair horizon. Although the contact of the lava is not well exposed, the Commenda flow is considered to be the final event of this cycle. Because the marker breccia contains accidental trachytic clasts (Palizzi lava?), but does not contain rhyolite fragments, the Commenda flow is assumed to be younger.

The parasitic cone of Forgia Vecchia I (Fig. 8e) was probably formed during this period. The ejection of a thick blanket of wet-surge deposits excavated a large crater (300 m diameter). Some ash-flow tuffs and lahars related to this explosion rest on fumarolized red-tuff beds of Fossa inside the Forgia Vecchia I crater. Dry-surge sandwave beds above the Forgia Vecchia I deposits contain abundant Pele's hair concentrated in layers. This horizon may correlate with a similar bed above the trachytic lava flow of the Palizzi cycle, although another Pele's hair horizon exists within Palizzi beds.

The next cycle (4), termed Pietre Cotte (Fig. 8d), may in fact consist of two cycles separated by a lava (sill?) that is only exposed in the gully between the Pietre Cotte lava flow and Forgia Vecchia crater. Because the relationship of this lava is uncertain due to lack of exposure, all the deposits between the Commenda and Pietre Cotte lavas are here interpreted as a single cycle. The Pietre Cotte cycle begins above this horizon with a widespread wet-surge deposit from Fossa that mantles the whole edifice and reaches the caldera wall at Pte. Luccia and Mte. Saraceno. After deposition of a lapilli-and-block fall sequence, the small (80 m diameter) parasitic cone of Forgia Vecchia II formed at the northern rim of the Forgia Vecchia I crater (Fig. 8e), producing thin deposits of lahar, wet-surge beds, and ash-flow tuffs within the crater. Above the Forgia II deposits lie ash-fall tuff and lahars from the Fossa vent. The final event of this cycle is the extrusion of the Pietre Cotte lava flow in 1739.

Deposits of the modern cycle (5) that post-date the Pietre Cotte lava are best exposed inside the southern crater rim of Fossa. Here a section of dark gray fall deposits consisting of pumice and weakly vesicular lapilli and blocks rests upon a thick wet-surge sequence (Fig. 9). Ash-fall deposits, surge beds, and erosional surfaces separate at least 10 lapilli-and-block fall horizons in the upper unit (Fossa III of Keller, 1980). The Pietre Cotte lava, which lies below the base of the modern cycle, is missing from the upper sequence at the rim. We assume that a prominent rhyolitic pumice layer that occurs 10 m below the rim is contemporaneous with the Pietre Cotte lava. Thus, all beds above this level belong to the modern cycle (5).

Products of the modern cycle consist of lapilli-and-block fall beds, coarse-and-fine deposits, and minor dry-surge beds. However, only products of the last active period (1888-90) can be stratigraphically identified. The 1888-90 eruption products presently constitute the upper 1.7 m of the 10 m thick modern cycle deposits at the crater rim. These beds mantle a collapse surface

Fig. 9. Southern crater rim of Fossa. Base of the section is composed of wet-surge deposits (Tufi Rossi). The overlying stratified beds belong to the Pietre Cotte and modern cycles. The 1888-90 deposits, forming the upper 1.7 m of the section, have quaquaversal dips over a collapsed rim of modern-cycle deposits.

formed with the present Fossa crater (Fig. 9). This eruption began with a thin yellow phreatic breccia (Sheridan and others, 1981) that was followed by coarse-and-fine beds with minor surge horizons that are restricted to channels on the Fossa rim. Subsequent water reworking produced lahars on the flanks and base of the cone. The middle period of activity produced distinctive composite breadcrust bombs that occur within no other cycle of the Fossa edifice.

Some controversy exists concerning the genesis of the breadcrust bombs in the 1888-90 deposit. Keller (1980, p. 50) interprets the last active period as "ultra-Vulcanian" in the sense of Mercalli (1907), because he believes that no juvenile material was erupted. He imagines the huge breadrust bombs to be a result of fusion of old material for several reasons including their wide range in chemical composition. Conversely, we concur with most of the other authors (Mercalli and Silvestri, 1891; Walker, 1979; Frullani, 1979) that the bombs are juvenile and consist of two components. Their glassy rhyolitic matrix is poor in crystals whereas the abundant porphyritic trachyte xenoliths contain plagioclase, clinopyroxene, and minor olivine phenocrysts. A similar mixed texture (porphyritic trachyte blebs within a glassy vitrophyre) is typical of the lava flows that terminate the Commenda and Pietre Cotte cycles.

An event like the 1888-90 eruption occurs nowhere else in the eruptive history of Fossa. We conclude that the modern cycle may still be in progress because the eruption of pumice-fall deposits and a final lava has not yet occurred. Conversely, this cycle may be irregular and the breadcrust bombs could correspond to the final lava of the other cycles.

Much of the modern-cycle deposits have now been eroded from the cone to expose underlying wet-surge deposits of other cycles at the surface in many places. Typical exposures of lahars at the base of the cone are thought to be a result of water-working of unconsolidated ash from higher on the cone. In support of this argument, modern lahars of small-scale presently form after heavy rains.

There are probably other eruptive cycles of Fossa that are not well exposed and therefore not described. For example, there is a series of wet-surge beds that occur above the Pte. Nere cycle (Fig. 8a) but below the Palizzi cycle (Fig. 8b). On the northern part of the cone this unit consists of three sequences of wet-surge beds that mantle a collapsed crater wall of the Pte. Nere cycle (Fig. 7). Above it are infilling deposits of ash-flow and wet-surge beds. Three sequences of wet-surge beds mantle the old crater rim of the Pte. Nere cycle and rest below the Pte. Nere cycles. The lower two layers are thicker and of local extent.

Keller (1970) called such wet-surge deposits "Tufi Rossi", and considered them to be a general stratigraphic horizon defining the middle sequence on Fossa. However, we find wet-surge beds (Tufi Rossi) at many stratigraphic intervals, not just within the middle sequence. For example, several levels of wet-surge deposits occur below the upper Pilato ash, whereas Keller only reports Tufi Rossi above this marker bed.

Products from later periods of activity are best exposed on the southern part of the volcano (Fig. 9). Two wet-surge units lie unconformably beneath products of the Palizzi cycle. These deposits are thought to be the main constructive units of the western volcanic edifice. Such wet-surge deposits probably constitute two or more unidentified cycles, but our present data is not sufficient to define them. The trachytic lava exposed in the Campo Sportivo may be a terminal event for one of these cycles.

GRAIN-SIZE ANALYSIS

Granulometric analyses performed on 66 samples characterize the clast-size frequencies in order to help interpret the origin of the deposits. Samples were only collected from beds with grains in the sand-size range. Therefore, the rare coarse deposits, such as pumice-fall beds and explosion breccias, are not represented. An attempt was made to reduce the adverse effect of multiple beds on size-population analysis. Most samples were collected from a 1 to 3 cm vertical interval with the intention of taking material from entire single beds. However, in a few cases with very fine laminae, grains from several beds had to be taken.

Dry sieving was used to determine the weight fraction of grains in the range of -2ϕ to 4.5ϕ using screen intervals of 0.5ϕ. This interval was adequate to characterize the populations. The resulting data were plotted as histograms as well as probability graphs in order to distinguish prominent modes. Thin section grain mounts were made of all samples to distinguish the type, shape,

TABLE 1

Grain-size data

SAMPLE	TEXT.	CLASS	Mdφ	σ	SAMPLE	TEXT.	CLASS	Mdφ	σ
VL45/1	C+F	3	0.98	0.98	VL79/1	SW	3	2.59	0.91
VL51/1	SW	3	2.03	1.48	VL79/2	SW	3	2.31	0.81
VL52/1	SW	5	2.19	0.54	VL81/5	AF	4	2.67	0.75
VL53/1	M	3	1.70	1.50	VL82/6	SW	3	2.19	1.44
VL54/1	M	1	1.86	0.53	VL82/7	SW	3	2.42	1.70
VL57/1	M	5	2.58	0.33	VL84/5	WS	3	2.35	1.53
VL60/1	SW	3	2.23	0.98	VL84/8	C+F	3	2.60	1.31
VL62/2	SW	3	1.77	1.63	VL84/9	C+F	1	0.03	0.97
VL62/3	SW?	3	1.85	1.23	VL84/10	C+F	3	1.86	1.54
VL62/4	SW	2	1.33	0.97	VL85/1	SW	4	2.11	1.32
VL63/1	SW	2	1.35	1.20	VL86/2	WS	5	3.76	0.82
VL70/3	WS	4	2.53	1.25	VL86/3	C+F	2	0.75	0.62
VL71/1	M	4	2.22	1.19	VL86/4	C+F	5	3.49	1.32
VL71/2	SW	3	1.70	0.91	VL90/2	SW	2	1.73	0.85
VL71/4	SW	4	2.94	0.40	VL90/4	SW	2	1.93	0.88
VL72/1	M	2	1.08	0.74	VL93/1	SW	3	1.92	1.12
VL72/5	SW	2	1.84	0.73	VL93/2	SW	3	2.00	1.29
VL72/6	SW	3	2.20	0.80	VL94/1	WS	3	2.54	1.36
VL73/1	SW?	3	1.10	1.20	VL94/2	WS	4	2.95	1.05
VL73/2	M	1	1.48	1.06	VL94/3	SW	3	1.73	0.98
VL75/1	M	2	2.00	1.30	VL95/1	SW	3	1.96	1.52
VL75/2	M	1	1.15	0.76	VL97/1	SW	2	1.16	1.09
VL75/3	M	2	1.17	0.94	VL97/2	WS	3	2.65	1.75
VL75/4	WS	3	2.95	1.85	VL97/3	F	2	1.65	1.25
VL75/5	M	2	1.09	1.92	VL99/1	WS	5	3.85	0.85
VL75/6	WS	3	2.31	1.59	VL99/2	SW?	3	1.68	1.13
VL75/16	F	1	-0.15	1.83	VL100/1	SW	4	2.38	1.10
VL75/17	SW	3	1.18	1.93	VL101/1	F	3	2.27	0.82
VL75/18	F	2	1.30	1.70	VL104/2	SW?	2	1.77	1.02
VL75/19	SW	2	1.80	1.50	VL105/1	SW	4	2.13	0.75
VL75/20	SW	3	1.62	1.70	VL119/1	SW	3	2.21	0.91
VL77/1	SW?	5	2.41	0.59	VL119/2	WS	5	3.34	0.56
VL77/3	SW	3	2.34	0.74	VL122/1	SW	5	2.49	0.57

and abundance of the constituent particles. In addition, approximately 900 SEM whole grain photographs were made to examine the surface morphology of crystals and glass fragments from the dominant size populations in selected samples. The detailed results of the SEM study will be reported in a later paper.

Grain-size data for the samples is summarized in Table 1. The sample number consists of section location, as shown in Figs. 1 and 2, and sample digit(s). Several types of outcrop textures are designated in the table: dry-surge sandwave (SW), dry-surge massive (M), wet-surge (WS), fall (F), coarse-and-fine (C+F), and ash flow (AF). The grain-size parameters are summarized in Table 2 for the three most common depositional types. In this table N refers to the number of samples used to compute the average. The sandwave beds are slightly coarser and the massive beds slightly finer than the average values reported for these types by Wohletz and Sheridan (1979). Both types are better sorted than values in the literature.

On detailed examination most of the samples could be represented by a combination of two populations with dominant modes respectively at 0.5-1.5 ϕ and 2.5-3.5 ϕ. We grouped the samples into 5 gradational classes (Table 1) ranging from a dominance of the coarse mode (class I) to a dominance of the fine mode (class V). Typical histograms of each class type are shown in Fig. 10. Samples in classes I and V are nearly unimodal. Those in class III are strongly polymodal with two dominant populations. Samples in classes II and IV show a dominance of one mode with a skew towards another. From this we conclude that mixing of the two main populations could reproduce 81% of the studied samples, excepting the extreme cases I and V. Because this sample suite is so strongly bimodal, the graphical parameters (Mdϕ and σ_ϕ) should not be interpreted in terms of strict population statistics. They are merely useful for comparative purposes. A better characterization of size distribution would be the location of specific modes and their relative contribution to the total population.

In addition to the two main size populations at 0.5-1.5 ϕ and 2.5-3.5 ϕ, two other minor populations were noted in some samples. Many deposits with a strong ash-fall component have an additional mode in the range of -2.0 to -1.0 ϕ. Conversely, most of the wet-surge deposits have a mode that is finer than 4.5 ϕ. The center of this mode was not accurately determined because only a dry screening technique was used.

TABLE 2

Average grain-size distribution parameters.

TEXTURES	N	Mdϕ	σ_ϕ
SW	35	1.96 ± .42	1.08 ± .37
M	10	1.63 ± .53	1.03 ± .47
WS	10	2.92 ± .56	1.26 ± .43

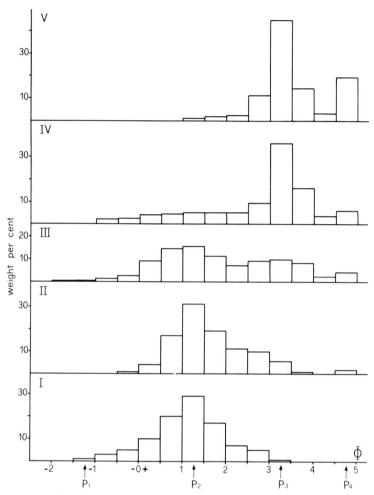

Fig. 10. Histograms of typical size distributions illustrating the bimodal character pyroclastic samples from Fossa. The five gradational classes of population types (classes I through V) represent a transition from dominantly ash-fall (P_2) through dominantly dry-surge (P_3) emplacement. The principal modes of the 4 main populations are also shown. P_1 is pumice fall, P_2 is ash fall, P_3 is dry surge and P_4 is wet surge.

The size parameter (Mdϕ and σ_ϕ) data for Vulcano fall distinctly outside of the main frequency fields of air fall and pyroclastic flow (Fig. 11) on the diagram of Walker (1971). Other typical surge deposits plot within the field defined by the Vulcano samples (for example data from Crowe and Fisher, 1973; Schmincke et al., 1973; Sheridan and Updike, 1975; Hoblitt et al., 1982). In addition, samples with similar size parameters are also produced by

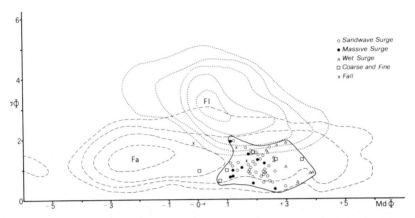

Fig. 11. Size parameters (Mdφ and σφ) for Fossa samples. The fall (dash) and pyroclastic flow (dot) fields of Walker (1971) are outlined. The surge samples of Fossa (dark line) define a tight field, distinct from the fall (Fa) and flow (Fl) modes.

phreatoplinian eruptions (Self and Sparks, 1978). It seems clear that size distributions produced during melt fragmentation and transportation by hydromagmatic processes are distinct from those of essentially magmatic origin (see Wohletz, this volume). The median diameter vs. sorting plot (Fig. 11) may prove useful for distinguishing products of hydrovolcanic explosions.

INTERPRETATIONS

The two main grain-size populations at 0.5-1.5 φ and 2.5-3.5 φ have different ratios of components. The finer population is enriched in juvenile glass fragments whereas the coarser population contains more crystals and lithic clasts. Fall deposits contain only the coarse mode plus even coarser lapilli. As the surge character of the bed texture increases, the fine component becomes more abundant. Wet-surge deposits are dominated by the fine component as well as an even finer ash population with a mode less than 4.5 φ. These data support a model for grain-size population development related to both the level of initial water/melt interaction and the effect of aerodynamic drag on the grains. The particle-size distribution in the eruptive column is related to initial water/melt ratios in the vent. Water/melt experiments at Los Alamos National Laboratory (Sheridan, 1982; Wohletz and Sheridan, 1982; Wohletz, this volume) have documented a direct relationship of water to grain size in hydromagmatic explosions. Relatively dry explosions produce coarser-grained fall deposits with a minor fine component. Relatively wet explosions produce a large component of the fine-ash population. Therefore, the initial size-populations for explosions at Fossa may be either unimodal or bimodal depending on mass flux rate of magma, hydrodynamics of the aquifer, and vent geometry.

Aerodynamic drag causes a distinct separation of population types into those that travel as single grains with their motion retarded by the atmosphere and those that move by mass flowage with their motion enhanced by gravity driven currents. Large particles ejected from the vent follow drag-modified parabolic paths but the small particles collapse in turbulent dust clouds and move laterally as surges. If there is a sufficient time interval between the arrival of the large (ballistic) and small (surge) grains produced by a single explosion, a distinctly layered coarse-and-fine deposit results (Fig. 4). If the two populations arrive at nearly the same time, typical surge structures (sandwaves) develop because the ballistic particles are incorporated into the denser portions of the turbulent surge clouds. This explains the presence of scoria imbedded in a matrix with surge textures in some of the proximal deposits. Likewise, dry-surge deposits with embedded fragile wet-surge clasts can result from this mechanism.

If the above assumptions are correct, the water/melt mixing ratio during eruption can be inferred from bedding structures, clast types, and grain-size populations. Pumice-fall beds represent eruptions with a low water/melt mixing ratio in which magmatic volatiles control the explosion and dispersal of clasts. Coarse-and-fine beds result from a significant mixing of water with melt in the vent that produces large explosions so that fall grains arrive significantly before the surge grains. With an increase in water/melt ratio the explosion energy is optimized causing all fall components to be incorporated within the dry-surge beds. The wet-surge horizons result from even more water in the vent so that the energy maximum is passed but the grain size is even finer due to greater melt fragmentation. Finally, phreatic breccias represent steam explosions with little or no juvenile magmatic component.

Using this scheme, the main Fossa cycles can be interpreted in terms of changes in degree of water/melt interaction. Both the mechanical energy of the eruption and fragmentation of the magma reach a maximum at a specific ratio of water to magma (Fig. 12a). For the Fossa vent of Vulcano, the fragmentation maximum appears to occur at a higher water content than the energy maximum. Proceeding from 'wet' to 'dry' the respective deposits include: phreatic breccias, slumped vesiculated tuffs, fresh sandwave surge beds, coarse-and-fine deposits, pumice-fall beds, and lava flows (Fig. 12b).

Using this framework, the cycles of Vulcano can be represented in a simple diagram (Fig. 12b) showing their approximate path on an energetics diagram (Sheridan and Wohletz, this volume). The Pte. Nere cycle (A) follows the path of dry surges, pumice falls, then a lava flow. The Palizzi cycle (B) appears to have two stages. The first stage (B_1) starts with wet surge and ends with pumice fall. The second stage (B_2) starts with wet surges, dry surges, then a lava flow. The Commenda cycle (C) begins with a phreatic explosion breccia and pyroclastic flow followed by a thin wet surge bed with accretionary lapilli followed by a thin dry surge. The cycle ends with a lava flow. The Pietre Cotte cycle (d) starts with widespread wet surges, and ends with a

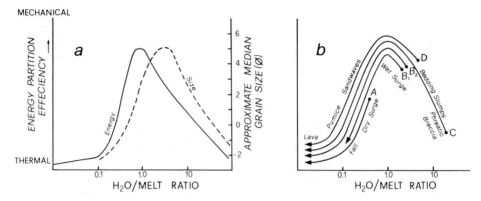

Fig. 12. Role of water/melt mixing ratio at Fossa of Vulcano. A. Relationship of conversion of thermal to mechanical energy (from: Sheridan and Wohletz, 1981; Wohletz, this volume) and particle size to water/melt ratio. B. Eruptive path followed by the various cycles [a) Pte. Nere, b) Palizzi, c) Commenda, and d) Pietre Cotte] on an energetics diagram following the method of Wohletz and Sheridan (in press). All cycles move from higher ratios of water to melt toward lower ratios (i.e. from wet to dry).

lapilli fall and lava. The record of other periods of activity are not depicted because of poor exposure (cycles between Pte. Nere and Palizzi), or incompleteness (modern cycle). In general, all of the cycles begin with either wet or dry surges and end with a lava flow. The general sequence of deposits follows a pattern that indicates a decrease in water/melt interaction with time.

MORPHOLOGICAL DEVELOPMENT OF FOSSA CONE

Our interpretation begins after emplacement of the tephritic lava flow of Pta. Luccia. The lower Pilato ash, dated at 11,000 to 8,500 y.b.p. (Keller, 1980, Bigazzi and Bonadonna, 1973) is not found within the Fossa edifice. Therefore, the Fossa cones are younger than this age. The rim of the first cone was constructed during the Pte. Nere cycle (Fig. 13). This first structure was a low tuff ring of dry-surge deposits followed by a cone-forming sequence of lapilli-and-block fall deposits. The lapilli and blocks are juvenile fragments of non-vesiculated glassy trachyte in layers separated by the thin surge beds. Finally the Pte. Nere trachytic lava flowed over the NE rim of the cone down the steep (30°) slope into the sea.

The period following the Pte. Nere cycle is not well documented because of a lack of good exposures. It is represented by wet surge deposits (Figs. 8a and 8b) which further develop the cone, possibly from a source to the west of the first vent (Fig. 13).

The next crater developed further to the west and was the source for the deposits of both the Palizzi and Commenda cycles. Wet surges from this vent

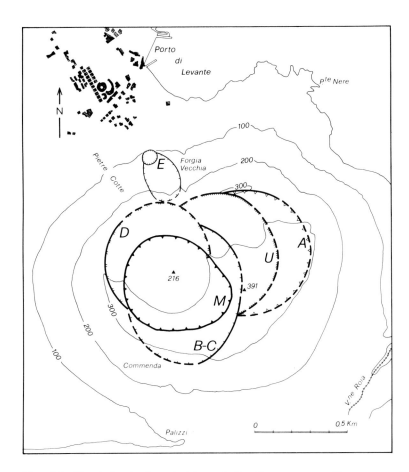

Fig. 13. Morphological evolution of the Fossa cone. The successive rims for cones of the various cycles are as follows: A. Pte. Nere cycle, B_1-B_2. Palizzi cycle, C. Commenda cycle, D. Pietre Cotte cycle, E. Forgia Vecchia parasitic cones, M. Modern cycles, and U. Unidentified cycles.

fill the previous crater and mantle the previous cone. The initial explosions of this cycle destroyed the western part of the first cone and concentrated the surge deposits in the western part of the Caldera. The lava of Palizzi spilled over the low southern edge of this new cone and the Commenda flowed over its SW edge.

The next crater to develop is related to the Pietre Cotte cycle that forms part of the present rim. This cycle began with the widespread red tuff that mantles much of the existing edifice. This wet-surge was followed by a fall deposit of poorly vesiculated clasts and interbedded surge horizons. The Pietre Cotte lava poured out on the northwestern flank of the cone to end this cycle.

The present crater produced by the explosive modern cycle, has somewhat modified the Pietre Cotte crater. The modern center of activity has moved to the south and hence the southern crater wall has been partly destroyed.

The Forgia Vecchia satellitic cone (Fig. 8c) was produced by two phreatic events that are not directly related to the main eruptive cycles. The first explosion formed the large crater of Forgia Vecchia I either contemporaneous with the latter part of the Palizzi cycle or with the Commenda cycle. The deposits are composed of local wet surge beds that are generally altered by fumarolic activity. The Forgia Vecchia II explosion in 1727 produced a very small crater superimposed on the rim of the Forgia Vecchia I crater. The wet-surge and pyroclastic-flow deposits produced by this event have a very local extent.

The distribution of modern cycle deposits is strongly modified by intense water erosion. These deposits are completely stripped from much of the Fossa Cone. For example, on the rim only 1.7 m of the original 5 m of the 1888-90 deposits now remain. Likewise, Mercalli and Silverstri (1891) reported 40 cm of ash on Mte. Sareceno, but no deposits can now be found there. Such intense erosion is supported by the frequent formation of small lahar deposits following intense rainfalls.

THE MODEL

A model for the mechanism of eruption is the basis for interpreting future activity. One critical factor at Fossa of Vulcano is the dominance of water/melt interaction over exsolution of magmatic volatiles in controlling melt fragmentation and explosive energy. Another factor could be the influence of mixing of the two melt components (mafic and acidic) that are observed in the rhyolitic lavas. A third factor is the role of active tectonism in the region.

Magma mixing is an important process in the Lipari-Vulcano eruptive system (De Rosa and Sheridan, this volume). On the Fossa of Vulcano, two types of magma have interacted during the rhyolitic cycles. One type is an evolved magma represented by the aphyric rhyolite glass of the late lavas. The other component could be a more basic melt in which phenocrysts of clinopyroxene, calcic plagioclase, olivine and magnetite are stable. The mafic melt inclusions in the rhyolite lavas have chill textures and strong reaction rims. Their outlines are fluidal and some blebs are vesiculated indicating a probable liquid condition just prior to eruption. Likewise, the pyroclastic materials contain pale brown aphyric glass (acid component) and abundant clinopyroxene and plagioclase crystals (mafic component).

Only textural evidence of mixing in the rhyolites is given here. Further investigation is being directed to ascertain whether or not the trachyte lavas are products of mixing as well. Accepting the mixing model, it is still unknown whether the two magmas existed in a single zoned chamber or two chambers at different depths. One obvious means to trigger mixing of the magmas, regardless of their settings, is tectonic faulting.

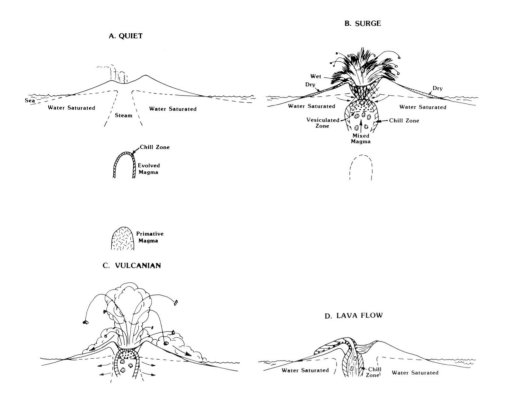

Fig. 14. Schematic diagram of the four stages of a typical Vulcanian cycle. A. Quiet. B. Surge. C. Vulcanian. D. Lava effusion.

The important stages in a typical Vulcanian cycle are schematicaly illustrated in Fig. 14. During a quiet period preceding an eruptive cycle two contrasting magmas are physically separated at an unknown depth beneath the vent (Fig. 14A). During this stage heat is rapidly transmitted to the surface through a chimney in the water table that is saturated with superheated steam.

A triggering event brings the hot primitive magma into contact with the cooler evolved liquid lowering its viscosity and promoting upward mobility. Effective mixing of the two magmas, perhaps associated with the separation of a magmatic volatile phase, occurs at this stage. The mixed magma rises to the point where it contacts a water-saturated zone near the surface (Fig. 14B). Dilation of the aquifer and fracturing associated with the rise of the magma allows water to contact hot congealed magma initiating fuel-coolant interactions (FCI) that could lead to sustained surge eruptions involving massive melt/water interactions (see Wohletz, this volume). The particles erupted at this stage show only moderate vesiculation because of the cooling effect of the water interaction. Wet- and dry-surge deposits construct a tuff cone with a wide crater at this stage.

Heat from the high-level magma body causes the steam chimney to expand reducing the direct contact of water with the melt. Eruptions at this stage (Fig. 14C) would be Vulcanian with vesiculated produced by magmatic exsolution as well as chilled juvenile fragments generated by FCI being ejected to relatively low elevations. Both fall and surge products are ejected, but the eruptive phenomena favors emplacement of coarse-and-fine deposits typical of the 1888-1890 eruptions. During this stage the steam halo is further widened and the vesiculated zone near the top of the magma body approaches the surface.

During the final stage of the cycle the pumiceous cap of the magma chamber erupts, either as a pumice-fall bed or as a pumiceous carpet for a lava flow (Fig. 14D). The degassed magma finally reaches the surface forming an obsidian-cored lava flow. The duration of a typical cycle or the repose between stages is still not known. However, this sequence has occurred repeatedly through the history of the Fossa vent and merits further study.

ACKNOWLEDGEMENTS

This work was financially supported by the National Council of Research of Italy (CNR) Geodynamics Project, volcanic risk, U.O. 3.4.2 and U.O. 3.6.6, publication No. 512. One of the authors (M.F.S.) was funded by NASA Grant NAGW-245. We would like to thank Tom Moyer and Ken Wohletz who carried out earlier field work on this program and who read this manuscript.

REFERENCES

Bigazzi, G. and Bonadonna, F., 1973. Fission track dating of the obsidian of Lipari Island (Italy). Nature, 242: 322-323.

Castellet y Ballara, G., Crescenzi, R., Pompili, A., and Trigila, R., 1981. Magma evolution of Vulcano eruptive complex: an approach for a deterministic model of volcanic activity. Mem. Soc. Astron. Ital., 52: 369-373.

Crowe, B. H., Fisher, R. V., 1973. Sedimentary structures in base surge deposits with special reference to cross bedding, Ubehebe craters, Death Valley, California. Geol. Soc. Am. Bull., 84: 633-682.

De Fiore, O., 1922. Vulcano (Isole Eolie). Monografia - Zeitsch. fur Vulk., 3: 3-393 (Rep. Swets and Zeitlinger N. V., Amsterdam, 1969).

De Rosa, R. and Sheridan, M.F., 1983. Preliminary data for magma mixing in the surge deposits of the Monte Guardia sequence, Lipari, In: M.F. Sheridan and F. Barberi (Editors), Explosive Volcanism. J. Volcanol. Geotherm. Res., 17: (this volume).

Frazzetta, G., Villari, A., and Villari, L., 1977. Ground deformation and geological structures on Vulcano-Lipari volcanic complex (Aeolian Islands, Italy). Abstract in IAVCEI-IASPEI Joint General Assemblies, Durham, 1977.

Frullani, A., 1979. Le eruzioni storiche di Vulcano (isole Eolie). Natura dei prodotti, meccanismo eruttivo ed implicazion per il rischio vulcanico. Tesi di Laurea Universita di Pisa.

Hoblitt, R. P., Miller, C. D. and Vallance, J. W., 1982. Origin and stratigraphy of the deposit produced by the May 18 directed blast. In: P.W. Lipman and D.R. Mullineaux (Editors), The 1980 eruptions of Mount St. Helens, Washington. U.S. Geol. Survey. Prof. Pap., 1250: 401-420.

Inman, D. L., 1952. Measures for describing the size distribution of sediments. J. Sediment. Petrol., 22: 125-185.

Keller, J., 1970. Die historischen Eruptionen von Vulcano und Lipari. Zeitschrift Deutsch. Geologie Geoch., 121: 179-185.

Keller, J., 1980. The Island of Vulcano. Rend. Soc. Ital. Min. Petrol., 36(1): 369-414.

Lo Giudice, E., Marino, C. M., and Tonelli, A. M., 1975. Le techniche di ripresa all'infrarosso termico applicate ai rilievi aerei dell'isola di Vulcano. Analisi dei risultati in rapporto alla situazione geovulcanologica locale. Boll. Geof. Teor. Appl., 17: 19-38.

Lorenz, V., 1974. Vesiculated tuffs and associated features. Sedimentology, 21: 273-291.

Mercalli, G. and Silvestri, O., 1891. Le eruzioni dell'isola di Vulcano, incominciate il 3 Agosto 1888 e terminate il 22 Marzo 1980. Relazione scientifica, 1891. Ann. Uff. Cent. Meteor. Geodin., 10(4): 1-213.

Schmincke, H.-U., Fisher, R. V., and Waters, A. C., 1973. Antidune and chute and pool structures in the base surge deposits of the Laacher-See area, Germany. Sedimentology, 20: 553-574.

Self, S. and Sparks, R. S. J., 1978. Characteristics of widespread pyroclastic deposits formed by the interaction of silicic magma and water. Bull. Volcanol., 41: 1-17.

Sheridan, M. F., 1981. Particles formed by fuel-coolant explosions. NASA Tech. Memor., 84211: 167-168.

Sheridan, M. F., and Updike, R., 1975. Sugarloaf Mountain tephra - a Pleistocene rhyolitic deposit of base-surge origin in northern Arizona. Geol. Soc. Am. Bull., 86: 571-581.

Sheridan, M. F., and Wohletz, K. H., 1981. Hydrovolcanic explosions: the systematics of water-pyroclast equilibration. Science, 212: 1387-1389.

Sheridan, M. F., Moyer, T. C., and Wohletz, K. H., 1981. Preliminary report on the pyroclastic products of Vulcano. Mem. Soc. Astron. Ital., 52: 523-527.

Sheridan, M. F. and Wohletz, K. H., 1983. Explosive hydrovolcanism: basic considerations. In: M.F. Sheridan and F. Barberi (Editors), Explosive Volcanism. J. Volcanol. Geotherm. Res., 17: (this volume).

Walker, G. P. L., 1971. Grain-size characteristics of pyroclastic deposits. Jour. Geol., 79: 696-714.

Walker, G. P. L., 1979. The breaking of magma. Geol. Mag., 106: 166-173.

Wohletz, K. H., 1979. Experimental modelling of hydromagmatic volcanism (abs.) EOS, Trans. Am. Geophys. Union, 61(6): 66.

Wohletz, K. H., 1983. Mechanisms of hydrovolcanic pyroclast formation: grain-size, scanning electron microscopy, and experimental data. In: M.F. Sheridan and F. Barberi (Editors), Explosive Volcanism. J. Volcanol. Geotherm. Res., 17: (this volume).

Wohletz, K. H. and Sheridan, M. F., 1979. A model of pyroclastic surge. In: C.E. Chapin and W.E. Elston (Editors), Ash-flow Tuffs. Geol. Soc. Amer. Sp. Pap., 180: 177-193.

Wohletz, K. H., and Sheridan M. F., 1980. Rampart crater ejecta: Experiments and analysis of melt-water interactions (abs.) NASA Tech. Memor. 82385: 134-136.

Wohletz, K. H., and Sheridan, M. F., 1981. Melt-water interactions: Series II experimental design. NASA Tech. Memor., 84211: 169-171.

Wohletz, K. H., and Sheridan, M. F., in press. Hydrovolcanic explosions II: Evolution of basaltic tuff rings and tuff cones. Am. Jour. Sci.

RECENT VOLCANIC HISTORY OF PANTELLERIA: A NEW INTERPRETATION

Y. CORNETTE[1], G. M. CRISCI[2], P. Y. GILLOT[1] and G. ORSI[3]

[1] Centre de Faibles Radioactivites, CNRS/CEA, Gif sur Yvette, (France)
[2] Dipartimento Scienze della Terra, Castiglione Scalo, Cosenza, (Italy)
[3] Instituto di Geologia e Geofisica, Napoli, (Italy) and Istituto di Mineralogia e Petrografia, Bari, (Italy)

(Received July 15, 1982; revised and accepted November 12, 1982)

ABSTRACT

Cornette, Y., Crisci, G.M., Gillot, P.Y., and Orsi, G., 1983. Recent volcanic history of Pantelleria: a new interpretation. In: M.F. Sheridan and F. Barberi (Editors), Explosive Volcanism. J. Volcanol. Geotherm. Res., 17: 361-373.

The island of Pantelleria is located in the Pantelleria Rift between Sicily and Africa. This rift is floored by a 20 km thick continental-type crust that is characterized by horsts and grabens formed by NW-SE tensional faults and NE-SW shear faults. Several faults and fractures that intersect the island follow these regional trends. Prominent volcano-tectonic features on the island include two calderas, the younger being related to the eruption of the green ignimbrite and the uplifted Montagna Grande block. Pantelleria mostly consists of peralkaline silicic rocks with minor trachytes and basalts. Volcanism on the island is characterized by both explosive and effusive activity.

Reconstruction of the volcanic history of the island since the eruption of the green ignimbrite is based on geological and geochronological data. The eruption of the green ignimbrite (50,000 y.b.p.) was followed by collapse of the Monastero caldera that has been the site of most of the subsequent acid volcanism. The Montagna Grande volcanic complex (35,000 y.b.p.) fills the central part while the lower pantelleritic lava flows (16,000 y.b.p.) erupted along its rim. Very probably the young Mt. Gelfiser and Khaggiar domes are also located on the caldera rim. The Montagna Grande block rose within the caldera at an estimated average rate between 1.0 and 1.5 cm/y for about 20,000 y. Subsequent pumice cones (9,000 y.b.p.), endogenous domes, and lava flows also erupted inside this caldera. Among the main acid volcanic units younger than the green ignimbrite, only the Mt. Gelkhamar endogenous dome is not linked to this caldera.

In the last 50,000 years basaltic activity only occurred in the northern part of the island. The basaltic vents are aligned parallel with the main tensional trends of the rift. At least two phases have been recognized; the older is about 29,000 y.b.p. and the younger is a few thousands of years old. A change occurred in the eruptive style after the eruption of the green ignimbrite and collapse of the Monastero caldera. Explosive activity diminished and extrusion of lava became dominant.

INTRODUCTION

The Island of Pantelleria is located in the Pantelleria Rift which is along the deepest part of the Strait of Sicily. According to Colombi et al. (1973) and Cassinis et al. (1979) the central part of this Strait is floored by a continental type crust which is about 20 km thick. Laterally the crust reaches 30-40 km under Sicily and Africa. The structural pattern of the strait is charaterized by horsts and grabens controlled by tensional and shear faults trending NW-SE and NE-SW, respectively (Colantoni and Zarudzki, 1973).

Faults and fractures following the regional trends are easily recognizable on the island (Fig.1). Easterly and northerly trending faults have also been identified. The main volcano-tectonic features of the island are: (1) two intersecting calderas of different ages (Orsi, in prep.) and (2) the uplifted Montagna Grande block with consequent endogenous domes, lava flows, and pumice cones (Fig.1). The two calderas located in the central part of the island are aligned in a NNW-SSE trend. The northern one, called Lago caldera, is older than the green ignimbrite. The southern one, called Monastero, is related to the eruption of the green ignimbrite (Orsi, in preparation). The results of gravity and geomagnetic prospecting (Agocs, 1959; Gantar el al., 1961) suggest the presence of a dome-shaped intrusion of $\rho=3.0$ g-cm^{-3} under the island. Its top, located under Montagna Grande and Mt. Gibele, should reach sea level.

The most recent comprehensive geological and geochemical study of the island was made by Villari (1974). The Island of Pantelleria mostly consists of acid peralkaline rocks with minor trachytes and basalts. Its volcanic history is characterized by both explosive and effusive activity. Most of the common endogenous domes have been vents for lava flows in their final stages of growth.

The aim of this paper is to reconstruct the volcanic history that post-dates the emplacement of the green ignimbrite of Pantelleria. The stratigraphy resulting from field work has been corroborated by means of K/Ar dating.

STRATIGRAPHY

The petrographic and geochemical designation of the Pantelleria rocks in this paper follows Rittmann (1967), Villari (1968; 1970; 1974) and Civetta et al. (1982). The rock units (Fig.2) are grouped from youngest to oldest as follows:
- Mursia basaltic lava flows and cinder cones
- Upper pantelleritic lava flows
- Pantelleritic domes and lava flows

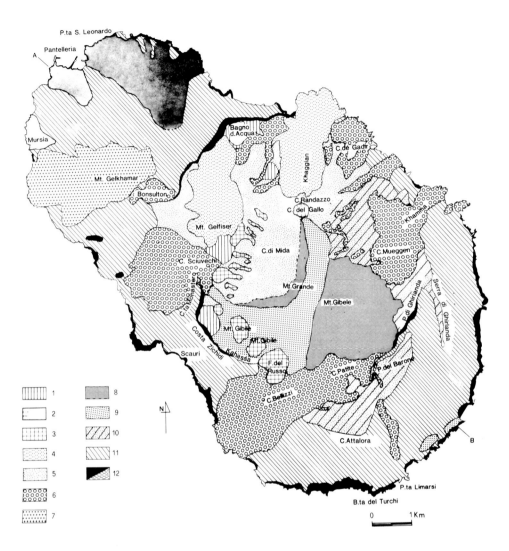

Fig. 1 Lithologic map of the island of Pantelleria (modified after Villari, 1967). (1) Recent sedimentary deposits; (2) Undifferentiated basaltic lava flows and cinder cones; (3) Upper pantelleritic lava flows; (4) Pantelleritic domes and lava flows; (5) Cuddia di Mida tephra; (6) Lower pantelleritic lava flows; (7) Gelkhamar pantelleritic endogenous dome; (8) Mt. Gibele lava flows; (9) Montagna Grande dome; (10) Serra di Ghirlanda tephra; (11) Green ignimbrite; (12) Volcanic units older than the green ignimbrite (a in this Figure, b in Fig. 3). Geological cross section of Fig. 3 is drawn through A-B.

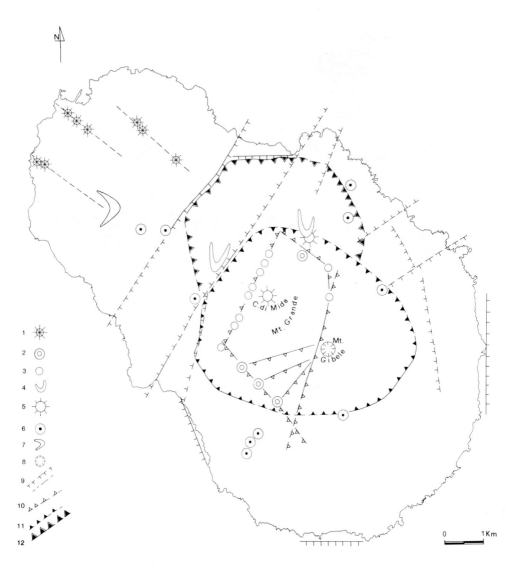

Fig. 2 Structural map of the island of Pantelleria (after Orsi, in prep.). (1) Cinder cones of the undifferentiated basaltic lava flows and cinder cones; (2) Domes of the upper pantelleritic lava flows; (3) Eruptive centers of the upper pantelleritic lava flows; (4) Domes of pantellerite; (5) Pumice cones of the Cuddio di Mida tephra; (6) Eruptive centers of the lower pantelleritic lava flows; (7) Gelkhamar pantelleritic endogenous dome; (8) Eruptive center of the Mt. Gibele lava flows; (9) Normal faults and fractures; (10) Volcano-tectonic faults; (11) Monastero Caldera rim. (12) Lago Caldera rim.

- Cuddia di Mida tephra
- Lower pantelleritic lava flows
- Gelkhamar pantelleritic endogenous dome
- Punta San Leonardo basaltic lava flows and cinder cones
- Montagna Grande soda trachytic volcanic complex
- Green ignimbrite
- Pre-green ignimbrite volcanic units.

The volcanic rocks older than the green ignimbrite (Villari, 1974; Wright, 1980) consist of a great variety of types including: lava flows, lava domes, pyroclastic-fall units, and pyroclastic-flow deposits. Their chemical composition includes basalts, trachytes, and rhyolites, the latter being dominant. These rocks are only exposed on the walls of the caldera depressions and along the coastal cliffs. At some places in the southern part of the island, these units attain a total thickness of about 300 m in sea cliff exposures. The ages of these rocks range between 50,000 y.b.p. (age of the green ignimbrite) and 140,000 y.b.p. (oldest age on Pantelleria determined by Bigazzi et. al., 1971). No attempt has yet been made to reconstruct the stratigraphy and the temporal evolution of these older products.

All of the above volcanic units are overlain by a pyroclastic sheet which was interpreted as an ignimbrite (Rittmann, 1967; Villari, 1969) and called "green ignimbrite" (Villari, 1974). This unit has also been considered to be a welded air-fall tuff (Korringa, 1971; Wright, 1980; Wolff and Wright, 1980) and was called "green tuff" (Wright, 1980). Although no special study has been performed on this formation by the present authors, the name "green ignimbrite" is used in this paper.

The green ignimbrite is a well-defined stratigraphic marker. It crops out over the entire island, except inside the calderas where it is buried by younger rocks. An ignimbrite found in a core taken at Bagno dell'Acqua (Villari, 1970; 1974) seems very likely to be the green ignimbrite. The thickness of the green ignimbrite, which varies from a few tens of centimeters up to 20 m, is linked to the previous morphology. It is thicker in topographic depressions and thinner on topographic highs. Because it covers an area of at least 80 km^2, which corresponds to the entire surface of the island, the dense rock volume is estimated at not less than 0.6 km^3 (Wright, 1980).

An ash layer correlative with the green ignimbrite has been found in deep sea cores from the Mediterranean (Keller et al., 1978; Keller, 1981). The interpolated strtigraphic age (45,000 y; Keller, 1981) is in a good agreement with the K/Ar ages of the green ignimbrite presented here (Table 1). Contrary to published papers, we consider the source of the green ignimbrite neither as fissural (Villari, 1969) nor as corresponding to the Cuddia di Mida pumice cone (Mahood and Hildreth, 1980). Rather, it should be from a central vent located inside the Monastero caldera depression that is now buried by younger rocks. The products of the Cuddia de Mida pumice cone overlie units younger than the green ignimbrite (Montagna Grande volcanic complex and lower pantelleritic lava flows). These products probably overlie the green ignimbrite as well. The different absolute ages (Table 1) and eruptive mechanisms for the Cuddia di Mida tephra and green ignimbrite clarify the relationships between the two formations.

The Montagna Grande complex consists of the Serra di Ghirlanda tephra, the Montagna Grande dome, and the Monte Gibele lava flows. All the rocks of this complex, contrary to previous interpretations (Rittmann, 1967; Villari, 1974; Mahood and Hildreth, 1980), are younger than the green ignimbrite as testified either by field relationships or by K/Ar age determinations (Table 1). This is

TABLE 1   Geochronological data of volcanic rocks from Pantelleria

Sample No.	K%	$^{40}Ar\hat{}$	$^{40}Ar\hat{}10^{11}at.g^{-1}$	Age (10^3y)
SIC-34	4.3	0.46	3.7	8.2 ± 3.6
SIC-34	"	0.33	4.4	9.7 ± 5.9
SIC-38	4.9	0.59	8.9	17.3 ± 5.8
SIC-38	"	0.80	7.8	15.1 ± 3.8
SIC-108	4.9	4.20	1.06	21.0 ± 1.5
SIC-108	3.9	1.55	1.06	26.0 ± 4.0
SIC-104	0.7	0.24	1.9	27.0 ± 20.0
SIC-104	"	0.48	2.2	30.5 ± 13.0
SIC-73	2.8	1.57	9.3	31.8 ± 4.0
SIC-73	"	0.59	9.7	33.0 ± 11.0
SIC-68	2.9	2.83	10.5	34.1 ± 2.4
SIC-68	"	2.81	10.9	35.3 ± 2.5
SIC-125	3.5	2.02	11.9	32.5 ± 3.2
SIC-125	"	2.20	13.2	35.8 ± 3.3
SIC-76	5.0	2.82	26.5	50.8 ± 3.6
SIC-20	5.0	2.89	26.1	50.0 ± 3.5
SIC-20	"	2.96	24.6	47.0 ± 3.2
SIC-111	4.7	7.3	2.85	59.0 ± 2.0
SIC-111	3.8	2.7	2.54	64.0 ± 5.0

$^{40}Ar\hat{}$ = radiogenic; The international constants were used for age calculation (Steiger and Jaeger, 1977).

TABLE 1 (continued)

Sample No.	Locality	Formation	Rock type	Analyzed phase
SIC-34	S. of Cuddia di Mida	Cuddia di Mida tephra	Peralkaline rhyolite	K-spar
SIC-38	Contrada Sciuvechi	Lower pantel. lava flows	Peralkaline rhyolite	K-spar
SIC-108	Punta dell'Alca	Galkhamar dome	Peralkaline rhyolite	K-spar
SIC-108	Punta dell'Alca	Galkhamar dome	Peralkaline rhyolite	gdmass
SIC-104	Punta San Leonardo	Basaltic lava flows	Basalt	gdmass
SIC-73	Monte Gibele	Monte Gibele lava flows	Trachyte	K-spar
SIC-68	Montagna Grande	Montagna Grande dome	Trachyte	K-spar
SIC-125	Serra di Ghirlanda	Serra di Ghirlanda tephra	Peralkaline rhyolite	K-spar
SIC-76	Punta Limarsi	Green ignimbrite	Peralkaline rhyolite	K-spar
SIC-20	Contrada Lago	Green ignimbrite	Peralkaline rhyolite	K-spar
SIC-111	Costa Zichidi	Pre-green ignimbrite	Peralkaline rhyolite	K-spar
SIC-111	Costa Zichidi	Pre-green ignimbrite	Peralkaline rhyolite	gdmass

the most significant evidence supporting our new interpretation of the recent volcanic history of the island. Trachytes of the Montagna Grande volcanic complex can be clearly demonstrated to be younger than both the green ignimbrite and the Monastero caldera collapse at two localities. At Contrada Kahassa the green ignimbrite forms the capping unit on the Monastero caldera rim, which has been faulted down towards the south. Trachytic lavas filled this caldera depression and flowed over the lowest parts of the rim. However, they butted against the higher parts of the rim. North of Piano del Barone the trachytic lavas butt against the Monastero caldera wall which is capped by the green ignimbrite. The Serra di Ghirlanda tephra mainly consists of peralkaline rhyolitic (Civetta et al., in prep.) pumice-fall deposits. At the western foot of Serra di Ghirlanda they overlie the green ignimbrite and at Piano di Ghirlanda they rest against the Monastero caldera wall. At Serra di Ghirlanda they accumulated either by primary fall or by subsequent downslope movement on steep surfaces (fall-and-roll). The large Montagna Grande soda-trachytic dome occupies most of the Monastero Caldera depression.

The Gelkhamar dome (Villari, 1968, 1974), on the northwestern part of the island, rests on the green ignimbrite. In places, this dome is covered by the younger basaltic lava flows of Contrada Mursia. Radiometric age determinations on separated phases and groundmass (Table 1) give concordant values. An age of 22,000 y.b.p. seems to be the most likely for this dome.

Two periods of volcanic activity produced pantelleritic lavas. Products of the older period are here called the lower pantelleritic lava flows. The most relevant features belonging to this volcanic activity are Cuddia Bonsulton, Cuddia Schiuvechi, Cuddia Bellizzi, Cuddia Patite, and Cuddia Mueggen, which are composed of lava flows erupted from central vents. At some of these vents the activity started with an explosive phase which formed pumice-fall deposits (Cuddia Patite). All products of this volcanic activity overlie the green ignimbrite and the trachytes of Montagna Grande volcanic complex. Cuddia di Bellizzi and Cuddia Sciuvechi are partly overlain by rocks of the upper pantelleritic lava flows.

The products of explosive volcanism, represented by Cuddia di Mida and Cuddia Randazzo pumice cones, are here called Cuddia di Mida tephra. Because the thickest pumice-fall deposits occur on the northwestern side of Montagna Grande, Cuddia di Mida probably produced the greatest eruptions. Pumice-fall deposits, locally only a few tens of centimeters thick, cover the lower pantelleritic lava flows (Cuddia Mueggen).

The two endogenous domes of Mt. Gelfiser and Khaggiar are here included as pantelleritic domes and lava flows. Their pantelleritic character is inferred on the basis of unpublished chemical data (Civetta et al., in prep.). These domes have been the vents for abundant lava flows in the latest stages of their growth. Khaggiar dome developed beneath the Cuddia Randazzo pumice cone. Pumice-fall deposits cover the flanks of the dome but none lie on the later lava flows. A good exposure at Contrada Caffefi shows the Mt. Gelfiser dome resting on the Cuddia di Mida tephra. The uppermost part of the pumice-fall deposits has been reddened by the Monte Gelfiser dome. Northwest of Cuddia di Mida a lava flow belonging to the upper pantelleritic lava unit lies on the Mt. Gelfiser dome.

Numerous small pantelleritic lava flows and domes are here grouped under the name of upper pantelleritic lava flows. They generally overlie the Cuddia di Mida tephra. A flow belonging to this formation rests on the Monte Gelfiser endogenous dome northwest of Contrada Siba Montagna. Three domes (two are Mts. Gibile and Fossa del Russo) which produced small lava flows in their final stages of growth occur on the southwestern side of Montagna Grande. The lava of the Fossa del Russo dome lies above rocks of the lower pantelleritic lava formation north of Cuddia di Bellizzi. The flows belonging to this formation are easily recognized because of their well preserved surficial features.

Basaltic lava flows and cinder cones crop out in the northwestern part of the island. Only those younger than the green ignimbrite are discussed here. Two phases of basic activity are distinguished in the proposed stratigraphic succession. The old flows dated at around 29,000 y.b.p. (Punta Sen Leonardo basaltic lava flows and cinder cones; Table 1) erupted in the northernmost part of the island. The young flows erupted in the Mursia area overlie the Mt. Gelkhamar volcanics (21,000 y.b.p.) and exhibit a strong ^{226}Ra excess (half life of 1,602 y) with respect to an equilibrium concentration of ^{238}U (L. Civetta, personal communication). More detailed stratigraphic, geochemical, and geochronological studies of these rocks by the authors are now in progress.

K/Ar DATING: ANALYTICAL TECHNIQUES AND RESULTS

The analytical techniques used by C.F.R. Gif sur Yvette for K/Ar dating of young volcanic rocks and minerals have been presented elsewhere (Cassignol et al., 1978; Gillot, 1978; Cassignol and Gillot, 1982). The range of application has been demonstrated by Gillot et al. (1979), Gillot and Nativel (1982) and Cassignol and Gillot (1982).

The present measurements have been done on mineral phases (feldspar), glass, or groundmass separated from the 250 to 500 µm fraction of the crushed rocks. These concentrations were made by either high-density liquids or magnetic separation. Potassium content was measured by atomic absorption spectrometry with a typical accuracy of 1.5%.

Sample weights of 2-8 g were used for the argon extraction. Pre-degassing was done at room temperature in order to avoid the fractionation effect as much as possible. The extraction line used an atmospheric argon blank of about 10^{11} atoms. Line blanks (including heating of the crucible at 1,500°C for 20 minutes and a purification step at 850°C for 1-3 hours) typically range between 0.5 and 1.5 x 10^{13} atoms. Mass analyses were made with a 180°, 6 cm radius mass spectrometer operated in the static mode at an accelerating potential of about 500 V.

The major uncertainty for such young age measurements is due to the predominance of atmospheric argon in the samples. The departure of ^{40}Ar/^{36}Ar ratio from the atmospheric value must be measured with maximum accuracy. The minimum departure value was determined to be 0.2% (2σ) for repeated analyses of atmospheric Ar and zero age samples (i.e., at 100% uncertainty for a sample with a level of 0.2% of radiogenic ^{40}Ar). The observed dispersion in the age results obtained for replicated measurements of the same samples always falls within the predicted error. To obtain such an accurate comparison, we operate under conditions that avoid any discrimination effects in the mass spectrometric measurements. Successive measurements of unspiked sample argon, atmospheric argon and standard argon are measured under the same conditions and at the same working point of the mass spectrometer (Cassignol and Gillot, 1982).

The agreement between replicated determinations appears to be good and is always better than the cited error which is calculated from the uncertainty in the correction for atmospheric contamination, the major uncertainty for samples with less than 10% of radiogenic argon. The location of the analyzed samples and the results are reported in Table 1. All of the obtained ages are in agreement with the stratigraphic succession and with either the available fission track (Bigazzi et al., 1971; Bigazzi et al., 1982) or interpolated stratigraphic ages (Keller et al., 1978; Keller, 1981). Fission track ages of 0.138± 0.08 and 0.072± 0.009 m.y. respectively correspond to one of the stratigraphically lowest formations of the island (exposed on the southern coastal cliff) and to a formation that crops out in the Lago caldera wall near Bagno dell'Acqua. Moreover, a good agreement exists between K/Ar (50,000 y) and interpolated stratigraphic ages (45,000 y) of the green ignimbrite.

VOLCANIC HISTORY

New stratigraphic and chronological data, together with previously published information, permit the construction of a geological cross section through the island of Pantelleria (Fig.3). The recent volcanic history considered in this paper starts with the eruption of the green ignimbrite (about 50,000 y.b.p.).

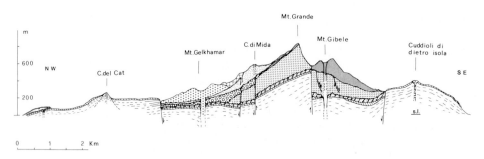

Fig. 3 Geologic cross section through the island of Pantelleria. See Fig. 1 for location and explanation of symbols.

This eruption was followed by the collapse of the Monastero caldera (Orsi, in prep.). After these events, the huge Montagna Grande volcanic complex erupted from the center of the Monastero caldera (35,000 y.b.p.). This feature could be interpreted as a resurgent dome (Fig. 1). The development of the Montagna Grande volcanic complex went through at least three stages. Volcanic activity started with the eruption of the Serra di Ghirlanda tephra and continued with the effusion of the great Montagna Grande dome. In the later stages of the formation of this dome a summit collapse took place with the subsequent development of the Monte Gibele volcanic cone.

Soon after the eruption of the Montagna Grande volcanic complex, fissure basalts formed lava flows and cinder cones (29,000 y.b.p.) in the northwestern part of the island (Punta San Leonardo basaltic lava flows and cinder cones). These basalts were erupted along NW-SE trending fractures which align with the Pantelleria rift direction. Following this fissure basaltic activity, the Gelkhamar dome was erupted about 22,000 y.b.p.

After a period of quiesence, volcanic activity started again with the extrusion of the lower pantelleritic lava flows (16,000 y.b.p.). This activity also seems to be linked to the volcano-tectonic pattern of the Monastero caldera collapse. Most of the centers of this phase are located along the rim of the Monastero caldera. Very few lie outside this caldera (Figs. 1 and 2).

After another period of quiescence, volcanic explosions formed the Cuddia di Mida tephra (9,000 y.b.p.). The main vents for this activity were Cuddia di Mida and Cuddia Randazzo. The emplacement of these pantelleritic domes and lava flows was followed by the upper pantelleritic lava flows. Pumice-fall deposits, lava flows, and domes occur along three sides of Montagna Grande. The geometric relationships among Montagna Grande, Monte Gibele, and the Monastero caldera rim, combined with the distribution of the vents of the Cuddia di Mida tephra and upper Pantelleritic lava flows (Fig. 2), suggest that after the formation of the Monte Gibele cone, the western part of Montagna Grande was uplifted about 300 m on its southeastern and southern sides and much less on its northeastern and northern flanks. The southeastern side is clearly defined by a NE-SW trending fault. The absence of volcanic vents along this fault line suggests that it is oriented perpendicular to the direction of maximum compression during uplift. The cause of the uplift must have been the intrusion of magma which refilled the feeding system of Montagna Grande volcanic complex and then produced the Cuddia di Mida tephra and the upper pantelleritic lava flows, because the uplift ended after the eruption of this magma. The crater rim of the Cuddia di Mida tephra cone is perfectly horizontal, also suggesting its formation after uplift. The uplift probably lasted about 20,000 years with an average rate between 1.0 and 1.5 cm/y.

Mt. Gelfiser and Khaggiar domes are younger than the Cuddia di Mida tephra. It is difficult to correlate this activity with any regional tectonic trend. A tentative hypothesis is that these domes are linked to the hidden northern rim of the Monastero caldera.

Very probably the most recent volcanic activity on Pantelleria is represented by the Mursia basaltic lava flows and cinder cones. At least two historical submarine eruptions are recorded. In 1831, the small and ephemeral Graham Island was formed 50 km northeast of Pantelleria (Washington, 1909). In 1891, an eruption took place about 7 km northwest of Pantelleria (Ricco, 1892; Washington, 1909).

In conclusion, the above information proves that a change occurred in the eruptive style on the island following the eruption of the green ignimbrite. Numerous eruptions of pyroclastic material took place between 140,000 and 50,000 y.b.p. as represented by the welded tuffs of this age (Villari, 1974; Wright, 1980). The dominance of lavas and domes in the last 50,000 y could be due to a change in the structural setting of the island induced by the Monastero caldera collapse.

ACKNOWLEDGEMENTS

This work was supported by the National Council of Research (C.N.R.S.) of Italy and by the National Council of Scientific Research of France; Programme interdisciplinaire des researches pour la prevision et la surveillance des eruptions volcaniques. The authors thank Professor M.F. Sheridan for helpful discussions in the field and for critical reading of the manuscript.

REFERENCES

Agocs, W. B., 1959. Profondita e struttura dell'orizzonte igneo fra Catania e Tunisi dedotta da un profilo aeromagnetico. Boll. Serv. Geol. It., 80, 1: 51-61.
Bigazzi, G., Bonadonna, F. P., Belluomini, G., and Malpieri, L., 1971. Studi sulle ossidiane italiane. IV. Datazione con il metodo delle tracce di fissione. Boll. Soc. Geol. It., 90: 469-480.
Cassignol, C., Cornette, Y., David, B., and Gillot, P. Y., 1978. Technologie potassium-argon. Rapport CEA-R-4908, CEN, Saclay.
Cassignol, C. and Gillot, P. Y., 1982. Range and effectiveness of unspiked potassium-argon dating. In: Odin (Editor), Numerical dating in stratigraphy. Wiley, New York, N.Y.
Cassinis, R., Franciosi, R., Scarascia, S., 1979. The structure of the Earth's crust in Italy. A preliminary typology based on seismic data. Boll. Geof. Teor. Appl., XXI, 81: 105-125.
Civetta, L., Crisci, G. M., Orsi, G., and Serri, G., 1982. Le vulcaniti basiche delle isole di Linosa, Pantelleria (Canale di Sicilia) e di Ustica: caratteristiche geochimiche delle loro regioni sorgenti. Rend. Soc. It. Miner. Petr. (Abstracts).
Civetta, L., Cornette, Y., Crisci, G.M., Gillot, P.Y., and Orsi, G., in preparation. Temporal compositional variation of the volcanic rocks of Pantelleria.
Colantuoni, P. and Zarudzki, E. F. K., 1973. Some principal sea floor features in the Strait of Sicily. Rapp. Comm. int. Mer. Medit., 22, 2a: 68-70, Monaco.

Colombi, B., Giese, P., Luongo, G., Morelli, C., Riuscetti, M., Scarascia, S., Schutte, K. G., Strowald, J. and de Visintini, G., 1973. Preliminary report on the seismic refraction profile Gargano-Salerno-Palermo-Pantelleria (1971). Boll. Geof. Teor. Appl., XV, 59: 225-254.

Gantar, C., Morelli, C., Segre, A. G., Zampieri, L., 1961. Studio gravimetrico e considerazioni geologiche sull'isola di Pantelleria. Boll. Geof. Teor. Appl., III, 12: 267-287.

Gillot, P. Y. 1978. K-Ar dating of the Laschamp magnetic excursion: a method for young volcanic rocks dating. Geol. Surv. Open File Rep., 78, 701: 140-142.

Gillot, P. Y., Labeyrie, J., Laje, C., Valladas, G., Guerin, G., Poupeau, G. and Delibrias, G., 1979. Age of the Laschamp magnetic excursion revisited. Earth Planet. Sci. Lett., 42: 444-450.

Gillot, P. Y. and Nativel, P., 1982. K-Ar chronology of the ultimate activity of Piton des Neiges volcano, Reunion Island, Indian Ocean. J. Volcanol. Geotherm. Res., 13: 131-146.

Keller, J., Ryan, W. B. F., Ninkovich, D., Altherr, R., 1978. Explosive volcanic activity in the Mediterranean over the past 200,000 years as recorded in deep-sea sediments. Geol. Soc. Am. Bull., 89: 591-604.

Korringa, M.K., 1971. Steeply-dipping welded tuff mantling the walls of the Pantelleria caldera. Conference on peralkaline acid volcanism, Catania, Abstracts.

Mahood, G. and Hildreth, W., 1980. Pantelleria, a new interpretation. EOS Trans. Am. Geophys. Union, 61: 1141.

Orsi, G.,. On the existence of two calderas on the island of Pantelleria. Submitted to Boll. Geol. Soc. It.

Ricco, A., 1892. Terremoto sollevamento ed eruzione sottomarina a Pantelleria. Ann. Uff. Cent. Meteorol. Geodin., II, III, XI: 5-23.

Rittmann, A., 1967. Studio geovulcanologico e magmatologico dell'isola di Pantelleria. Riv. Miner. Sicil., 106-108: 147-204.

Villari, L., 1967. Isola di Pantelleria. Carta Geologica. In: A. Rittmann (Editor), Studio geovulcanologico e magmatologico dell'isola di Pantelleria. Riv. Miner. Sicil., 106-108: 147-204.

Villari, L., 1968. On the geovolcanological and morphological evolution of an endogenous dome (Pantelleria, Mt. Gelkhamar). Geol. Rundsch., 57(3): 784-795.

Villari, L., 1970. Studio Petrologico di alcuni campioni dei pozzi Bagno dell'Acqua e Gadir (Isola di Pantelleria). Rend. Soc. It. Miner. Petrol., XXVI, 1: 353-376.

Villari, L., 1974. The island of Pantelleria. Bull. Volcanol., 33(3): 1-24.

Washington, H. S., 1909. The submarine eruptions of 1831 and 1891 near Pantelleria. Am. J. Sci., 27: 131-150.

Wolff, J. A. and Wright, J. V., 1980. Rheomorphism of welded tuffs. J. Volcanol. Geotherm. Res., 10: 13-34.

Wright, J. V., 1980. Stratigraphy and geology of the welded air-fall tuffs of Pantelleria, Italy. Geol. Rundsch. 69,1: 263-291.

ORIGIN AND EMPLACEMENT OF A PYROCLASTIC FLOW AND SURGE UNIT AT LAACHER SEE, GERMANY

RICHARD V. FISHER[1], HANS-ULRICH SCHMINCKE[2] and PAUL VAN BOGAARD[2]

[1] Department of Geological Sciences, University of California, Santa Barbara, CA 93016 (U.S.A.);
[2] Ruhr Universität Bochum Institut für Mineralogie, (W. Germany)

(Received June 26, 1982; revised and accepted December 20, 1982)

ABSTRACT

Fisher, R.V., Schmincke, H.-U., and Van Bogaard, P., 1983. Origin and emplacement of a pyroclastic flow and surge unit at Laacher See, Germany. In: M.F. Sheridan and F. Barberi (Editors), Explosive Volcanism. J. Volcanol. Geotherm. Res., 17: 375-392.

Within the pyroclastic ejecta sequence of Laacher See volcano, West Germany, are many repeated phreatomagmatic depositional cycles. A complete cycle begins with a thin, low aspect ratio, fines-depleted breccia layer, followed by a sequence with several coarse- to fine-grained, cross-bedded units, in turn overlain by a massive pyroclastic flow layer. Each sequence is interpreted in terms of water/magma interaction and effects upon the eruption column. The breccia represents the triggering eruption with a high eruption column where fines were winnowed out and fallback from high altitudes produced a low aspect ratio layer. The cross-bedded sequence containing more fine-grained ash resulting from an increase in amount of water, represents a fluctuating eruption column of sufficient height to produce large antidunes within 700 meters of the rim. The massive pyroclastic flow layer which contains more fine-grained ash than the underlying layers, represents the culmination of the eruption where an increase in amount of water interacting with the magma resulted in a dense and very low eruption column producing a dense, laminar moving pyroclastic flow.

STRATIGRAPHY AND SEDIMENTOLOGY

Introduction

Laacher See, located in the Eifel District of West Germany (Fig. 1) is a large maar volcano, about 2.5 km in diameter, which has had a complex history (Schmincke, 1977). The volcanic basin is encircled by basanitic and tephritic scoria cones (270,00 y.b.p., Schmincke and Mertes, 1979). Laacher See ejecta (~11,000 y.b.p.) have filled valleys between the cones and have largely covered most of them. Laacher See tephra is found as far as coastal Sweden and Northern Italy (Bogaard, in prep.). The ejecta consist of (1) juvenile and cognate, pumiceous to poorly vesiculated phonolitic ash, blocks, and bombs, (2) accidental older volcanics, and (3) slate and graywacke of Devonian age which form the regional basement.

The Laacher See tephra sequence has been divided into three major units (Schmincke, 1974) called lower (LLST) middle (MLST) and upper (ULST) Laacher See tephra. The sequence is chemically and mineralogically zoned from highly evolved, nearly crystal-free phonolite at the base to strongly phyric crystal-rich phonolite on top (Wörner and Schmincke, 1983). The lower two units are dominated by Plinian fallout and flow deposits (Freundt, 1982; Bogaard, in prep.) while ULST is interpreted to have been produced by phreatomagmatic mechanisms (Schmincke, 1970; Schmincke et al., 1973; Bogaard, 1978). Several different aspects of these deposits will be published elsewhere. Here, we discuss eruptive and transport mechanisms of some layers of ULST. ULST consists largely of unconsolidated deposits of pyroclastic flows, surges and some fallout deposits with fallout predominating at distances exceeding 3 km. The ULST deposits locally exceed 30 m in thickness near the rim but decrease to less than 1 m 15 km from the rim.

The flow and surge unit discussed here is but one out of many within this upper sequence and occurs on the Mendig Fan (Fig. 1), one of 5 or 6 pyroclastic fans radiating outward mostly from low parts of the rim of the volcano.

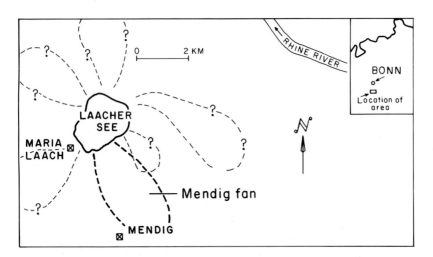

Fig. 1 Locality map. Laacher See, Eifel District, West Germany. Map shows pyroclastic fans. Units described herein are from the Mendig fan.

Most layers within the upper Laacher See tephra consist of three kinds of depositional units: breccia (B-layers), cross-bedded layers (D-layers) and massive flow deposits (M-layers) (Schmincke et al., 1973; Bogaard, 1978). These different kinds of depositional units occur repeatedly throughout the sequence. They differ in sedimentary textures, structures, volumes and areal distribution but there are several transitions between beds, especially between B and D, and D and M. Sequences of B-, D- and M-layers in that order seem to recur throughout the sequence although this type of cycle is not unique to the entire sequence (Fig. 2). Complete B-D-M cycles occur in several places in the

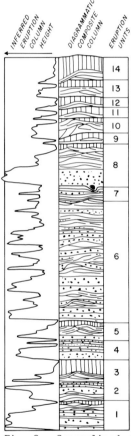

Fig. 2 Generalized column, upper Laacher See tephra sequence, from the Mendig fan (see Fig. 1), Laacher See. Solid dots = breccia beds (B-layers); cross and wavy lines = duned layers (D-layers); vertical lines = massive layers (M-layers). Inferred relative eruption column heights at left; highest for B-layers, lowest for M-layers. Simplified from Schmincke et al., (1973) and Bogaard (1978).

sequence, but there are also B-D, B-M or D-M sequences. We interpret these cyclic sequences in terms of eruption column dynamics, a topic which is discussed following the description of a single B-D-M sequence (B5, D2, M14, see Schmincke et al., 1973) which we infer to be a single eruption unit. However, rather than use the letters B, D, and M for the studied depositional sequence, we refer to them as eruption subunits I, II, and III, respectively, corresponding to stages of eruption as discussed later.

First in sequence is subunit I (Fig. 3) which generally is a single bed of gray breccia with a sandy matrix (lithic-rich, coarse-grained ash). Its base has fewer large fragments than its central and upper parts and therefore is considered to be inversely graded. Second in sequence is subunit II which contains numerous coarse- to fine-grained cross-bedded layers. At several proximal localities, subunit II is composed of a finer-grained lower part and a coarser-grained upper part, both of which are duned and cross-bedded. Breccia beds occur within subunit II but these differ from the breccia bed of subunit I by being yellowish in color owing to the greater abundance of yellowish silt. Where subunit II thins out to a few centimeters thick, it consists of a yellowish, sandy breccia with crude plane parallel bedding. Third in the sequence is subunit III which is a generally massive, yellowish bed with blocks and lapilli supported in a matrix of coarse- to fine-grained ash. In places, crude bedding occurs where there are trains of large fragments.

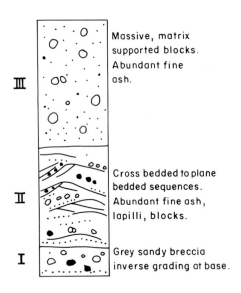

Fig. 3 Generalized eruption unit. Filled areas = lithics; open circles = pumiceous fragments; dots = sand-size ash; cross and wavy lines = dunes and cross-beds. Roman numerals refer to eruption subunits discussed in text.

Distribution, thickness and volume

The reference line for distance measurements, hence area and volume estimates, is the present-day crestal divide between the Laacher See crater and its outer slopes. The total estimated volume of the studied eruption unit is 0.015 km^3 inside the 10 cm isopach (Fig. 4) (uncorrected for equivalent solid-rock density), although to determine this estimate, isopach lines are extrapolated to the crest from the most proximal exposures. This estimated

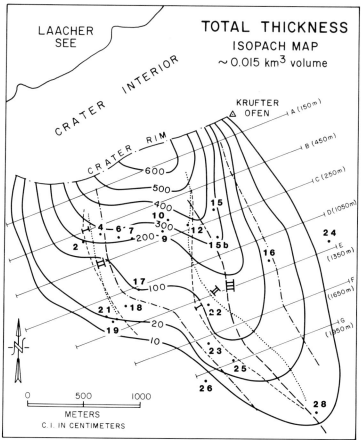

Fig. 4 Generalized total thickness isopach map of studied eruption unit constructed from isopachs of individual subunits. Short dashes, dots, and long dashes are thickness axes for designated subunits I, II and III (in Figs. 5, 6, and 7). Dash-dot lines are shoulders of "valley" sides as shown in Fig. 9. Numbered dots are locations of measured sections. Lines A-G are location of sections shown in Fig. 8: distances from rim (in meters) are shown in parentheses.

volume represents only the amount of material which escaped by flow from the crater onto the Mendig Fan. It does not include material which may have (1) fallen back into the crater and remained there, (2) escaped during the same eruptive phase in other directions onto other pyroclastic fans or (3) the amount of downwind fallout. Of the total volume, subunit I = 11%, subunit II = 38%, and subunit III = 51%.

Subunit I, ranging from 25 cm to 5 cm thick, is spread out in a sheet-like form at the base of the eruption unit, but deposition was in part controlled by topography as shown by Fig. 5. It has axial patterns consistent with the overall axis of the entire unit (Fig. 4), but contributes little to the total thickness pattern of the unit except in distal areas.

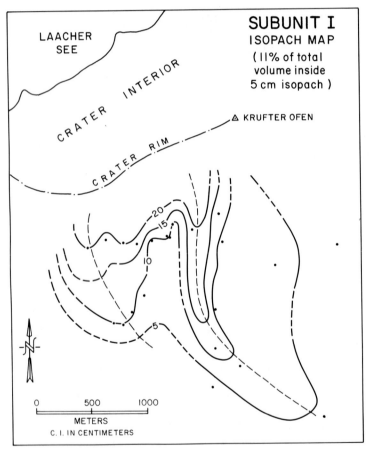

Fig. 5   Isopach map of subunit I (breccia layer). Thickness axes are thin dashed lines, which are also shown in Fig. 4.

Subunit II is a wedge-shaped body ranging from 250 cm to less than 5 cm thick. Its thickness pattern (Fig. 6) has weaker axial patterns than either subunit I or III suggesting less topographic control upon deposition. Subunit II appears to form a significant part of the volcano's rim on the Mendig Fan.

Subunit III, ranging in thickness from more than 300 cm to 5 cm, shows a strong axial trend (Fig. 7) indicating that deposition was strongly controlled by topography. In detail, subunit III fills minor irregularities between dune crests at the top of subunit II.

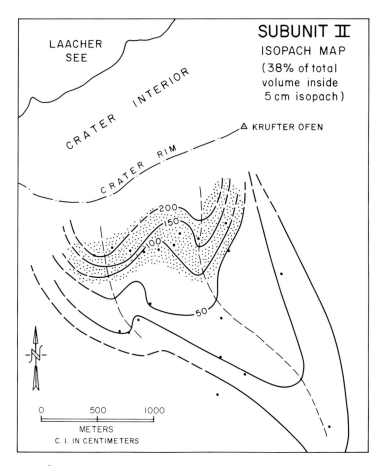

Fig. 6 Isopach map of subunit II (duned layer). Stipples show area of large antidunes. Thin dashed lines are axes of thickness maxima. Outer two isopach lines are 10 cm and 5 cm.

Several stratigraphic cross sections (Fig. 8) constructed from the isopach maps strongly suggest that the eruption unit filled a gentle swale or valley on the side of the volcano. Such a valley is probably exaggerated, however, because it is assumed that the tops of subunits II and III are flat and horizontal, whereas they could be bowed upward, especially subunit II. The thickness of subunit II is skewed to the northwest suggesting that it followed and partly filled the west part of the original valley near the source. Subunit III later filled the more restricted remaining part of the valley.

Using the thicknesses from the stratigraphic cross section which were averaged across the fan at regular intervals (Fig. 8), it was possible to construct graphs showing rate of average thickness decrease down the fan (Fig. 9), and cumulative percent of thickness versus distance (Fig. 10). Both

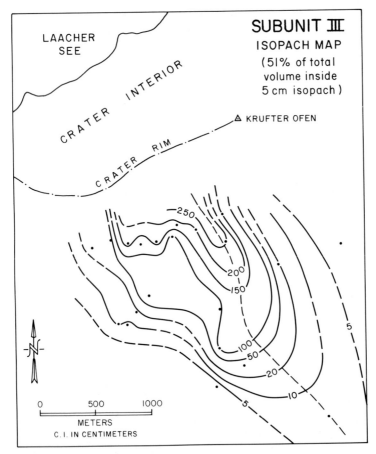

Fig. 7 Isopach map of subunit III (massive layer). Dashed line shows axis of thickness maxima.

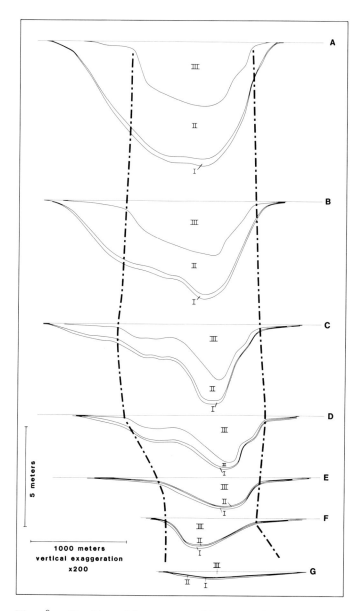

Fig. 8 Stratigraphic cross sections constructed from isopach maps projected to crater rim. See text for explanation. Dash-dot line connects shoulder of "valley" (also shown in Fig. 4).

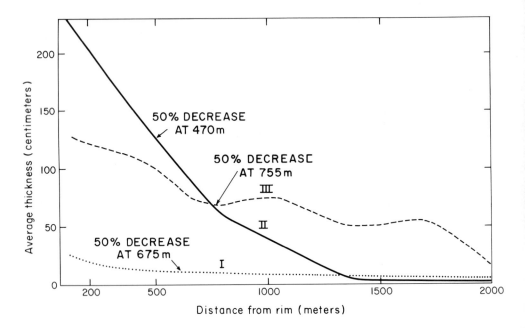

Fig. 9 Average thickness decrease of subunits I, II and III down the Mendig Fan. Average thickness is calculated across the fan at each stratigraphic cross section (A-G) shown in Figs. 4 and 8. Values for 50% decrease in thickness are taken from cumulative percent graph (Fig. 10).

figures show that the rate of overall thickness decrease for subunit II is faster than for subunits I and III. Subunit II begins with the greatest average thickness, but at nearly 1500 m from the rim, it becomes thinner than subunit I (Fig. 9). Fig. 10 also shows that 50 percent of the cumulative thickness of all the subunits occurs at less than about 800 m from the rim. These differences are believed to result from different flow mechanisms as discussed later.

Bedforms related to distribution and thickness

The inversely graded breccia bed (subunit I) generally consists of a single layer, but in some proximal locations it contains one or two poorly defined 1-2 cm thick sandy layers within its central part, and at one locality (Loc. 2, Fig. 4), the breccia is underlain by up to 30 cm of steel-gray, essentially silt-free sand-size ash which shows crude wave-like and subtly cross-bedded internal layering grading up into the breccia at the top.

In subunit II there are large dune structures in the region where greater than 50% of its volume is deposited and where its thickness exceeds 50 cm (Fig. 6). These are interpreted as chute and pool structures (Schmincke et al.,

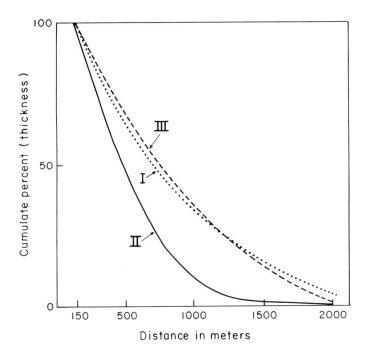

Fig. 10 Cumulative percent of average thickness down the Mendig fan for designated subunits.

1973). There appear to be at least two major duned sequences within subunit II, the lower of which is finer-grained, but each consists of coarse- to fine-grained beds.

Subunit III is mostly massive with matrix supported lapilli and blocks in areas of thickest expression, but contains fragment trains in places which gives it a crude bedding. In proximal areas where it is thick, it is composed of two flow units and it has a thin, cross-bedded top which is in sharp contact with the underlying flow. This is interpreted to be an ash-cloud surge deposit similar to ash-cloud surges described elsewhere (Fisher, 1979). Where subunit III is thin, as at locality 2 (Fig. 4), the entire unit is cross-bedded and is interpreted to be an ash-cloud surge deposit which sheared off the massive layer and flowed laterally over higher ground as the massive flow moved downslope. At distal localities, a 2-3 cm thick accretionary lapilli ash bed overlies the massive tuff (Loc. 23, Fig. 4), but farther down the depositional fan out of the area of Fig. 4, the massive layer is represented only by this thin accretionary lapilli layer.

DISCUSSION

Granulometry

Median grain size (Md_ϕ), Inman (1952) sorting coefficients (σ_ϕ) and percentage of silt were determined at several localities for the three subunits (Table 1). Median diameter and sorting coefficient plots show that the median grain size decreases in the range from -0.5ϕ to 2.5ϕ from the basal breccia layer (subunit I) to the upper massive layer (subunit III). The cross-bedded layer (subunit II), consisting of base surge units (Schmincke et al., 1973) overlaps both the other subunits. Except for three samples from the base of the massive layer, all samples range in σ_ϕ between 1.2 and 1.9. Greater discrimination between subunits, however, is given by the $<1/6$ mm ($>4\phi$) versus Md_ϕ plot (Fig. 11). Placing emphasis upon the fine-grained component of the

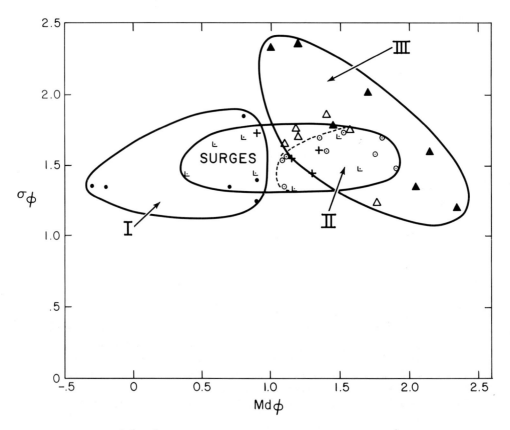

Fig. 11 Md_ϕ-$<1/16$ mm% plot showing fields for subunits I, II (in dashed line), III and various samples from surge deposits.

TABLE 1
Size Parameters, Laacher See Eruption Unit

Subunit	Locality*	Md$_\phi$	σ_ϕ	% Silt	Comments
I	6	-0.20	1.35	2.0	Sandy base
	9	0.80	1.85	2.7	" "
	15	0.70	1.35	2.3	" "
	16	-0.30	1.35	2.0	" "
	19	0.90	1.25	2.1	" "
	23	0.90	1.40	2.4	" "
II	4	1.35	1.69	3.4	Silty layer
	4	1.10	1.35	3.1	Sandy layer
	6	1.75	1.58	3.4	Base, massive layer
	6	1.08	1.53	4.9	Top, " "
	6	1.52	1.73	7.3	Silty layer, base
	6	1.80	1.69	4.7	" " , top
	6	1.90	1.48	2.8	Silty breccia
	6	1.40	1.60	5.4	Sandy, laminated
	15	1.10	1.56	7.1	Silty, sandy layer
III	4	1.76	1.24	3.4	Top)
	4	2.35	1.20	6.5	Base) Massive layers
	6	1.56	1.70	5.1	Top)
	6	2.05	1.35	5.8	Base) " "
	9	1.20	1.70	10.4	Top)
	9	1.45	1.78	10.8	Base) " "
	15	1.18	1.75	8.0	Top)
	15	1.70	2.05	17.9	Base) " "
	16	1.00	2.33	12.4	Base) " "
	19	1.10	1.65	8.4	Top)
	19	1.20	2.35	6.5	Base) " "
	23	1.40	1.85	17.4	Top)
	23	2.15	1.60	23.8	Base) " "
Miscellaneous Surges	4	1.15	1.55	4.8	Surge beneath Subunit III
	6	0.90	1.73	4.2	" " " "
	7	1.30	1.45	4.1	" " " "
	9	1.35	1.60	10.0	" " " "
	4	1.65	1.48	5.0	Surge above Subunit III, base
	6	1.70	1.50	4.6	" " " " , top
	6	1.18	1.33	3.4	" " " " , base
	9	0.60	1.65	5.3	" " " " , top
	15	0.40	1.45	6.8	" " " " , base
	19	0.90	1.45	6.5	" " " " , top

*Localities shown on Figure 4.

deposits has interpretive significance with regard to comminution processes, eruption column height and density, and on transport processes as discussed below. At each locality, the percentage of silt is least in subunit I and greatest in subunit III. Averages for silt of the subunits are I = 2.3%, II = 4.7%, and III = 10.4%.

The height, volume, sediment-size distribution, and particle concentration (hence density of the column) are determined by magmatic volatiles, the presence or absence of interaction of magma with phreatic water, changing radius of vents, and other factors (Wilson, 1976; Sparks and Wilson, 1976; Sparks et al., 1978). There is evidence that many of the eruptions at Laacher see were phreatomagmatic (Schmincke et al., 1973) as reflected in the deposits by bedding sags and accretionary lapilli layers along with development of cross-beds and dunes which are characteristic of base-surge deposits (Fisher and Waters, 1969, 1970; Waters and Fisher, 1971; Crowe and Fisher, 1973; Sheridan and Updike, 1975; Wohletz and Sheridan, 1979). The greater the volume of water in the erupted mixture, the denser and shorter the eruption column (Sparks et al., 1978), and the greater the amount of fine-grained ash from more efficient comminution of the magma (Self and Sparks, 1978). From experiments with water and thermite explosions, Wohletz (1980; this volume) has shown that the conversion from thermal energy to mechanical energy is a function of the water/melt ratio. Fragmentation of the melt reaches a maximum when the water/melt ratio is about 0.5 for basaltic eruptions and could be higher in silicic magmas with higher initial partial pressures. Thus, initial ("vent-coring") eruptions with small water/melt ratios could result in the formation of coarse-grained tuff and breccia. Increasing ratios result in more abundant fines and wetter and more dense eruption columns giving rise to base surges with consequent development of finer-grained, laminated and cross-bedded deposits. Still wetter eruptions and greater amounts of fines produce heavier eruption columns, collapse, and consequent mass flows (Fig. 12).

The breccia subunit I is interpreted as representing the initial or "triggering" stage I of eruption (Fig. 13). Paucity of silt can result from either one of three causes or a combination of each: (1) Little fine-grained material was produced during the eruption because comminution of the magma was "inefficient" owing to a low water/melt ratio or else explosive energy was supplied only by expansion of magmatic volatiles. (2) The eruption had sufficient energy to produce a high eruption column from which fines were elutriated out. (3) Fines were elutriated out during flow. Emplacement of the breccia by flow is suggested by presence of inverse grading at its base and the absence of bedding sags beneath large blocks which in places rest upon fine-grained ash. The 30 cm thick wavy laminated coarse-grained ash bed beneath and gradational with the unit in proximal locations (Loc. 2, Fig. 4) is interpreted as a ground surge which is not laterally continuous across or down the fan. The low thickness to area ratio (low aspect ratio) of the breccia suggests that it was emplaced with high kinetic energy which further suggests that it was derived from the collapse of a high column. The breccia layer clearly is not a ground layer produced from the advancing head of a pyroclastic flow by ingestion of air from subunit II (e.g. Wilson, 1980; Walker et al., 1981). This is shown by the fact that the breccia layer is reduced in thickness much more slowly and becomes thicker than the overlying subunit II at distal locations (see Fig. 9).

Subunit II is interpreted to represent the second stage of the eruption (Fig. 13). As the volume (related to duration) of the eruption increased, more fines were produced. At least two episodes of dune formation contain

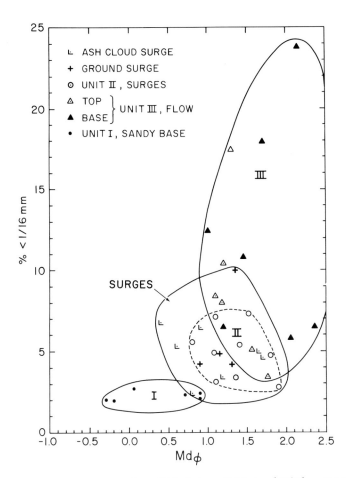

Fig. 12 Adapted and modified from Wohletz (1980). Wohletz showed explosive energy versus water/melt ratio for hydroclastic eruptions. This figure qualitatively relates column height deduced from aspect ratios of layers, percent silt, and flow mechanics determined by particle concentration as explained in text. Percentages are volumes of total eruption unit. Digrammatic eruption column height is also shown in Fig. 2.

alternating coarse- to fine-grained beds. Rapid deposition occurred near the crater rim where large-scale dunes were formed. The interpretation of this sequence is that water gained access to the vent closely following, or continuous with state I of the eruption giving rise to the low aspect ratio breccia. The higher water/magma ratio caused more efficient comminution of the magma, faster cooling, and therefore a more dense and lower eruption column.

Alternating coarse- to fine-grained layers which occur together in the dune structures suggest that the build-up of explosive energy followed a continuous but pulsating pattern. We envision an eruption where the column was maintained by staccato-like energetic blasts with intervening very short pauses so that clasts were continuously being supplied to the eruption column. Abundant fines and increase in water vapor resulted in a heavy eruption column where collapse of its outer margin was relatively continuous while the interior of the column was being fed by repeated eruptions. At localities 2, 4, 6 and 7 (and at other proximal localities shown in Fig. 4) subunit II is composed of a lower sequence of alternating thin, fine-grained tuff layers and grit layers (coarse-sand size tuff to fine lapilli layers) and a coarser upper sequence. This suggests a crescendo-like build-up of activity. The upper and lower sequences of subunit II are both duned. In places they are duned together; in others they form separated (but migrating) dune sequences.

Over fifty percent of the volume of subunit II was deposited within 1 km of the volcano crest in the area where dunes are largest and have the longest wavelengths. The formation of large dunes indicates that kinetic energy was high suggesting that the eruption column was also high, but its relative low aspect ratio suggests a lower eruption column than during Stage I. The formation of dunes is interpreted to be the result of a "hydraulic jump" (see Schmincke et al., 1973) caused by rapid deceleration and creation of great turbulence as the vertically collapsing column struck ground near the rim of the volcano. Formation of dunes and cross-beds is attributed to turbulence in the area of deposition. These structures progressively get smaller down the fan and die out

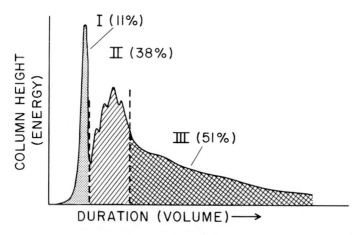

Fig. 13   Summary diagram relating eruption columns to the major kinds of deposits and facies as explained in text.

completely within about 1.5 km of the crest. Distal emplacement of the planar, relatively thin, silty-sandy breccia of subunit II is interpreted to have been accomplished by laminar flow.

Subunit III, emplaced as a fluidized laminar-moving grain flow, is interpreted to represent Stage III (Fig. 13) of the eruption during which extruded products were most voluminous and the greatest amount of fine-grained ash was produced. Abundance of silt suggests an increase in the water/magma ratio over Stage II. This in turn increased the density of the eruption column to such an extent that it merely "boiled-over" low places in the rim rather than forming a high column. The mobility of the flow, its ability to fill low places leaving a flat top, and its massive character is an indication that it was fluidized or partially fluidized, and was able to flow more easily downslope, unlike subunit II which was rapidly deposited near the rim. Gaseous fluidization is also suggested by the presence of a thin cross-bedded unit which lies above it in proximal locations and on channel margins. Stage III ended the eruption, but build-up of pressure for the next "throat-clearing" triggering magmatic eruption (the silt-poor breccia which lies above Subunit III) proceeded rapidly.

ACKNOWLEDGEMENTS

The senior author (R.V.F) sincerely thanks the Institut für Mineralogie, Ruhr-Universität Bochum, West Germany for providing research facilities and the Alexander von Humboldt Foundation for their generous financial support from the U.S. Senior Scientist Award (1980-81).

REFERENCES

Bogaard, P.v.d., 1978. Stratigraphie und Verbreitung der Oberen Laacher Pyroklastika. Dipl.-Arbeit, Ruhr-Universität Bochum, 1-184 pp.
Crowe, B.M., and Fisher, R.V., 1973. Sedimentary structures in base-surge deposits with special reference to cross-bedding, Ubehebe Craters, Death Valley, California. Geol. Soc. Amer. Bull., 84: 663-682.
Fisher, R.V., and Waters, A.C., 1969. Bed forms in base-surge deposits: lunar implications. Science, 165: 1349-1352.
Fisher, R.V., and Waters, A.C., 1970. Base surge bed forms in maar volcanoes. Amer. J. Sci., 268: 157-180.
Fisher, R.V., Smith, A.L. and Roobol, M.J., 1980. Destruction of St. Pierre, Martinique by ash cloud surges, May 8 and 20, 1902. Geology, 8: 472-276.
Freundt, A., 1982. Stratigraphie des Brohltaltrass und seine Entstehung aus pyroklastischen Strömen des Laacher See Vulkans. Dipl.-Arbeit, Ruhr-Universität Bochum, 1-319 pp.
Inman, D. L., 1952. Measures of describing the size distribution of sediments. Jour. Sed. Petrol. 22: 125-145.

Schmincke, H.-U., 1970. "Base surge" - Ablagerungen des Laacher-See-Vulkans. Aufschluss. 21: 359-364.

Schmincke, H.-U., 1974. Laacher-See Tephra: Eruptionszentren, Eruptions-und Transport mechanismen (Abst.). Dt. Geol. Ges., p. 126, Jahrestagung Bonn.

Schmincke, H.-U., 1977. Phreatomagmatische Phasen in Quartären Vulkanen de Osteifel. Geol. Jahb., 39: 3-45.

Schmincke, H.-U. and Mertes, H., 1979. Pliocene and Quaternary volcanic phases in the Eifel volcanic fields. Naturwissenschaften, 66: 614-615.

Schmincke, H.-U., Fisher, R.V., and Waters, A.C., 1973. Antidune and chute and pool structures in the base surge deposits of the Laacher See area, Germany. Sediment., 20: 553-574.

Self, S. and Sparks, R.S.J., 1978. Characteristics of wide spread pyroclastic deposits formed by the interaction of silicic magma and water. Bull. Volcanol., 41: 1-17.

Sheridan, M.F. and Updike, R.G., 1975. Sugarloaf Mountain tephra - A Pleistocene rhyolitic deposit of base-surge origin in northern Arizona. Geol. Soc. Amer. Bull., 86: 571-581.

Sparks, R.S.J. and Wilson, L., 1976. A model for the information of ignimbrite by gravitational column collapse. Jour. Geol. Soc. London, 132: 441-451.

Sparks, R.S.J., Wilson, L., and Hulme, G., 1978. Theoretical modeling of the generation, movement, and emplacement of pyroclastic flows by column collapse. Jour. Geophys. Res., 83: 1727-1739.

Walker, G.P.L., Self, S. and Froggatt, P.C., 1981. The ground layer of the Taupo ignimbrite: a striking example of sedimentation from a pyroclastic flow. J. Volcanol. and Geotherm. Res., 10: 1-11.

Waters, A.C., and Fisher, R.V., 1971. Base surges and their deposits: Capelinhos and Taal Volcanoes. Jour. Geophys. Res., 76: 5596-5614.

Wilson, C.J.N., 1980. The role of fluidization in the emplacement of pyroclastic flows: An experimental approach. J. Volcanol. and Geotherm. Res., 8: 231-249.

Wilson, L., 1976. Explosive volcanic eruptions - III. Plinian eruption columns. Geophys. J. R. Astr. Soc., 45: 543-556.

Wohletz, K.H., 1980. Explosive hydromagmatic volcanism. Ph.D dissertation, Arizona State Univ., 303 pp.

Wohletz, K.H. and Sheridan, M.F., 1979. A model of pyroclastic surge. In: C.E. Chapin and W.E. Elston (Editors), Ash-flow Tuffs. Geol. Soc. Amer. Sp. Paper 180: 177-194.

Wohletz, K.H., 1983. Mechanisms of hydrovolcanic pyroclast formation: grain-size, scanning electron microscopy, and experimental data. In: M.F. Sheridan and F. Barberi (Editors), Explosive Volcanism. J. Volcanol. Geotherm. Res., 17: (this volume).

Wörner, G. and Schmincke, H.-U., 1983. Geochemical evolution of the Laacher See magma chamber. J. Petrol., (in press).

VOLCANISM IN THE EASTERN ALEUTIAN ARC: LATE QUATERNARY AND HOLOCENE CENTERS, TECTONIC SETTING AND PETROLOGY

JUERGEN KIENLE and SAMUEL E. SWANSON

Geophysical Institute, University of Alaska, Fairbanks, Alaska 99701 (U.S.A.)

(Received October 11, 1982; revised and accepted February 15, 1983)

ABSTRACT

Kienle, J. and Swanson, S.E., 1983. Volcanism in the eastern Aleutian arc: Late Quaternary and Holocene centers, tectonic setting, and petrology. In: M.F. Sheridan and F. Barberi (Editors), Explosive Volcanism. J. Volcanol. Geotherm. Res., 17: 393-432.

Calc-alkaline volcanism and oceanic plate subduction are intimately linked in the eastern Aleutian arc. The volcanic arc is segmented: larger caldera-forming volcanic centers tend to be located near segment boundaries. Intrasegment volcanoes form smaller stratocones. Ten of the 22 volcanoes that make up the 540 km long volcanic front in the eastern Aleutian arc have erupted in recorded history and another six show hydothermal activity.

The geometry of the Benioff zone in the eastern Aleutian arc has been defined by earthquake data from a local, high-gain short-period seismograph network. The Benioff zone dips at an angle of about 45° beneath the volcanic arc and reaches a maximum depth of 200 km. Based on the alignment of volcanoes, the eastern Aleutain arc can be subdivided into two main segments, the Cook and Katmai segments. A misorientation of 35° of the two segments reflects a change in strike of the underlying Benioff zone and implies a lateral warping of the subducting plate. The Cook segment volcanoes line up closely on the 100 km isobath of the Benioff zone. The Katmai segment volcanoes lie on a crosscutting trend with respect to the strike of the underlying Benioff zone. Depths to the dipping seismic zone beneath volcanoes of the Katmai segment vary by 25% from 100 to 75 km. In the Katmai segment there is also good geophysical evidence that crustal tectonics plays an important role in localizing volcanism. Narrowly spaced linear groups of volcanoes appear to be positioned over a deep crustal fault that underlies the volcanic front. Transverse arc elements divide the arc into subsegments and localize larger magma reservoirs at shallow levels in the crust.

Intrasegment volcanoes in both the Cook and Katmai segments erupt andesite and minor dacite of remarkably uniform composition despite differences in depths to the Benioff zone. Segment boundary volcanoes erupt lavas with a wider range of compositions (basalt to rhyolite) but are still calc-alkaline, in contrast to volcanoes in similar tectonic settings near segment boundaries in the central Aleutains. Greater crustal thickness in the eastern Aleutian arc, coupled with structural traps in the crust, allow magma ponding at shallow crustal levels. Differentiation at shallow depths yields dacite and even rhyolite.

INTRODUCTION

The Aleutian arc stretches 2,600 km westward from Hayes Volcano (152.5° W) (Miller and Smith, 1976) to a small island volcano, Buldir, in the western Aleutians (176° E). Fig. 1 shows the location of the Quaternary volcanoes that make up the Aleutian arc (Coats, 1962; Smith and Shaw, 1975; Simkin et al., 1981). Forty of these volcanoes have erupted in recorded history beginning in 1741. Many Aleutian volcanoes have not been studied and in fact new Quaternary volcanic centers are still being discovered. Because of the remoteness of many of the volcanoes, eruptions may go unnoticed.

VOLCANOES OF THE EASTERN ALEUTIAN ARC

The eastern Aleutian arc, as arbitrarily defined here, includes principally the volcanoes of Katmai and Cook Inlet from Ugashik Caldera (names of volcanic centers are unofficial and are given in Tables 1 and 2) to Hayes Volcano (Fig. 2). This section is 540 km long and contains 22 Late Quaternary and Holocene volcanoes. Young volcanoes are still being identified in this heavily glaciated region (Miller and Smith, 1976). Ugashik Caldera has also only recently been recognized as a separate volcanic center (T. P. Miller, personal communication). Kienle et al. (1982) completed a reconnaissance field study in Katmai National Park and Cook Inlet. They identified Late Quaternary and Holocene volcanic centers and collected a first suite of rock samples from many of these remote volcanoes. In the course of this work we positively identified five new volcanic centers in Katmai National Park: Fourpeaked Mtn., Devils Desk, Mt. Stellar, Snowy Mtn. and Kejulik Volcano. The first four peaks were previously thought to be of volcanic origin because of their morphology, but no mapping or sampling had been done.

Volcanoes of Cook Inlet and Katmai

In the Cook Inlet region, reconnaissance and photogeologic maps have been or are being prepared for Redoubt (A. Till, U.S. Geol. Survey, in prep.), Iliamna (Juhle, 1955) and Augustine (Detterman, 1973; Kienle and Swanson, 1980), but Spurr remains unmapped. A series of papers have dealt with eruptions of Spurr (Juhle and Coulter, 1955), Redoubt (Wilson et al., 1966; Wilson and Forbes, 1969; Riehle et al., 1980) and Augustine (Davidson, 1884; Detterman, 1968; Kienle and Forbes, 1976; Meinel et al., 1976; Stith et al., 1977; Johnston, 1978; Kienle and Shaw, 1979; Kienle and Swanson; 1980).

In Katmai National Park much of the previous work has been concentrated

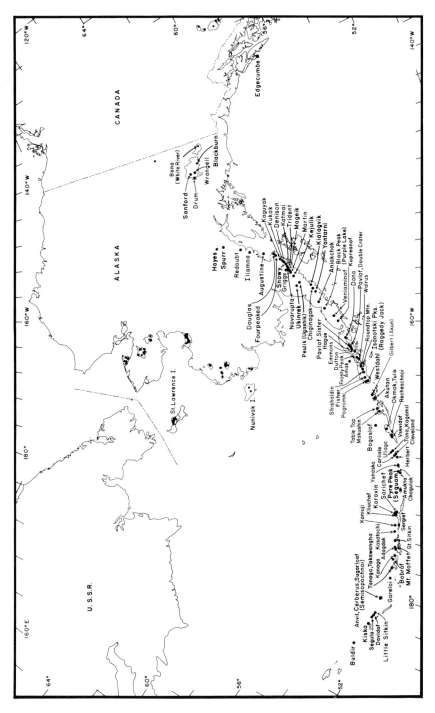

Fig. 1 Quaternary volcanic centers in the Aleutian arc.

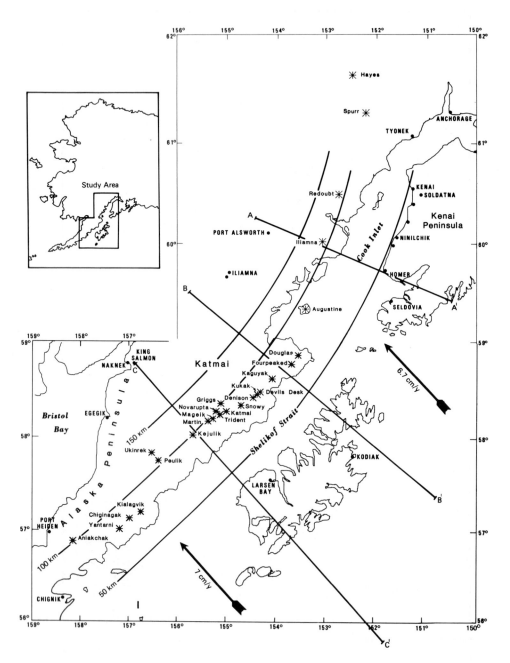

Fig. 2 Location of Late Quaternary and Holocene volcanoes in the eastern Aleutian arc. Relative motion vectors between Pacific and North American plates after Minster and Jordan (1978, model RM2). Line AA', BB' and CC' indicate location of seismic cross sections shown in Fig. 8.

TABLE 1

VOLCANIC CENTERS IN COOK INLET

Name Latitude (N) Longitude (W) (of indicated feature) Highest Point (ft)	Type of Feature	Last Eruption	Current Activity	Historic Eruptions	Comments	Petrology/SiO2 Range (n: No. of samples analyzed)
Hayes Volcano 61°37.2' 152°28.9' (remnant) 5,400	remnant of unconsolidated ash fall and avalanche deposit	no historic activity	not active	none	young vent, probably Holocene, original vent covered by Hayes Glacier	hornblende-biotite dacite (Miller and Smith, 1976)
Mt. Spurr (Crater Peak) 61°16.0' 152°12.7' (highest peak) 11,070	complex volcano, covering about 25 km²	1953, major eruption, ash spread over Anchorage, damage to equipment (Juhle and Coulter, 1955; Wilcox, 1959)	50 m diameter, hot, acid lake within Crater Peak; fumaroles on south rim of crater at boiling point in 1982 (94.1°C) (R. Motyka, pers. comm.)	1953	heavily glaciated, deeply dissected	?
Iliamna Volcano 60°01.9' 153°05.41 (highest peak) 10,016	complex volcano, consisting of four peaks aligned in a N-S direction (Juhle, 1955)	1876, major eruption spreading ash to Kenai Peninsula (Rymer and Sims, 1976)	two vigorous, sulfurous fumaroles, one just below the summit of Iliamna on upper east face above Red Glacier, the other near the saddle between Iliamna and North Twin	eruptions in 1876, 1867, 1843, depositing ash 160 km to the ENE in Skilak Lake, ash eruption in 1778/79; "smoke" reported in 1947, 1933, 1793 1786, 1779, 1778, 1768. Active in 1978, 1952/53.	heavily glaciated, deeply dissected	olivine basalt and two-pyroxene andesite; SiO2 = 51 to 61% (n = 9) (Juhle, 1955; Kienle et al., 1982)
Redoubt Volcano 60°29.3' 152°45.5' (highest peak) 10,197	stratocone	1966/67/68, major eruptions (Wilson et al., 1966; Wilson and Forbes, 1969) producing 2 flash floods in Drift River Valley because of excessive ice melt in summit crater (Post and Mayo, 1971)	heated ground; weak steam vents in ice caves at northern rim of summit crater	eruptions in 1966-68, 1902 (ash 200 km to the E, to Hope, Knik Sunrise); smoke reported in 1933 and 1819; active in 1778.	extensive lahar deposits in Crescent River Valley (Riehle et al., 1980); heavily glaciated	two-pyroxene andesite, widespread hornblende; SiO2 = 55 to 60% (n = 9) (Forbes et al., 1965; Kienle et al., 1982)
Augustine Volcano 59°21.7' 153°25.9' (highest peak) 4,100	island, stratocone and dome complex with pyroclastic flow apron	1976, major eruption, eruption columns to 14 km (Kienle and Shaw, 1979); emplacement of pyroclastic flows; ash dispersal 380 km to the NNE (Talkeetna) and 1100 km to the ESE (Sitka).	cooling lava dome, extruded in 1976, vigorous summit fumaroles, 754°C in 1979 (0. Johnston, pers. comm.), 464°C in 1982 (R. Motyka, pers. comm.)	major eruptions in 1976, 1963/64, 1935, 1883, 1812, producing high eruption columns and pyroclastic flows; eruptions often end with dome intrusions, the 1883 eruption was tsunamigenic (Doroshin, 1870; Davidson, 1884; Detterman, 1968, 1973; Johnston, 1978; Kienle and Swanson, 1980); ash to Skilak Lake, 200 km to the NE, 1963/64, 1935, 1812 (Rymer and Sims, 1976)	not glaciated	two-pyroxene andesite and dacite, minor hornblende and olivine; SiO2 = 57 to 69% (n = 21) (Detterman, 1973; Kienle and Forbes, 1976; Kienle et al., 1982)

397

TABLE 2

VOLCANIC CENTERS IN KATMAI NATIONAL PARK AND ON THE UPPER ALASKA PENINSULA

Name / Latitude (N) / Longitude (W) (of indicated feature) / Highest Point (ft)	Type of Feature	Last Eruption	Current Activity	Historic Eruptions	Comments	Petrology/SiO2 range (n: No. of samples analyzed)
Mt. Douglas 58°51.3' 153°32.4' (crater lake) 7,020	stratocone	no historic activity	200 m diameter crater lake at 25°C, pH < 1, fumaroles at boiling point in 1982 (93.4°C) (R. Motyka, pers. comm.)	none	heavily glaciated	two-pyroxene andesite, some olivine; SiO_2 = 59 to 62% (n = 7) (Kienle et al., 1982)
Fourpeaked Mtn. 58°46.1' 153°40.8' (center of four peaks) 6,903	stratocone	no historic activity	not active	none	hydrothermally altered summit, but no apparent solfataric activity, heavily glaciated and dissected	two-pyroxene andesite, minor hornblende; SiO_2 = 57 to 58% (n = 5) (Kienle et al., 1982)
Kaguyak Crater 58°36.8' 154°03.5' (island in crater lake) 2,956	caldera, 2.4 km diameter, contains 189 m deep lake	no historic activity	weak solfataric activity on central island, no elevated temperature in lake	none	pyroclastic flow apron, several lava domes within and outside caldera; not glaciated	two-pyroxene dacite, and minor andesite, some hornblende and quartz; SiO_2 = 59 to 66% (n = 14) (Swanson et al., 1981; Kienle et al., 1982)
Devils Desk 58°28.5' 154°17.9' (highest peak) 6,410	central vent and dike complex	no historic activity	not active	none	deeply dissected, few small glaciers	two-pyroxene andesite, minor olivine and hornblende; SiO_2 = 55 to 61% (n = 4) (Kienle et al., 1982)
Kukak Volcano 58°27.6' 154°20.9' (fumarole field on northernmost peak) 6,710	stratocone	no historic activity	vigorous fumarole field, with steam escaping through holes in ice	none	heavily glaciated	two-pyroxene andesite, and dacite; SiO_2 = 59 to 66% (n = 6) (Kienle et al. 1982)
Mt. Denison 58°25.1' 154°26.9' (highest peak) 7,520	stratocone, includes Mt. Steller, an erosional remnant	no historic activity	not active	none	heavily glaciated and dissected	two-pyroxene andesite; SiO_2 = 57 to 63% (n = 2) (Kienle et al., 1982)
Snowy Mtn. 58°20.1' 154°41.0' (highest peak) 7,090	complex volcano, consisting of 3 peaks aligned in a NE-SW direction	no historic activity	heated ground and weak steam vents on peak 6,875, 89°C in 1982 (R. Motyka, pers. comm.)	none	heavily glaciated	olivine, two-pyroxene andesite, hornblende dacite; SiO_2 = 56 to 64% (n = 4) (Kienle et al., 1982)
Mt. Katmai 58°15.8' 154°58.5' (center of crater lake) 6,715	caldera, 2 km x 2.5 km, ~ 600 m deep, contains 200 m deep lake	1912, caldera formation	lake has elevated temperature, there is a zone of upwelling gas, indicating fumarolic activity at lake bottom; lake level rising at ~ 3 m/yr (average 1975-1977, Motyka, 1978)	1912	glaciated	andesite dominant, minor basalt, dacite and rhyolite (Fenner, 1926; Kosco, 1981b)
Mt. Griggs 58°21.2' 155°06.2' (summit crater) 7,650	stratocone	no historic activity	summit crater with heated ground and weak steam vents, high pressure fumarole on SW-flank	none	heavily mantled with 1912 tephra from Novarupta; few small glaciers	andesite, minor dacite (Fenner, 1926; Kosco, 1981b)

Volcano	Description	Historic activity	Thermal activity	Last major eruption	Glaciation	Composition
Novarupta 58°16.0' 155°09.4' (dome) 2,760	dome, ~250 m diameter, located within buried, ~2.8 km diameter, caldera (Hildreth, 1981)	1912, paroxysmal explosion, source vent of Valley of 10,000 Smokes ash flow	heated ground, weak steam vents	1912		rhyolite and dacite (Hildreth, 1982)
Trident Volcano 58°13.8' 155°07.2' (summit of new flank cone) 3,800	cinder cone with lava flows and block avalanches, formed in 1953/54, 57, 58, 59/60, 63 (Snyder 1954; Decker, 1963), located on SW flank of ancestral group of 3 volcanoes, the Tridents	1968, normal explosion	fumaroles at boiling point within and outside summit crater	vent plugs emplaced in 1961, 66, 74 (74 plug subsided again by 1975); normal explosions in 1953, 60, 61, 62, 63, 64, 67, 68 with plumes to 6-14 km, former vent plugs destroyed during 62, 67 eruptions no vent plug in 1982.		two-pyroxene andesite; minor dacite (Ray, 1967; Kosco, 1981b)
Mt. Mageik 58°11.8' 155°14.6' (summit crater lake) 7,140	stratocone	1936, explosion	small crater lake at 70°C, $P_H \sim 1$ (Hildreth, personal communication, 1981), weak steam plume can be seen frequently	1936; questionable event in 1946, 1929, small explosion; 1927, small explosion (Jaggar, 1927)	heavily glaciated	dacite and andesite (Fenner, 1926; Kosco, 1981b)
Mt. Martin 58°10.2' 155°21.0' summit crater solfatara field 6,110	stratocone	1951, questionable ash eruption	almost continuous steam plume fed from vigorous solfatara field in crater floor, small acid lake	none	glaciated	two-pyroxene andesite and dacite, some olivine; $SiO_2 = 61$ to 64% (n = 2) (Fenner, 1926)
Kejulik Volcano 155°40.0' (center of complex) 5,510	central vent and dike complex	no historic activity	not active	none	deeply dissected, few small glaciers	two-pyroxene andesite; SiO_2 56 to 60% (n = 6) (Kienle et al., 1982)
Mt. Peulik 58°45.0' 156°22.2' (highest peak) 5,030	stratocone	1814, normal explosion	not active	1814; "smoke" reported in 1852	summit crater occupied by large dacitic dome, not glaciated	andesite (T. Miller, pers. comm.)
Ukinrek Maars 57°50.1' 156°23.2' (eastern maar) 250	2 maars, 170 and 600 m in diameters	formed March 30 to April 9, 1977	hot springs in dry west maar, east maar occupied by ~70 m deep lake	none		alkali-olivine basalt (normative Ne = 1.2%); $SiO_2 = 48\%$ (n = 2) Kienle et al., 1980)
Ugashik Caldera 57°42.9' 156°23.2' (dome in center of caldera) 2,600	caldera, 4 km diameter, containing nested dome complex (T. Miller, pers. comm.)	no historic activity	not active	none	not glaciated	dacite to rhyolite (T. Miller, pers. comm.)

around Mt. Katmai and, specifically on the 1912 eruption that formed the Valley of Ten Thousand Smokes (VTTS). In their study of the 1912 eruption Griggs (1920, 1922) and Fenner (1920, 1922, 1923, 1926, 1930, 1950) identified several volcanic centers in the area including Mt. Katmai, Novarupta, Knife Peak (now renamed Mt. Griggs), Mt. Mageik, Mt. Martin, Trident Volcano and Mt. Peulik. Curtis (1968) also studied the 1912 eruption and made a detailed examination of the tephra deposits. Hildreth (1982) has recently been working on the petrology of the 1912 ejecta.

Few studies of other volcanic centers in this region are available. Muller et al. (1954) reported on the volcanic activity in the Katmai area during the summer of 1953, which included the 1953 eruption of Trident. Muller et al. also noted fumarole activity on six other volcanoes: steam issued in 1953 from several vents on the side of a crater lake on Douglas and steam vents were observed at Kukak, Griggs, Novarupta, Mageik and Martin. Snyder (1954) gave the first detailed report of the new activity at Trident. Ray (1967) studied the mineralogy and petrology of the lavas formed during the 1953-1963 eruptions of Trident. Kosco (1981a, and 1981b) reported that the lavas of Griggs, Katmai, Trident and Mageik are dominantly dacites and andesites (SiO_2 = 54 to 68%) and that the disequilibrium phenocryst assemblage (olivine + quartz + orthopyroxene) suggests magma mixing in some of these lavas. Keller and Reiser (1959) presented a regional geologic map for the area from Cape Douglas to the Kejulik Mts. and showed the distribution of volcanic rocks, but most of their effort was devoted to the Mesozoic and Tertiary sedimentary rock units. Keller and Reiser (1959) did comment that Martin, Mageik, Novarupta, Knife, Kukak and Douglas were steaming with varying intensities and that Trident was steaming and extruding blocky lava. An important summary of volcanic activity in Katmai National Monument from 1870 to 1965 was prepared by Ward and Matumoto (1967). Just outside the Park boundary Kienle et al. (1980) and Self et al. (1980) described the formation (in 1977) of two maars near Mt. Peulik, which they named Ukinrek Maars. No IAVCEI volcano catalogue currently exists for the Aleutian arc and much of the previous work on volcanoes in Cook Inlet and Katmai is generally only available as abstracts, unpublished theses or not easily accessible reports. We have, therefore, summarized the principal facts about the volcanoes of Cook Inlet and Katmai in Tables 1 and 2.

Ten of the 22 volcanic centers of probably Late Quaternary and Holocene age in the eastern Aleutian arc have erupted in recorded history. Figs. 3 and 4 summarize the most important eruptive events in Cook Inlet and Katmai in historic times. Spurr, Redoubt, Iliamna, Augustine, Trident, Mageik and Peulik had Vulcanian eruptions. A Plinian eruption of Novarupta in 1912 led to the formation of the Valley of Ten Thousand Smokes ash flow and the collapse of the summit of Mt. Katmai, producing ~20 km^3 of ash-fall tephra and 11-15 km^3 of ash-flow tuff within ~60 hours (Hildreth, 1982). The Ukinrek eruption was phreatomagmatic/Strombolian. Six volcanoes that have not erupted in historic times show hydrothermal activity (Douglas, Kaguyak, Kakak, Snowy, Griggs, Martin). Four of the eastern Aleutian volcanoes have advanced to the caldera

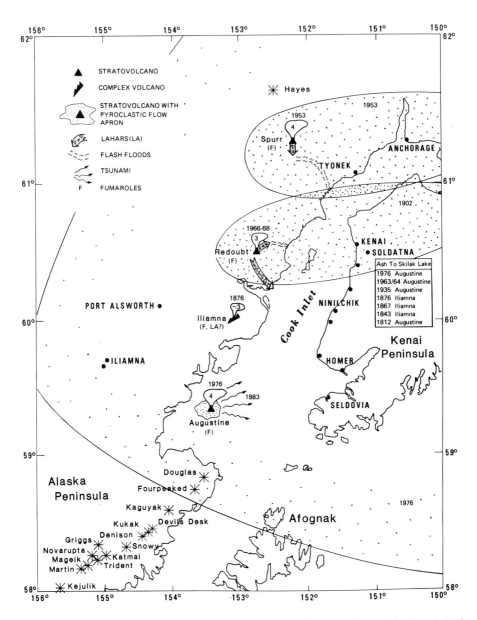

Fig. 3 Types of volcanic edifices, major historic eruptions, limits of tephra dispersal, and generalized volcanic hazards in Cook Inlet. Numbers within plume symbols refer to the Volcanic Explosivity Index (VEI), described by Simkin et al. (1981). Skilak Lake lies 18 km ESE of Soldatna.

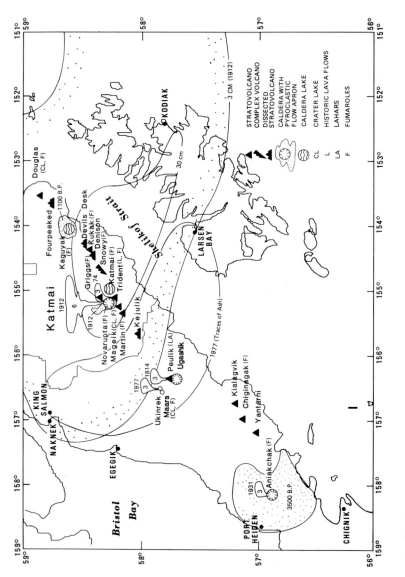

Fig. 4 Types of volcanic edifices, major historic eruptions and generalized volcanic hazards in Katmai National Monument and on the upper Alaskan Peninsula. Numbers within plume symbols refer to the Volcanic Exposivity Index (VEI), described by Simkin et al. (1981).

stage: Kaguyak, Katmai, Novarupta (a possibility proposed by Hildreth, 1981), and Ugashik (Miller, personal communication). Both Kaguyak and Katmai have well developed summit calderas occupied by deep (~200 m) lakes. Most of the other volcanoes in Katmai and Cook Inlet are composite cones. Domes and pyroclastic flows are common at Augustine Volcano, Kaguyak Crater and Novarupta.

Glaciers are found on all the eastern Aleutian volcanoes except for the very young centers such as Augustine, Kaguyak, Novarupta and Ukinrek. Glacial dissection can be used as a means to crudely classify the volcanic centers with regard to their relative age. Extensive erosion of Kejulik, Denison, Devils Desk, Fourpeaked and Hayes, and current lack of hydrothermal activity, suggest that these volcanoes have not been active since the last period of extensive glaciation, ending ~10,000 years ago in this region (Karlstrom, 1964; Péwé, 1975). On the other end of the spectrum Griggs, Kaguyak, Augustine, and of course Novarupta and Ukinrek have not been affected by glacial erosion and have all been recently active or show hydrothermal activity. The bulk of these volcanic structures was most likely built in the past 10,000 years. The remaining volcanoes are still active, have erupted recently, or show surface hydrothermal activity; but these centers have also been extensively modified by glacial erosion. Here, construction of the volcanic edifices could not keep up with glacial dissection.

PRESENT-DAY TECTONIC FRAMEWORK

Quaternary volcanism in the Aleutians is the result of plate convergence between the American and Pacific plates. Fig. 5 shows directions and rates of relative plate motion calculated from Minster and Jordan's (1978) model RM2 which describes present-day global plate motions from an average over the past 3 m.y. Slip rates increase regularly from east to west along the Aleutian volcanic arc as the angular distance to the pole of rotation for the Pacific-American plates increases. However, the component of convergence normal to the arc decreases from east to west as plate motion becomes more and more oblique to the arc. Also shown in Fig. 5 is the seismicity based on worldwide teleseismic data between 1962 and 1969 and the approximate 150 km isobath to the Benioff zone (after Jacob et al., 1977). The Aleutian volcanic front is developed over a distance of 2,600 km between Hayes and Buldir Volcanoes along that section of the arc which is underlain by a dipping seismic zone. This illustrates the intimate relationship between active arc volcanism and the presence of a downgoing oceanic slab in the Aleutian arc. Quaternary volcanism ceases in the Kommandorsky section of the arc (between Buildir Volcano and Sheveluch Volcano in Kamchatka), where the slip vector between the two plates becomes tangential to the arc. Lack of intermediate depth seismicity and strike slip mechanisms for large earthquakes (McKenzie and Parker, 1967; Stauder, 1968) imply that the Pacific plate is currently not subducting beneath the Kommandorsky section of the arc and that this section acts as a transform plate boundary (Cormier, 1975). Sheveluch Volcano, the northernmost Quaternary Volcano in Kamchatka,

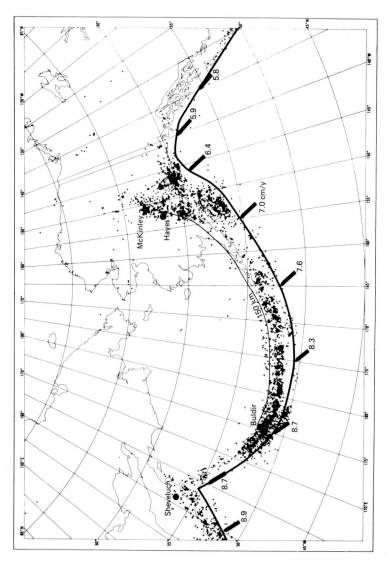

Fig. 5 Relative motion vectors between Pacific and North American plates after Minster and Jordan (1978, model RM2), and seismicity of the Aleutian arc (WWSSN teleseismic locations 1962-69). Pacific-North American plate boundary and 150 km depth contour to Benioff zone shown by solid lines (Figure modified from Jacob et al., 1977).

marks the next transition back to near normal plate convergence against the
Kamchatka Peninsula with its well developed chain of active volcanoes.
Kamchatka is again underlain by a seismically well defined down-going oceanic
slab (e.g. Fedotov and Tokarev, 1973).

The volcanic front in the Aleutian arc lies generally over the 100 to 150 km
depth contour to the Benioff zone (see Fig. 2). Marsh and Carmichael (1974)
presented the interesting hypothesis that the regular spacing of 60 to 70 km
between large volcanic centers observed along the Aleutian arc could be
explained by regularly spaced mantle plumes rising from a gravitationally
unstable ribbon of melt that has developed at some critical depth (100-150 km)
along the strike of the down-going plate. Bogoslof and Amak Islands are the
only volcanic centers that have developed behind the primary volcanic front
above greater depths to the dipping seismic zone. These two volcanoes also
erupt more potash-rich lavas (Marsh, 1979). Both Bogoslof and Amak are very
young, probably less than 10,000 years old. Marsh and Leitz (1979) and Marsh
(1979, 1982) proposed that this embryonic secondary volcanic front may be the
result of diapirism above a ribbon of melt that has now developed down dip from
the main volcanic front along the surface of the subducting plate.

At the eastern end of the Aleutian arc there are two apparent anomalies in
the relationship of volcanism to plate subduction. (1) No major Late Quaternary
or Recent volcanic center has yet been found north of Hayes Volcano (Fig. 5),
even though northwesterly subduction is well defined by intermediate depth
earthquakes into the interior of Alaska, extending 300 km beyond the last
Aleutian arc volcano. (2) A group of voluminous Pleistocene volcanoes, which
represent the culmination of past mid-Miocene andesite volcanism in the Wrangell
Mountains (Nye, 1982) lies some 350 km east of the Aleutian trend (Fig. 1).
This group of volcanoes is not underlain by a currently active Benioff zone.

SUBDUCTION DEFINED BY SEISMICITY

The geometry of the subducting Pacific plate in the eastern Aleutian arc is
now quite well defined through earthquake studies. A series of theses (Davies,
1975; Lahr, 1975; Estes, 1978), various reports (Pulpan and Kienle, 1979; Lahr
et al., 1974), and a recent paper by Kienle et al. (1982) have analyzed the
configuration of the Benioff zone in the area.

Since 1976, the University of Alaska has operated a high resolution
seismographic network with some 30 short-period stations in the lower Cook
Inlet-Katmai-Kodiak Island region, where the relationship of volcanism to a well
developed down-going plate is quite clear. All stations have single component,
vertical, short-period (natural period 1 sec) seismometers. Fig. 6 shows the
current configuration of the seismic network and the location of an additional 4
stations operated by other agencies and routinely used for our earthquake
locations. Data from all stations are transmitted by radio to three data-
gathering points, located at King Salmon, Kodiak, and Homer. Details of our
procedure for locating hypocenters have been discussed by Pulpan and Kienle

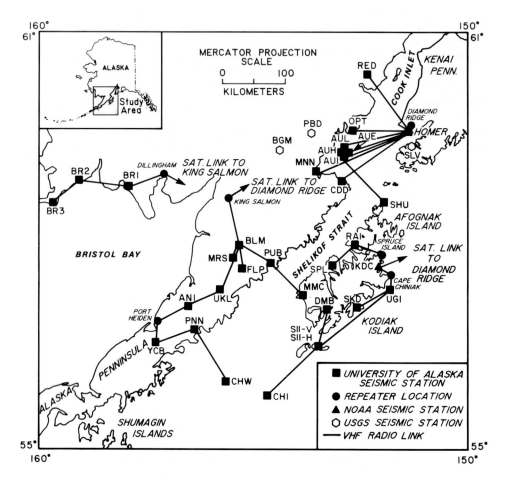

Fig. 6 Cook Inlet-Alaska Peninsula-Bristol Bay-Kodiak seismic network, of the University of Alaska, 1981 configuration (from Kienle et al., 1982).

(1979) and Kienle et al. (1982).

Epicenters

Fig. 7 is a map of epicenters in the eastern Aleutian arc. For this figure we selected a data file of 1,881 events from the period July 1977 to June 1981. These events were recorded on at least 6 stations, had relative vertical and

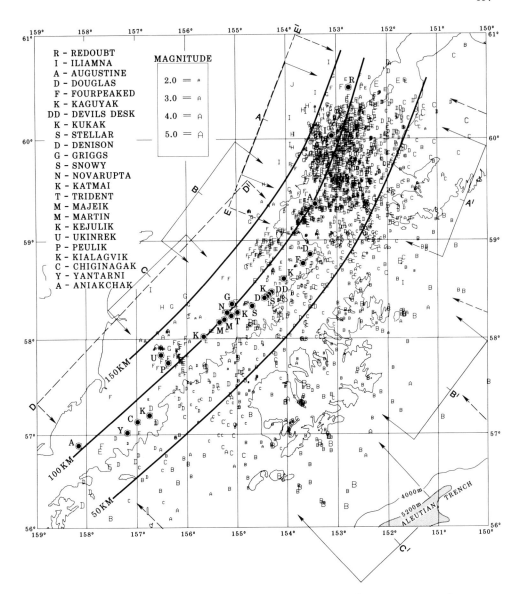

Fig. 7 Epicenters, volcanoes, isobaths of Benioff zone (50, 100, 150 km) and location of projection volumes for cross sections and frontal views shown in Figs. 8 and 9. Epicenters from local network data, July 1977 to June 1981; selection criteria; Recorded on at least 6 stations, RMS travel time residual ≤ 0.4 sec, relative vertical and horizontal location error (ERZ and ERH) ≤ 10 km. Epicenters are coded according to magnitude and depth range, A: 0-25 km, B: 26-50 km, C: 51-100 km, etc. (from Kienle et al., 1982).

horizontal location errors of less than 10 km, and had a root mean square travel time error of less than 0.4 sec.

The absolute hypocentral location errors are unknown because the velocity model used does not account for lateral velocity inhomogeneities in the mantle associated with the down-going slab. However, location errors due to an inadequate velocity structure would be systematic and would not affect the general conclusions reached here. The earthquake detection threshold within the network is about magnitude 2, but can be as low as magnitude 1 in some portions.

Hypocentral Cross Sections

In Fig. 8 earthquake hypocenters within 3 individual blocks shown on the epicenter map (Fig. 5) have been projected on vertical planes indicated by lines A-A', B-B' and C-C'. The orientations of the planes were chosen by trial and error to minimize the thickness of the Benioff zone and thus correspond to the best-fit plunge direction of the subducting plate. Two types of data are plotted on Fig. 8: hypocenters for events with a depth greater than 50 km are closely constrained (crosses). We required that an event be recorded on at least 6 stations, have a vertical (ERZ) and horizontal (ERH) location error of less than 10 km, have a root mean square (RMS) travel time error of less than 0.3 sec and the distance from the epicenter to the nearest station was within twice the depth to the hypocenter. The last requirement insures fairly good depth control. Since these constraints filtered out too many of the shallow events (0 to 50 km depths), we relaxed the selection criteria, requiring that the event be recorded on at least 5 stations and have an RMS travel time residual of less than 0.5 sec, and an ERH and ERZ of less than 10 km (triangles)

The seismicity shown in the cross section defines four distinct tectonic regions:

(1) A shallow thrust zone dips at an angle of about 10° landward from the Aleutian trench over a distance of several 100 km. This thrust zone widens from southwest to northwest (bottom to top in Fig. 8). Great earthquakes (M > 7.8), such as the 1964 Alaskan earthquake, rupture this plane of generally strong plate coupling.

(2) A Benioff zone, about 20 to 30 km thick, reaches a maximum depth of 200 km, and dips at an angle of about 45° beneath the volcanic arc.

(3) An aseismic wedge of asthenosphere is located above the cold subducting slab and below the overriding plate.

(4) A brittle top layer of the overriding plate is characterized by generally diffuse seismicity that does not correlate with mapped faults. Cross sections BB' and CC' show shallow clusters of earthquakes beneath certain volcanic centers, e.g. Douglas-Fourpeaked and Ukinrek. The pipe-like cluster of shallow events beneath Ukinrek is associated with the formation of the two maars in the course of the April, 1977 eruptions (Kienle et al., 1980). The cluster beneath Douglas-Fourpeaked may be associated with hydrothermal activity (Pulpan and Kienle, 1979). Had we used data recorded before 1977, extending back to 1975, a dense cluster of very shallow (0-5 km) seismicity would have been plotted directly beneath Augustine

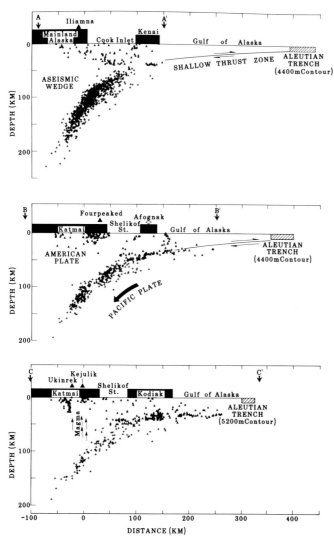

Fig. 8 Cross sectional views of Benioff zone along lines A-A', B-B', C-C' shown in Figs. 2 and 7. No vertical exaggeration. Landmasses (solid black), the position of the trench and location of volcanoes are also shown. Shallow thrust zone is added schematically. Selection criteria for events at 0-50 km depth (triangles): Recorded on at least 5 stations (STA $\geq$ 5), relative horizontal and vertical location error (ERZ and ERH) $\leq$ 10 km, RMS travel time residual $\leq$ 0.5 sec. Selection criteria for event > 50 km depth (crosses): STA $\geq$ 6, ERH and ERZ $\leq$ 10 km, RMS $\leq$ 0.3 sec, and distance from epicenter to nearest station $\leq$ 2 x depth. (Modified from Kienle et al., 1982).

Volcano, which erupted explosively in 1976.

Frontal View

Fig. 9 is a sectional view of the hypocenter distribution looking toward the volcanic arc. All the data shown in Fig. 8 were projected onto two vertical planes, one oriented parallel to the strike of the volcanic front in Katmai (DD', for location of projection, see Fig. 7), the other oriented parallel to the strike of the volcanic front in Cook Inlet (EE', compare Fig. 7). As in Fig. 8, the data set is split into shallow events (0-50 km depth, triangles) and deeper events (depths greater than 50 km, crosses).

Shallow events in the 30 to 50 km depth range give a good picture of the seismicity in the thrust zone. Very shallow earthquakes generally scatter in a diffuse manner but are sometimes concentrated beneath certain volcanoes as we have just discussed.

Earthquakes in the 50 to about 100 km depth range are fairly uniformly distributed along the arc. However, earthquakes deeper than about 100 km appear to cluster near the lower end of the Benioff zone behind certain volcanoes. Fairly well defined clusters of intermediate depth earthquakes occur beneath Peulik (100-150 km deep), landward of Kaguyak-Douglas-Fourpeaked (100-150 km deep), and slightly south of Augustine (100-150 km deep). A very intense, persistent cluster of intermediate depth earthquakes occurs in the vicinity of Iliamna. This 100 km diameter and 70 to 180 km deep cluster frequently produces earthquakes in the magnitude 3-4 range and some larger events up to magnitude 5. We have as yet no explanation why these deeper clusters of events occur spatially beneath and behind certain active volcanoes but such clusters have also been observed beneath some of the volcanoes in the central Aleutians (Engdahl, 1977).

TECTONIC SETTING OF VOLCANISM

Segmentaion of the Volcanic Front

The front of Late Quaternary and Holocene volcanoes in the eastern Aleutian arc is clearly segmented as shown in Fig. 10. A similar segmentation of the volcanic front has been observed by other workers in the Aleutian arc (Marsh, 1979, 1982; Kay et al., 1982), and also in other circum-Pacific areas such as the Cascades (Hughes et al., 1980) and the Central American arc (Stoiber and Carr, 1973; Carr et al., 1982). Linear segments of volcanoes are often separated by transverse offsets, which have been correlated in the Aleutians with fracture zones or breaks in the down-going oceanic plate (Van Wormer et al., 1974; Spence, 1977) or with apparent transverse arc boundaries mapped out by the boundaries of aftershock sequences of great thrust earthquakes (Holden and Kienle, 1977). In the Cascades, intrasegment volcanoes are characterized by calc-alkaline fractionation patterns and commonly erupt two-pyroxene andesite (Hughes et al., 1980). Volcanoes near the offsets often erupt a variety of lavas that range from basalt to rhyolite (Hughes et al., 1980) and may show tholeiitic fractionation trends (Kay et al., 1982).

411

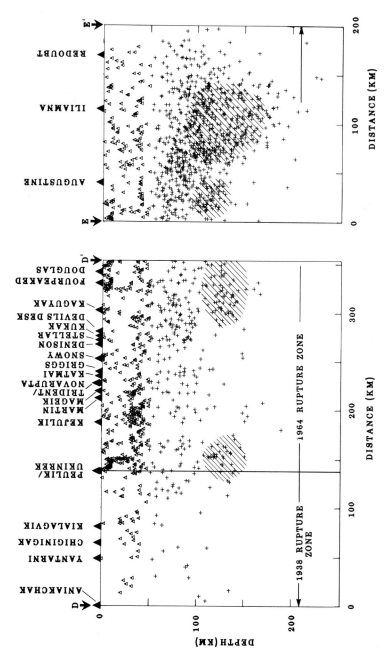

Fig. 9 Same data as in Fig. 8 projected on two vertical planes oriented along the strike of the Benioff zone (along lines D-D' and E-E', Fig. 7). Position of volcanoes also shown. Deep clusters of seismicity cross-hatched. Vertical line in section D-D' marks rupture zone boundary of 1964 (M_w 9.2) shallow thrust earthquakes.

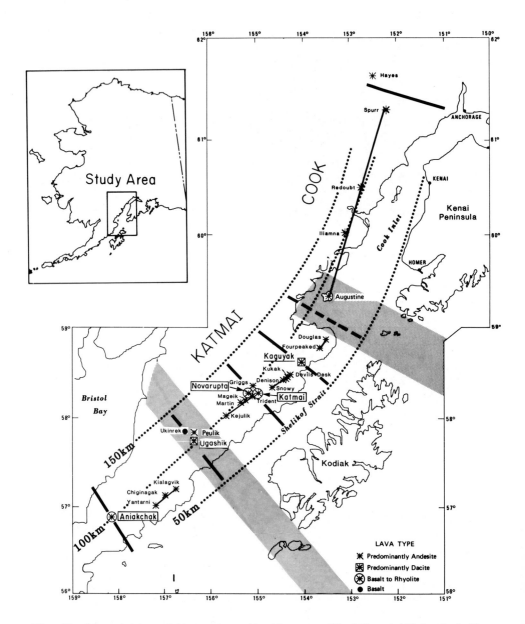

Fig. 10 Segmentation of the eastern Aleutian arc. 50, 100 and 150 km isobaths to the Benioff zone are also shown. Arc-transverse tectonic boundaries shown as stippled zones (after Fisher et al. (1981)).

In Fig. 10 two major arc segments are defined on the basis of tectonic boundaries and volcano alignment (Fisher et al., 1981) and we call these the Cook and Katmai segments. A third segment formed by Kialagvik, Chiginagak and Yantarni Volcanoes is offset seaward by 40 km from the Katmai front across a transverse boundary that also marks the southern boundary of the 1964 great Alaskan earthquake rupture zone (compare Figs. 5 and 9). That boundary passes through Ugashik, Peulik, and Ukinrek that have erupted a variety of lavas ranging from basalt to rhyolite (Table 2).

The Katmai segment consists of several en echelon subsegments, each one characterized by a very close volcano spacing. Volcanoes with more fractionated magmas occur near the subsegment boundaries (Novarupta/Katmai, Kaguyak). On the average, volcanoes of the Katmai segments have a narrow spacing of 13 ± 7 km (11 pairs between Kejulik and Douglas).

Volcanoes within in the Cook segment form a single N 20°E trending front that diverges from the average Katmai trend of N 55°E by 35°. In marked contrast to the Katmai segment, volcano spacing in Cook Inlet is much wider, 70 ± 18 km (3 pairs, Spurr to Augustine). The northern end of the Cook segment arbitrarily has been chosen to pass between Spurr and Hayes, because Hayes is offset to the west from the Cook Inlet trend. Van Wormer et al. (1974) discussed seismologic evidence for a break in the down-going slab that coincides with the end of the Aleutian volcanic front near Hayes. The location of the southern boundary of the Cook segment is not clear. Kienle et al. (1982) chose a line south of Augustine (dashed line in Fig. 10) which coincides with a broad zone of deformation in the overriding plate (stippled zone in Fig. 10) proposed by Fisher et al. (1981). Based on volcanic alignment alone, one could equally well group the Douglas/Fourpeaked pair of volcanoes with the Cook segment, thus drawing the Cook segment boundary through Kaguyak (Marsh, 1982). If we include the Douglas/Fourpeaked pair of volcanoes in the Cook segment, volcano spacing becomes 66 ± 18 km (4 pairs, Spurr to Douglas).

Subduction and Volcanism

Based on the epicentral data presented in Fig. 7, it is possible to contour isobaths to the Benioff zone. Three isobaths (50, 100 and 150 km) are shown in Fig. 10. Comparing volcano positions to the Benioff zone isobaths we find: (1) The 35° change in volcanic alignment between the Katmai and Cook segments reflects a lateral change in orientation of the down-going oceanic plate to a more northerly strike in Cook Inlet. (2) There is no evidence for a break or hinge fault in the subducting plate in the vicinity of Douglas Volcano, a result that corroborates the findings of Estes (1978) who analyzed an earlier earthquake data set for the region. The lateral warping of the subducting plate in lower Cook Inlet appears to occur smoothly without failure of the slab. (3) Volcanoes of the Cook segment are closely aligned above the 100 km isobath of the Benioff zone. Volcanoes of the Katmai segment lie on a trend that is misoriented by about 20° with respect to the strike of the Benioff zone. In Katmai the depth to the Benioff zone decreases by 25% from Kejulik to Douglas Volcanoes, from 100 to 75 km.

Crustal Tectonics and Volcanism

Fisher et al. (1981) have recently defined crustal tectonic boundaries at the northeastern and southwestern end of the Kodiak Islands that extend further inland across the Aleutian volcanic arc. These boundaries are not single faults but consist of broad zones of disruption that began to form during the late-Miocene or Pliocene (stippled zones in Figure 10). The southern zone of disruption coincides with the southern boundary of the Katmai Volcano segment as we defined it based on the 40 km offset of the volcanic front across an arc-tranverse line drawn through Ugashik, Peulik and Ukinrek. According to Fisher et al. (1981) the northern zone of crustal disruption is less clearly defined than the southern one and passes between the Kenai Peninsula and the Kodiak Islands into Lower Cook Inlet. This zone coincides with the Cook-Katmai segment boundary proposed by Kienle et al. (1982) and includes Augustine as a segment boundary volcano.

Evidence cited by Fisher et al. (1981) to define the transverse arc boundaries includes regional changes in geology, terminations of structural trends, cross-arc trends of young tectonic elements on the continental shelf at the two boundaries, changes in sign of long wavelength gravity anomalies representing regional changes of crustal structure, and the discrete nature of the southern boundary of the aftershock zone of the 1964 great Alaskan earthquake (compare Fig. 5). Fisher et al. (1981) also regard offsets in the volcanic front as strong indicators of transverse tectonic boundaries.

The question arises whether or not the transverse tectonic boundaries in the overriding plate can be correlated with tectonic features on the down-going plate such as seamount chains and fracture zones. In the central Aleutians, Marsh (1979) correlated offsets in the volcanic front with subducting fracture zones. However, since subduction is oblique in the eastern Aleutian arc tectonic elements on the down-going plate would sweep along the continental margin with time and hence could not give rise to any localized zones of disruption in the overriding plate. Specifically, Fisher et al. (1981) calculated that the intersection of the Kodiak seamount chain with the continental margin near the southern transverse arc boundary must have swept ~180 km northward in the past 3 m.y. (based on Jordan and Minster's 1978 model RM2). Similarly oceanic fracture zones near the northern boundary would have swept ~160 km northward for the same time period.

Crustal control of volcanic alignment in the Katmai segment

There is geologic and geophysical evidence that the northeasterly alignment of the volcanic front between Mageik and Kaguyak may be controlled by near-surface structures. Two anticlines mapped by Keller and Reiser (1959) on or near the crest of the volcanic arc between Martin and Katmai and between Kaguyak and Devils Desk clearly trend toward each other. Keller and Reiser propose that these anticlines may be the manifestation of a deep-seated fault that provides a zone of crustal weakness for magma migration.

There is geophysical evidence for such a fault. Kubota and Berg (1967) and Berg et al. (1967) found that the active volcanic front is located along a steep

35 mgal gravity gradient. Kienle (1969) showed that this gravity gradient extends from about Martin to Kaguyak along 90 km of the volcanic front (Fig. 11, top). Fig. 11 (bottom) shows a two-dimensional crustal density model derived from the gravity data for section A-A' which crosses the volcanic range between Snowy Mtn. and Mt. Denison. At least part of the steep gravity gradient can be attributed to a deep-seated crustal fracture beneath the volcanic front. Based on seismic data, Berg et al. (1967) also found distinctly different average crustal thicknesses on either side of the volcanic front. This implies the existence of a deep crustal boundary or fault. To the northeast of the Katmai volcanoes the crust has an average thickness of 38 km, southwest average crustal thickness decreases to 32 km.

In addition to these regional trends there is also evidence from gravity and seismic data that major magma accumulation may exist in the crust beneath several volcanoes of the Katmai segment. Matumoto (1971) used wave attenuation patterns from local earthquakes to show that large magma bodies may underlie Martin-Mageik, Griggs-Katmai-Trident-Novarupta, and Snowy at depths of less than 10 km. Beneath the Katmai group of volcanoes Matumoto (1971) identified two magma reservoirs. A shallow chamber lies at a depth of less than 10 km and has a cross-sectional dimension of about 10 x 20 km. A deeper magma chamber beneath the shallow one has a similar size and lies near the base of the crust at a depth of 20-30 km. A local negative gravity anomaly, about 7-8 km wide, with an amplitude of nearly 20 mgal has been mapped by Decker (1963), Berg and Kienle (1966), and Kienle (1969) centered in the volcanic triangle defined by Mageik-Trident-Novarupta (Fig. 11, top). The anomaly has been interpreted by Kubota and Berg (1967) and Kienle (1969) as a shallow magma body beneath the recent Trident cone (Table 2). It may be part of the Katmai body defined by Matumoto (1971). The crustal density model and gravity profile shown in Fig. 11 (bottom) also shows a small gravity low beneath the volcanic axis near Snowy, modelled as a shallow magma body with a low density of 2.46 g-cm^{-3}. That body coincides with the shallow magma chamber identified by Matumoto (1971) near Snowy.

Another piece of evidence that suggests that fairly large magma bodies and associated hydrothermal systems may be present in the shallow crust beneath the Katmai volcanic front is microearthquake activity. Microearthquake studies carried out in Katmai National Park at various times during the past 15 years consistently showed a fairly intense clustering of microearthquakes in the section of the Katmai volcanic front between Martin and Snowy (Kubota and Berg, 1967; Berg et al., 1967; Matumoto and Ward, 1967; Matumoto, 1971). Continuous earthquake data acquired by the University of Alaska over the past few years clearly demonstrated the volcanic nature of the shallow seismicity beneath the Katmai volcanic front. Earthquakes typically occur in swarms of short duration. Pulpan and Kienle (1979) and Kienle et al. (1982) proposed that this seismicity may be associated with volcanic and hydrothermal activity beneath several Katmai segment volcanoes. In recent years Snowy Mtn. has been the most consistent source region for shallow earthquake swarms.

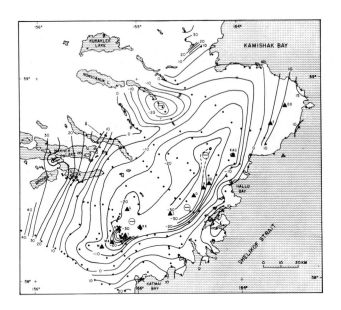

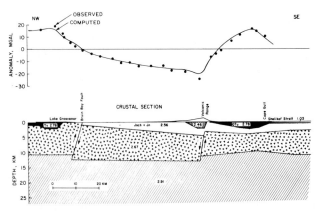

Fig. 11 Top: Simple Bouguer (2.67) gravity map of Katmai National Monument, showing gravity stations (solid dots) and volcanoes (modified from Kienle, 1969).

Bottom: Two-dimensional crustal density section (g-cm^{-3}) along profile AA' (see top of figure). Terrane corrected and geologically corrected gravity data and computed anomalies are also shown. Jnch + Jn: Upper Jurassic sandstone and shales (Naknek Form.); Qtu: Tertiary/Quaternary igneous rocks undifferentiated; Tv: Eocene volcanic rocks (all units after Keller and Reiser, 1959). Upper crustal layer: 2.67 g-cm^{-3}; lower crustal layer: 2.91 g-cm^{-3} (modified from Kienle, 1969).

Stoiber and Carr (1973) found that calderas tend to occur on segment boundaries in Central America. Calderas in the eastern Aleutian arc are found at Aniakchak, Ugashik, Katmai and Kaguyak (Fig. 4) and these volcanoes are on our proposed segment boundaries (Fig. 10).

PETROLOGY OF THE EASTERN ALEUTIAN LAVAS

Petrologic studies of volcanic centers in the eastern Aleutian arc range from none or only the briefest survey to detailed studies that utilize major and minor element analyses, electron microprobe analyses, and isotopic studies. Tables 1 and 2 include a summary of rock types in the eastern Aleutian arc and also identify the data source for each volcanic center. Conclusions based on a few samples from a volcanic center (such as Denison or Martin, Table 2) may obviously change as more samples are collected. The discussion that follows is based on a somewhat limited and uneven distribution of data. However, the data base used here is the largest and most complete for volcanic centers in this region and should be useful in identifying regional patterns of variation.

Porphyritic andesite containing abundant phenocrysts of plagioclase and lesser amounts of hypersthene and augite is the most common rock type in the eastern Aleutian arc. Dacite, often containing hornblende, is found as a minor constituent at several of the volcanic centers and a quartz-bearing dacite is the dominant lava type at Kaguyak Crater (Swanson et al., 1981). Dacite is also an important component of the 1912 VTTS deposits (Hildreth, 1982). Quartz-bearing rhyolite is present in the basal portions of the VTTS deposits and at Novarupta (Fenner, 1950; Curtis, 1968; Hildreth, 1982). Basalt has been collected from boulders on Katmai (Fenner, 1926; Kosco, 1981b) and olivine-bearing basalt has been found as rare hyaloclastite on Iliamna, Augustine (Johnston, 1979) and at Ukinrek (Kienle et al., 1980). Overall, basalts are rare in this region.

Mineralogy

Plagioclase is the most common phenocryst in the eastern Aleutian arc. These phenocrysts are complexly twinned and zoned. Normal compositional zoning from Ca-rich cores to Na-rich rims is common, but reversed zoning is also widespread, especially in grains with abundant glass inclusions and a sieve texture. Plagioclase commonly shows a wide range in compositions within a particular volcanic center, often almost the entire compositional range is found in a single thin section. Hildreth (1982) found the plagioclase in the VTTS deposits ranged from An_{25} to An_{94}. A less extreme, but probably more typical range of plagioclase compositions (An_{37} to An_{86}) is shown in the dacites and andesites of Kaguyak.

Hypersthene is more abundant than augite in most of the lavas from the eastern Aleutian arc. The orthopyroxene forms slightly pleochroic (light green to colorless) phenocrysts that range in composition from Wo_3 En_{45} Fs_{52} (VTTS; Hildreth, 1982) to Wo_3 En_{80} Fs_{17} (Kaguyak). Some hypersthene grains are rimmed by clinopyroxene.

Augite is present in moderate to trace amounts in most lavas of the eastern Aleutian arc. Clinopyroxene shows a more restricted compositional range than the coexisting hypersthene. Kosco (1981b) reports a range of augite compositions from the volcanic centers around the VTTS of Wo_{45} En_{51} Fs_6 to Wo_{44} En_{36} Fs_{20}.

Olivine is an accessory phase in some of the andesites of the eastern Aleutian arc. Reaction rims of pyroxene often surround the olivine grains. Kosco (1981b) reports a range of olivine compositions in the volcanoes around VTTS of Fo_{68} to Fo_{80}. Olivine phenocrysts are a major component in the basalts from Augustine (Johnston, 1979) and Iliamna.

Amphibole, either edenite or edenitic hornblende according to the classification of Leake (1978), is a minor accessory phase in some andesites and dacites and is a major phase in some dacites from Kaguyak. The amphibole shows the common green to brown pleochroic color scheme. Oxidation of some hornblende grains has resulted in opaque pseudomorphs after amphibole. Other amphiboles are surrounded by reaction coronas of pyroxene and plagioclase, apparently formed by dehydration of the amphibole at low pressure.

Quartz is found as phenocrysts in the rhyolite and dacite of the VTTS deposits (Fenner, 1920; Hildreth, 1982) and in the dacites of Kaguyak (Swanson et al., 1981). Reaction coronas of pyroxene and plagioclase are found around some quartz grains and the rounded and embayed form of some quartz crystals also suggests the quartz may be unstable.

Fe-Ti oxides are common accessory phases in the lavas of the eastern Aleutian arc. Titanomagnetite is generally more abundant the ilmenite. Kosco (1981b) used Fe-Ti oxide compositions to estimate temperature (780°-940°C) and oxygen fugacity (log f_{02} = -10.7 to -14.8) in lavas from the region around the VTTS. Similar results are reported by Hildreth (1982) for the VTTS deposits and reconnaissance studies we have done at Kaguyak are also consistent with this estimate.

Chemistry

Lavas of the eastern Aleutian arc are calc-alkaline and their pattern of chemical variability is similar to other calc-alkaline volcanic provinces. FeO*, MgO and CaO decrease with increasing SiO_2, while Na_2O and K_2O increases (Fig. 12). Al_2O_3 shows little change with increasing SiO_2 until bulk compositions reach basaltic andesite (about 57% SiO_2) then there is a progressive decrease in Al_2O_3 with increasing SiO_2. Scatter in the variation patterns of some basalts indicates fractionation of olivine and plagioclase. Most of the volcanic centers show similar patterns of variation and are grouped together in Fig. 12. Augustine is slightly higher in MgO and CaO than other Cook Inlet volcanoes. Kaguyak is higher in FeO* than most of the volcanoes, lower in MgO and CaO than the Cook Inlet volcanoes and lower in K_2O than the Katmai volcanoes. Griggs is higher in Na_2O and K_2O than other Katmai volcanoes and this is consistent with its position behind the main volcanic front. The alkaline basalts (nepheline = 1.2 in the norm) of Ukinrek do not seem to follow the same compositional trend as the other Katmai volcanoes (Fig. 12) and this is

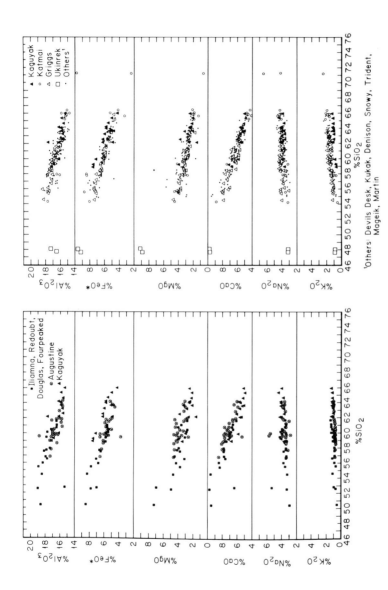

Fig. 12 Harker variation diagrams for the Cook Inlet volcanoes (Iliamna, Redoubt, etc.) and the Katmai volcanoes (Kaguyak, Katmaie, etc.). Note that Kaguyak is given on both diagrams for comparison. Data from Becker (1898), Fenner (1926, 1950), Juhle (1955), Forbes et al. (1965), Ray (1967), Detterman (1973), Kienle and Forbes (1976), Kosco (1981b) and Kienle et al. (1982).

consistent with the anomalous character of Ukinrek relative to the calderas and composite volcanoes of Katmai.

Katmai shows a large range of lava compositions, but other volcanoes show a much more restricted range of compositions (Fig. 12). Fig. 13 further illustrates the restricted compositional range for lavas of the eastern Aleutian arc. The diversity of the Katmai lavas and the 1912 pumices from the VTTS suggests some common source for both sites. Fenner (1930), Curtis (1968) and Hildreth (1982) all suggest a magmatic connection between Novarupta (the supposed source of the 1912 pumices) and Mt. Katmai. The 1912 trends in Fig. 13 are consistent with this argument. Other volcanoes in the eastern Aleutian show a much more restricted compositional range. It might be argued that the restricted compositional ranges shown on Fig. 13 are an artifice of the limited number of available analyses. This may be true for centers such as Martin (n = 2) or Devils Desk (n = 4). However, volcanoes such as Augustine (n = 21) and Trident (n = 36) also show restricted compositional ranges. The restriction here is probably not related to a limited number of samples, but may be related to a limited number of separate rock units.

Igneous fractionation in the Aleutian arc follows either a calc-alkaline or tholeiitic trend. Kay et al. (1982) have used FeO*/MgO vs. SiO_2 plots to distinguish between these differentiation trends and Fig. 14 shows the results for the eastern Aleutian lavas. Increases in FeO*/MgO with increasing SiO_2 are characteristic of tholeiitic lavas while calc-alkaline differentiation shows little, if any, increase in FeO*/MgO with increasing SiO_2. A number of volcanic centers in the eastern Aleutian arc show similar overlapping trends on Fig. 14 and these centers have been combined. None of these centers show FeO*/MgO increases with increasing SiO_2. In fact, only Kejulik could be characterized as tholeiitic based on Fig. 14. Other volcanic centers (such as Kaguyak or Katmai) have individual analyses that plot in the tholeiitic field, but lack the pattern of iron-enrichment with increasing SiO_2. Volcanoes of the Katmai and Cook segments appear to be uniformly calc-alkaline with the exception of Kejulik, based on the FeO*/MgO vs. SiO_2 plots. This trend is quite different from the central and western Aleutian arc where segment-boundary volcanoes show an increase in FeO*/MgO with increasing SiO_2 (tholeiitic), while the calc-alkaline intrasegment volcanoes show little increase in FeO*/MgO with increasing differentiation (Kay et al., 1981). The FeO*/MgO vs. SiO_2 diagram is not a useful discriminator of volcanoes in the eastern Aleutian arc.

PETROLOGY AND VOLCANIC ARC GEOMETRY

Magmas of the eastern Aleutian arc show little compositional variation along the trace of the arc. The two major segments, Katmai and Cook (Fig. 10), are quite similar in their patterns of variation (Figs. 12, 13 and 14). Both arc segments are dominated by andesites that show a rather restricted range of variation (Fig. 14). Cook lavas are slightly higher in CaO and MgO, but lower in Na_2O and K_2O relative to the Katmai samples (Fig. 12). The variation of K_2O

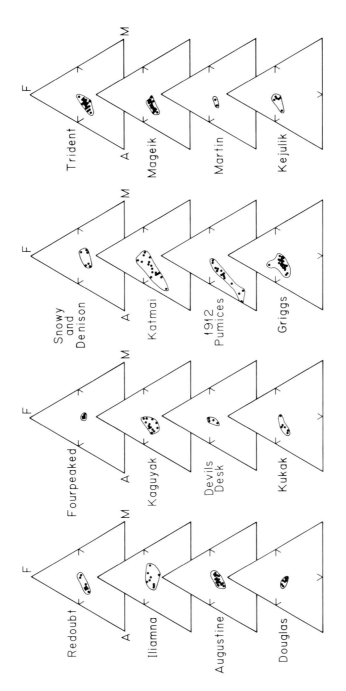

Fig. 13 A ($Na_2O + K_2O$), F ($FeO + 0.899\ Fe_2O_3$), M (MgO) diagrams for volcanic centers in the eastern Aleutian arc. Data sources listed in the caption for Fig. 12.

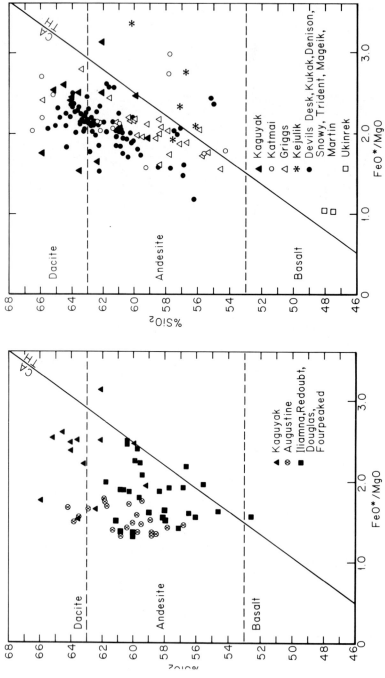

Fig. 14 FeO*/MgO vs. SiO$_2$ variation diagram, where FeO* = FeO 0.899 Fe$_2$O$_3$. CA/TH is the calc-alkaline/tholeiitic trend for island arcs from Miyashiro (1974). Data from Cook Inlet (Kaguyak, Augustine, etc.) and Katmai (Kaguyak, Katmai, etc.) are on separate diagrams with Kaguyak being repeated for comparison.

(at a particular SiO$_2$ value) along the eastern Aleutian arc is shown on Fig. 15. Generally volcanic centers in the Katmai segment are richer in K$_2$O than the Cook Inlet volcanic centers. Kaguyak, located at the segment boundary has the lowest K$_2$O content in the eastern Aleutian arc, while Mt. Katmai, another volcano that may lie on an arc segment boundary, is as low or lower in K$_2$O than other nearby volcanoes. Augustine is also relatively low in K$_2$O (Fig. 15). This correlation of low K$_2$O contents with segment-boundary volcanoes is similar to the pattern in Central America where low K$_2$O volcanic rocks are found (more or less) on segment boundaries (Carr et al., 1979).

Volcanic centers in the Katmai segment are aligned at an oblique angle to contours on the Benioff zone seismicity (Fig. 10). Attempts to relate the chemical variation to the depth to Benioff zone have not been successful (Kienle et al., 1982). In some volcanic arcs a correlation is found between K$_2$O content (at a particular SiO$_2$ value) and depth to the seismic zone (Dickinson, 1975). However, only a poor correlation between seismicity and K$_2$O variation in lava chemistry could be found for the eastern Aleutian arc (Kienle et al., 1982). No correlation between depth to Benioff zone and other chemical components can be demonstrated for the eastern Aleutian lavas.

Volcanic centers that lie along segment boundaries in the eastern Aleutian arc seem to be characterized by a higher degree of fractionation within the lava suite, resulting in a diversity of rock types or more evolved mineral assemblages. Kaguyak is characterized by dacites that form domes, flows, and ash flows containing phenocrysts of quartz and relatively abundant hornblende. The prevalence of this mineral assemblage is unique in the eastern Aleutian arc. Katmai and Ugashik, volcanic centers that seem to mark segment boundary

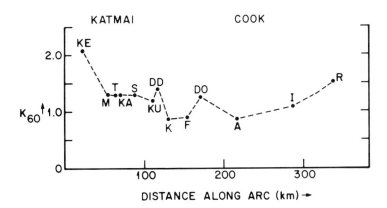

Fig. 15 Variation of K$_{60}$ (K$_2$O content at 60% SiO$_2$) along the strike of the eastern Aleutian arc. KE = Kejulik, M = Mageik, T = Trident, KA = Katmai, S = Snowy, KU = Kukak, DD = Devils Desk, K = Kaguyak, F = Fourpeaked, DO = Douglas, A = Augustine, I = Iliamna and R = Redoubt.

volcanoes along the arc, are characterized by a diverse suite of lavas (Tables 1 and 2, Fig. 13). Andesite is the most common lithology at Augustine, but some dacite and rare basaltic hyaloclastite are also found on this island.

DISCUSSION

Tectonic segmentation in the eastern Aleutian arc is generally well defined and there are two 250 km long segments (Fig. 10). Volcanic segmentation is less well defined and suggests that there are subsegments approximately 100 km long. Based upon the alignment of the volcanic front, Kienle et al. (1982) drew the boundary between Augustine and Douglas. Fisher et al. (1981) propose a segment boundary in the same region based on anomalies in crustal geology and tectonics. Based solely on volcanic alignment, Marsh (1982) proposed the segment boundary is through Kaguyak Crater, a model which explains the anomalous character of the dacites of Kaguyak. These models can be brought into agreement if boundaries are located at Augustine and also through Kaguyak resulting in a subsegment that contains Fourpeaked and Douglas (Fig. 10). Thus, the tectonic segment boundaries pass through Ugashik and Augustine (Fig. 10), while volcanic subsegment boundaries include Aniakchak, Katmai, Kaguyak, and Hayes.

The boundary volcanoes, Augustine, Kaguyak, Novarupta/Katmai, and Ugashik appear to be characterized by calc-alkaline fractionation that has taken place at shallow levels in the crust. This accumulation of magma at shallow depths promotes caldera formation. Kaguyak lavas are dominated by quartz- and hornblende-bearing dacite prior to the caldera-forming eruption. Dacitic pumices erupted during the caldera-forming event contain quartz and hornblende in a glassy groundmass. Post-caldera andesites at Kaguyak contain neither quartz nor hornblende. Swanson et al. (1981) interpret this eruptive sequence as the draining of a zoned magma chamber under Kaguyak. The quartz- and hornblende-bearing dacites represent the top of the chamber while the post-caldera andesites represent a lower level of the magma chamber. Hildreth (1981) notes a similar eruptive sequence from many zoned magma chambers. Fractionation within the Kaguyak lavas apparently took place at shallow levels because the phenocrysts and groundmass minerals have similar compositions (Fig. 16).

Hildreth (1982) studied the stratigraphy and petrology of the 1912 VTTS deposits and concluded that the 1912 Novarupta eruption partially drained a shallow zoned magma chamber. The VTTS deposits have a rhyolite at the base and become dacitic and andesitic toward the top. Magma mixing, as evidenced by extensive banded pumices, is common in the upper portions of the deposits. Hildreth (1982) reports distinct compositional breaks between the rhyolite, dacites and andesite in the VTTS deposits, suggesting the presence of three distinct lavas in the magma chamber prior to eruption. Temperatures calculated from Fe-Ti oxides in the VTTS deposits show a continuum from rhyolite through dacite, to andesites suggesting that the 1912 lavas were comagmatic. Therefore, the mixing evidenced in the 1912 products is probably the result of internal mixing of more- and less-fractionated magmas prior to eruption. Results of

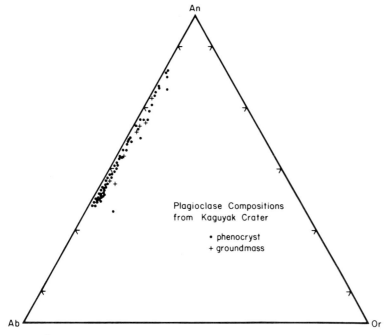

Fig. 16 Comparison of phenocryst and groundmass plagioclase compositions from Kaguyak Crater.

seismic and gravity studies (Kubota and Berg, 1967; Berg et al., 1967; Matumoto and Ward, 1967; Kienle, 1969; Matumoto, 1971) are consistent with the presence of a large shallow-level (< 10 km deep) magma chamber under the entire Griggs-Novarupta-Trident-Katmai area.

Augustine lavas do not show an Fe-enrichment trend (Figs. 13 and 14). The Augustine and Kaguyak lavas show similar patterns of variation on AFM diagrams (Fig. 14) and both lava suites have relatively low K_2O contents (Fig. 15). Johnston (1978) studied the mineralogy and fluid inclusions of the 1976 Augustine eruptive products and estimated 3-8 km as the depth of magma fractionation prior to eruption.

Ugashik (Table 2) is another segment boundary volcano (Fig. 10) that has erupted a suite of lavas that range from rhyolite to andesite. However, little is known about the petrology of this volcanic center. Only the abundance of dacite and rhyolite suggests a similarity to the Kaguyak and Novarupta/Katmai systems.

Segment-boundary volcanoes in the eastern Aleutian arc (Kaguyak and Katmai) are characterized by calc-alkaline differentiation in contrast to the tholeiitic segment-boundary volcanoes in western Aleutian arc (Kay et al., 1982). Based upon the results of trace element analyses, parental magmas for the calc-alkaline and tholeiitic lavas appear to be the same (Kay et al., 1982), but the fractionation mechanisms are different. Kay et al. (1982) suggest that tholeiitic lavas fractionate at shallower levels than calc-alkaline lavas. Perhaps the ascent of tholeiitic lavas to shallower levels is aided by zones of crustal weakness at transverse arc boundaries. As discussed previously, transverse arc boundaries in the eastern Aleutian arc appear to be related to changes in crustal geology, but here these zones of weakness do not promote tholeiitic volcanism.

Volcanic centers of the Cascade Range are segmented (Hughes et al., 1980) and the segment boundary volcanoes are not associated with tholeiitic differentiation. Instead, the segment-boundary volcanoes of the Cascades show calc-alkaline fractionation as do the intrasegment volcanoes. However, the intrasegment volcanoes are typically andesitic while the boundary volcanoes range from rhyolite to basalt (McBirney, 1968; Hughes et al., 1980). This same pattern of chemical variation is found in the eastern Aleutian arc.

A rather thick section of continental crust underlies both the Cascade Range and the eastern Aleutian arc, while the central Aleutian arc is underlain by either oceanic crust or relatively thin continental crust. The presence of thick continental crust seems to control whether the segment boundary volcanoes are calc-alkaline or tholeiitic. Marsh (1982) related the presence of rhyolite on the Alaska Peninsula to the presence of continental crust and suggested the rhyolite formed by the melting of the continental crust. However, Sr-isotope studies cited by Hildreth (1982) on the rhyolite of the VTTS deposits show no evidence for crustal contamination. Thus, whatever role the continental crust has in controlling fractionation in the segment-boundary volcanoes, it apparently does not involve partial melting or chemical interaction of the crust with the rising magmas.

Kosco (1981b) recognized that the continental crust was not involved in magma generation in the Katmai area. Instead he felt that the density similarity between the continental crust and the rising magma would stagnate the magma near the base of the crust (about 30-35 km under Katmai) and fractonation would begin. Fractionation at such moderate pressure would produce hornblende. Kay et al. (1982) believe this fractionation would follow a calc-alkaline trend. Structural traps higher in the crust beneath the Katmai volcanoes may result in fractionation at still shallower levels, thus accounting for the textural disequilibrium observed for some phenocryts (decomposition of amphibole, reaction coronas on some quartz and hypersthene).

Geophysical evidence indicates a discontinuity (offset of layers of different density) in the crust that extends along the strike of the volcanic arc from the VTTS region to about Kaguyak. This discontinuity may provide a structural trap for the accumulation of magma at shallow depths in the crust. Since the discontinuity extends laterally along the volcanic arc, magma may accumulate to form elongate bodies that feed several closely spaced volcanic vents. This may account for some of the close spacing of volcanic vents in the Katmai segment (such as Devils Desk-Kukak-Denison/Stellar and the line of 3 peaks forming Snowy). Intersection of this lateral structural discontinuity with a subsegment boundary near the VTTS may explain the abnormally large magma chamber that exists at shallow depths in this region. The differences between the eruptive products at Katmai (greater diversity, more K_2O) along with Augustine and Kaguyak (less diversity and K_2O) shown on Fig. 13 and 15 may be related to the shallow level, structurally controlled differentiation at Katmai as opposed to fractionation of the other lavas deeper in the crust.

CONCLUSIONS

Subduction in the eastern Aleutain arc is marked by a smooth, continuous Benioff zone and a segmented volcanic arc. Transverse arc boundaries offset the volcanic front and correspond to some change in crustal topography and regional geology. Back-arc volcanic centers (Ukinrek and Griggs) are sometimes found behind the main volcanic front along the transverse boundaries. The strike of the Benioff zone and trend of the volcanic arc cross-cut in the Katmai segment, but are generally parallel in the Cook Inlet segment.

Despite the difference in the tectonic setting between the Katmai and Cook segments, the chemistry of the eruptive products between the two areas is remarkably uniform andesites. Only the segment boundary volcanoes Augustine, Ugashik, Katmai/Novarupta, and Kaguyak are significantly different and contain a wider range of compositions from rhyolite to basalt. Mineralogic and petrologic trends suggest initial fractionation of the magmas near the base of the crust. Intrasegment volcanoes tap this lower crust magma body and erupt the magma with little further fractionation. Segment boundary volcanoes tap the same magma source, but discontinuities higher in the crust allow ponding of the magma promoting further fractionation, resulting in a wider diversity of lava types.

ACKNOWLEDGEMENTS

Daniel Kosco generously made his analytical data on Griggs, Katmai, Trident and Mageik available to us and without this excellent data base our speculations would have been even less constrained. Wes Hildreth sent us a preprint of his work in the VTTS for which we are grateful. Cooperation of the National Park Service personnel, especially David Morris and Bruce Kaye, was invaluable during our work in Katmai National Park. We also thank William Witte who calculated plate motion vectors and rates for Fig. 5. This study was supported by the U.S. Bureau of Land Management (BLM) through an interagency agreement with the U.S. National Oceanic and Atmospheric Administration (Contract NOAA-03-04-022-55) under which a multi-year program responding to the need of petroleum development of the Alaska Continental Shelf is managed by the Outer Continental Shelf Environmental Assessment Program (OCSEAP) office. The operation of a part of the seismic network is supported by the office of Basic Energy Science of the U.S. Department of Energy (Contract EY-76-S-06-02229, Task No. 6, Dr. Hans Pulpan, principal investigator).

REFERENCES

Becker, G.F., 1898. Reconnaissance of the goldfields of southern Alaska, with some notes on the general geology, U.S. Geol. Surv., 18th Ann. Rept., Part 3: 28-31, 50-58.

Berg, E., and Kienle, J., 1966. Gravity measurements in the Katmai volcano area, Alaska. Univ. Alaska, Geophys. Inst. Rep., UAG R-176: 13 pp.

Berg, E., Kubota, S., and Kienle, J., 1967. Preliminary determination of crustal structure in the Katmai National Monument, Alaska. Bull. Seismol. Soc. Am., 57: 1367-1392.

Carr, M.J., Rose, W.I., and Mayfield, D.G., 1979. Potassium content of lavas and depth to the seismic zone in Central America. J. Volcanol. Geotherm. Res., 5: 387-401.

Carr, M.J., Rose, W.I., and Stoiber, R.E., 1982. Central America. In: R. S. Thorpe (Editor), Andesites. John Wiley and Sons, New York, N.Y.: 149-166.

Coats, R., 1962. Magma type and crustal structure in the Aleutian arc. In: G. MacDonald and H. Kuno (Editors), The Crust of the Pacific Basin. Am. Geophys. Union, 6: 92-109.

Cormier, V.F., 1975. Tectonics near the junction of the Aleutian and Kuril-Kamchatka arcs and a mechanism for Middle Tertiary magmatism in the Kamchatka Basin. Geol. Soc. Am. Bull., 86: 443-453.

Curtis, G.H., 1968. The stratigraphy of the ejecta of the 1912 eruption of Mount Katmai and Novarupta, Alaska. Geol. Soc. Am. Mem., 116: 153-210.

Davidson, G., 1884. Notes on the volcanic eruption of Mount St. Augustine, Alaska, October 6, 1883. Science, 3: 186-189.

Davies, J.N., 1975. Seismological investigations of plate tectonics in south-central Alaska. Ph.D. Thesis, University of Alaska, Fairbanks: 193 pp.

Decker, R.W., 1963. Proposed volcano observatory at Katmai National Monument, a preliminary study. Unpublished report, Dartmouth College, Hanover, N.H.: 54 pp.

Detterman, R.L., 1968. Recent volcanic activity on Augustine Island Alaska. U.S. Geol. Surv. Map, GQ-1068.

Dickinson, W.R., 1975. Potash-depth (K-h) relations in continental margin and intra-oceanic magmatic arcs, Geology, 3: 53-56.

Doroshin, P., 1870. "Some volcanoes, their eruptions, and earthquakes in the former Russian holdings in America". Verhandlungen der Russisch-Kaiserlichen Mineralogischen Gesellschaft zu St. Petersburg, Zweite Serie, Fünfter Band: 25-55 (in Russian).

Engdahl, E.R., 1977. Seismicity and plate subduction in the central Aleutians. In: M. Talwani and W. C. Pitman III (Editors), Island Arcs, Deep Sea Trenches, and Back-Arc Basins. Am. Geophys. Union, Maurice Ewing. Ser., 1: 259-272.

Estes, S.A., 1978. Seismotectonic studies of lower Cook Inlet, Kodiak Island and the Alaska Peninsula areas of Alaska. M.S. Thesis, University of Alaska, Fairbanks, 142 pp.

Fedotov, S.A. and Tokarev, P.I., 1973. Earthquakes, characteristics of the upper mantle under Kamchatka, and their connection with volcanism (according to data collected up to 1971). Bull. Volcanol., 37-2: 245-257.

Fenner, C.N., 1920. The Katmai Region, Alaska, and the great eruption of 1912. J. Geol., 28: 596-606.

Fenner, C.N., 1922. Evidences of assimilation during the Katmai eruption of 1912. Bull. Geol. Soc. Am., 33: 129 pp.

Fenner, C.N., 1923. The origin and mode of emplacement of the great tuff deposit of the Valley of Ten Thousand Smokes. National Geograph. Soc. Contributed Technical Papers, Katmai Series, 1: 1-74.

Fenner, C.N., 1926. The Katmai magmatic province, J. Geol., 35: 673-772.

Fenner, C.N., 1930. Mount Katmai and Mount Mageik. Z. Vulkanologie, 13: 1-24.

Fenner, C.N., 1950. The chemical kinetics of the Katmai eruption. Am. J. Sci., 248: 593-627.

Fisher, M.A., Bruns, T.R., and Von Huene, R., 1981. Transverse tectonic boundaries near Kodiak Island, Alaska. Geol. Soc. Am. Bull., 92: 10-18.

Forbes, R.B., Kay, D.K., Katsura, T., Matsumoto, H, Haramura, H., Furst, M.J., 1969. The comparative chemical composition of continental vs. island arc andesites in Alaska. In: A. R. McBirney (Editor), Proceedings of the Andesite Conference, Int. Upper Mantle Project, Scientific Rep. 16, State of Oregon, Dept. of Geol. and Min. Res., Bull. 65: 111-120.

Griggs, R.F., 1920. Scientific results of the Katmai expedition of National Geographic Society. Ohio State University, compilation of other papers: 492 pp.

Griggs, R.F., 1922. The Valley of Ten Thousand Smokes. National Geograph. Soc., Washington, D.C.: 340 pp.

Hildreth, W., 1981. Gradients in silicic magma chambers: implications for lithospheric magmatism. J. Geophys. Res., 86: 10153-10192.

Hildreth, W., 1982. The compositionally zoned eruption of 1912 in the Valley of Ten Thousand Smokes, Katmai National Park, Alaska, J. Volcanol. Geotherm. Res., in press.

Holden, J.C. and Kienle, J., 1977. Geometry of a subducted plate and anti-arc segmentation of the Aleutian volcanic chain. Trans. Am. Geophys. Union, 58: 168.

Hughes, J.M., Stoiber, R.E., and Carr, M.J., 1980. Segmentation of the Cascade volcanic chain. Geology, 8: 15-17.

Jacob, K.H., Nakamura, K., and Davies, J.N., 1977. Trench-volcano gap along the Alaska-Aleutian arc: Facts, and speculations on the role of terrigeneous sediments. In: M. Talwani and W. C. Pitmann III (Editors), Island Arcs, Deep Sea Treanches and Back-Arc Basins. Am. Geophys. Union, Maurice Ewing, Ser., 1: 243-258.

Jaggar, T.A., 1927. Eruption of Mageik in Alaska. The Volcano Letter, Hawaiian Volcano Observatory, No. 147.

Johnston, D.A., 1978. Volatiles, magma mixing, and the mechanism of eruption of Augustine Volcano, Alaska. Ph.D. Thesis, Univ. of Washington, Seattle: 177 pp.

Johnston, D.A., 1979. Onset of volcanism at Augustine Volcano, Lower Cook Inlet, U.S. Geol. Surv. Circ. 804B: B78-B80.

Juhle, W., 1955. Iliamna Volcano and its basement. U.S. Geol. Surv. Open File Rep.: 74 pp.

Karlstrom, T.N.V., 1964. Quaternary geology of the Kenai Lowland and glacial history of the Cook Inlet region, Alaska. U.S. Geol. Surv. Prof. Paper, 443: 69 pp.

Kay, S.M., Kay R.W., and Citron, G.P., 1982. Tectonic controls on tholeiitic and calc-alkaline magmatism in the Aleutian Arc. J. Geophys. Res., 87: 4051-4072.

Keller, A.A., and Reiser, H.N., 1959. Geology of the Mount Katmai area, Alaska. U.S. Geol. Surv. Bull., 1058-G: 261-298.

Kienle, J., 1969. Gravity survey in the general area of Katmai National Monument, Alaska. Ph.D. Thesis, University of Alaska, Fairbanks, 151 pp.

Kienle, J. and Forbes, R.B., 1976. Augustine - evolution of a volcano. Geophys. Inst., Univ. of Alaska, Fairbanks, Ann. Rep. 1975/76: 26-48.

Kienle, J. and Shaw G.E., 1979. Plume dynamics, thermal energy and long distance transport of Vulcanian eruption clouds from Augustine Volcano, Alaska. J. Volcanol. Geotherm. Res., 6: 139-164.

Kienle, J., Kyle, P.R., Self, S., Motyka, R.J., and Lorenz, V., 1980. Ukinrek Maars, Alaska, I. April 1977 eruption sequence, petrology and tectonic setting. J. Volcanol. Geotherm. Res., 7: 11-37.

Kienle, J., and Swanson, S.E., 1980. Volcanic hazards from future eruptions of Augustine Volcano. Univ. Alaska, Geophys. Inst. Rep., UAG R-275: 122 pp., appendix and map.

Kienle, J., Swanson, S.E., and Pulpan, H., 1981. Volcanic centers in the Katmai area, Alaska. EOS, Trans. Am. Geophys. Union, 62: 430 (abstract).

Kienle, J., Swanson, S.E., and Pulpan, H., 1982. Magmatism and subduction in the eastern Aleutian arc. In: D. Shimozuru and I. Yokoyama (Editors), Arc Volcanism. Advances in Earth and Planetary Sciences, Ctr. for Acad. Publ., Japan: in press.

Kosco, D.G., 1981a. Characteristics of andesitic to dacitic volcanism at Katmai National Park, Alaska. Geol. Soc. Am. Abstracts with Programs, 13: 490.

Kosco, D.G., 1981b. Part I: The Edgecumbe volcanic field, Alaska: an example of tholeiitic and calc-alkaline volcanism; Part II: characteristics of andesitic to dacitic volcanism at Katmai National Park, Alaska. Ph.D. Thesis, University of California, Berkely, California: 249 pp.

Kubota, S., and Berg, E., 1967. Evidence for magma in the Katmai volcanic range. Bull. Volcanol., 31: 175-214.

Lahr, J.C., Page, R.A. and Thomas, J.A., 1974. Catalog of earthquakes in south central Alaska, April–June 1972. U.S. Geol. Surv. Open File Rep., Menlo Park, California: 30 pp.

Lahr, J.C., 1975. Detailed seismic investigation of Pacific-North American plate interaction in southern Alaska. Ph.D. Thesis, Columbia University, New York: 140 pp.

Leake, B.E., 1978. Nomenclature of amphiboles. Can. Min., 16: 501-520.

Marsh, B.D., 1979. Island arc volcanism. Am. Sci. 67: 161-172.

Marsh, B.D., 1982. The Aleutians. In: R. S. Thorpe (Editor), Andesites. John Wiley and Sons, New York, N. Y.: 99-114.

Marsh, B.D. and Carmichael, I.S.E., 1974. Benioff zone magmatism. J. Geophys. Res., 79: 1196-1206.

Marsh, B.D. and Leitz, R.E., 1979. Geology of Amak Island, Aleutian Islands, Alaska. J. Geol., 87: 715-723.

Matumoto, T., and Ward, P.L., 1967. Microearthquake study of Mount Katmai and vicinity. J. Geophys. Res., 72: 2557-2568.

Matumoto, T., 1971. Seismic body waves observed in the vicinity of Mount Katmai, Alaska, and evidence for the existence of molten chambers. Geol. Soc. Am. Bull., 82: 2905-2920.

McBirney, A.R., 1968. Petrochemistry of Cascade andesite volcanoes. Andesite Conf. Guidebook, Ore. Dept. Geol. and Min. Ind. Bull. 62: 101-107.

McKenzie, D. and Parker, R.L., 1967. The North Pacific: An example of tectonics on a sphere. Nature, 216: 1276-1280.

Meinel, A.B., Meinel, M.P. and Shaw, G.E., 1976. Trajectory of the Mt. St. Augustine 1976 eruption ash cloud. Science, 193: 420-422.

Miller, T.P. and Smith, R.L., 1976. "New" volcanoes in the Aleutian volcanic arc. U.S. Geol. Survey Circular 733, U.S. Geol. Survey in Alaska, Accomplishments during 1975: 11.

Minster, J.B., and Jordan, T.B., 1978. Present-day plate motions. J. Geophys. Res., 83: 5331-5354.

Miyashiro, A., 1974. Volcanic rock series in island arcs and active continental margins. Am. J. Sci., 274: 321-355.

Motyka, R.J., 1978. Surveillance of Katmai caldera and crater lake, Alaska: 1977. Univ. of Alaska, Geophys. Inst. Rep.: 19 pp.

Muller, E.H., Juhle, W., and Coulter, H.W., 1954. Current volcanic activity in Katmai National Monument. Science, 119: 319-321.

Nye, C.J., 1982. Voluminous volcanism accompanying microplate accretion. Geol. Soc. Am., Abstracts with Programs, 14(4): 221 (abst.).

Péwé, T.L., 1975. Quaternary geology of Alaska. U.S. Geol. Surv. Prof. Paper, 835: 145 pp.

Post, A. and Mayo, L.R., 1971. Glacier-dammed lakes and outburst floods in Alaska. U.S. Geol. Surv. Hydrol. Inv. Atlas HA-455.

Pulpan, J., and Kienle, J., 1979. Western Gulf of Alaska seismic risk. Proc. 11th Offshore Tech. Conf., April 30-May 3, 1979, Houston, Texas: 2209-2218.

Ray, D.K., 1967. Geochemistry and Petrology of the Mt. Trident andesites, Katmai National Monument, Alaska. Ph.D. Thesis, University of Alaska, Fairbanks: 198 pp.

Riehle, J.R., Kienle, J., and Emmel, K.S., 1981. Lahars in Crescent River Valley, Lower Cook Inlet, Alaska. Alaska Div. of Geol. and Geophys. Surveys, Geologic Report, 53: 10 pp.

Rymer, M.J. and Sims, J.D., 1976. Preliminary survey of modern glaciolacustrine sediments for earthquake-induced deformational structures, south-central Alaska. U.S. Geol. Surv. Open File Rep., 76-373: 20 pp.

Self, S., Kienle, J., and Huot, J.-P., 1980. Ukinrek Maars, Alaska, II. Deposits and formation of the 1977 crater. J. Volcanol. Geotherm. Res., 7: 39-65.

Simkin, T., Siebert, L., McClelland, L., Bridge, D., Newhall, C., and Latter, J. H., 1981. Volcanoes of the World. Smithsonian Institution. Hutchinson Ross, Stroudsburg, Pennsylvania: 232 pp.

Smith, R.L., and Shaw, H.R., 1975. Igneous-related geothermal systems. In: D. E. White and D. L. Williams (Editors), Assessment of Geothermal Resources of the United States - 1975. U.S. Geol. Surv. Circ. 726: 58-83.

Snyder, G.L., 1954. Eruption of Trident Volcano, Katmai National Monument, Alaska, February-June, 1953. U.S. Geol. Surv. Circular 318: 7 pp.

Spence, W., 1977. The Aleutian arc: tectonic blocks, episodic subduction, strain diffusion, and magma generation. J. Geophys. Res. 82: 213-230.

Stauder, W., 1968. Tensional character of earthquake loci beneath the Aleutian trench with relation to sea-floor spreading. J. Geophys. Res., 73: 7693-7701.

Stith, J.L., Hobbs, P.V. and Radke, L.F., 1977. Observations of a nuee ardente from St. Augustine Volcano. Geophys. Res. Lett., 4: 259-262.

Stoiber, R.E., and Carr, M.J., 1973. Quaternary volcanic and tectonic segmentation of Central America. Bull. Volcanol., 37: 304-325.

Swanson, S.E., Kienle, J., and Fenn, P.M., 1981. Geology and petrology of Kaguyak Crater, Alaska. EOS, Trans. Am. Geophys. Union, 62: 1062 (abstract).

Van Wormer, J.D., Davies, J., and Gedney, L., 1974. Seismicity and plate tectonics in south central Alaska. Bull. Seismol. Soc. Am., 64: 1467-1475.

Ward, P.L., and Matumoto, T., 1967. A summary of volcanic and seismic activity in Katmai National Monument, Alaska. Bull. Volcanol., 32: 107-129.

Wilcox, R.E., 1959. Some effects of recent volcanic ash falls with special reference to Alaska. U.S. Geol. Surv. Bull., 1028-N: 409-476 and plates.

Wilson, C.R., and Forbes, R.B., 1969. Infrasonic waves from Alaskan volcanic eruptions: J. Geophys. Res., 74: 4511-4522.

Wilson, C.R., Nichparenko, S., and Forbes, R.B., 1966. Evidence for two sound channels in the polar atmosphere from infrasonic observations of the eruption of an Alaskan volcano: Nature, 211: 163-165.

LARGE-SCALE PHREATOMAGMATIC SILICIC VOLCANISM: A CASE STUDY FROM NEW ZEALAND

STEPHEN SELF*

[1]Department of Geology, Arizona State University, Tempe, AZ 85287, (U.S.A.)

(Received November 5, 1982; revised and accepted May 3, 1983)

ABSTRACT

Self, S., 1983. Large-scale phreatomagmatic silicic volcanism: a case study from New Zealand. In: M. F. Sheridan and F. Barberi (Editors), Explosive Volcanism. J. Volcanol. Geotherm. Res., 17: 433-469.

Pyroclastic deposits of the 20,000 year old Wairakei Formation occur in the region surrounding Lake Taupo in the central North Island of New Zealand. Recent advances in the correlation of both proximal and distal equivalents of the Wairakei deposits show that they were the result of one large eruption. The ash layers once covered most of New Zealand; they are also present in deep sea cores from the SW Pacific and on the Chatham Islands. The dispersal and grain-size parameters of the ash-fall members indicate that their vent area was within the Taupo volcano-tectonic depression which, for a long period, has contained a large lake.

The Wairakei eruption is interpreted to have been phreatomagmatic throughout. Each phase of the eruption sequence generated its own characteristic deposits. The sequence of events can be summarized as follows: (1) Eruption and deposition of a white, fine-grained, bedded, fall deposit; (2) eruption and deposition of a medium-grained, pumice-fall deposit; (3) generation of ultra-fine, wet, ash clouds, with a fall deposit accumulating largely as aggregates and accretionary lapilli accompanied by base surges; (4) formation of water-cooled pyroclastic flows and deposition of an extensive ignimbrite; (5) return to a dominantly vertical eruption column and deposition of a medium-grained bedded ash fall, immediately followed by another wet, ultra-fine fall deposit charged with accretionary lapilli. Finally, (6) an ignimbrite less widespread than that of phase (4) was produced. Co-ignimbrite fall deposits of fine vitric dust and medium-grained pumice were laid down over a huge area. There may have been a short hiatus in activity between phases (4) and (5).

---

*Present address: Department of Geology, University of Texas at Arlington, P. O. Box 19049, Arlington, TX 76019, U.S.A..

0377-0273/83/$03.00 © 1983 Elsevier Science Publishers B.V.

The ash-fall members are all of the phreatoplilnian type. The two accretionary lapilli-rich layers represent some of the most completely fragmented, finest, ash beds yet documented. Violent rain-flushing, particle aggregation and accretionary lapilli formation were largely responsible for deposition of these layers. Individual fall units have somewhat irregular thickness distributions reflecting both their mode of deposition from the wet eruption clouds and, perhaps, some contemporaneous fluvial erosion due to torrential rains.

Pyroclastic flows generated by the collapse of phreatoplinian eruption columns were steam-laden, cool, thick (up to 50 m) and highly mobile. The ignimbrite occurs as valley ponds on the distal side of mountains rising 600 m above Lake Taupo, and as a veneer of pyroclastic-flow material up to 65 km from source. The area covered by ignimbrite may have exceeded 10,000 km^2. Above the pyroclastic flows and beyond the distal ends, billowing, damp, ash-laden clouds were generated. Fine ash and accretionary lapilli fell from these clouds.

During this remarkable event the changing eruptive style was probably due to fluctuations in the mixing ratio of lake water to vesiculating magma, reflecting a complex interplay between water access, mixing processes and morphology of the vent area. The inconsistent primary thickness of all members, and extensive fluvial and aeolian erosion after the event, make volume estimates for the deposits very approximate. The fine fall deposits (phreatolinian and co-ignimbrite ashes) may exceed 90 km^3 (dense rock equivalent); the ignimbrites probably totalled at least 65 km^3.

This study demonstrates that large-scale phreatomagmatic eruptions produce a variety of ash-fall and pyroclastic-flow deposits of exceptionally wide dispersal. Future rhyolitic eruptions from the Taupo volcanic center could possibly be of this type. An understanding of processes involved in such events can aid volcanic hazards planning.

INTRODUCTION

Fine-grained deposits of rhyolitic pumice, variously known as the Oruanui, Kawakawa, and Wairakei Formations in the Taupo region, are probably one of the largest volume pyroclastic deposits erupted from the central North Island of New Zealand during the past 100,000 years. These units are the product of the greatest magnitude eruption from the Taupo Volcanic center (Healy, 1962) within this time period (Froggatt, 1982). In the Taupo Volcanic Zone (Fig. 1), there have been two phases of mid-late Quaternary explosive rhyolitic activity (Healy, 1964); (1) an early phase from about 2.1 m.y. to 100,000 years ago, from which the welded ignimbrites of New Zealand are the best known products (Kohn, 1973; Murphy and Seward, 1981), and (2) a more recent phase starting 100,000 years ago, typified by Plinian eruptions and the formation of nonwelded ignimbrites. During this latter phase many of the eruptions produced extrusive volumes of 5 to 30 km^3 (expressed as dense rock equivalent). Two, however, approached or may have exceeded 100 km^3. One of these, the Rotoiti/Rotoehu event from Okataina

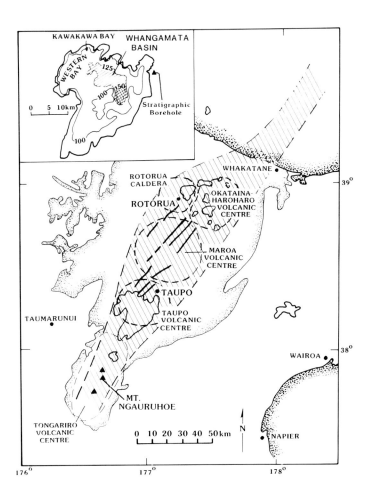

Fig. 1 The Taupo Volcanic Zone, North Island, New Zealand (after Cole and Nairn, 1975) showing the 4 main rhyolitic volcanic centers (dashed circles), major faults and andesite volcanoes of Tongariro Volcanic Center (triangles). Thin line encloses outcrop of Quaternary ignimbrites. Shaded area is active part of zone. Inset shows Lake Taupo (357 m above sea level) with locations mentioned in text and bathymetry in meters (after Irwin, 1972).

Caldera (Fig. 1), is dated tentatively at >42,000 y.b.p. and has been described by Nairn (1971, 1972), and Walker (1979). The other event, from the Taupo Volcanic Center, is the subject of this paper.

The Oruanui Pumice Formation was described by Vucetich and Pullar (1964), who first mapped its distribution in New Zealand (see also Pullar and Birrell, 1972). Recent unpublished work by the author and J. Healy show that the Oruanui Pumice Formation is a correlative of the earlier-named Wairakei Breccia (Grindley, 1965). The two were produced by the same eruptive episode and here the name Wairakei Formation is adopted. The Wairakei deposits are dated with some confidence at 20,000 y.b.p. by an average of seven independent ^{14}C dates on soils or carbonaceous material above and below the Formation (see Vucetich and Howorth, 1976a, their Table 1). A composite section (Fig. 2) shows that the Wairakei Formation can be divided into six members. Detailed tephrostatigraphic mapping has allowed the recognition and correlation of the six members across the North Island of New Zealand (Fig. 3). Individual members are mappable units distinguished either by their lithological characteristics or by bounding erosion surfaces, or both. In very distal locations, for example the Chatham Islands, where the Wairakei deposit is locally called Rekohu Ash (Mildenhall, 1976), stratificaiton is generally absent, perhaps due to bioturbation of the ash layers.

Deposits of the Wairakei Formation are typical rhyolites of the Taupo Volcanic Zone as described by Ewart (1968) and Ewart et al. (1968, 1975). The magma has a phenocryst content of 12-14% by volume, consisting of plagioclase (andesine), quartz, hypersthene, hornblende and titanomagnetite (Ewart, 1968; Briggs, 1973; Howorth, 1976), in order of decreasing modal abundance.

Conventionally, the deposits of large rhyolitic explosive eruptions are widespread pyroclastic-fall layers that are also coarse-grained near the source, their median grain-size decreases downwind and sorting generally improves away from the vent. Associated with such deposits are coarse, welded to nonwelded ignimbrites in the region surrounding the source. In stark contrast are the deposits of the Wairakei event; both ash-fall and pyroclastic-flow deposits with an exceptionally fine grain size were produced. The very widespread fall units have unusual characteristics such as locally irregular thickness distributions and a high content of accretionary lapilli. Data presented here indicate that the Wairakei eruption was phreatomagmatic, the source of water being in the Taupo basin, which has been a volcano-tectonic depression containing a lake for most of the past 100,000 years. In recent times the lake capacity has been about 28 km^3 of water. In an earlier stratigraphic study of the lowermost three members of the Wairakei Formation, Vucetich and Howorth (1976b) proposed that the source vent was located in an area of Lake Taupo near Kawakawa Bay (Fig. 1).

Throughout this paper, it is assumed that magma-water interaction produces a fine-grained assemblage of clasts in the eruption column (Walker and Croasdale, 1972; Wohletz, 1980 and this volume) and that varying ratios of magma to water change the grain-size characteristics of the resulting deposits. Self and Sparks (1978) based their recognition of phreatoplinian deposits - ash-fall beds

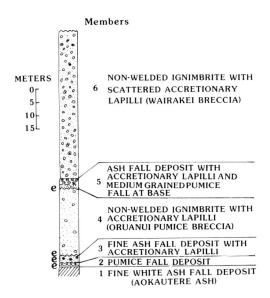

Fig. 2 Composite section through Wairakei Formation based on locations within 20 km from source: e indicates erosion surface. Previous names for members 1, 4 and 6 are given in parentheses (see text).

of Plinian extent but with very fine grain size and other features typical of phreatomagmatic tephra - to a large extent on the Wairakei deposits, then called the Oruanui Formation. They published a generalized isopach map (their Fig. 2) of the fall members. This paper gives a detailed account of both the ash-fall and pyroclastic-flow deposits. It attempts to reconstruct the eruption sequence and provide a mechanism to explain the characteristics of the deposits and the changing eruptive style.

STRATIGRAPHY AND PYROCLASTIC DEPOSITS

The Wairakei Formation consists of interbedded fine-grained, pyroclastic-fall and flow deposits in the proximal regions (Fig. 2). Characteristically, large Plinian deposits underlie ignimbrites, a sequence shown in many silicic eruptions (Smith, 1960; Sparks et al., 1973). Both phreatoplinian phases in the Wairakei event were followed by an ignimbrite-forming phase. The lower, widespread, nonwelded ignimbrite (member 4), occurs between two phreatoplinian deposits and could thus be called an intra-plinian ignimbrite, as described by Wright (1981), but it is of a simple, rather than compound type. In distal locations fine- to medium-grained ash falls (co-ignimbrite ash of Sparks and Walker, 1977; Sparks and Huang, 1980), and perhaps another type of ash fall

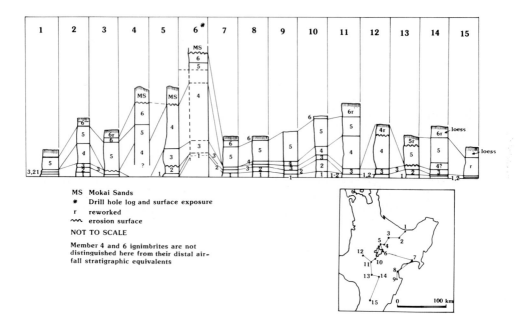

Fig. 3 Regional correlation diagram. Members are indicated by number against section. Wavy contacts are prominent intraformation erosion breaks. Inset is North Island of New Zealand showing location of correlated sections.

associated with secondary explosions in ignimbrites (Walker, 1979), are time correlatives of the nonwelded ignimbrite members. They are interstratified with and overlie the phreatoplinian ash-fall units. The six-member division given below is possibly a simplified representation of the deposits from this complex eruption. Future work may unravel more intricacies of the deposits representing detail in the timing of events between ash-fall and pyroclastic-flow phases.

Phreatoplinian fall deposits

Member 1 is a distinctive white ash which is fine-grained even near source, where it totals up to 24 cm thickness (Fig. 3, location 5 and Fig. 4). It has up to 12 inversely-graded fall units and is mantle-bedded upon older soils, loess, and eroded pyroclastic deposits from the Taupo Volcanic Center. Vucetich and Howorth (1976b) discuss the pyroclastic series underlying the Wairakei Formation. The ash contains small pumice clasts and shards, fragmented crystals of quartz, plagioclase and hypersthene, and tiny lithic fragments. The average grain size parameters of Md_ϕ = 4.0 and σ_ϕ = 2.1 for eight proximal samples show it to be exceptionally fine-grained. Very few pumice clasts are greater

Fig. 4 a) to c). Exposure of lower part of Wairakei Formation on Whangumata Road near Taupo (see also section 5, Fig. 3): a) members 1-4 with base-surge beds in member 3 showing pinch-and-swell structure. Members 2 and 1 are by base of shovel which is 90 cm long; b) close-up of a) showing members 1, 2 and the base of 3; c) close-up of an accretionary lapilli-rich fall unit in 3 overlain by one with few lapilli: knife is 8 cm long. d) exposure of Wairakei Formation near Wairoa, 130 km from source area on each coast of North Island. Scale is 50 cm long: members 1 through 6 are present and white band of member 1 is visible, bottom right; members 4 and 6 are co-ignimbrite ash.

than 1 mm in diameter. The overall grain-size population is so fine that the fall-off in median diameter downwind (Fig. 5) is almost imperceptible. In the southern area of the North Island member 1 is 2 cm thick and forms a prominent white bed. In these regions it was called the Aokautere Ash by Cowie (1964) and

440

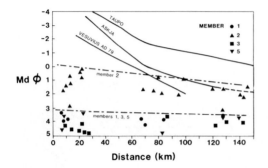

Fig. 5 Median diameter (Md$_\phi$) plotted against distance from vent for samples from fall members 1, 2, 3 and 5. The trends of the Taupo ultraplinian deposit (Walker, 1980) and two plinian deposits are shown for comparison (A.D. 79, Lirer et al., 1973; Askja, Sparks et al., 1981). Analyses from members 3 and 5 do not include accretionary lapilli. Limiting trendlines are shown for medium-grained member 2 and for the fine phreatoplinian fall members 1, 3 and 5.

Rhea (1968). This distinctive white ash, with an age of 20,000 years, is an important stratigraphic marker horizon over much of the North Island.

Member 2 is a thin, fine- to medium-grained pumice-fall deposit separated from member 1 by an erosion surface. In some proximal locations member 1 is completely missing and either member 2 or 3 is at the base of the formation. A maximum thickness of 8 cm is found in exposures 8-10 km north of Lake Taupo. In the most proximal locations it has a variable Md$_\phi$ of 0.2 to 1.7 and σ_ϕ of 1.0 to 2.0. The median diameter changes gradually downwind (Fig. 5). In distal regions along the dispersal axis of the lower fall members (Fig. 6), this member is up to 6-10 cm thick with 3 fall subunits distinguishable. In the southern part of the North Island it is a barely discernible layer < 1 cm thick and slightly coarser than member 1. The largest single pumice found in all of the Wairakei fall deposits comes from this layer; it is 3.3 cm in diameter. This is a remarkably small maximum pumice size for such widespread ash-fall deposits. An isopleth map of maximum pumice diameter shows the distribution of the unit (Fig. 7). Like member 1, member 2 is always partly, and often completely, eroded away in areas < 30 km from the presumed source area.

Member 3 is an extremely fine-grained fall deposit, light olive in color, with up to four distinct fall units totaling 2.6 m maximum exposed thickness. It contains accretionary lapilli in great abundance (Fig. 4c), up to 33 wt% in some ash layers. Concentrations almost entirely composed of accretionary lapilli mark the base of some fall units. The grain-size distribution of the ash and that of accretionary lapilli is similar (Table 1). A range of Md$_\phi$ (4.0-5.0)

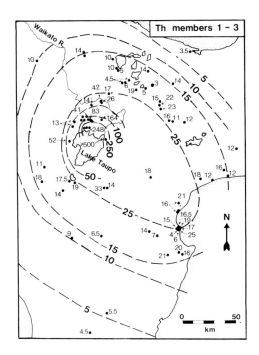

Fig. 6 Isopach map of measured thickness (in cm) of members 1 through 3 on North Island, N.Z. Note that deposits have variable thickness within small areas: isopachs are drawn in include greatest values in each area and thus generalize the thickness distribution. Some data points North of Lake Taupo have been omitted for clarity.

and σ_ϕ (1.5-3.0) has been determined for samples collected near Lake Taupo. A Coulter Counter was used for grain-size analyses of the fraction smaller than 63 µm(4ϕ). Sparse 'large' pumices 1-3 cm in size are occasionally found. Generally 98 wt% of the deposit is less than 1 mm in grain size and more than 65% is in the sub-50 µm size range. This is true of the deposit from proximal to distal regions. The median diameter of samples does not change significantly downwind (Fig. 5), even along 140 km of dispersal toward the east coast of the North Island.

The size distribution of accretionary lapilli in member 3 deceases downwind (Fig. 8a). Away from the dispersal axis of the lower three fall deposits, member 3 is generally devoid of recognizable accretionary lapilli. It is often severely eroded in regions nearer to source than 30 km; in many places the thickness varies from a few centimeters to more than 1 m between outcrops less than 1 km apart. It thins to about 6 cm along the east coast, 140 km downwind (Fig. 3, sections 8 and 9). Irregular, lens-like, bedforms on a centimeter

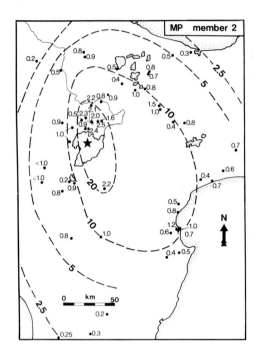

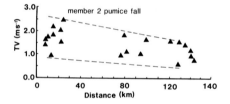

Fig. 7 Isopleth map of average maximum diameter of 3 largest pumice clasts in member 2 fall deposit: location values in cm, isopachs in mm. Star marks possible vent area. Below: median sea-level terminal fall velocity (after Walker et al., 1971) for sieved samples of member 2 plotted against distance from source area.

scale may be indicative of rain-splatter (see also Walker, 1981a, his Fig. 4).

In a few exposures on Whangamata Road near Taupo, (map ref. N93/373458)[1], 4-5 km from the north shore of Lake Taupo, base-surge deposits are found at the top of member 3 (Fig. 4). They show low-angle cross-bedding, pinch-and-swell forms,

---

[1]Map grid references refer to N.Z. Topographical Map Series 1, 1:63360, 1971 Edition.

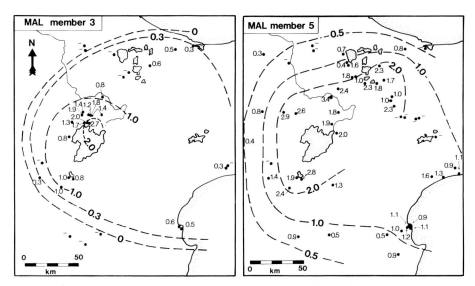

Fig. 8 a) Isopleth map of average maximum diameter (in cm) of 5 largest accretionary lapilli in member 3 ash-fall deposit; dash denotes accretionary lapilli absent in ash: b) as a) for member 5 fall deposit.

and contain accretionary lapilli. The presence of surge beds suggests that these exposures are some of the nearest to source (Self and Sparks, 1978).

Member 5 closely resembles member 3 in color, grain-size, and content of accretionary lapilli. The basal unit is a fine- to medium-grained, bedded, pumice-fall deposit similar to member 2. It is best seen in distal regions, but is so seldom exposed in proximal areas that it is not here designated as a separate member.

Member 5 exceeds 2 m in thickness in the few exposures near Lake Taupo. Evidence from a stratigraphic hole drilled near Taupo township (N94/355 571) in 1975 for purposes of solving correlation problems of the Wairakei Formation suggests it may be more than 6 m thick. It is composed of three prominent fall units, each one marked by a concentration of accretionary lapilli. This is especially apparent in the Pukeonake exposure (N112/065824) near Mt. Ngauruhoe (Fig. 1), where the whole of the Wairakei sequence is preserved in a small basin within the crater of a scoria cone (Topping, 1974). Here member 5, some 50 km south of source and south of the dispersal axis, is 65 cm thick (Fig. 9). On the east coast of the North Island, some 140-180 km downwind, member 5 is up to 40 cm thick near Wairoa, (Fig. 4d), where it contains 1.0 cm accretionary lapilli and has three prominent fall units. Maximum accretionary lapilli diameters are up to 4.5 cm in an exposure 20 km north of Lake Taupo and fall off to 1.6 cm on the east coast. An isopleth map of accretionary lapilli size distribution (Fig. 8b) shows a gradual decrease with distance from source.

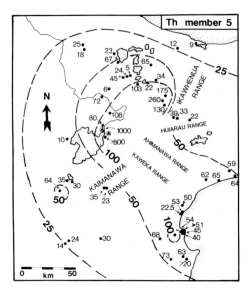

Fig. 9 Isopach map of ash-fall member 5, thicknesses in cm. For discussion of dispersal pattern, see text. Note that thicknesses are variable within small area; isopachs are drawn to include greatest value in each area and thus generalize the thickness distribution. Locations where thicknesses derived from borehole data shown as triangles. The main mountain ranges to the east of the Taupo region are indicated by their names.

Member 5 is probably the most voluminous of the ash-fall members and is the thickest member at locations on the east coast of the North Island (Fig. 9). In rare exposures in proximal locations, member 5 has a variable thickness; an eroded upper surface is assumed to occur, but cannot be proven directly in the field. Member 5 may contribute significantly to the undifferentiated exposures of distal upwind (westward) ashes, and to ashes found to the north and south more than 80-90 km from source. Typical exposures in these distal areas are decimeters thick but the only recognizable member is the white band of member 1.

Ignimbrite members

Member 4 is a nonwelded ignimbrite, originally known as the Oruanui Breccia (Vucetich and Pullar, 1969). It rests on member 3 disconformably, separated by an erosion surface in most near source localities. It is widespread north, northwest and southwest of Lake Taupo (Fig. 10), where it generally occurs in areas up to 600 m above sea level, and fills topographic lows and valleys. Everywhere the ignimbrite has a reworked top, generally represented by an aeolian deposit locally called the Mokai Sands (Vucetich and Pullar, 1969). These sands are up to several meters thick, are compositionally the same as the ignimbrite, and have low-angle ripple cross-laminations up to 30 cm in amplitude

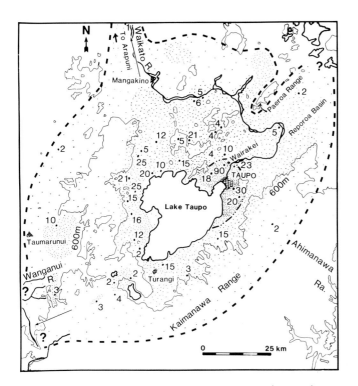

Fig. 10 Distribution of ignimbrite members 4 and 6 in area around Lake Taupo. Basins lower than 600 m (contour shown on map) have fill of ignimbrite (moderate stipple): thickest ignimbrite in Taupo-Wairakei area shown by heavy stipple. Areas between 600 m and 1500 m and distal regions probably had a cover of ignimbrite veneer deposit (light stipple). Thicknesses (in m) shown for selected locations must be minima, due to erosion of ignimbrite. Note basin and valley fills to NE and SW implying that pyroclastic flows topped mountain ranges to arrive in these areas. Major routes of fluvially reworked material in Waikato and Wanganui river systems are shown by arrows. Thick dashed line shows approximate limit of Wairakei Formation ignimbrites. Member 4 is widespread: member 6 dispersal is much reduced by erosion and is shown by dash-dot line.

and 90 cm in wavelength.

The ignimbrite has a pinkish hue, is nonwelded and has sparse, coarse pumice clasts. Typically only 5 wt% is coarser than 4 cm, unlike many other ignimbrites (Fig. 11a). Concentrations of large pumice, occasionally up to 42 cm in diameter, and coarse lithics are seen in a few areas north of the lake. Plots of maximum clast diameters in the body of the ignimbrite show coarser zones which may represent loci of the main pyroclastic flow routes, as suggested

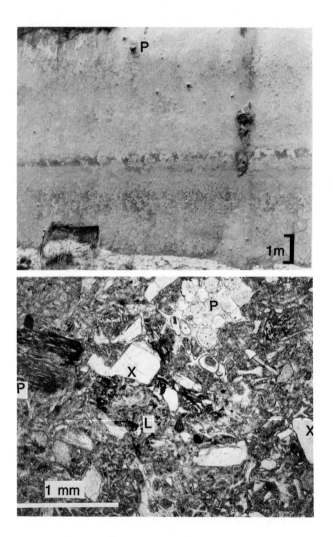

Fig. 11 a) Featureless member 4 ignimbrite (Oruanui Pumice Breccia) exposed in road-cut on Western Bay Road, 5 km west of Lake Taupo. Note lack of coarse pumice clasts (P). b) Photomicrograph of indurated member 6 ignimbrite from 8 km northwest of Taupo township on Poihipi Road, showing an undeformed shard matrix, pumice clasts (P), plagioclase phenocrysts (X) and a lithic fragment (L). Grey groundmass is fine, partly devitrified glass.

by Kuno et al. (1964) and Sparks (1975) for other ignimbrites (Fig. 12). The lithic assemblage consists largely of andesite and graywacke cobbles, highly suggestive of coarse fluvial deposits. Sparse angular spherulitic and banded rhyolite clasts and clasts of Whakamaru Ignimbrite (Briggs, 1976) also occur. The ignimbrite also contains up to 2 wt% of accretionary lapilli, which are sometimes concentrated into small depositional lenses. Some fumarolic pipes in the ignimbrite display concentrations of accretionary lapilli.

Component analyses of pumice, crystals and lithics show that the ignimbrite has a small lithic content but is slightly crystal enriched (an average of 19 wt% in 6 samples collected near source) over the magmatic content of crystals (14-15 wt%), determined by the mass of crystals from crushed pumice clasts. This indicates that some loss of fine dust from an eruption column and pyroclastic flows could have contributed towards a co-ignimbrite ash component.

In places a finer basal layer (layer 2a of Sparks et al., 1973) occurs at the contact with member 3, but pumice or lithic concentration zones of lateral extent are uncommon. Two flow units are recognized; a thick one which is up to 25 m in exposures, but undoubtedly was originally thicker before it was eroded, and a thinner, lower one that is poorly exposed. Up to 45 m of ignimbrite at the level of member 4 was cored in the Wairakei geothermal boreholes (Grindley, 1965).

A thin (<8 cm thick) lithic-rich bed of small clasts occurs at the base of the ignimbrite in many near-source locations and may be a ground layer (Walker et al., 1981b). Beds of large lithics (up to 1 m in diameter) interstratified in the ignimbrite are found near the north shore of Lake Taupo (best exposed at N93/413275). They resemble proximal co-ignimbrite lithic breccia deposits (Druitt and Sparks, 1982). The occurrence of these beds is indicatd on Fig. 12b, and provides further evidence that this is the source region. Near to this location, lithic-rich mudflow deposits overlie the member 4 ignimbrite.

Pyroclastic flows were widely dispersed and are thought to have climbed to considerable heights (300-600 m above the present level of Lake Taupo) to reach outlying regions such as Taurewa (N112/050896), the Taumarunui area, the Reporoa and Paeroa grabens towards Rotorua, and possibly valleys in the Kaimanawa Range, implying ascent of 900 m high ranges (C. J. N. Wilson, personal communication). The ignimbrite is also found in the Pukeonake scoria cone indicating that pyroclastic flows crossed the plateau (800-900 m above sea level) on which the presently active composite volcanoes of Tongariro, Ngauruhoe and Ruapehu sit. The upper parts of these volcanoes largely post-date the Wairakei Formation (Topping, 1973). The distribution of member 4 ignimbrite (Fig. 10) suggests that the pyroclastic flows covered a wide area with much topographic irregularity, but the thin ignimbrite 'veneer' deposits, characteristic of the passage of a pyroclastic flow (Walker et al., 1981b; Walker and Wilson, 1982a), have been largely eroded away.

It is proposed that the member 4 ignimbrite resulted from the collapse of a phreatomagmatic eruption column. Evidence for collapse is the proximal breccias and the widespread nature of the ignimbrite. Some criteria for recognition of

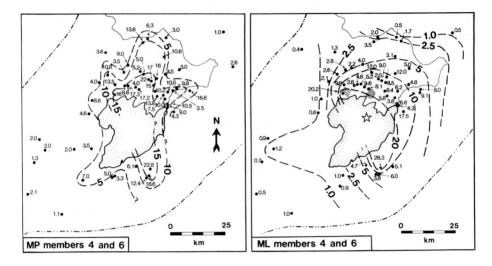

Fig. 12 a) Isopleth map of average maximum diameter of 5 largest pumice clasts in Wairakei Formation ignimbrites (members 4 and 6); isopleth lobes may follow main pyroclastic flow routes; b) isopleth map as a), for lithic clasts: shaded areas are locations of proximal lithic breccias in member 4. Star shows possible source area. Dash-dot line is edge of known dispersal of Wairakei Formation ignimbrites.

phreatomagmatic origin are: (1) the distinct lack of an abundant coarse pumice component compared to other ignimbrites; (2) the presence of rounded greywacke and andesite pebbles, possibly derived from deposits of such material that were on the bed of Lake Taupo; (3) the accretionary lapilli found in proximal exposures, unusual in most ignimbrites, but present in the eruption column immediately prior to collapse. Accretionary lapilli could have survived in the collapsing column or could have fallen into the pyroclastic flows from the ash clouds above.

Member 6 is another nonwelded ignimbrite, originally known as both the Wairakei (Pumice) Breccia (Grindley, 1965) and the Waitahauni (Pumice) Breccia (Grindley, 1960). Vucetich and Pullar (1969) suggested a correlation between the Oruanui and Wairakei Breccias (members 4 and 6), as did Kohn (1970). Although it superficially resembles member 4, member 6 is rather coarser-grained (typically 8-10 wt% greater than 4 mm), having a higher proportion of large pumice. It also has a more varied lithic assemblage including, in addition to rounded greywacke and andesite clasts, angular fragments of welded ignimbrite, rhyolite, chert and conglomerate. It also contains some basalt fragments in a coarse-grained, lithic-rich ground layer exposed near Taupo township (N94/355572). Again, accretionary lapilli are present and are particularly noticeable where the ignimbrite is indurated in the region of the Wairakei and Broadlands hydrothermal areas. The induration is probably due to replacement and cementation by secondary minerals (Steiner, 1953, 1977), not to welding or vapor-phase alteration. Thin section examination reveals an undeformed shard matrix and the absence of vapor phase minerals (Fig. 11b). Wood (charcoal) remains are notably absent from members 4 and 6. This may be due either to the coolness of the flows or to the lack of significant surface vegetation at the time of the eruption (Vucetich and Pullar, 1969).

The distribution of this ignimbrite member is mainly northeast to east from Lake Taupo, overlapping with the member 4 ignimbrite in the Taupo-Wairakei area. Thicknesses of 100 m of the Wairakei Formation have been recovered from boreholes drilled for gethermal exploration in the Wairakei region (see Grindley, 1965, Appendix III, Fig. 62). As with member 4, member 6 everywhere has an eroded top, reworked in many places to form the aeolian Mokai Sands. Accretionary lapilli occur in fumarolic pipes in the ignimbrite in several areas. Member 6 is also thought to be an ignimbrite generated from a phreatomagmatic eruption.

Wairakei ignimbrite exposures up to about 40 km from source are typical valley ponds of poorly-sorted nonwelded pumice and ash. Generally, the proximal ignimbrite was of a high aspect ratio type (see Wright, 1981). Beyond the valley ponds, the ignimbrite is thin (<5 m), lacks any coarse clasts and in some locations has an appreciable content of accretionary lapilli. This distal deposit might be interpreted as a type of low aspect ratio ignimbrite (Walker et al., 1979), deposited by pyroclastic flows which had lost much of their content of coarse clasts. Above the flows both co-ignimbrite and fine phreatoplinian ash mingled together to produce an eruption cloud that probably extended from ground level to great altitudes.

Co-ignimbrite and distal ash fall deposits

The stratigraphic position of member 4 is occupied by a thin fall unit of fine- to medium-grained, moderately well-sorted pumice (Fig. 3). The distal equivalent of member 6 is three pumice-fall units similar to those of member 4, which are especially prominent along the east coast of the North Island. The grain-size characteristics of these units change little over the area of outcrop. The median grain size and the moderately well-sorted nature show that they are not like most fine vitric co-ignimbrite ashes (Sparks and Walker, 1977). Walker (1979) as proposed that deposits of a similar nature formed from secondary explosions as hot pyroclastic flows entered water. In the Wairakei case, the explosions generating the pyroclastic flows were themselves within water and the distal ash beds could be a well-sorted, co-ignimbrite ash fall resulting from such explosions. Beds of very fine, vitric-enriched co-ignimbrite ash were also deposited totalling up to 2.5 to 3 m in some locations. They were later largely obliterated by erosion in the higher areas near source, but are preserved in distal regions where they appear to be very similar to the fine ashes of members 3 and 5. In the northeastern part of the dispersal, accretionary lapilli occur in the co-ignimbrite ash equivalent of member 6. On Fig. 13 these fine-grained ashes are included with the coarser ashes to show the distribution of co-ignimbrite fall deposits, which thicken away from Lake Taupo. These widespread ashes may account for a substantial fraction of the volume of material that fell at sea.

The regional correlation and distribution of the Wairakei fall members is summarized on Figs. 3 and 14. Wairakei ash occurs over much of the North Island of New Zealand, in the Amberley area of the South Island (Kohn, 1979), in submarine cores taken off eastern New Zealand (Lewis and Kohn, 1973; Kohn and Glasby, 1978) and in the Chatham Islands 800 km downwind of Lake Taupo (Fig. 15). Here, the Rekohu Ash, an average of 21 cm thick, has been dated confidently by C-14 at 20,100 ± 400 years (Mildenhall, 1976). It is a a vitric-rich, non-layered ash and may contain a significant proportion of co-ignimbrite ash. SEM examination of ash from mainland members and the Rekohu Ash yielded no clear clues as to which member the Chatham Island ash layer most resembled. Scanning electron photographs of the phreatoplinian fall deposits show blocky shards and 'micro-pumices', typical of phreatomagmatic ejecta (Fig. 16). Samples of ash grains from the mainland and the Chatham Islands are similar in both size and morphology.

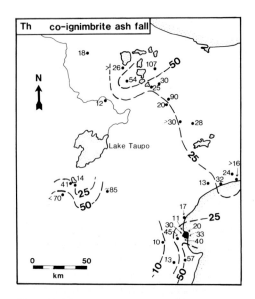

Fig. 13 Dispersal of co-ignimbrite ash-fall deposits shown by thickness distribution map (in cm). Fine- and medium-grained ash fall equivalents of members 4 and 6 are plotted together, giving a total thickness of co-ignimbrite ash. Note thickening away from Lake Taupo. Lack of data near source is due to widespread erosion and younger covering tephra deposits.

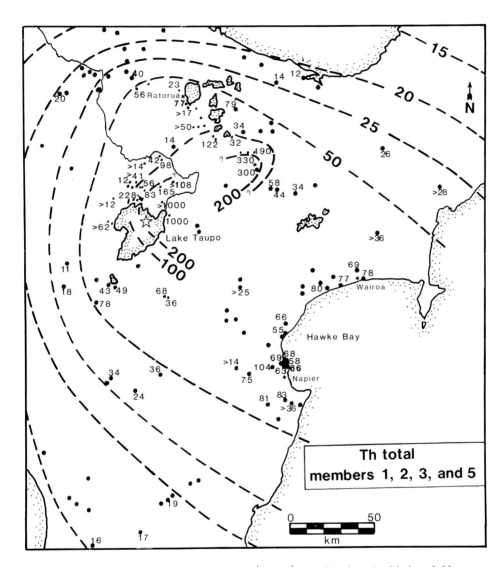

Fig. 14 Isopach map of total thickness (in cm) of all phreatoplinian fall deposits in Wairakei Formation. Locations without thickness values fall below the maximum values shown on the map and are compiled from work by Pullar and Birrel (1973) and the author.

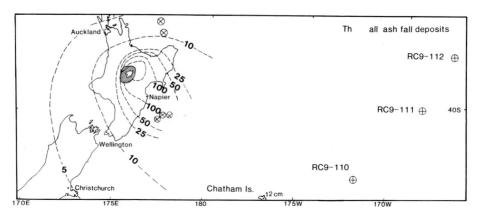

Fig. 15 Map of N.Z. and Pacific Basin showing Chatham Islands and core locations where Wairakei Formation has been tentatively identified; x = cores of Lewis and Kohn (1973) and Kohn and Glasby (1978). Isopachs for all primary Wairakei members, including co-ignimbrite ash, are shown in cm. RC9110-112 are Lamont-Doherty Geological Observatory deep-sea core sites (see Ninkovich, 1968). Shaded region around Lake Taupo is area covered by Wairakei Formation ignimbrites.

Ash layers from the Lamont-Doherty Geological Observatory deep sea core collection (see Ninkovich, 1968, for core numbers in the Pacific east of New Zealand), were sampled and minor elements in the glass analysed by electron microprobe. Correlation matrices (Borchardt et al., 1978) were constructed to test for similarity to ashes from the Wairakei Formation on land. Other Quaternary New Zealand ashes were also tested. Fig. 15 shows core locations where tentative correlatives of the Wairakei ash were found. Samples from the Rekohu Ash also correlates well on minor element glass chemistry with the Wairakei Formation glasses on land.

Dispersed and prominent ash-beds were described from deep-sea cores taken downwind from the Balleny Islands (68°S) by Watkins and Huang (1977) and some are in the 20,000 year age-range. These rhyolitic ashes, ascribed to a Balleny Island source by Watkins and Huang, may come from New Zealand and are in line with a southeasterly dispersal from the North Island. The Balleny Islands volcanoes are basaltic (Hatherton et al., 1965) and probably did not produce the ashes (P. R. Kyle and D. Seward, personal communication). It is feasible that the Wairakei ash could be recognized in South Pacific cores.

Reworked deposits

Vast amounts of Wairkei Formation ash were remobilized after deposition. The climate 20,000 years ago at 300-700 m elevation in the Taupo Region was cool, as this period was a stadial in the Otiran (Wisconsin) Glaciation (Barry and Williams, 1973). The presence of a periglacial-tundra type terrain with little vegetation may partly eplain the lack of vegetal remains in the Wairakei Formation deposits. Much material was removed by wind to form the Mokai Sands, local to the Taupo Area (Vucetich and Pullar, 1969), and to add to the North

Fig. 16 Scanning electron photomicrographs of untreated samples of Wairakei Formation: a) Member 3 fall deposit from exposure 4.5 km N. of Lake Taupo at location of Fig. 4. Note bimodal size distribution, vitric, platy 100-200 μm shards and lithics and sub 30 μm matrix. The fine matrix is mainly composed of vesicle wall fragments and clay particles. b) Rekohu ash (=Wairakei Formation) from Chatham Is., 800 km downwind from source. Sample courtesy of NZDSIR Soil Bureau. Note similar size of clasts to larger clasts in (a), 100-200 μm, and much better sorting. Shard character is similar. Vesicle walls are thicker than 10 μm and grains are fracture-bounded giving blocky, angular shards. c) Detail of shards in b) showing poorly vesicular 'micro-pumices' (suggestive of arrested vesiculation) and platy, but blocky vitric bubble wall shards. Photos courtesy of G. Heiken, Los Alamos, National Laboratories.

Island tephric loess deposits further afield (Pullar and Birrell, 1973; Cowie and Milne, 1973; Kohn, 1979). Fluvial equivalents of the Wairakei Formation in the Waikato River catchment are prominent in the Hinuera Formation (Schofield, 1965; Hume et al., 1975), which underlies the plains of the Hamilton Basin 75-100 km to the NW of Taupo. Pain and Pullar (1975) attribute terraces and lake deposits of post-20,000 year age in the Reporoa Basin, NE of Taupo (Fig. 10), to ash-fall and pyroclastic-flow deposition. The pyroclastic flows dammed the basin causing temporary lakes. Pyroclastic flows also reached the upper catchment of the Wanganui River system to the SW of source (Fig. 10). Campbell (1973) describes cross-bedded pumice sands forming terraces in that region and correlates these with the primary deposits of the Wairakei Formation. Much of the material in these voluminous post-20,000 year pumice sands was derived from the erosion of valley-filling ignimbrite. Water-reworked material probably choked all drainage systems leading south, west, and north from the Taupo Volcanic Zone after the eruption.

INTERPRETATION

Grain size characteristics of the deposits

A plot of median diameter (Md_ϕ) vs. graphic standard deviation (σ_ϕ, sorting) (Fig. 17a) shows that the phreatoplinian fall members (1, 3 and 5) are very fine-grained and, considering that the samples represent distances from source ranging from about 8 km to 180 km, show little change in grain size with distance downwind (see also Self and Sparks, 1978). Granulometric analysis of crushed accretionary lapilli from members 3 and 5 are consistently finer than the ashes they are found in. This is due to a more restricted size population in the coarser size range and a greater amount of fine dust in the fine size range, probably "scavenged" from the eruption cloud to form the fine outer rims of the accretionary lapilli. The ignimbrites (member 4 and 6) are at the finer-grained size range on the Md_ϕ vs. σ_ϕ plot (Fig. 17a) due to the paucity of coarse clasts. Samples notably richer in coarse pumice plot in the middle of the range for known ignimbrite size distributions (see e.g., Walker, et al., 1980, Fig. 5). On a plot of size parameters that compares the coarse and fine fractions of the size population (Fig. 17b), the Wairakei ignimbrites can be seen to be lacking in fine material. Many of the samples plotted are from proximal locations, where there may be little development of a fine population by attrition. Accretionary lapilli from proximal member 6 are distinctly finer than the ignimbrite that they are found in, and similar in median grain size, but better sorted, to those from the phreatoplinian ashes. This suggests that they were formed in the eruption column, by the same processes that formed the accretionary lapilli in the phreatoplinian clouds, and that they were incorporated into the pyroclastic flows during flow formation, or fell into the flows from the clouds above.

The Wairakei Formation fall deposits are of 'phreatoplinian' type (Self and Sparks, 1978), formed by explosive rhyolitic eruption through a voluminous body

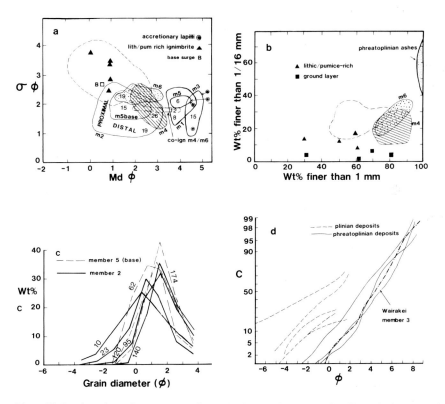

Fig. 17 Grain-size characteristics of the members of the Wairakei Formation a) Median diameter plotted against σ_ϕ: fields for various members shown, with number of analyzed samples represented by each field. Dashed field is 4% contour for pyroclastic flow deposits (see Wright et al., 1980). Circled points are crushed accretionary lapilli; tie lines connect lapilli to members from which they were sampled B = base surge bed in member 3: P = coarse, pumice-rich variant of ignimbrite of member 4. Member 5 plots are for fine, phreatoplinian ash only. b) Plot of weight percent finer than 1/16 mm against weight percent finer than 1 mm for ignimbrite members 4 and 6, and also for phreatoplinian members 1, 3 and 5. Dashed field is 4% contour of Walker et al. (1980) for pyroclastic flow deposits. Coarse variants are shown as individual symbols. c) Frequency curves for Wairakei fall members 2 and 5 (medium-grained base only) collected at increasing distances from source (numbers on curves refer to distance (in km)). Note more or less stationary mode at 250 μm and gradual loss of coarse tail. d) Plot of total grain size distributions for plinian and phreatoplinian deposits, after Carey and Sigurdsson (1982) and Walker (1981). A total grain size distribution representative of the finest Wairakei Formation fall deposit, based on analyses from member 3, is not unusually fine. Aggregate formation must occur to allow premature flushing of such deposits from eruption column. C is cumulative weight percent greater than size stated.

of water. A phreatomagmatic origin is suggested by the abundance of accretionary lapilli, interpreted to be formed in wet, fine-grained eruption columns (Self and Sparks, 1978; Sigurdsson, 1982; Brazier et al., 1982), the exceptionally fine grain-size, the general closure of isopachs and isopleths (Fig. 6-8) upon Lake Taupo (a lake basin of long history), and the presence of diatoms in the ash of member 3 (C. G. Vucetich, personal communication). Erosion of the ash was probably contemporaneous with deposition in some cases due to torrential rains from the water-laden eruption clouds. The most striking feature of the ultra-fine ash-fall deposits, especially members 1, 3 and 5, is the very high degree of magma fragmentation that can be implied to have occurred by magma-water explosive interaction.

Dispersal and deposition of ash

The great dispersal of even the coarsest members implies high eruption columns and high mass fluxes of magma (Wilson et al., 1978). The median terminal fall velocity of member 2 pumice is only 1-2 m s^{-1} (Fig. 7), giving ample time for the WNW winds to disperse the ash plume. Premature fallout with rain must have occurred in order to account for the deposition of proximal fine ash. Deposition occurred from all levels of the eruption cloud, probably in the form of mud rains and aggregates (e.g., pellets to accretionary lapilli), and was possibly influenced by local rainstorms or other meteorological phenomena, contributing to the uneven thicknesses of the fall units. The NE-SW elongation of isopachs of member 5 (Fig. 9) parallel to the main axial mountain ranges of the North Island supports this conclusion. A very approximate estimation of column height can be obtained from the accretionary lapilli. Using the data on accretionary lapilli size distribution on Fig. 8, the measured density of wet lapilli of 1600 kg m^{-3} and a drag coefficient of 0.5, and assuming release of lapilli from an eruption cloud into a stably-stratified atmosphere, a range of fall times can be calculated (Walker et al., 1971). Lapilli of 0.5 cm diameter are found at locations 130 km from source in member 3: an estimate for fall from 20 km is 33 minutes, implying wind speeds of 77 ms^{-1}. This is a reasonable upper tropospheric wind speed and is less than that estimated for dispersal of pumice during the Taupo event by Walker (1980). The above data suggests eruption columns in excess of 20 km were likely.

The prevalence of fine particles in the eruption cloud aided wide dispersal. A total grain-size distribution representative of the whole deposit for fall member 3 has been estimated by an average of 6 samples collected from near Taupo to a distance of 130 km downwind (Fig. 17d and Table 1). Due to the flushing of material from the column, the whole deposit is not greatly size-graded downwind. This assumption is supported by the consistent mode in the fine-size range for sieved samples at varying distances downwind (Figs. 5 and 17c). The size distribution determined for whole, carefully crushed accretionary lapilli is not dissimilar to the total deposit grain-size distribution (Table 1).

TABLE 1

Total grain-size distributions of phreatoplinian ashes

mm	ø	WAIRAKEI MEMBER 3		WAIRAKEI FORMATION ACCRETIONARY LAPILLI[c]				ROTONGAIO ASH[a]		MT. ST. HELENS 1980[b]	
				Member 3		Member 5					
		wt%	cum.wt%	wt%	cum.wt%	wt%	cum.wt%	wt%	cum.wt%	wt%	cum.wt%
2	−1	0.5	0.5	–	–	–	–	1.1	1.1	1.3	1.3
1	0	3.1	3.6	–	–	–	–	1.5	2.6	1.7	3.0
0.5	1	3.9	7.5	4.1	4.1	5.7	5.7	2.1	5.0	4.5	7.5
0.25	2	5.0	13.4	11.0	15.1	7.5	13.2	7.1	12.1	8.5	16.0
0.125	3	16.6	30.0	11.3	26.4	8.8	22.0	14.9	17.0	9.0	25.0
0.063	4	14.6	44.6	16.4	42.8	16.8	38.8	26.5	53.5	11.0	36.0
0.032	5	11.4	56.0	22.0	64.8	24.9	63.7	15.2	68.7	14.0	50.0
0.016	6	22.4	78.4					16.6	85.3	21.0	71.0
0.008	7	16.0	94.4					5.4	90.7	14.0	85.0
0.004	8	4.5	98.9	35.2	100.0	36.3	100.0	5.9	96.6	13.2	98.2
<0.004	>8	1.1	100.0					3.4	100.0	1.8	100.0

[a] after Walker, 1981a
[b] after Carey and Sigurdsson, 1982
[c] crushed lapilli from members 3 and 5

Vent location

Due to the fine grain-size of the fall deposits, accurate location of source by isopleth maps has proved difficult, but a general source area in the northern part of Lake Taupo is indicated (Fig. 7). One method adopted to represent the dispersal of the lowermost fall deposits is to sum the thicknesses of members 1-3 (Fig. 6). This leads to some oversimplification and an underestimate of thickness in the near-source area, but the isopachs again generally indicate the northern part of the lake as a source area. In this region of the lake is a large depression over 100 m deep, the Whangamata Basin (Fig. 1), largely filled by Taupo Ignimbrite and sediment (D. Northey, personal communication). That this may be the source region is further supported by the isopleth maps of accretionary lapilli (Fig. 8), the presence of base-surge deposits north of the lake (Fig. 4d), and by the size distribution of lithic fragments and of proximal breccias in the ignimbrites (Fig. 12b).

Volume of the Deposits

The volume of ash-fall deposits can be estimated by integrating under a curve of area (km^2) plotted against isopach thicknesses (see Walker, 1980). The curves are extrapolated to the 1 mm isopach, so estimates do not include deposits beyond this thickness. Following the method of Rose et al. (1973), a bulk volume of 150 km^3 for members 1-3 and 5 is obtained. This volume does not take into account the distal co-ignimbrite (layer 3) fall beds, which fell largely at sea. Walker (1971) and Sparks and Walker (1977) suggest that co-ignimbrite ashes may represent a substantial proportion (30-50%) of the volume erupted as pyroclastic flows in an ignimbrite-forming eruption. The volume of ignimbrite (members 4 and 6) is roughly estimated by totalling the area covered by various thicknesses, giving 130 km^3. This, again, may be an underestimate due to eroded material and the very uneven topography upon which the ignimbrites were deposited. Co-ignimbrite ash-fall deposits are conservatively estimated to be 30 km^3. The total volume obtained by summing these estimates is 310 km^3 bulk volume. Taking an average tephra density of 1.3 g cm^{-3} (the average of member 3, member 5 and the two ignimbrites) and expressing the volume as dense rock (2.6 g cm^{-3}), the volume is very approximately 155 km^3.

DISCUSSION

Eruption Sequence

The stratigraphy of the Wairakei Formation, together with field characteristics and grain-size distributions of the pyroclastic deposits, enables the eruption sequence and mechanisms to be evaluated. Twenty thousand years ago a large rhyolitic magma body under the northwestern part of the Taupo basin begin to erupt. The vent, or vents, were under proto-Lake Taupo and water had access to the vesiculating magma. The event began with the deposition of fine-grained, phreatoplinian ash (member 1) from a high eruption column. This widely dispersed ash shows a remarkably small decrease in thickness with distance from source. Its fine grain-size is a result of explosive mixing of vesiculating magma with lake water (Self and Sparks, 1978). The assemblage of grain-sizes ejected from the vent gradually coarsened and pyroclastic fall deposition continued, laying down a fine- to medium-grained pumice-fall deposit (member 2), which also shows a very gradual decrease in thickness downwind. The change of grain-size could be explained by a lower degree of magma fragmentation, perhaps implying that less water was reaching the magma at this stage of the eruption.

This condition was, however, temporary and a return to full-scale phreatoplinian activity took place, giving an extremely fine-grained fall deposit. This layer (member 3) fell as mud, particle aggregates, and accretionary lapilli, giving a patchy thickness distribution. The phreatoplinian ash clouds may have extended from ground level to the top of the eruption column, spreading laterally away from source with deposition occurring from a range of altitudes. Some base surges occurred and spread to about 10 km from the presumed source area. Up to this horizon, no ignimbrite flow units have been found in the stratigraphic succession, but there are no exposures of the lower members nearer than 10 km from the proposed source area. It is feasible that early pyroclastic flows occurred but were confined to the lake basin.

The wet eruption column then went into a period of collapse and voluminous pyroclastic flows swept out from source with great mobility, traversing a rough landscape, surmounting topographic barriers up to at least 600 m and perhaps as much as 1000 m above source, and leaving low regions (<600 m) full of ponded ignimbrite (member 4) (Fig. 10). The pyroclastic flows are interpreted to have been substantially water-cooled. Co-ignimbrite ashes from the convecting part of the eruption column fell over a wide region.

The eruption then reverted to a dominantly convecting column, and deposited a fine-grained pumice-fall similar in grain-size to member 2. This, the lower part of member 5, indicates a decrease in the amount of water entering the vent. The fundamental change in the eruption that accounts for this sequence of events is discussed later. The eruption column then became wetter and finer-grained, depositing a very fine ash bed consisting of 3 major fall units each charged with accretionary lapilli. This bed, like member 3, probably fell as damp aggregates and accretionary lapilli. It has a locally variable thickness distribution but an ispoach map drawn around maximum thickness values (Fig. 9), delineates an elongate zone enclosed by the 1 m isopach. This zone is parallel to the main mountain ranges to the east of Lake Taupo and suggests topographic and, perhaps, meteorologic influence on deposition.

After a period of phreatoplinian activity the column began to collapse and a second major set of pyroclastic flows deposited ignimbrite (member 6) mainly to the northeast and east (Fig. 10). Substantial thicknesses of ash fell in distal areas at this phase of the eruption (Fig. 13). They probably represent both deposition from the convective part of the eruption column and from possible secondary explosions in the lake basin. At this stage no more widespread pyroclastic deposits were produced and collapse of part of the Taupo depression probably ensued. A cartoon explaining this sequence is shown on Fig. 18.

Controls on eruption sequence

A simple setting for this large phreatomagmatic eruption is a lake-filled basin with a subaqueous vent supplying continuous access of water to the conduit. The eruption could then begin in a phreatomagmatic style and continue in that mode until the water supply is exhausted. At this point it would then revert to magmatic activity. However, the Wairakei eruption must have been more complex than this because alternations in style and in magma fragmentation occurred. These changes are interpreted to reflect differing mixing ratios of water and magma. Several workers have attempted to explain the role of magma-water interactions on volcanic phenomena using fuel-coolant models (e.g., Colgate and Sigurgeirsson, 1973; Peckover et al., 1973; Wohletz, 1980). Recently, Sheridan and Wohletz (1981) and Wohletz (this volume), using data from

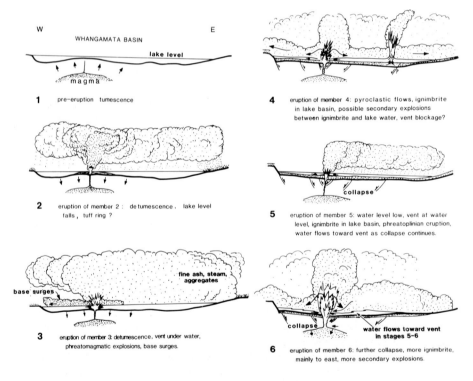

Fig. 18 A cartoon representing sequential stages in the eruption of the Wairakei Formation, illustrating interplay between lake water and vent area. See text for discussion.

explosion experiments, have shown that thermal to mechanical energy transfer is most efficient at mass mixing ratios (coolant: fuel) of 0.3 - 0.5. Since such maximum energy conditions can be equated to efficiency of magma fragmentation (Wohletz, this volume), maximum fragmentation may be produced by similar mixing ratios. Self and Sparks (1978) interpret that in phreatoplinian events the rhyolite magma is first disrupted in the conduit and then encounters water at the surface. Explosive interaction takes place in the vent and an assemblage of fine particles vigorously rises with the vaporized steam and exsolved magmatic volatiles.

The rise of a large (>100 km^3) body of magma under the Taupo depression probably caused regional tumescence. The scale of the tumescence cannot be estimated but the potential vent site was likely located on the bulge (Fig. 18). Lake water would have been displaced laterally from the area. The eruption could then have begun with a vent under shallow water and member 1, a fine-grained, well-bedded phreatoplinian deposit, was erupted. The lake level dropped as water was vaporized; proximal member 1 fall units show a slight overall increase in grain size upwards (Md$_\phi$ 4.8 to 3.4) which may represent a slow decrease in the water/magma mixing ratio. At the start of the eruption of member 2 the vent was at lake level, or perhaps a tuff ring was built around the vent, limiting free access of water. Member 2 thus has an overall lower degree of fragmentation and clasts perhaps fell from a dryer eruption column (Fig. 18). The deposit is generally better sorted than member 1 and lacks accretionary lapilli or very fine ash.

Member 3 represents sudden incursion of water to the vent, as indicated by the radical change to high fragmentation and a much wetter eruption column, producing water (rain) and accretionary lapilli. Laterally moving base surges were generated, perhaps indicating more energetic explosions and/or heavier eruption columns, more prone to collapse. The influx of water may have been due to collapse of the vent area (or perhaps destruction of a tuff ring, previously built around the vent), either locally or as part of the beginning of regional (caldera) collapse. Water could then have filled the depression created. Eventually the vent widened or the eruption column became greatly overloaded with steam and fine particles generated in the vent, and column collapse began (Sparks and Wilson, 1976, Sparks et al., 1978; Wilson et al., 1981). Water interacting with the magma would also act as a heat sink, further reducing the ability of the column to convect (Wilson et al., 1978; Sheridan et al., 1981).

After the initiation of column collapse, during the emission of member 4, mass flux rates of magma were probably so high that water-magma interaction was less efficient, yielding a larger coarse component to the grain-size distribution. Member 4 was laid down by very mobile pyroclastic flows covering a wide area. Uncollapsed, convecting parts of the column rose and continued to supply fine ash to the atmosphere. The Taupo basin was at least partially inundated with ignimbrite and some flows were probably deposited into water causing large secondary phreatic explosions (Walker, 1979), thus adding to the high altitude eruption column. Walker (1981b) has already suggested that medium-grained pumice-fall deposits found on the east coast of the North Island (the fall equivalent of member 6) in the Wairakei Formation sequence are of this secondary explosion origin.

The eruption continued by reverting to the ejection of a coarser phreatoplinian eruption column and the deposition of the coarse base of member 5, indicating once again that water did not have completely free access to the vent. Several explanations regarding the cessation of free access of water are: exhaustion of water in the lake basin (or in the part containing the vent), a lack of free flowing water due to pyroclastic flow fill, and/or vent blockage due to collapse after the eruption of several 10's of km^3 of fall deposits and ignimbrite. However, after a period of relatively coarse tephra deposition (base of member 5), the eruption returned to very fine, wet, ash production, suggesting that water again had free access to the vent (member 5).

Whether or not there was a hiatus in activity between the pyroclastic flow phase of member 4 and the beginning of phase 5 is a question that cannot be easily resolved. There is a prominent erosion break between members 4 and 5 in many distal locations but the time represented is indeterminable. Moreover, there is evidence that high intensity rhyolite eruptions, once initiated, continue until conditions change in the magma body -- for example, until volatile-poor magma is tapped (e.g., Hildreth, 1979).

The eruption ended with the production of an ignimbrite with four flow units, presumably in a similar style to that described for member 4. It can be evisaged that by this stage of the eruption, which so far had yielded >100 km^3 of magma, caldera collapse was occurring. Water in the whole Taupo lake basin could have been flowing into the depression. Again, secondary phreatic explosions may have yielded a coarser co-ignimbrite ash deposit, as well as finer grained beds of co-ignimbrite ash. This conjectural scheme to explain the complex Wairakei Formation deposit sequence is summarized on Fig. 18.

CONCLUSIONS

Evidence has been presented for a phreatomagmatic eruption 20,000 years ago of exceptional volume and dispersive power. A total bulk volume in excess of 300 km^3 was probably erupted in a short period of time (up to a few days but perhaps much less). The whole event was phreatomagmatic including the production of ignimbrites and associated co-ignimbrite ash falls. Possibly, most of the proto-Lake Taupo was vaporized during the eruption. A complex history of water access to the vent is inferred to have produced the varying styles of phreatomagmatic explosive activity interpreted from the deposits.

The ash-fall deposits formed are exceptionally fine-grained and contain up to 33 wt% accretionary lapilli. Some layers are totally formed of such lapilli. The name 'phreatoplinian' (Self and Sparks, 1978) was earlier proposed for fine, widespread, silicic ash beds erupted through water. The great dispersal of very fine ash (either individual clasts or aggregates) was facilitated by eruption columns possibly up to 30 km high, and a low mean terminal fall velocity (0.5 m s^{-1}) for the entire assemblage of particles produced. Ash was dispersed over a wide area of the South Pacific and southern hemisphere. Some 60% of the ash produced fell off mainland New Zealand. If the plot of thickness vs. area covered by all fall members is extrapolated to the 1 mm isopach, the area covered is greater than 10 million km^2, about 10% of the area of the southern hemisphere.

Much of the fine ash prematurely fell from the eruption clouds by aggregate formation mechanisms possibly involving electrostatic and other cohesive forces. Generation of accretionary lapilli in the vertical column further contributed to the removal of much ash from the eruption clouds by the adhesion of the finest suspended ash onto aggregate clusters of fine particles. Accretionary lapilli also have formed in ignimbrite fumarole pipes due to escape of steam through fine ash particles.

The ignimbrites generated by this event were cool, but voluminous and highly mobile. The magma of the Wairakei Formation is a typical Taupo Volcanic Zone rhyolite, containing phenocrysts of orthopyroxene, hornblende, and clinopyroxene for which pre-eruption equilibration temperatures of 725-800°C have been estimated (Ewart et al., 1971; Ewart et al., 1975; Howorth, 1976). Howorth also demonstrates a temperature increase (750 to 800°C) in the magma body from the initial products (member 1) to the first ignimbrite (member 4), based on co-existing titanomagnetite and ilmenite pairs and phenocryst chemistry. Welding temperatures for ignimbrites can be as low as 535°C (Smith, 1960). The Wairakei Formation ignimbrites are all nonwelded, even where thickness exceeds 100 m (e.g., in the Wairakei Basin). Water perhaps cooled the erupted material during initial contact with the melt in the vent which led to low-temperature pyroclastic flows. No carbonized plant material occurs, so that temperatures may have been as low as 350°C.

Large-scale pyroclastic flows formed from the collapse of phreatomagmatic eruption columns have not been described before. Their mobility, despite their coolness, underlines the fact that the runout of pyroclastic flows is largely controlled by momentum derived from eruption column collapse height, and the volume (mass) of material in the flow unit (Francis and Baker, 1977; Sheridan, 1979; Walker and Wilson, 1982). The area covered by pyroclastic flows was about 10,000 km^2, less than the Taupo ignimbrite (Walker et al., 1981b), but still large for an ignimbrite that has a high-aspect ratio near its source.

Finally, the environmental implications of this type of eruption should be considered, especially with respect to future volcanic activity in New Zealand. Walker (1980, 1981a, 1981b, 1982) and Froggatt (1981a, 1981b) have shown that all recent rhyolitic eruptions of the Taupo Volcano have been in the lake basin. Therefore a future eruption could be of this type. The scale of eruption exemplified by the Wairakei event, involving torrential mud-rains and, perhaps, eruption-induced storms, may in fact set up its own set of unusual meterological conditions. Mud falls and widespread pyroclastic-flow devastation occurred over a vast area (e.g., all of New Zealand was covered by at least 1 cm of ash). Reports of fallout from phreatomagmatic eruptions note that the ash may fall dry or wet, become sticky after deposition due to absorbed moisture, and then harden. Such deposits would not be easy to remove by conventional methods. The large amounts of reworked material testify to the disruption of drainage systems over an area of 10^5 km^2. Blockage of much of the sun's radiation is likely during this type of eruption due to the great optical depth perturbation in the lower atmosphere and stratosphere. The recurrence of such an event would be of severe consequence anywhere in the world.

ACKNOWLEDGEMENTS

This work was begun at the suggestion of Colin G. Vucetich and his encouragement and stimulation are much appreciated. I would like to acknowledge the following colleagues for valuable discussion, constructive criticism and helpful information: G. Grindley, J. Healy, R. Howorth, J. D. Milne, I. A. Nairn, V. E. Neal, W. A. Pullar, W. W. Topping, G. P. L. Walker and C. J. N. Wilson. The paper was reviewed and improved by M. F. Sheridan and K. H. Wohletz. F. McCoy (Lamont-Doherty Geological Laboratory) made deep-ash samples available to the author. Field studies on the Wairakei Formation were made during the tenure of a Victoria University of Wellington post-doctoral fellowship during 1974-1975.

REFERENCES

Barry, R.G. and Williams, J., 1973. The climate of the southern hemisphere at the last glacial maximum and an interhemispheric comparison. Abstracts, Ninth Congress, International Union for Quaternary Research, New Zealand, 1973: 12-13.
Borchardt, G.H., Haward, M.E., and Schmitt, R.A., 1978. Correlation of volcanic ash deposits by activation analysis of glass separates. Quatern, Res., 1: 247-260.
Brazier, S., Davis, A.N., Sigurdsson, H., and Sparks, R.S.J., 1979. Fall-out and deposition of volcanic ash during the 1979 explosive eruption of the Soufriere of St. Vincent. J. Volcanol. Geotherm. Res., 14: 335-360.
Briggs, N.D., 1973. Investigations of New Zealand pyroclastic flow deposits. Unpublished Ph.D. thesis. Victoria University of Wellington, New Zealand: 438 pp.
Briggs, N.D., 1976. Welding and crystallization in the Whakamaru Ignimbrite, Central North Island, New Zealand, N.Z.J. Geol. Geophys., 19: 189-212.
Campbell, I.B., 1973. Late Pleistocene alluvial pumice deposits in the Wanganui Valley. N.Z.J. Geol. Geophys., 16: 717-721.
Carey, S.N. and Sigurdsson, H., 1982. Influence of particle aggregation on deposition of distal tephra from the May 18, 1980 eruption of Mount St. Helens volcano. J. Geophys. Res., 87: 7061-7072.

Cole, J.W. and Nairn, I.A., 1975. Catalogue of Active Volcanoes of the World: 22, New Zealand. International Association of Volcanology, Rome, 156 pp.

Colgate, S.A. and Sigurgeirsson, T., 1973. Dynamic mixing of water and lava. Nature, 244: 552-555.

Cowie, J.D., 1964. Aokautere Ash in the Manawatu District, New Zealand, N.Z.J. Geol. Geophys., 7: 67-77.

Cowie, J.D. and Milne, J.D.G., 1973. Maps and sections showing the distribution and stratigraphy of North Island loess and associated cover deposits, New Zealand. 1: 1,000,000. N.Z. Soil Survey Report 6, map and explanatory notes. N.Z. D.S.I.R., Wellington, New Zealand.

Druitt, T. and Sparks, R.S.J., 1982. A proximal ignimbrite breccia facies on Santorini, Greece. J. Volcanol. Geotherm. Res., 13: 147-171.

Ewart, A., 1968. The petrography of the Central North Island rhyolitic lavas part 2 - regional petrography including notes on associated ash-flow pumice deposits. N.Z.J. Geol. Geophys., 11: 478-545.

Ewart, A., Taylor, S.R. and Capp, A.C., 1968. Trace and minor element geochemistry of the rhyolite volcanic rocks, Central North Island, New Zealand, Contrib. Mineral. Petrol., 18: 76-104.

Ewart, A., Green, D.C., Carmichael, I.S.E. and Brown, F.H., 1971. Voluminous low temperature rhyolitic magmas in New Zealand. Contrib. Mineral. Petrol., 33: 128-144.

Ewart, A., Hildreth, W. and Carmichael, I.S.E., 1975. Quaternary acid magma in New Zealand. Contrib. Mineral. Petrol., 51: 1-27.

Francis, P.W. and Baker, M.C.W., 1977. Mobility of pyroclastic flows. Nature, 270: 164-165.

Froggatt, P.C., 1981a. Karapiti Tephra Formation: a 10,000 year B.P. rhyolitic tephra from Taupo. N.Z.J. Geol. Geophys., 24: 95-98.

Froggatt, P.C., 1981b. Motukere Tephra Formation and redefinition of Hinemaiaia Tephra Formation, Taupo volcanic centre, New Zealand. N.Z.J. Geol. Geophys., 24: 99-105.

Froggatt, P.C., 1982. Review of methods of estimating rhyolitic tephra volumes: application to the Tampo Volcanic Zone, New Zealand, J. Volcanol. Geotherm. Res., 14: 301-318.

Grindley, G.W., 1960. Sheet 8 - Taupo (1st Edition). Geological Map of New Zealand 1: 250,000. N.Z. Dept. of Sci. and Industr. Res. Wellington.

Grindley, G.W., 1965. The geology, structure and exploitation of The Wairakei Geotherm Field, Taupo, New Zealand. N.Z. Geological Survey Bull., n.s. 75: 131.

Hatherton, T., Dawson, E.W. and Kinsky, F.C., 1965. Balleny Islands reconnaissance expedition 1964. N.Z.J. Geol. Geophys., 8: 164-179.

Healy, J., 1962. Structure and volcanism in the Taupo Volcanic Zone, New Zealand. In: Crust of the Pacific Basin. Am. Geophys. Union, Geophys. Monogr., 6: 151-157.

Healy, J., 1964. Volcanic mechanisms in the Taupo Volcanic Zone. N.Z.J. Geol. Geohpys., 7: 6-23.

Hildreth, W., 1979. The Bishop Tuff: evidence for the origin of compositional zonation in silicic magma chambers. Geol. Soc. Amer. Spec. Paper 180: 43-75.

Howorth, R., 1976. Late Pleistocene tephras of the Taupo and Bay of Plenty regions. Unpublished Ph.D. thesis, Victoria University of Wellington, N.Z., 192 pp.

Hume, T.M., Sherwood, A.M., Nelson, C.S., 1975. Alluvial sedimentology of the Upper Pleistocene Hinuera Formation, Hamilton Basin, New Zealand. J. Roy. Soc. N.Z., 5: 421-462.

Irwin, J., 1972. Lake Taupo, Provisional bathymetry, 1: 50,000. N.Z. Oceanographic Institute Chart, Lake Series.

Kohn, B.P., 1970. Identification of New Zealand tephra layers by emission spectrographic analysis of their titanomagnetites. Lithos, 3: 361-368.

Kohn, B.P., 1973. Some studies of New Zealand Quaternary pyroclastic rocks. Unpublished Ph.D. thesis. Victoria University of Wellington, New Zealand, 340 pp.

Kohn, B.P. and Glasby, G.P., 1978. Tephra distribution and sedimentation rates in the Bay of Plenty, New Zealand. N.Z.J. Geol. Geophys., 21: 49-70.

Kuno, H., Ishikaswa, T., Katsui, Y., Yaki, K., Yamasaki, M. and Taneda, S., 1964. Sorting of pumice and lithic fragments as a key to eruptive and emplacement mechanisms. Japan J. Geol. Geog., 35: 223-238.

Lewis, K. and Kohn, B.P., 1976. Ashes, turbidites and rates of sedimentation on the continental slope off Hawkes Bay. N.Z.J. Geol. Geophys., 16: 439-54.

Lirer, L., Pescatore, T., Booth, B. and Walker, G.P.L., 1973. Two plinian pumice fall deposits from Somma Vesuvius, Italy. Geol. Soc. Amer. Bull., 84: 759-772.

Mildenhall, D.C., 1976. Exotic pollen rain on the Chatham Islands during the Late Pleistocene, N.Z.J. Geol. and Geophys., 19: 327-333.

Murphy, R.P. and Seward, D., 1981. Stratigraphy, lithology, palaeomagnetism, and fission track ages of some ignimbrite formations in the Matahana Basin, New Zealand. N.Z.J. Geol. Geophys., 24: 325-331.

Nairn, I.A., 1971. Studies of the Earthquake Flat Breccia Formation and other unwelded pyroclastic flow deposits of the Central Volcanic Region, New Zealand. Unpublished MSc thesis, Victoria University of Wellington, New Zealand.

Nairn, I.A., 1972. Rotoehu Ash and the Rotiti Breccia Formation, Taupo Volcanic zone, New Zealand. N.Z.J. Geol. Geophys., 15: 251-261.

Ninkovich, D., 1968. Pleistocene volcanic eruptions in New Zealand recorded in deep-sea sediments. Earth Planet. Sci. Lett., 4: 89-102.

Pain, C.F. and Pullar, W.A., 1975. Chronology of 'paleosurfaces' and present land surfaces in the Reporoa Basin, North Island, New Zealand. N.Z.J. Sci., 18: 313-322.

Peckover, R.S., Buchanan, D.J. and Ashby, D.E.T.F., 1973. Fuel-coolant interactions in submarine volcanism. Nature, 245: 307-308.

Pullar, W.A. and Birrell, K.S., 1973. Age and distribution of late Quaternary pyroclastic and associated cover deposits of the Rotorua and Taupo Area, North Island, New Zealand, Parts I and II. Maps of isopachs and volumes of tephra and subsurface loess, central North Island, New Zealand 1: 1,000,000. New Zealand Soil Survey Report 1, NZDSIR, Wellington, New Zealand.

Rhea, K.P., 1968. Aokautere Ash, loess and river terraces in the Dannevirke District, New Zealand. N.Z. Journ. Geol. Geophys., 11: 685-692.

Rose, W.I., Bonis, S., Stoiber, R.E., Keller, M. and Bickford, T., 1973. Studies of volcanic ash from two recent Central American eruptions. Bull. Volcanol., 37: 338-364.

Schofield, J.C., 1965. The Hinuera Formation and associated Quaternary events. New Zealand Jour. Geol. Geophys., 8: 772-791.

Self, S., and Sparks, R.S.J., 1978. Characteristics of widespread pyroclastic deposits formed by the interaction of silicic magma and water. Bull. Volcanol., 41: 196-212.

Sheridan, M.F., 1979. Emplacement of pyroclastic flows: A review. Geol. Soc. Amer. Spec. Paper 180: 125-136.

Sheridan, M.F. and Wohletz, K.H., 1981. Hydrovolcanic explosions: the systematics of water-pyroclast equilibration. Science, 212: 1387-1389.

Sheridan, M.F., Barberi, F., Rosi, M. and Santacroce, R., 1981. A model for plinian eruptions of Vesuvius. Nature, 289: 282-285.

Sigurdsson, H., 1982. Tephra from the 1982 Soufriere explosive eruption. Science, 216: 1106-1108.

Smith, R.L., 1960. Ash Flows. Geol. Soc. Amer. Bull., 71: 795-842.

Sparks, R.S.J., 1975. Stratigraphy and geology of the ignimbrites of Vulsini volcano. Central Italy. Geol. Rundsch., 64: 497-523.

Sparks, R.S.J., Self, S., and Walker, G.P.L., 1973. Products of ignimbrite eruptions. Geology, 1: 115-118.

Sparks, R.S.J., and Wilson, L., 1976. A model for the formation of ignimbrite by gravitational column collapse. J. Geol. Soc. London, 132: 411-452.

Sparks, R.S.J. and Walker, G.P.L., 1977. The significance of vitric-enriched air-fall associated with crystal-enriched ignimbrites. J. Volcanol. Geotherm. Res., 2: 329-341.

Sparks, R.S.J., Wilson, L. and Hulme, G., 1978. Theoretical modeling of the generation, movement and emplacement of pyroclastic flows by column collapse. J. Geophys. Res., 83: 1727-1739.

Sparks, R.S.J. and Huang, T.C., 1980. The volcanological significance of deep sea layers associated with ignimbrites. Geol. Mag., 117: 425-436.

Sparks, R.S.J., Wilson, L. and Sigurdsson, H., 1981. The pyroclastic deposits of the 1875 eruption of Askja, Iceland. Phil. Trans. Roy. Soc., A299: 241-273.

Steiner, A., 1953. Hydrothermal rock alteration at Wairakei, New Zealand. Econ. Geol., 48: 1-12.

Steiner, A., 1977. The Wairakei geothermal area, North Island, New Zealand. N.Z. Geological Survey Bull., n.s. 90: 136 pp.

Topping, W.W., 1973. Tephrostratigraphy and chronology of late Quaternary eruptions from the Tongariro Volcanic Center, New Zealand. N.Z.J. Geol. Geophys., 16: 397-423.

Topping, W.W., 1974. Some aspects of the Quaternary history of Tongariro Volcanic Centre. Unpublished Ph.D. thesis, Victoria University of Wellington, New Zealand.

Vucetich, C.G. and Howorth, R., 1976a. Proposed definition of the Kawakawa Tephra, the c 20,000 years B.P. marker horizon in the New Zealand region. N.Z.J. Geol. Geophys., 10: 43-50.

Vucetich, C.G. and Howorth, R., 1976b. Late Pleistocene tephrostratigraphy in the Taupo district, New Zealand. N.Z.J. Geol. Geophys., 19: 51-69.

Vucetich, C.G. and Pullar, W.A., 1964. Stratigraphy and chronology of late Quaternary volcanic ash in Taupo, Rotorua and Gisborne district. Pt 2. N.Z. Geol. Surv. Bulletin., n.s. 73: 43-88.

Vucetich, C.G. and Pullar, W.A., 1969. Stratigraphy and chronology of Late Pleistocene volcanic ash beds in Central North Island, New Zealand. N.Z. Jour. Geol. Geophys., 12: 784-821.

Walker, G.P.L., Grain size characteristics of pyroclastic deposits. J. Geol., 79: 696-714.

Walker, G.P.L., 1982. Crystal concentration in ignimbrites. Contrib. Mineral. Petrol., 36: 135-146.

Walker, G.P.L., 1979. A volcanic ash generated by explosions where ignimbrite entered the sea. Nature, 281: 642-646.

Walker, G.P.L., 1980. The Taupo Pumice: product of the most powerful known (ultraplinian) eruption? J. Volcanol. Geotherm. Res., 8: 69-94.

Walker, G.P.L., 1981a. Characteristics of two phreatoplinian ashes, and their water-flushed origin. J. Volcanol. Geotherm. Res., 9: 395-407.

Walker, G.P.L., 1981b. New Zealand case histories of pyroclastic studies. In, Tephra Studies, Self, S. and Sparks, R.S.J., Eds. Reidel, Dordrecht, Holland: 317-330.

Walker, G.P.L., 1982. The Waimihia and Hatepe plinian deposits from the Taupo rhyolitic volcano, New Zealand. N.Z.J. Geol. Geophys. 74: 305-324.

Walker, G.P.L. and Croasdale, R., 1972. Characteristics of some basaltic pyroclastics. Bull. Volcanol., 35: 303-317.

Walker, G.P.L., Wilson, L. and Bowell, E.L.G., 1971. Explosive volcanic eruptions. I. The rate of fall of pyroclasts. Geophys. J. Roy. Astron. Soc., 22: 377-383.

Walker, G.P.L., Heming, R.F. and Wilson, C.J.N., 1979. Low-aspect ratio ignimbrites. Nature, 282: 286-287.

Walker, G.P.L., Wilson, C.J.N. and Froggatt, P.C., 1980. Fines depleted ignimbrite in New Zealand - The product of a turbulent pyroclastic flow. Geology, 8: 245-249.

Walker, G.P.L., Self, S. and Froggatt, P.C., 1981a. The ground layer of the Taupo Ignimbrite: A striking example of sedimentation from a pyroclastic flow. J. Volcanol. Geotherm. Res., 10: 1-11.

Walker, G.P.L., Wilson, C.J.N. and Froggatt, P.C., 1981b. An ignimbrite veneer deposit: the trail marker of a pyroclastic flow. J. Volcanol. Geotherm. Res., 9: 490-521.

Walker, G.P.L. and Wilson, C.J.N., 1982. Ignimbrite depositional facies: the anatomy of a pyroclastic flow. J. Geol. Soc. London, 139: 581-592.

Watkins, N.D. and Huang, T.C., 1977. Tephras in abyssal sediments east of the North Island, New Zealand: chronology, paleowind velocity, and paleoexplosivity. N.Z.J. Geol. Geophys., 20: 179-188.

Wilson, L., Sparks, R.S.J., Huang, T.C. and Watkins, N.D., 1978. The control of volcanic column heights by eruption energetics and dynamics. J. Geophys. Res., 83: 1829-1836.

Wilson, L., Sparks, R.S.J. and Walker, G.P.L., 1981. Explosive volcanic eruptions IV. The control of magma properties and conduit geometry on eruption column behavior. Geophys. J. Roy. Astron. Soc., 63: 117-148.

Wohletz, K.H., 1980. Explosive hydromagmatic volcanism. Unpublished Ph.D. Thesis. Arizona State University, 303 pp.

Wohletz, K.H., 1983. Mechanisms of hydrovolcanic pyroclast formation: grain-size, scanning electron microscopy, and experimental data. In: M.F. Sheridan and F. Barberi (Editors), Explosive Volcanism. J. Volcanol. Geotherm. Res., 17: 31-63.

Wright, J.V., 1981. The Rio Caliente Ignimbrite: Analysis of a compound intra-plinian ignimbrite from a major late Quaternary Mexican eruption. Bull. Volcanol., 44: 189-212.

Wright, J.V., Smith, A.L. and Self, S., 1981. A working terminology of pyroclastic deposits. J. Volcanol. Geotherm. Res., 8: 315-336.

SUBJECT INDEX

Aeolian Islands	314,321,327
Agnano eruption	278
Aira Caldera	76
Akaroa Ash	76
Alban Hills	220
Alution Arc	393-432
Amak Island	405
Aniakchak Caldera	417,424
Aokautere Ash	439
Arenal Volcano	126,150,152
Aso 4 flow	77
Aso Caldera	77
Augustine Volcano	394,400,403,408,410,413,414, 417,418,420,423-425,427
Avellino Pumice	239,255
Baccano Caldera	221-223,230,234
Benioff zone	314,405,408,410,413,423,427
Bishop Tuff	75,76
Bogoslof Island	405
Breccia Museo	274,275,283,284,293
Bromo Volcano	178
Buldir Volcano	394,403
Campanian Ignimbrite	75,274,275,280,284,290,293,305, 308
Campanian Y-5 Ash	89-109
Chiginagak Volcano	413
Cimino Ignimbrite	72,204,207,211,213
Cimino Volcano	204,206
Devils Desk Volcanic Center	394,403,414,420,426
Eifel District	375
El Chichon eruption, 1982	149
El Tajo Tephra	126
Fossa of Vulcano	329-360
Fourpeaked Mountain	394,403,408,410,413,424
Fuego eruption, 1974	139,145-151
Hawaii eruption, 1790	19
Hawaii eruption, 1924	19
Hawaiian activity	7
Hayes Volcano	394,403,405,413,424
Hekla Volcano	178
Iliamna Volcano	394,400,410,417,418
Ito Ignimbrite	75
Kaguyak Crater	403,410,413,414,417,418,423, 424-427

Karymsky eruption	124
Katmai Volcano	394,400,403,413-415,417,418, 420,423-425,427
Katmai eruption, 1912	112
Kejulik Volcano	394,403,413,420
Kiaglagvik Volcano	413
Kikai Caldera	76
Koya Flow	77
Koya Ignimbrite	67,72,75
Krakatau Caldera	76
Krakatau eruption, 1883	74,76,77
La Soufriere of Guadeloupe	160,190
Laacher See	375-392
Laacher See Tephra	376
Lewotobi Lakilaki eruption	124
Lipari	189,195,313-328
Los Chocoyos Ignimbrite	75,76
Mageik	414
Magone Plinian Pumice	71
Merapi eruption, 1883-1915	123,126
Merapi-type pyroclastic flows	123,171
Monte Guardia Sequence	313-328
Morrinsville Ignimbrite	73-75
Mt. Denison	403,415,426
Mt. Douglas	400,408,410,413,424
Mt. Griggs	400,403,415,418,425,427
Mt. Kukak	400,426
Mt. Mageik	400,415
Mt. Martin	400,415,420
Mt. Nuovo eruption, 1538	278,281,293
Mt. Pelée	159-185
Mt. Pelée eruption, 1902	72,79,81,172,173
Mt. Peulik	400,410,413,414
Mt. St. Helens Ash	92,94,106
Mt. St. Helens eruption, 1980	72,79,81,82,90,123,127,133-157, 188,192,198,199
Mt. Stellar Volcanic Center	394
Neapolitan Tuff	277,278,282
Ngauruhoe eruption, 1975	81
Novarupta Volcano	400,403,413,415,424,425
Novarupta eruption, 1912	400,424
Oruanui Ash	76
Oruanui Ignimbrite	73,75,434,436
Pantelleria	361-373
Pantelleria eruptions	372

Phlegraean Fields	273-311
Phlegraean Fields Caldera	90
Pilato Ash	332,342,344
Plinian activity	7,9,19,72,73,163,168,169,175, 177,198,238,241,243,244,247, 253,278,280,282,400,434
Plinian eruption column	90,106-108
Plinian eruption cycle	238
Plinian pumice fall	20,68,75,76,172,174,196,376,437
Pompeii Pumice	239,256
Popocatepetl	178
Rabaul Ignimbrite	72
Redoubt Volcano	394,400
Rekohu Ash	436,451
Rio Caliente Ignimbrite	71,72,74
Roccamonfina Caldera	233
Roman Volcanic Province	220,293
Roseau Ash	77
Rotoehu Ash	76
Rotoiti Ignimbrite	75
Ruapehu eruption, 1945	124
SEM analysis	39-42,229
SEM images	16,40,316,318,326
Sabatini Volcanic Complex	219,225
Sacrofano Caldera	220,225,228-231
Santiaguito eruption	124
Sheveluch Volcano	403
Skessa Tuff	72
Snowy Mountain Volcanic Center	394,400,415,426
Somma-Vesuvius	197,237-248
Soufriere of St. Vincent	123
Spurr Volcano	394,400,413
St. Vincent type eruption	163
Stromboli	314,321
Strombolian activity	7,9,47,225,241,280,281,400
Surtseyan activity	7,9,16,33,47
Taal eruption, 1965	79
Tarumai Volcano	178
Taupo Caldera	76
Taupo Ignimbrite	67,68,72,458,463
Taupo Volcanic Center	434
Toba Ash	75
Toba Tuff	75
Trident Volcano	400,415,420,425
Trident eruption, 1953	123

Ugashik Caldera	394,413,414,417,423-425,427
Ukinrek Maars	400,403,408,413,414,417,418, 420,427
Valley of Ten Thousand Smokes	67,400,418,424,426,427
Vesuvius	178,189,190,196,198,230
Vesuvius eruption, 1631	241,247
Vesuvius eruption, 1944	243
Vesuvius eruption, A.D. 472	197,199,240,241,247,249-271
Vesuvius eruption, A.D. 79	20,238,250
Vico Volcano	204,207
Volmero Yellow Tuff	278
Vulcanello	326,327,330
Vulcanian activity	7,77,400
Vulcano	193,199,329-360,314,318,321, 327,329-360
Vulcano eruption	20,189,190
Vulcano eruption, 1888-1890	333,345
Vulsini D Ignimbrite	71
Vulsini Volcanic Complex	220,233
Wairakei Formation	433-469
Walcott Tuff	72
Whakamaru Ignimbrite	447
Wickman diagram	178
Yantarni Volcano	413
accretionary lapilli	11,17,20,71,76,145,225,229,281, 281,342,344,385,388,436,440, 441,443,447-449,454,456,458, 459,461-463
acoustic waves	6
aggregation of particles	106,138,142,152,456,459,463
alteration of pyroclasts	14,16,33
andesite	162,165,168-172,195,315,325, 326,327,341,400,417,420,424, 425,447,448
ash fall	18,71,90,106,107,400,436-438, 440,454,458,462
ash flow	20,71,74-79,238,423
ash-cloud deposit	169,385,448
ash-cloud surge	79,82,168,173,253
ash-flow deposit	171,344
ash-flow tuff	66,251,315,339
aspect ratio	67
atmospheric ash fractionation	139
atmospheric hazards	133-157
avalanche	123

banded pumice	164,171,172,315,318,424
basalt	35,327,362,365,368,369,371,372, 410,413,417,418,426,448
basaltic andesite	164,172,418
basanite	238,243,375
base surge	8,77,79,82,195,196,209,210,282 388,442,458,459
bed transitions	377
bimodal volcanism	327
block-and-ash flow	163,165,168,172,174,175
block-and-ash surge	168
breadcrust	74,172,175
breccia	11,163,173,209,283,292,318,342 377,378,384,388,389,458
bubble walls	41,95
calc-alkaline series	314,327,410,417,420,424,425,426
caldera	214,274,278,282,284,290,292, 305,362,365,368,371,400,403, 420,424,461,462
carbonized wood	106,169,171,172,251,254,449
central volcano	212
chlorine emission	145-152
cinder cone	238,241,369,371,372
co-ignimbrite ash cloud	106,108,142,143,164,238,459
co-ignimbrite ash fall	75,90,101,105,106,437,447,449, 450,451,458,462
co-ignimbrite lithic breccia	447
coarse-and-fine deposits	338,342,344,345,351
coarsetail grading	75
column collapse	73,74,77,83,123,173,175,390
component analysis	66,139,447
composite volcano	315
compostional gap	294,327
computer graphics	187-202
computer models	96,99,107,188
computer simulation	89-109
cone sheet	275
convective plume	77
cross-bedded layers	20,377,378,384,385,388,391,442, 444,454
cryptodomes	112
crystal enrichment	75,172
crystal fractionation	245,303,308,418,420,424-427
dacite	35,95,126,135,400,417,423-425
deflation	82,83

detonation wave	35,53,59
differentiation	270
directed-blast eruption	79,81,82,123,145,172-176,188, 199,281
discharge rate	68,83,247
dome	212,213,274,315,327,330,365, 369,371,403,423
dry surge	17,20,190,193,327,332,338,342, 343,344,351
dune bedding	79,81,82,210,378,384,385, 388-391
effusion rate	242
ejecta plume	8
endogenous dome	123,206,362,368
energetics diagram	351
energy cone model	188,189,192,197,199
epicenter	406
eruption column	73,90,99,169,282,378,387-391, 389,390,447,456,458,459,461,463
eruptive cycles	19,244,247,332-334,342,378
exogenous dome	123
explosion breccia	3,17,315,341,343
explosive energy	6,37
fallout deposit	376
feeding rate	270
feeding system	243,371
fine ash	134
fines-depleted deposits	69-71,77,79-82,172,173,275,277, 283,284,293
flow head	70,73,82
flow velocity	68
fluid instabilities	56
fluidization	69,73,79,82,283,391
fluidization experiments	66
fragmentation vaporization	36
fuel-coolant interactions	4-7,4,10,34,37,145
fumarolic pipes	17,74,274,283,342,344,447,449, 463
geothermal	18,145,274,282,283,290,447,449
glass enrichment	142,449
glass shards	95
globules	41,44
glowing cloud	173,175,251,255
grain flow	79
grain size	76,90,108,136,139

grain-size distribution	91-94,105,387,439,454
granulometric analysis	66,346,441,454
granulometric data	10,33,37,38,347-349,386
green ignimbrite, Pantelleria	362,365,366,368-370
ground layer	70,447,448
ground surge	82,171,238,253,389
growth of domes	111-131,168,175
hazard zones	193
hazards from L.A.R. ignimbrite	71
hazards from ash fall	76
heat transfer	48
high aspect ratio ignimbrite	67,463
high potassium series	243,255,314,321,325
high-grade ignimbrite	72-74
hot avalanche	241
hyaloclastite	9,17,19,20,32,417
hyalotuff	32
hydroexplosion	2,7-9,222,228
hydrofracture	145
hydromagmatic activity	2,16,144,230-234,239,241,253, 281
hydromagmatic breccia	224
hydromagmatic explosions	282
hydromagmatic tuff	274,292
hydromagmatism	228,229
hydrothermal metamorphism	278
hydrovolcanic ash	33,40
hydrovolcanic cycles	19
hydrovolcanic vents	18
hydrovolcanism	2,8,20,34,37,42,188,193
ignimbrite	66,71-76,82,91,142,206,437,447, 438,445,449,454,458,459,462,463
impact sags	229,388
inverse grading	378,384
kinematic properties	192
lag breccia	83
lahar	20,74,75,196,229,340,344-346
laminar flowage	75,168
lapilli fall	11,18,20,315
latite	207,274,278,280,294,330
lava flow	20,71,74,162,196,225,238,241, 252,327,334,342,362,365,366, 368,369,371,423
lava fountain	74
layer 1 deposits	69-71,82

leucite tephrite	326,327
leucitite	228,238,243
leucititic phonolite	265
limit of destruction	192
low aspect ratio ignimbrite	67,71,72,81,449
low-grade ignimbrite	72,73
maar volcano	9,228,375
magma chamber	2,284,290,308,310,326,327,415,427
magma column	172
magma mixing	325,327,334,354,400,424
magma replenishment	247
magmatic explosions	111
mantle plume	405
massive bed	17,38,251,339,342,377,385
melt fragmentation	32,43,46,139,286,327,332,338,456,459,461
microseismic activity	308
missing volume of ash	134-136
mobility	391,459,463
mud flow	74,238,240,241,447
mud hurricane	238,251
nonwelded ignimbrite	74,434,436,437,444,448
nuée ardente	79,81,163,164,168,169,174,178
palagonite	3,16,17,252
paleosoil	254
pantellerite	368,369,371
particle aggregation	101,106-108
particle morphology	13,14
peralkaline rhyolite	362,368
phonolite	207,243,255,376
phonolitic leucitite	241,255
phonolitic tephrite	209-211
phonolitic trachyte	284,293
phreatic activity	3,77,145,190-194,228,230,233,234,461,462
phreatic breccia	339,345,351
phreatic explosion pits	18
phreatomagmatic	3,73,77,139,144,145,177,238,376,388,400,436,447-449,459,460,464
phreatoplinian activity	7,16,33,56,76,145,282,436,459,461,462
phreatoplinian ash	73,438,449,451,454,458,459
pillow	20

pillow breccia	3,19,33
pillow lava	3,9,17,19,67
planar bed	17,38,378
plate motion	403
plug domes	123
polymodal grain size	38,90,94,99,136,139
potassic series	220,243,293
pumice cone	362,365,368
pumice fall	20,163,195,240,250,256,280,282, 327,337,341,351,368,371,440, 443,449,459,462
pumice flow	20,163,168-171,238
pumice-and-ash flow	171,172,174
pyroclast shape	33,55-58,315-317
pyroclast size	46
pyroclast types	40
pyroclastic fall	11,34,228,337,365,436,459
pyroclastic fan	376
pyroclastic flow	11,34,77,81-83,162,164,168,169, 172,188,191,196,207,209,224, 225,228-230,232,240,241,251, 252-254,256,274,280,282,339, 341,343,365,376,403,436-438, 447-449,454,458,459,462,463
pyroclastic surge	11,16,34,77,79,82,83,172,173, 187-192,196,198,210,232,238, 240,251,280-282,376,443
red tuff with black scoria	209,211,228
repose period	178,181,238,242-245,246,247,270
resurgent dome	371
rheoignimbrite	74
rheomorphic structures	74
rhyodacite	206,213,316
rhyolite	35,195,315-318,320,321,325-327, 330,334,365,410,413,417,424, 425,426,435,448,454,458,461,462
ring fracture	212,275,284
risk zones	199
sandwave bed	11,17,18,37,38,79,210,229,250, 251,339,342,344,384
scoria cone	196,228,229,376,443
scoria flow	163
segment boundary volcano	425-427
segregation features	70
seismicity	403,405,408,415,423

shock propagation	54
shoshonite series	314,318,327
steam-blast explosion	7,19
stochastic models	178
stratified fall	20
stratovolcano	196
sub-Plinian deposit	195,280,315,318
subaqueous	3,16,17,460
subduction	403,413,414,427
subglacial	3,17,19
submarine activity	36,274,292,372
sulfur emission	145-152
superheat vaporization	34,59
surge facies	17,251,281
survivor function	178
tectonic segmentation	424
tephrite	207,211,238,327,330,334,352,375
tephritic leucitite	232,241,265
tephritic phonolite	207,209,228
terminal settling velocity	96
thermal fragmentation	53
thermite	9,10,42,388
tholeiitic series	410,420,425,426
total grain-size distribution	139,456
trachyandesite	149
trachybasalt	278,280,293,294,308
trachyte	35,95,207,209,210,243,274,278, 280,281,284,293,308,318,320, 321,326,327,330,334,341,342, 362,365,368
tuff breccia	252
tuff cone	2,18,34,222,229,281,330,332,356
tuff ring	2,18,34,281,332,461
underflow	168,169,173
valley pond deposits	81,449
vapor-film collapse	36,52
velocity	192
veneer deposits	70,71,447
vesiculated tuff	17,20,75,251,281,342
volcanic explosivity index	123
volcanic front	403,405,410,414,415,424,427
volcanic hazards	286
water/melt experiments	4,9,10,388
water/melt interaction	4,16,33,46,229,231-233,247,280, 281,282,286,315,327,333,350, 354,356,388,389,436,456

water/melt ratio	332,351,391,460,461
wavy-bedded deposits	82
welded ignimbrite	434,448
welded scoria	281,283
welded tuff	67,72,74,277
wet pyroclastic flow	233,252
wet surge	17,20,190,327,332,339,342-344, 346,350,352
wind profile	98,105
xenolith	207,327
yellow tuff	277,278,281,284,293,306